HARMONIC ANALYSIS ON HOMOGENEOUS SPACES

Symposium in Pure Mathematics, XXI Williams College, 1972

PROCEEDINGS OF SYMPOSIA
IN PURE MATHEMATICS
VOLUME XXVI

HARMONIC ANALYSIS ON HOMOGENEOUS SPACES

AMERICAN MATHEMATICAL SOCIETY
PROVIDENCE, RHODE ISLAND
1973

PROCEEDINGS OF THE SYMPOSIUM IN PURE MATHEMATICS
OF THE AMERICAN MATHEMATICAL SOCIETY

HELD AT WILLIAMS COLLEGE
WILLIAMSTOWN, MASSACHUSETTS
JULY 31-AUGUST 18, 1972

EDITED BY
CALVIN C. MOORE

Prepared by the American Mathematical Society
with the partial support of National Science Foundation Grant GP-33391

Library of Congress Cataloging in Publication Data CIP

Symposium in Pure Mathematics, Williams College, 1972.
Harmonic analysis on homogeneous spaces.

(Proceedings of symposia in pure mathematics, v. 26)
Includes bibliographical references.
1. Locally compact groups. 2. Lie groups.
3. Functions, Special. 4. Harmonic analysis.
I. Moore, Calvin C., 1936– ed. II. Williams
College. III. Title. IV. Series.
QA387.S94 1972 512'.55 73-10456
ISBN 0-8218-1426-5

AMS (MOS) subject classifications (1970). Primary 22D10, 22E25, 22E30, 33A75, 22E35, 22E40, 12A85, 22E50, 22E55; Secondary 44A25, 31B10, 31B25, 60J45, 60J50

Printed in the United States of America

CONTENTS

Indexes

PREFACE

This book is an outgrowth of the nineteenth Summer Research Institute of the American Mathematical Society which was devoted to the topic Harmonic Analysis on Homogeneous Spaces. The Institute was held at Williams College in Williamstown, Massachusetts from July 31 to August 18, 1972, and was supported by a grant from the National Science Foundation. It was the aim of the Institute to concentrate mainly on the more analytic aspects of the subject and, of course, the theory of unitary group representations for groups not only over the reals and complex numbers but also for p-adic groups plays the fundamental role. Part of the program was devoted to automorphic functions, but it was decided not to go exhaustively into the very rich and deep connections of representation theory with automorphic functions nor to explore the arithmetic and algebraic geometric aspects of the theory of automorphic forms. This is certainly a most fruitful direction of research, but this is more properly the subject of another conference or future Summer Institute organized with this as its major goal.

The scientific program of the Institute consisted first of all of six major lecture series, each devoted to relatively broad areas within the subject and consisting of from four to six lectures. The lecturers and their topics were: C. C. Moore, Representations of solvable and nilpotent groups and harmonic analysis on nil and solvmanifolds; V. S. Varadarajan, Character theory and the Plancherel formula for semisimple groups; S. Helgason, Function theory on symmetric spaces; B. Kostant (jointly with R. Blattner), Quantization and representation theory; Harish-Chandra, Representation theory of p-adic groups; and H. Furstenberg, Boundary theory and stochastic processes on homogeneous spaces. The first part of this volume contains articles prepared by the lecturers based on these lecture series. The manuscript for the lectures on Quantization and representation theory was prepared by R. Blattner. These are intended to be surveys of various aspects of the subject of the Institute and it is hoped that they would provide material not only for advanced graduate students and postdoctoral mathematicians who want to get into the subject, but also for more senior mathematicians who work in related areas and who need a survey of what is known in the subject, and what techniques are available.

The second part of the scientific program of the Institute consisted of five research seminars each under the leadership of an invited chairman. These seminars and chairmen are as follows: Representation theory of solvable groups and harmonic analysis on solvmanifolds (Chairman, L. Auslander); Irreducibility and realization of various series of representations of semisimple groups (Chairman, B. Kostant); Boundary behavior, special functions, and integral transforms in group representations (Chairman, E. Stein); Representations of p-adic groups (Chairman, Harish-Chandra); and a seminar devoted to $L^2(G/\Gamma)$ and automorphic functions (Chairman, R. Godement). These seminars were devoted to the presentation of talks by participants in the Institute on current research in the various fields. The second part of this volume contains short summaries of these talks in the format of Research Announcements. Formal papers containing complete proofs will appear in the regular journals. These short notes are arranged under the five seminar headings. It should be noted that Varadarajan's survey lectures contain a description of the discrete part of the spectrum of a semisimple group; a description of the continuous spectrum seemed more properly to fit into one of the seminars, and is in fact contained in the two articles by Trombi and Wolf listed under Kostant's seminar. It is hoped that this combination of lecture series devoted to broad areas and short summaries of current research results and problems will facilitate access to the subject and foster further progress.

CALVIN C. MOORE

PART I: INVITED LECTURES

REPRESENTATIONS OF SOLVABLE AND NIL-POTENT GROUPS AND HARMONIC ANALYSIS ON NIL AND SOLVMANIFOLDS

CALVIN C. MOORE*

Contents

1. Introduction. Our object is to sketch the recent history and the current state of development of the theory of unitary representations of nilpotent and solvable Lie groups and of the associated harmonic analysis on nilmanifolds and solvmanifolds. This study of unitary representations has many historical antecedents, beginning with theory of finite groups and the discovery, for instance, that every irreducible representation of a finite nilpotent group is induced by a one-dimensional representation of some subgroup – or, equivalently, is a monomial representation. The Kirillov theory has as one of its starting points the analogue of this for Lie groups. It is true that generally speaking the unitary representation theory of semisimple groups has received more attention than has the theory for solvable and nilpotent groups. The solvable theory has now developed to a certain fullness and has revealed many interesting theorems and unsolved problems. Many of the techniques used are in a sense adapted from the abelian case, and one constant theme in the proofs is an inductive argument on the dimension of the group.

The present moment appears to be a good point in time at which to survey

AMS (MOS) subject classifications (1970). Primary 22D10, 22E25, 22E40, 22D30.

* Supported by NSF contract GP-30798.

this subject for now one has in hand theorems which completely describe the dual space of a solvable Lie group, at least if that group is type I, together with simple necessary and sufficient conditions for a solvable Lie group to be type I. One has in addition a great deal of information about the existence and properties of the global characters of the irreducible representations. Moreover, one has a concrete description of the Plancherel formula in general and also a description of the primitive ideal space of the group C^*-algebra of a solvable Lie group. It is hoped that a complete understanding of the solvable and nilpotent theory will have an impact on the semisimple theory such as exploiting the relations between representations of a semisimple group and those of its maximal unipotent group or of its Borel subgroup. Finally the study of harmonic analysis on nil and solvmanifolds is, in a sense, in its infancy, but there is already a good understanding of the nilpotent case. This theory which is a natural generalization of the study of multiple Fourier studies poses many interesting problems and should have many equally interesting applications.

Let us turn to some of the more specific points. The first step was taken by Dixmier in a series of papers [**13**] in which he analyzed the representation theory of nilpotent groups in detail. Soon thereafter, these results were amplified and placed in a more systematic context by Kirillov, and indeed all future developments stem from Kirillov's idea of associating irreducible representations with orbits in the dual of the Lie algebra. At about the same time, an alternate realization of these representations on spaces of holomorphic functions was given in [**7**], a development whose significance did not fully emerge until later. Takenouchi [**55**] and then later Bernat [**8**] studied the representation theory of the simplest kind of nonnilpotent solvable groups, namely the groups of type E, and obtained very similar results. There were other contributions to the theory for various classes of solvable groups, but these ultimately led to the work of Auslander and Kostant ([**4**], [**5**]) which gives a full generalization of the Kirillov theory to solvable groups (at least those to type I), and gives a necessary and sufficient condition for such a group to be type I. Simultaneously the Kirillov construction was vastly generalized geometrically by Kostant [**33**] with the suggestion that this process should work in many nonsolvable groups too. In a subsequent paper Pukanszky [**45**] found an analogue of the results in [**5**] in the non-type I case and gave a description of enough primary representations à la Kirillov to decompose the regular representation. The existence or nonexistence and properties of global characters for irreducible representations have been treated in a series of papers by Pukanszky and Duflo. The theory of harmonic analysis on nil and solvmanifolds is just beginning; there are fairly complete initial results in the nilpotent case which we shall describe in the final section.

Several quasi-expository works have appeared including the book of Pukanszky [**47**] which covers, among other topics, nilpotent groups. Very recently the book [**9**] has appeared, and we have leaned heavily on the treatment in this excellent work in preparing this paper and we are deeply grateful – see also the lecture notes by S. R. Quint [**48**], which is a useful source for much of the same material. In this paper we shall assume that the reader is familiar with the basic facts about unitary representation theory, including, as necessary introductory material, the theory of Borel spaces, and then induced representations, irreducible, primary, and CCR representations, projective representations, and perhaps most importantly the Mackey little group method. All of this material is contained in the Introduction of [**6**] among a number of other places. As general background, the reader should know something of the algebraic and geometric properties of solvable groups and solvmanifolds, and a discussion of these points may be found in [**1**] or [**2**].

Let us add one comment on induced representations; when H is a closed subgroup of G, the quotient space G/H may or may not admit a G-invariant measure. Indeed, if Δ_H and Δ_G are the modular functions of H and G respectively, the condition for existence of such a measure is that $m(h)=\Delta_H(h)/\Delta_G(h)=1$ for $h\in H$. However, in general, if we consider the space of all continuous functions C on G with compact support mod H such that $f(gh)=m(h)\,f(g)$, then C admits a unique positive G-invariant linear functional denoted suggestively by $\int f\,d\mu$ which replaces the invariant integral on G/H when $m=1$. We shall use this linear functional in forming our function spaces for induced representations.

2. Nilpotent groups. We will begin with a discussion of the Kirillov theory – a remarkable contribution which presents us with a complete description of the dual space, i.e., the space of equivalence classes of irreducible representations, of a simply connected nilpotent group N. If such an N is given, let $\mathfrak{n}$ be its Lie algebra and let exp denote the exponential map and log its inverse (exp is a global diffeomorphism by a classical result of Birkhoff). Now N acts on $\mathfrak{n}$ via the adjoint representation, and hence by duality, N acts on $\mathfrak{n}^*$, the dual vector space to $\mathfrak{n}$, via what we call the coadjoint representation, and we write $\mathrm{ad}^*(g)$ for the linear transformation on $\mathfrak{n}^*$ corresponding to $g\in N$.

Suppose that $f\in\mathfrak{n}^*$ is a linear functional on $\mathfrak{n}$, and that H is a closed connected Lie subgroup of N with Lie algebra $\mathfrak{h}$. We shall say that $\mathfrak{h}$ (or H) is subordinate to f if $f([X, Y])=0$ for $X, Y\in\mathfrak{h}$, or, in other words, that f vanishes on the commutator subalgebra of $\mathfrak{h}$. Then we may define a homomorphism λ_f of H into the circle group T as follows: $\lambda_f(x)=\exp(2\pi i\log(x))$. We view λ_f as a one-dimensional unitary representation of H and we denote by $\pi(\mathfrak{h}, f)$ the representation of N induced by this one-dimensional representation. In terms of this we may state the Kirillov results beginning with an irreducibility criterion:

THEOREM 1. *The representation $\pi(\mathfrak{h}, f)$ is irreducible if and only if $\mathfrak{h}$ has maximal dimension among all subordinate subalgebras.*

It will turn out presently when we study solvable groups that the condition of maximality above is the same as insisting that $\mathfrak{h}$ have maximal dimension among all subspaces V such that $f([X, Y])=0$ for $X, Y \in V$ whether or not V is a subalgebra. Now that we have produced a number of irreducible representations of N, we seek to determine when they are distinct.

THEOREM 2. *For fixed* $f \in \mathfrak{n}^*$, $\pi(\mathfrak{h}, f)$ *is independent of the choice of* $\mathfrak{h}$ *provided that it is of maximal dimension. We denote the corresponding representation by* $\pi(f)$.

THEOREM 3. *The irreducible representation* $\pi(f_1)$ *is unitarily equivalent to* $\pi(f_2)$ *if and only if there is an* $\mathfrak{n} \in N$ *with* $\mathrm{ad}^*(\mathfrak{n})\, f_1 = f_2$, *or in other words if and only if* f_1 *and* f_2 *are in the same N-orbit in* $\mathfrak{n}^*$.

This last theorem says that we may attach to each orbit $O \subset \mathfrak{n}^*$, an irreducible unitary representation which we denote by $\pi(O)$, by letting $\pi(O)=\pi(f)$ for any $f \in O$. The final result of Kirillov is a completeness statement which is an analogue of a known theorem for finite nilpotent groups.

THEOREM 4. *Every irreducible representation of N is of the form* $\pi(O)$, *and therefore we may identify* $\hat{N}$, *the dual space of N, with the orbit space* $\mathfrak{n}^*/\mathrm{ad}^*(N)$.

Of course both $\hat{N}$ and the orbit space $\mathfrak{n}^*/\mathrm{ad}(N)^*$ carry natural topologies, the former, the Fell topology [**24**], and the latter the quotient topology from $\mathfrak{n}^*$. It is still unknown whether the natural isomorphism between these spaces is a homeomorphism; however it is known that the map from $\mathfrak{n}^*/\mathrm{ad}^*(N)$ onto $\hat{N}$ is continuous (this is a special case of Proposition 2 of [**42**]). The natural map is therefore a Borel isomorphism from $\mathfrak{n}^*/\mathrm{ad}^*(N)$ to $\hat{N}$. Before going any further it would be wise to look at an example, which is commonly called the Heisenberg group or algebra. Let V be a k-dimensional complex vector space with a nondegenerate positive definite Hermitian inner product $(\ ,\ )$. Let $\mathfrak{n}=\mathfrak{n}_k=\boldsymbol{R}\oplus V$ ($\boldsymbol{R}$=real line) and define the structure of real Lie algebra on this $(2k+1)$-dimensional real vector space as follows

$$[t+v,\ t'+v'] = \mathrm{Im}(v, v') \qquad (t \in \boldsymbol{R}).$$

If X_j is an orthonormal basis of V, and $Y_j = iX_j$ and if $Z=(1+0)\in \mathfrak{n}_k$, then these $2k+1$ vectors are a basis for $\mathfrak{n}_k$ and we have $[X_i, Y_j]=\delta_{ij}Z$ with all other brackets between the basis vectors equal to 0. This is in a sense the simplest kind of non-abelian nilpotent Lie algebra – its name coming from the familiar Heisenberg commutation relations between the momentum and position observables in elementary quantum mechanics which this algebra reproduces. Let (Z^*, X_i^*, Y_j^*)

be the dual basis in $\mathfrak{n}^*$; one may observe that if $f \in \mathfrak{n}^*$, and $f(Z)=0$ that then the orbit of f is one point in $\mathfrak{n}^*$. Any subalgebra is subordinate and so the maximal ones are $\mathfrak{n}_k$ itself and $\pi(f)$ is one dimensional or more precisely $\pi(f)(\exp(t+v)) = \exp(2\pi i f(v)) \cdot 1$ where 1 is the unit operator. If on the other hand $f(Z) \neq 0$, then one may easily verify that the orbit through f is a $2k$-dimensional hyperplane described by $O = \{h \mid h(Z) = \alpha = f(Z)\}$. (It is evident that O is contained in this hyperplane, since $h(Z)$ is constant as h runs over any orbit.) We may therefore select representatives for the orbits, i.e., parameterize them, by taking linear functionals of the form αZ^*, $\alpha \neq 0$. The corresponding representation will be denoted by π_α and is infinite dimensional. Furthermore one sees that $\mathfrak{h}_1 = \{Z + \sum t_i X_i, t_i \in \boldsymbol{R}\}$ and $\mathfrak{h}_2 = \{Z + \sum t_i Y_i, t_i \in \boldsymbol{R}\}$ are subordinate subalgebras, and indeed one can in fact determine all subordinate subalgebras of maximal dimension as follows: Any such $\mathfrak{h}$ must contain $\mathfrak{z} = (Z)$ which is the center of $\mathfrak{n}_k$ and is hence of the form $\mathfrak{h} = \mathfrak{z} + W$ where $W \subset V$. It is evident that $\mathfrak{h}$ is a subalgebra, and it is subordinate if and only if W is an isotropic subspace of V for the skew bilinear form $\operatorname{Im}(\cdot, \cdot)$. It is of maximal dimension if and only if W is a maximal isotropic subspace, and in particular is of dimension k.

If we choose $\mathfrak{h}_1$ as our maximal subordinate subalgebra, we may realize π_α as a representation on a certain space of functions on N_k transforming under $H_1 = \exp(\mathfrak{h}_1)$. Since there is a natural cross section of H_1 in N_k given by $\exp(V_2)$, where V_2 is the space of linear combinations of the Y's, we may indeed realize π_α on a space of functions on a real k-dimensional vector space V_2. By computing invariant measures, we see that this space is $L_2(V_2)$ and that π_α is given by

$$(\pi_\alpha(\exp(t+w_1+w_2))\,\phi)(v_2) = \exp(2\pi i\alpha(t+\operatorname{Im}(w_1, v_2) - \tfrac{1}{2}\operatorname{Im}(w_1, w_2)))\,\phi(v_2 - w_2)$$

where $w_2 \in V_2$ and $w_1 \in V_1$ which is the span of the X's. If on the other hand we use the maximal subordinate subalgebra $\mathfrak{h}_2$, we get by Theorem 2 the same representation but now realized on $L_2(V_1)$ as follows:

$$(\pi'_\alpha(\exp(t+w_1+w_2))\,\psi)(v_1) = \exp(2\pi i\alpha(t+\operatorname{Im}(w_2, v_1) + \tfrac{1}{2}\operatorname{Im}(w_1, w_2)))\,\psi(v_1 - w_1).$$

It is of interest to know the unitary operator from $L_2(V_2)$ to $L_2(V_1)$ which implements the unitary equivalence of these representations, and one may see that it is the Fourier transform where we note that V_1 and V_2 are in natural duality with each other via the bilinear form $\operatorname{Im}(\cdot, \cdot)$ restricted to $V_1 \times V_2$.

The proofs of the Kirillov theorems above proceed by induction on the dimension of the group N as follows. If the center $\mathfrak{z}$ of $\mathfrak{n}$ has dimension greater than one, the linear functional f vanishes on a codimension one subspace $\mathfrak{z}_0$ of $\mathfrak{z}$ which is an ideal in $\mathfrak{n}$. One can replace $\mathfrak{n}$ by $\mathfrak{n}/\mathfrak{z}_0$ which has smaller dimension and use the inductive hypothesis. If $\mathfrak{z}$ is one dimensional, we can select a Y not in $\mathfrak{z}$ such that the centralizer of Y is a codimension one ideal $\mathfrak{n}_0$. Then the span of $\mathfrak{z}$ and Y, $\mathfrak{z}_1$, is an abelian ideal, and if $Z_1 = \exp(\mathfrak{z}_1)$, one applies Mackey's little group method

[35] to Z_1. Questions about representations of N are reduced to questions about representations of $N_0 = \exp(\mathfrak{n}_0)$ which is of smaller dimension, and the inductive hypothesis is used. This pattern of induction on dimension is used constantly in the study of nilpotent groups.

We have now a complete description of the irreducible representations of N, and we note now that we can obtain additional information about these representations. For instance, such a group N is known to be CCR (which we recall means that if π is any irreducible representation of N, and if ϕ is any function in $L_1(N)$, then the integrated or "smoothed" operator $\pi(\phi) = \int_N \phi(s)\, \pi(s)\, ds$ is a compact operator). This was first proved by Dixmier [13], and later a shorter proof was given by Fell [25]. Not only is the operator $\pi(\phi)$ compact, but for suitable ϕ it is also a trace class operator, and indeed the representation π has a distribution character. Let us make this more precise as follows:

THEOREM 5. *Let ϕ be a C^∞ function with compact support on N, then the operator $\pi(\phi)$ has a trace and the map $\phi \to \mathrm{Tr}(\pi(\phi)) = \chi_\pi(\phi)$ is a distribution on the manifold N.*

This distribution χ_π is called the global character of the representation π, and it generalizes in an obvious way the notion of the character of a finite-dimensional representation which is simply the function on the group given by $\chi_\pi(s) = \mathrm{Tr}(\pi(s))$. In our case the distribution χ_π turns out to be tempered in the following obvious sense: The exponential map is a diffeomorphism from N to $\mathfrak{n}$ and so χ_π may be transported back to a distribution $\bar{\chi}_\pi$ on $\mathfrak{n}$ which is then tempered in the usual sense as a distribution on the vector space $\mathfrak{n}$.

In addition we would of course like to be able to compute the distribution χ_π explicitly or, what is the same thing, the distribution $\bar{\chi}_\pi$. This is done as follows: Associated with π, there is an orbit O in $\mathfrak{n}^*$, and this orbit is of course a homogeneous space for N; since N and any subgroup are unimodular, it follows that O has an N-invariant measure μ_0, unique up to positive scalar multiples. This measure may be viewed then as a measure on the vector space $\mathfrak{n}^*$.

THEOREM 6. *The measure μ_0, as a measure on $\mathfrak{n}^*$, is tempered in that it defines a tempered distribution on $\mathfrak{n}^*$. Finally μ_0 may be normalized so that its Fourier transform is precisely $\bar{\chi}_\pi$.*

This was first proved by Kirillov [30], but one may refer to Pukanszky [43] for a more delicate discussion. Indeed one may wonder what the required normalization of the measure μ_0 is so that Theorem 6 holds, and Pukanszky gives a very precise answer as follows in [43]. This involves anticipating some of our future discussion and for this discussion $\mathfrak{n}$ can be an arbitrary finite-dimensional Lie algebra, not necessarily nilpotent. We note that O is a manifold and that if $f \in O$, let $N_f=\{g:\mathrm{ad}^*(g)f=f\}$ be the isotropy subgroup at f. The isotropy algebra $\mathfrak{n}_f$ is the Lie subalgebra of $\mathfrak{n}$ corresponding to N_f and is described as $\mathfrak{n}_f=\{X: f([X,Y])=0 \text{ for all } Y\in\mathfrak{n}\}$. We then observe that if we let $B_f(X,Y)=f([X,Y])$ which is a skew symmetric bilinear form on $\mathfrak{n}$, then $\mathfrak{n}_f$ is precisely the singular subspace for B_f. Therefore B_f may be viewed as a nondegenerate alternating two form on $\mathfrak{n}/\mathfrak{n}_f$ which is therefore even dimensional. Now we note that the tangent space to O at f is, by general theory, canonically isomorphic to $\mathfrak{n}/\mathfrak{n}_f$, and so the tangent space to O at f carries a nondegenerate skew two form ω_f. (Note that this implies that O is even dimensional, say of dimension $2k$.) Now the kth exterior power $(\omega_f)^k$ of ω_f defines a volume form on the tangent space to O at f, and hence defines a volume form on O, and hence a measure which we denote by ν_0. Note that this measure is God-given to us, and there are no normalizations. We return to the nilpotent case.

THEOREM 7. *The proper normalization of* μ_0 *in Theorem* 6 *is* $\mu_0=(k!\,2^k)^{-1}\nu_0$.

We noted for later reference that the discussion of N_f, $\mathfrak{n}_f$, B_f and the construction of ω_f is quite general and works in any Lie group G (not just in the nilpotent case). Thus, in general, any G orbit in the dual of its Lie algebra $\mathfrak{g}^*$ is even dimensional and carries a canonically determined differential two form ω which is nondegenerate at every point. It may be checked that ω is closed and invariant under the action of G so that O is in fact a symplectic manifold [**33**]. In this context it is helpful to mention the very striking result of Duflo and Vergne [**23**] on the generic structure of the isotropy algebra $\mathfrak{g}_f$ – namely it is almost always abelian for any Lie algebra $\mathfrak{g}$.

PROPOSITION. *Let* $\mathfrak{g}$ *be any finite-dimensional Lie algebra and let* m *be the minimal dimension of* $\mathfrak{g}_f$ *as* f *varies in* $\mathfrak{g}^*$. *Then* $H=\{f \mid \dim\mathfrak{g}_f=m\}$ *is a Zariski open set in* $\mathfrak{g}^*$ *and* $\mathfrak{g}_f$ *for* $f\in H$ *is abelian.*

As an example of Theorem 6, one may compute explicitly the characters of the irreducible representations of the Heisenberg group N_k. The infinite-dimensional representations π_α have characters which as distributions are supported on the one-dimensional center Z of N_k and as a distribution on Z which is simply a line, the character is a function, $C_\alpha\exp(2\pi i\alpha z)$ times Haar measure where C_α is a constant and where z is the natural parameter on Z.

The global characters χ_π satisfy partial differential equations on the group N, as such characters do whenever they exist. More precisely if $U(\mathfrak{n})$ is the universal enveloping algebra of the Lie algebra $\mathfrak{n}$, and if $\mathscr{Z}$ is its center, we may view elements of $\mathscr{Z}$ as bi-invariant differential operators on N. If π is an irreducible unitary representation, it is known [**51**] that there is a corresponding infinitesimal representation $d\pi$ of $\mathfrak{n}$ as unbounded essentially skew adjoint operators all defined on a common invariant dense domain, say the space of all C^∞ vectors of the representation. The representation extends to a representation of $U(\mathfrak{n})$, and, for $z \in \mathscr{Z}$, $d\pi(z)$ is known to be a scalar operator $\lambda_\pi(z) \cdot 1$ where 1 is the unit operator ([**51**], [**53**]). The object $z \to \lambda_\pi(z)$ is a homomorphism of $\mathscr{Z}$ into the complex numbers and is called the infinitesimal character of π. Then one checks that χ_π satisfies the equations

$$z \cdot \chi_\pi = \lambda_\pi(z)\, \chi_\pi, \qquad z \in \mathscr{Z},$$

where the left-hand side is the result of applying the differential operator z to the distribution χ_π.

The infinitesimal character λ_π associated with π can be viewed in a slightly different way as follows: We consider the symmetric algebra $S(\mathfrak{n})$ on the vector space $\mathfrak{n}$, which may equivalently be viewed as the polynomial functions on $\mathfrak{n}^*$. The Birkhoff-Witt theorem asserts that $S(\mathfrak{n})$ and $U(\mathfrak{n})$ are isomorphic as N-modules, an isomorphism being specifically the symmetrization map s of $S(\mathfrak{n})$ onto $U(\mathfrak{n})$. Now clearly $S(\mathfrak{n})$ and $U(\mathfrak{n})$ are not isomorphic as algebras, but s does carry the ring of N-invariants $I(\mathfrak{n})$ in $S(\mathfrak{n})$ onto the N-invariants in $U(\mathfrak{n})$, that is, the center $\mathscr{Z}$. The remarkable fact is that s is an algebra isomorphism of $I(\mathfrak{n})$ onto $\mathscr{Z}$ [**20**], when $\mathfrak{n}$ is nilpotent. In fact we may remark that Duflo has shown that $I(\mathfrak{n})$ and $\mathscr{Z}$ are always isomorphic as algebras [**22**]. In the solvable and semi-simple cases an obvious modification of the map s accomplishes this isomorphism, but whether this modification of s works in general is unknown.

Now if π is an irreducible representation of N, it is associated to some orbit $O \subset \mathfrak{n}^*$ according to the Kirillov theory, but now an orbit O defines a homomorphism λ_0 of $I(\mathfrak{n})$ to the complex numbers simply by evaluating invariant polynomial functions at any point of O. Then λ_0 may be transported via s^{-1} to give rise to a homomorphism $s_*(\lambda_0)$ of $\mathscr{Z}$ into the complex numbers. The following ties this together with the previous discussion.

THEOREM 8. *We have $s_*(\lambda_0) = \lambda_\pi$, the infinitesimal character as defined above.*

Finally to round out the picture of harmonic analysis on N, we turn to the Plancherel formula, which from one point of view describes the decomposition of the regular representation of N as a direct integral of irreducible representations, and from a slightly different point of view is an expansion of the Dirac delta measure δ concentrated at the identity element of N in terms of the characters of the irreducible representations. (That this is really very much the same, see [**50**].) The characters χ_π are parameterized by $\pi \in \hat{N}$ or, equivalently, let us write χ_a where $a \in \mathfrak{n}^*/\mathrm{ad}^*(N)$ is the corresponding point in the quotient space. The Plancherel formula is then $\delta = \int \chi_a \, d\mu(a)$ where μ is Plancherel measure, a measure on $\mathfrak{n}^*/\mathrm{ad}^*(N)$. Now this orbit space carries a uniquely determined family of measures, all equivalent to each other in the sense of mutual absolute continuity which we shall call rational measures. One describes them as follows: Let F be the field of fractions of the integral domain $I(\mathfrak{n})$ so that F is the set of all N-invariant rational functions on $\mathfrak{n}^*$, and may be viewed as functions on $\mathfrak{n}^*/\mathrm{ad}^*(N)$. The field F has finite transcendence degree, say d, over the complex numbers (equal to the algebraic dimension of $\mathfrak{n}^*/\mathrm{ad}^*(N)$, and equal to the codimension of a generic orbit O in $\mathfrak{n}^*$). Now pick generators $\phi_1, \ldots, \phi_s$ for F and then the map $a \to (\phi_1(a), \ldots, \phi_s(a))$, $a \in \mathfrak{n}^*/\mathrm{ad}^*(N)$, is one-to-one outside of the image in $\mathfrak{n}^*/\mathrm{ad}^*(N)$ of a Zariski open set. The image outside of this set is a submanifold in R^s and we take any Euclidean metric on R^s and this gives rise to a volume form and hence a measure on this submanifold. We pull this volume form back to a measure μ on almost all of $\mathfrak{n}^*/\mathrm{ad}^*(N)$ and declare the complement to have measure zero. If we change the generators or the metric, the resulting measure μ is changed into a new measure μ^1 but the Radon-Nikodym derivative $d\mu^1/d\mu$ is a function in F. Thus we have a whole family of measures differing only by multiplication by elements of F, and we call these rational measures. There are alternate and more elegant ways of describing such measures such as by specifying a d-differential Ω on the field F in the sense of algebraic geometry, and then using Ω to get a measure on $\mathfrak{n}^*/\mathrm{ad}^*(N)$ which is an algebraic variety having F as its function field, as in **[60]**.

THEOREM 9. *Plancherel measure as a measure on* $\mathfrak{n}^*/\mathrm{ad}^*(N)$ *is rational.*

In this connection let us note that Kirillov in **[31]** has announced without proof a more explicit form of the Plancherel theorem from which Theorem 9 would follow. In the case of the Heisenberg group N_k of dimension $2k+1$, the orbit space $\mathfrak{n}^*/\mathrm{ad}^*(N)$ consists of a $2k$-dimensional linear space consisting of orbits reduced to single points, together with a one-parameter family O_α of orbits where O_α is the set of linear functionals taking the value $\alpha \neq 0$ on a fixed element of the center of N_k. It is evident that the orbits reduced to a point form a null set for any rational measure (taken all together these points form what would be O_0) so that these measures live on the remaining orbits and α running over $R - \{0\}$ is a parameterization of these orbits. Rational measures are evidently precisely those of the form $f(\alpha)\, d\alpha$ where f is any rational function. For N_k one may explicitly compute the Plancherel measure μ_k for N_k and it is given by $d\mu_k(\alpha) = C_k |\alpha|^k \, d\alpha$ where C_k is a constant.

Our discussion has now given a complete description of $\hat{N}$, the characters of the representations and the rudiments of harmonic analysis on these groups. Before going onto the theory for solvable groups, we must investigate an alternate method for constructing irreducible representations which was discovered by Bargmann and Segal; see **[7]**.

We consider the Heisenberg group N_k of dimension $2k+1$; let Z be the one-dimensional center, and note that N/Z which is abelian can be identified with

$\mathfrak{n}_k/\mathfrak{z} \simeq V$ (recall that the Lie algebra $\mathfrak{n}_k$ is written as $\mathfrak{z}+V$, where V is a k-dimensional complex Hermitian vector space). Thus N/Z has the structure of a k-dimensional complex Hermitian vector space and we let $|\cdot|$ denote the norm in N/Z. Let $\lambda>0$ and consider the space $H(\lambda)$ of all holomorphic functions ϕ on $N/Z \approx V$ such that

$$|\phi|^2 = \int_V \exp(-\pi\lambda|z|^2)\,|\phi(z)|^2\,dz < \infty$$

where dz indicates Haar measure. We are going to define a representation T_λ of $N=N_k$ on this space of functions as follows: Any group element $g \in N$ can be written uniquely as $g=\exp(tZ)\exp(w)$ with $L \in \boldsymbol{R}$ and $w \in V$, and we identify g with the pair (t, w). Then

$$(T_\lambda(g)\,\phi)\,(v) = \exp(2\pi i\lambda t)\,\exp\left(-\pi\left(\frac{\lambda}{2}|w|^2 + \lambda(v, w)\right)\right)\phi(v-w)$$

where $g=(t, w)$.

The following result holds for these representations.

THEOREM 10. *The space $H(\lambda)$ is a Hilbert space and is nonempty if and only if $\lambda>0$. The representation T_λ is unitary, and is equivalent to the representation π_λ defined previously corresponding to a linear functional $f=\lambda Z^* \in \mathfrak{n}_k^*$.*

We shall not give the proof but refer the reader to [7], [**48**] for more details. For the sake of completeness let us write down the unitary operator implementing the equivalence of T_λ with π_λ. This should be a unitary operator from $H(\lambda)$ to $L_2(V_1)$ where V_1 is the span of the X_i's in our previous discussion. Let us formally write down

$$(*) \qquad (U\phi)\,(s) = \int_{V_1} \exp(-i\pi sy\lambda)\,\exp(-\pi\lambda(s^2+y^2)/2)\,\phi(s+iy)\,dy$$

with $s \in V_1$ and where the integration is carried out over the real k-dimensional vector space V_1.

THEOREM 11. *The formula (*) is well defined for any ϕ which is a linear combination of functions of the form $\exp((z, \alpha))\,p(z)$ where $\alpha \in V$ and where p is a polynomial, and these form a dense subspace of $H(\lambda)$. Moreover $U(\phi)$ is in $L_2(V_1)$ and U extends to a unitary equivalence of the representation T_λ and the representation π_λ.*

The reader may consult [7] or [**48**] for the details of this calculation. This result then gives us an alternate way of realizing the representations of this particular nilpotent group at least for $\lambda > 0$. One may reasonably ask about the representations for $\lambda < 0$, and the answer quite simply is that they can be realized on similarly defined spaces of antiholomorphic functions on V. These realizations at first glance appear to be vastly different from the previous realization given by Kirillov. The point is, however, that both kinds of realizations may be subsumed under a common construction to which we shall turn presently. However, first we want to recast the realization of these spaces of holomorphic functions. Essentially what we have done is to look at functions on $V \approx N/Z$; however, what we should really be looking at are functions on N transforming according to a certain character of Z, or in other words, we want to look at sections of a certain line bundle on $N/Z \simeq V$. We are going to express the transformation properties and the holomorphy of such sections by using the left invariant vector fields on N.

More precisely let W be any element of the Lie algebra of $\mathfrak{n}_k = \mathfrak{n}$, so that W can be viewed as a left invariant vector field on N. If ϕ is C^∞ function on N, we write the action of W on ϕ by $\phi * W$; now if $W = W_1 + iW_2$ is in the complexified Lie algebra $\mathfrak{n}_C$ of $\mathfrak{n}$, we define $\phi * W = \phi * W_1 + i\phi * W_2$. Now let $\mathfrak{h}$ denote the complex vector subspace of $\mathfrak{n}_C$ spanned by $X_j + iY_j$ for $j = 1, \ldots, k$, and by Z. We note that $\mathfrak{h}$ is a subalgebra of $\mathfrak{n}_C$ and in fact is abelian. Furthermore let us denote by f a linear functional on $\mathfrak{n}$ describing the representation π_λ (for instance $f = \lambda Z^*$) and extend f to $\mathfrak{n}_C$ so as to be complex linear. Now clearly $f([\mathfrak{h}, \mathfrak{h}]) = 0$ so that $\mathfrak{h}$ is subordinate to f in the sense of Kirillov above except that it is a complex subalgebra instead of a real one. Moreover it has the same dimension as the real ones of maximal dimension, and indeed $\mathfrak{h}$ has the largest possible dimension of any such complex subordinate subalgebra. Let us now consider the conditions

$$(*) \qquad \phi * X = -2\pi i f(x)\,\phi \quad \text{for } X \in \mathfrak{h}$$

and for a C^∞ function ϕ on N. For $X = Z$, this says that $\phi(\exp(tZ)\,g) = \exp(-2\pi i\lambda t)\,\phi(g)$ for all $g \in N$. This says in particular that $|\phi|^2$ is a function on N/Z. Moreover since $X_j + iY_j \in \mathfrak{h}$, we have then that $\phi * X_j + i\phi * Y_j = 0$ or in other words a kind of holomorphic condition. We shall make this more precise in a moment, but now let $H(f, \mathfrak{h})$ denote the space of all C^∞ functions ϕ on N satisfying (*) above and such that $|\phi|^2$ is integrable over N/Z, which makes sense by the remark above. We norm this space by using

$$|\phi|^2 = \int_{N/Z} |\phi|^2 \, dz.$$

We now define a representation S_λ of N on this space by $(S_\lambda(g))\,(\phi)\,(u) = \phi(g^{-1}u)$, i.e., action by left translations, which is isometric for the above norm. This representation then looks very much like an induced representation, but instead of inducing a character from an honest subgroup H of N, we are "inducing" a character from a complex subgroup of the complexification of N. Indeed if $\mathfrak{h}$ were a real subalgebra of $\mathfrak{n}$, conditions (*) above are easily seen to be the conditions for a C^∞ function to belong to the usual Hilbert space for an induced representation, i.e., that it transforms in a certain way under the right action of H.

We have now constructed a representation S_λ which appears to be quite similar to T_λ and is easily seen to be equivalent to it. Indeed for $\phi_0 \in H(\lambda)$, the Hilbert space of T_λ, define $(V\phi_0)(\exp(tZ)\exp(v)) = \exp(-2\pi i\lambda t)\exp(-\pi|v|^2/2)\,\phi_0(v)$.

THEOREM 12. *The map V is a unitary equivalence of the unitary representation T_λ on $H(\lambda)$ with the representation S_λ on $H(f, \mathfrak{h})$ (and so in particular this last space is complete as it stands).*

This discussion completes our treatment of the nilpotent case and shows the way in a sense to a discussion of unitary representations of general solvable groups. This holomorphic realization which we have described in detail for the Heisenberg groups plays an essential role in the general theory, and, as we shall see, the construction is far from the seemingly accidental phenomenon it might appear to be.

Before passing on let us stop to remark on some extensions of the Kirillov theory for nilpotent groups. First of all, it is a rather routine matter to extend the description of $\hat{N}$ to the case when N is a unipotent algebraic group over a p-adic field. One has a Lie algebra $\mathfrak{n}$, a coadjoint representation, and again $\hat{N} \approx \mathfrak{n}^*/\mathrm{ad}^*(N)$ [**36**]. The additional information on characters should carry over also.

There is also a natural generalization of this to the adele group of a rational unipotent group over a number field [**36**]. In a different direction, Howe [**27**] has shown that the Kirillov theory works for discrete subgroups Γ of nilpotent Lie groups in a suitably modified form. Of course $\hat{\Gamma}$ is terrible, since Γ is not type I, but the point is that the primitive ideal space of the C^* group algebra of Γ is still nice and one could hope, and indeed it is the case, that this has a Kirillov type description at least in some cases. Howe has also extended this reasoning to more general nilpotent topological groups [**29**].

3. Exponential groups. Let us now turn our attention to the case of a general simply connected solvable Lie group G and its representation theory, and we begin with the case which in one sense most resembles the nilpotent case, and that is the case of exponential solvable groups or Lie algebras or as they are sometimes called groups of type E. In order to clarify this, recall the definition of a root of a solvable Lie algebra $\mathfrak{g}$; we triangulate the adjoint action of $\mathfrak{g}$ on its complexification $\mathfrak{g}_C$ using Lie's theorem, and then the diagonal entries of the action, viewed as functions of $X \in \mathfrak{g}$ are linear maps $\varrho_i(X)$ of $\mathfrak{g}$ into the complex numbers. Taken as a set, this finite family of linear maps is independent of the choice of triangulation, and are called the roots of $\mathfrak{g}$. One has the following characterization of exponential groups G and their Lie algebras $\mathfrak{g}$.

THEOREM 13. *The following are equivalent:*
(1) $\exp: \mathfrak{g} \to G$ *is surjective,*
(2) $\exp: \mathfrak{g} \to G$ *is injective,*
(3) $\exp: \mathfrak{g} \to G$ *is a diffeomorphism,*
(4) *the imaginary part of every root is a multiple of its real part.*

Now let G be an exponential solvable Lie group; Takenouchi [**55**] established the analogue of Theorem 4 in this case by showing that any irreducible unitary representation could be realized as the representation induced by a one-dimensional representation of some connected subgroup H of G. (He also showed that the group was type I.) Bernat refined this by putting it into the Kirillov framework. Given $f \in \mathfrak{g}^*$, the dual of the Lie algebra of G, one says as before that a subalgebra $\mathfrak{h}$ or the corresponding subgroup $H = \exp(\mathfrak{h})$ is subordinate to f if f vanishes on $[\mathfrak{h}, \mathfrak{h}]$ and one defines the character λ_f of H as in the nilpotent case, and we form the induced representation $\pi(f, \mathfrak{h})$. The result quoted above asserts that every $\pi \in \hat{G}$ is of the form $\pi(f, \mathfrak{h})$; the next result to be established by Bernat [**8**] was that, for a given f, there exists an $\mathfrak{h}$ such that $\pi(f, \mathfrak{h})$ is irreducible; and finally he determined when two such irreducible representations were equivalent. We summarize these results as follows:

THEOREM 14. *For each* $f \in \mathfrak{h}^*$, *there exists a subordinate subalgebra* $\mathfrak{h}$ *such that* $\pi(f, \mathfrak{h})$ *is irreducible; such subalgebras have maximal dimension among subordinate subalgebras. Moreover* $\pi(f, \mathfrak{h})$ *depends only on* f, *and is denoted by* $\pi(f)$ *and* $\pi(f)$ *depends only on the orbit* $O \subset \mathfrak{g}^*$ *to which* f *belongs. Finally* $f \to \pi(f)$ *sets up a bijection between* $\mathfrak{g}^*/\mathrm{ad}^*(G)$ *and* $\hat{G}$.

Thus we get a nearly complete analogue of the Kirillov theory, the only missing piece being the characterization of those $\mathfrak{h}$ for which $\pi(f, \mathfrak{h})$ is irreducible, and Pukanszky in [**41**] settled this question very beautifully as follows: For a subalgebra $\mathfrak{h}$ of $\mathfrak{g}$ let $\mathfrak{h}^\perp$ denote its annihilator in $\mathfrak{g}^*$.

THEOREM 15. *The representation* $\pi(f, \mathfrak{h})$ *is irreducible if and only if* $\mathfrak{h}$ *is of maximal dimension among subordinate subalgebras and the orbit* O *of* f *in* $\mathfrak{g}^*$ *contains the affine subspace* $f + \mathfrak{h}^\perp$. *The maximality condition is equivalent to the condition that* $\dim \mathfrak{h} = \dim \mathfrak{g} - \frac{1}{2} \dim(O)$.

Kostant will discuss in greater detail in his lectures the intrinsic geometric meaning of the condition $O \supset f+\mathfrak{h}^{\perp}$. (See Blattner's article in this volume.) This is called the Pukanszky condition and is equivalent to the assertion that $\exp(H)\cdot f = f+\mathfrak{h}^{\perp}$. It is automatic in the presence of maximality of dimension in the nilpotent case and also in the special case of type E groups when no nonzero root is real [**41**]. The behavior described in Theorem 15 is well illustrated by the $ax+b$ group which is of type E and has Lie algebra $\mathfrak{g}$ with a basis of two elements A and B with $[A, B]=B$. Let A^* and B^* be the dual basis and it is easy to calculate the orbits in $\mathfrak{g}^*$ in terms of these. The elements tA^*, $t\in \boldsymbol{R}$, are orbits reduced to single points and correspond to the one-parameter family of one-dimensional representations of G [**34**]. There are two other orbits $O^{\pm}=\{tA^*+sB^* \text{ with } s>0 \text{ (or } s<0)\}$ which are of dimension two. These correspond to the well-known two infinite-dimensional representations $\pi^{\pm}$ of G [**34**].

We may conveniently choose representatives $f^{\pm}=\pm B^*$ and we note that any one-dimensional subspace of $\mathfrak{g}$ is a subordinate subalgebra of maximal dimension. It is well known by Mackey's theory that the "correct" subalgebra to take is $\mathfrak{h}=(B)$ so that $\mathfrak{h}^{\perp}=(A^*)$. In this case $\pi(f^{\pm}, \mathfrak{h})$ is irreducible and this is in accord with Theorem 15 for $O^{\pm}$ consists of the upper (lower) half plane in the A^*-B^* plane and $f^{\pm}+\mathfrak{h}^{\perp}$ is a horizontal line lying entirely within a single orbit! On the other hand, any other $\mathfrak{h}$ fails to satisfy the condition of Theorem 15 for $f+\mathfrak{h}^{\perp}$ is a diagonal line meeting both $O^{\pm}$, and it is known that $\pi(f, \mathfrak{h})$ is not irreducible but is in fact $\pi^{+}\oplus\pi^{-}$. This observation suggests at once a conjecture which was established by M. Vergne [**59**]. We first remark that Pukanszky in [**42**] showed that $\pi(f, \mathfrak{h})$ is a finite direct sum of irreducible representations if $\mathfrak{h}$ satisfies the condition on maximality of dimension. (This was done at the same time as he showed that $f\to\pi(f)$ is a continuous map from $\mathfrak{g}^*/\mathrm{ad}^*(G)$ onto $\hat{G}$ for exponential groups – a fact we mentioned in the special case of nilpotent groups in our earlier discussion.) Vergne's result is as follows:

THEOREM 16. *Let G be the type* E, *$f\in\mathfrak{g}^*$ and $\mathfrak{h}$ of maximal dimension. Then the affine subspace $f+\mathfrak{h}^{\perp}$ is up to a closed set of smaller dimension the intersection of $f+\mathfrak{h}^{\perp}$ with a finite number of orbits $O_1, \dots, O_k$, each $O_i\cap(f+\mathfrak{h}^{\perp})$ being open in $f+\mathfrak{h}^{\perp}$, and having a finite number n_i of connected components. Let $\pi_1, \dots, \pi_k$ be the irreducible representations corresponding to the O_i; then the decomposition of $\pi(f, \mathfrak{h})$ is given by $\sum\bigoplus n_i\pi_i$.*

With these additional facts one has as virtually a complete picture as one had in the nilpotent case. Let us now turn to a discussion of the existence of global characters and related topics for exponential solvable groups. One of the first questions that may be asked is that of determining when an irreducible representation π associated with an orbit is CCR. There is an obvious conjecture that an irreducible representation is CCR if and only if the corresponding orbit is closed.

As evidence for this one may quote Pukanszky's result in [**42**] that establishes that if π is CCR, then the orbit is closed – a fact that follows from the continuity of the map from $\mathfrak{g}^*/\mathrm{ad}^*(G)$ onto $\hat{G}$. Intuitively this conjecture is correct, since CCR representations are the closed points in $\hat{G}$ in the Fell topology and closed orbits are the closed points in the orbit space. Moreover, in the final theorem in [**42**], Pukanszky proves the converse of the first statement and hence establishes the validity of the conjecture in the case when the adjoint group of G is algebraic. However, the issue appears to be open in general. One might go even further and pose as a general principle the following (which involves anticipating some of our later discussion, and we defer detailed discussion of it until then):

PRINCIPLE. *To the extent that irreducible representations of a general Lie group G correspond to orbits in the dual of the Lie algebra of G, CCR representations correspond to closed orbits.*

Let us return to exponential solvable groups and global characters of their representations. In the case when $\mathrm{ad}(G)$ is algebraic and the representation π associated to a closed orbit O in $\mathfrak{g}^*$, Pukanszky in [**44**] produces a distribution χ_π on the group G, and shows that $\mathrm{Tr}(\pi(\phi))=\chi_\pi(\phi)$ at least for ϕ of the form $\psi^* * \psi$ with ψ C^∞ (or in fact C^n for some n) and of compact support. As Duflo points out, any C^∞ ϕ of compact support can be written as a finite linear combination of such functions (in fact $\phi=\phi_1*\phi_2+\phi_3*\phi_4$, and then polarize) and so Pukanszky's χ_π is the global distribution character. Pukanszky gives a formula for it which is a direct generalization of Theorem 6; Duflo amplified and extended this formula in [**20**] and [**8**], and we shall use his formulation, slightly modified so that integrals can be taken over the group rather than the Lie algebra. Specifically let O be an orbit and $f\in O$, and let $\mathfrak{g}_f$ be the isotropy algebra. Then $\mathfrak{g}_f$ acts symplectically on the quotient space $\mathfrak{g}/\mathfrak{g}_f$ and hence the roots of this action are of the form $\pm\mu_i$, $1\leq i\leq d$, where $2d=\dim\mathfrak{g}/\mathfrak{g}_f$. We extend these linear functionals on $\mathfrak{g}_f$ to $\mathfrak{g}_f+[\mathfrak{g},\mathfrak{g}]$ by making them zero $[\mathfrak{g},\mathfrak{g}]$ (which is consistent), and note that they are restrictions of roots of $\mathfrak{g}$. Specifically let $\lambda_1,\dots,\lambda_d$ be roots of $\mathfrak{g}$ restricting to $\mu_1,\dots,\mu_d$ on $\mathfrak{g}_f+[\mathfrak{g},\mathfrak{g}]$ and let F be the remaining roots. For a root λ we define the function $S_\lambda(x)$ to $\sinh(\lambda(x)/2)/\lambda(x)$ on $\mathfrak{g}$, and we define

$$L_O(\exp(x))=\Delta(\exp(x))^{-1/2}\prod_{\lambda\in F} S_\lambda(x)$$

where Δ is the modular function, and $\Delta(\exp(x))=\prod\exp(\lambda(x))$ (for all roots λ). Let F_1 be the roots $\{\lambda_i\}$ and define a similar function on $\mathfrak{g}$ by

$$K_O(x)=\prod S_\mu(x) \qquad (\mu\in F_1).$$

We note that as a consequence of the fact that G is exponential solvable, $S_\lambda(x)$ is never zero for any λ and x (and this is equivalent to being exponential solvable). The function L_O above depends on the orbit O together with some choices; however, the restriction of L_O to $\exp(\mathfrak{g}_f+[\mathfrak{g},\mathfrak{g}])$ which is all we ever want is uniquely determined by O.

With this preparation, we proceed to Duflo's formula which is proved under broader assumptions than Pukanszky's on the orbit O. More specifically, we say that an orbit O in $\mathfrak{g}^*$ (for any Lie algebra $\mathfrak{g}$) satisfies condition (H) if there is an integer $m>0$ such that

$$\text{(H)} \qquad \int_O (1+|h|)^{-m}\, d\nu_O<\infty$$

where $|h|$ is some norm on $\mathfrak{g}^*$ and where ν_O is the canonical measure on O discussed in Theorem 7, which, as noted there, always exists. Note that condition (H) assures us that the measure ν_O viewed as a distribution on $\mathfrak{g}^*$ is tempered; Duflo notes that if G is ad-algebraic (not necessarily even exponential) and if O is closed, then condition (H) is satisfied. Since Duflo proves his result assuming condition (H), his results contain Pukanszky's.

THEOREM 17. *Let G be exponential solvable and let π be the representation of G associated to an orbit O in $\mathfrak{g}^*$ satisfying condition* (H). *Then π has a global distribution character χ_π given as follows: For a C^∞ function ϕ of compact support on G, let $\phi_1(x)=\phi(\exp(x))\, L_O(\exp(x))$ where L_O is as above. Then* $\operatorname{Tr}(\pi(\phi))=\chi_\pi(\phi)$ $=\int \hat{\phi}_1\, d\mu_O$ *where* ^ *denotes Fourier transform and where the invariant measure μ_O on O must be normalized relative to ν_O just as in Theorem* 7.

We note that such a π is CCR and, according to our philosophy, O should be closed – or in other words condition (H) should imply that the orbit is closed, but we do not know if this is true yet. This statement of Theorem 17 differs from Duflo's only in that an integral has been transferred from $\mathfrak{g}$ to G via the exponential map. If we write the formula in $\mathfrak{g}$ instead of on G, as will be more convenient when dealing with nonexponential groups, it becomes the following: Let ϕ be C^∞ of compact support on $\mathfrak{g}$, and then

$$\tilde{\chi}_\pi(\phi)=\operatorname{Tr}\left(\int \phi(x)\, \pi(\exp(x))\, dx\right)=\int (\phi K_O^{-1})^\wedge\, d\mu_O$$

where K_O is as described above.

It is somewhat annoying that the function L_O that enters into the definition of χ_π depends on the orbit; however, Duflo has shown that one can fix things up so that it does not at least for a generic set of orbits. To be more precise, let us introduce the function $L(\exp(x))=\Delta(\exp(x))^{-1/2}(\det(\sinh(\operatorname{ad}(x)/2))/(\operatorname{ad}(x)/2))^{1/2}$;

note that the determinant inside is always positive since G is exponential solvable. The following theorem gives a uniform formula valid for almost all orbits (recall from above that $\mathfrak{g}_f$ is abelian for f in a Zariski open set in $\mathfrak{g}^*$).

THEOREM 17$'$. *Let G be exponential solvable; if O satisfies condition* (H) *and if $\mathfrak{g}_f$ for $f \in O$ is nilpotent, then the global character χ_π of the corresponding representation π has the form $\chi_\pi(\phi) = \int \hat{\phi}_1 \, d\mu_O$ where $\phi_1(x) = \phi(\exp(x)) \, L(\exp(x))$ and where $\hat{}$ is as usual the Fourier transform and L is as above.*

If we introduce the function $K(x) = \varrho(x)^{1/2}$ where

$$\varrho(x) = \det(\sinh(\mathrm{ad}(x/2))/\mathrm{ad}(x/2))$$

then this formula may be rewritten on $\mathfrak{g}$ as follows for $\phi \in C_c^\infty(\mathfrak{g})$:

$$\tilde{\chi}_\pi(\phi) = \mathrm{Tr}\left(\int \phi(x) \, \pi(\exp(x)) \, dx\right) = \int (\phi K^{-1})\hat{} \, d\mu_O .$$

This completes our discussion of exponential groups; one of the most important points to note is that the irreducible representations could always be obtained by inducing from real subordinate subgroups. This is not true in general and one should perhaps first look at an example where such complex subordinate subalgebras are necessary (as opposed to just being an alternate way of doing things as for the Heisenberg groups). We consider the (four-dimensional) oscillator group whose Lie algebra $\mathfrak{g}$ has a basis denoted by (R, X, Y, Z) with $[X, Y] = Z$, $[R, X] = Y$, $[R, Y] = -X$ and all other brackets zero. Note that (X, Y, Z) span an ideal isomorphic to the three-dimensional Heisenberg algebra $\mathfrak{n}_1$. One might remark now that in some sense one understands the representation theory of solvable Lie groups once one understands it for three examples, namely the three-dimensional Heisenberg group N_1, the $ax+b$ group, and the oscillator group. All of the essential phenomena are displayed, and, although it is a vast simplification, the rest of the theory is one of compounding these examples.

It is false for this oscillator group that every irreducible representation is induced by a one-dimensional representation of a subgroup. Indeed Z is in the center of $\mathfrak{g}$ and any irreducible representation π is scalar on Z, and suppose that $\pi(\exp sZ) = \exp(2\pi i \lambda s)$ with $\lambda \neq 0$. We may apply the Mackey little group method to the normal subgroup N with Lie algebra $\mathfrak{n}$ spanned by (X, Y, Z), and with the knowledge we have of the representations of this Heisenberg group, we can deduce that the restriction of π to N is irreducible. Now this says that if H is the subgroup from which π is induced, then $H \cap N$ is two dimensional and $HN = G$, and hence that H must be three dimensional. However, a moment's inspection shows that there can be no such H. Another way of looking at this is that we start with $f \in \mathfrak{g}^*$ which would give rise to such a π, say $f = Z^*$, and that we would try to construct H or its Lie algebra $\mathfrak{h}$, a subalgebra subordinate to f, by first constructing $\mathfrak{h}\cap\mathfrak{n}$ which is subordinate to f restricted to $\mathfrak{n}$. To do this properly the point is that we would have to select $\mathfrak{h}\cap\mathfrak{n}$ in such a way that it is invariant under the subgroup generated by the element R. Since this subgroup acts by rotations in the X-Y plane, it is clear that there is no way to pick $\mathfrak{h}\cap\mathfrak{n}$ as a real subalgebra. But on the other hand it is clear that there is a complex subalgebra subordinate to f restricted to $\mathfrak{n}$ which does the trick, namely the subalgebra spanned by Z and $X+iY$. (The algebra spanned by Z and $X-iY$ is also invariant, and these are the only two.) Finally, the complex subalgebra $\mathfrak{h}$ of $\mathfrak{g}_C$ spanned by Z, $X+iY$, and R is subordinate to f, and is what should be used for "holomorphic" induction. Indeed if we follow exactly the prescription given for the Heisenberg group in our previous discussion using this $\mathfrak{h}$, one arrives at an irreducible representation of G, and the space of functions turns out to be exactly the same as the space of functions for the nilpotent group $N=N_1$ when we use the subalgebra spanned by Z and $X+iY$.

This example and our previous discussion set the stage for the introduction of the general notion of polarization, and for the Auslander-Kostant theory of representations of general solvable groups. We have all the examples and special cases worked out and one may easily see the simplicity and elegance and strong unifying features which this way of looking at things yields.

4. Solvable groups in general. We turn now to the case of a general solvable Lie group and its representation theory. For the moment let us consider a quite general connected Lie group with Lie algebra $\mathfrak{g}$. Let $f\in\mathfrak{g}^*$ be a linear functional on $\mathfrak{g}$ and note that then B_f defined by $B_f(X, Y)=f([X, Y])$, $X, Y\in\mathfrak{g}$, is a skew symmetric bilinear form on $\mathfrak{g}$. Now G operates on $\mathfrak{g}^*$ via the coadjoint representation and let O denote the orbit of f. Let $G_f=\{g:\mathrm{ad}^*(g)f=f\}$ be the isotropy group of f and let $\mathfrak{g}_f$ be the corresponding isotropy subalgebra, $\mathfrak{g}_f=\{X:f([X, Y])=B_f(X, Y)=0, \forall Y\}$. We note then that $\mathfrak{g}_f$ is precisely the singular subspace of B_f, and that B_f may be viewed as a nondegenerate form on $\mathfrak{g}/\mathfrak{g}_f$, which is necessarily even dimensional. Now O, the orbit of f, is a manifold and, as we have remarked, its tangent space at f may be canonically identified with $\mathfrak{g}/\mathfrak{g}_f$ and hence B_f may be transported to the tangent space to O at f, and we denote it by ω_f. The assignment of ω_f to each $f\in O$ is nothing other than a differential two form on O which we denote by ω. One sees that ω is closed and G-invariant so that O is a homogeneous symplectic manifold for G. Now O is even dimensional say $2d$, and ω^d is a G-invariant volume form on O – a fact we have used previously on several occasions in special cases.

If n is the dimension of G and m is the dimension of $\mathfrak{g}_f$, then $n=m+2d$. We now complexify everything, obtaining $\mathfrak{g}_C$, and an extension of B_f to $\mathfrak{g}_C$. For $Z=X+iY\in\mathfrak{g}_C$, let $\bar{Z}=X-iY$. Now the skew two form B_f has maximal isotropic complex subspaces V (i.e., $B_f(u, v)=0$ for all $u, v\in V$) and any such V contains

$\mathfrak{g}_f$ by maximality, and, moreover, has dimension $m+d$, or put another way, "V is halfway between $(\mathfrak{g}_f)_C$ and $\mathfrak{g}_C$." In accord with our previous discussion we want to consider such complex subspaces $\mathfrak{h}=V$ which in addition are subalgebras. We note that $\mathfrak{h}\cap\bar{\mathfrak{h}}$ is also a subalgebra which is real in the sense that it is invariant under $x\to\bar{x}$, or, equivalently, it is the complexification of a real subalgebra, which we denote by $\mathfrak{d}$, so $\mathfrak{d}=\mathfrak{h}\cap\mathfrak{g}$ and $\mathfrak{h}\cap\bar{\mathfrak{h}}=\mathfrak{d}_C$. If we consider $\mathfrak{h}+\bar{\mathfrak{h}}$, it is also a real subspace, but there is no *a priori* reason why it should be a subalgebra, unless of course $\mathfrak{h}=\bar{\mathfrak{h}}$. In general let $\mathfrak{e}_C=\mathfrak{h}+\bar{\mathfrak{h}}$. We are now ready to introduce the notion of a polarization.

DEFINITION. A subalgebra $\mathfrak{h}$ of $\mathfrak{g}_C$ is said to be a polarization for f if
(1) as a subspace, $\mathfrak{h}$ is maximally isotropic for B_f;
(2) $\mathfrak{h}+\bar{\mathfrak{h}}=\mathfrak{e}_C$ is a subalgebra.
The polarization is said to be real if $\mathfrak{h}=\bar{\mathfrak{h}}$.

We have seen many examples of such objects in our discussion of nilpotent and exponential solvable groups, and the reader may well pause for a moment to reflect on the previous discussion. One may ask whether for a given $f\in\mathfrak{g}^*$, polarizations always exist. It is easy to see that the answer is 'no' in general; specifically let $\mathfrak{g}$ be a semisimple Lie algebra split over $\boldsymbol{R}$, and not Sl_n, for instance the Lie algebra in the higher symplectic groups. The Killing form identifies $\mathfrak{g}$ with $\mathfrak{g}^*$ in a G-equivariant manner, and lets us take $f=X^*$ an element corresponding to a highest root vector. These elements fall into one G-orbit which is minimal in the sense that the image of this G-orbit in the projective space $P(\mathfrak{g})\approx P(\mathfrak{g}^*)$ is the unique compact G-orbit. One sees that $\mathfrak{g}_f$ is codimension one in a maximal parabolic subalgebra, and from this it is relatively easy to see that there is no polarization, real or complex for f. We note that $\mathfrak{g}_f$ for this f is a large as it can be for $f\neq 0$. On the other hand, Dixmier has established [**19**] a positive result for f in a Zariski open set in $\mathfrak{g}^*$ where $\mathfrak{g}_f$ is small.

PROPOSITION. *Let $f\in\mathfrak{g}^*$, any finite-dimensional Lie algebra, be such that* dim$(\mathfrak{g}_f)$ *is minimal* (*so that $\mathfrak{g}_f$ is in particular abelian*). *Then there exists a* (*possibly complex*) *polarization $\mathfrak{h}$ for f. Moreover, $\mathfrak{h}$ can be chosen to be solvable as a Lie algebra.*

Let us return to the definition of a polarization $\mathfrak{h}$, and examine some consequences of this definition. We note that $\mathfrak{h}$ being maximal isotropic is equal to its own orthogonal complement with respect to B_f. From this it follows immediately that $\mathfrak{d}_C$ and $\mathfrak{e}_C$ are orthogonal complements of each other and hence so are $\mathfrak{d}$ and $\mathfrak{e}$, back in $\mathfrak{g}$, and in particular the restriction of the bilinear form B_f to $\mathfrak{e}$ or $\mathfrak{e}_C$ has $\mathfrak{d}$ or $\mathfrak{d}_C$ as a singular subspace. Therefore $\mathfrak{e}/\mathfrak{d}$ has a canonical alternating two form, also denoted by B_f, and is hence even dimensional.

Our next goal (as in the Heisenberg algebra $\mathfrak{n}_k$ where, for the complex polarization we discussed earlier, we have $\mathfrak{z}=\mathfrak{d}$, and $\mathfrak{n}_k=\mathfrak{e}$) is to get a complex and, even better, a Hermitian structure on $\mathfrak{e}/\mathfrak{d}$. That is, we need a real linear map J such that $J^2=-1$. We define J on $\mathfrak{e}_C$ first by noting that any $z\in\mathfrak{e}_C$ may be written as $z=x+\bar{y}$ with $x, y\in\mathfrak{h}$ and these are unique mod $\mathfrak{d}_C$. Note that $z\in\mathfrak{e}$ if and only if $y=x$, and now define $J(z)=-ix+i\bar{y}$ and since $(\overline{-ix})=i\bar{x}$, we see that $J(z)$ is real if z is, and this defines J on $\mathfrak{e}/\mathfrak{d}$ and makes it into a complex vector space. The next step is to show that B_f on $\mathfrak{e}/\mathfrak{d}$ is invariant under J, and this is a simple calculation. One may then define a nondegenerate symmetric bilinear form S_f by the formula $S_f(x, y)=B_f(x, Jy)=B_f(y, Jx)$, and a nondegenerate Hermitian form $H_f=S_f+iB_f$, thereby making $\mathfrak{e}/\mathfrak{d}$ into a Hermitian vector space.

It is sometimes convenient to replace $\mathfrak{e}/\mathfrak{d}$ with another space $\mathfrak{h}/\mathfrak{d}_C$, and we note that the map $p(x)=x+\bar{x}$ from the latter to the former is an isomorphism from the natural complex structure on $\mathfrak{h}/\mathfrak{d}_C$ (multiplication by $-i$) to $\mathfrak{e}/\mathfrak{d}$ with the complex structure J. The Hermitian form H_f may then be transported back to $\mathfrak{h}/\mathfrak{d}_C$ and in this space it is given by $H_f(x, y)=2iB_f(x, \bar{y})$, $x, y\in\mathfrak{h}/\mathfrak{d}_C$.

The following distinguishes a most important property of polarizations, and indeed those failing to satisfy it are of no use to us in constructing representations.

DEFINITION. One says that the polarization $\mathfrak{h}$ is *positive* if H_f is positive definite, i.e., if $2iB_f(x, \bar{x})\geqq 0$ for $x\in\mathfrak{h}$ (or equivalently if the symmetric form S_f is positive definite).

One sees by an immediate calculation that the complex polarization of $\mathfrak{n}_k$ which we discussed is positive provided in the notation there that $\lambda>0$, where $f=\lambda Z^*$. This brings out an important point; for that polarization is not positive (in fact it is negative) for $f=-\lambda Z^*$, $\lambda>0$, and in general $\mathfrak{h}$ is not positive for f and $-f$. We denote the set of polarizations for f by $P(f, \mathfrak{g}_C)$ and the set of positive ones by $P^+(f, \mathfrak{g}_C)$.

Recall that G_f is the isotropy subgroup of f, and this group acts on the set of polarizations for a fixed linear functional. It is a natural additional condition on a polarization to demand that it be invariant under G_f. Note that it is automatically invariant under the connected component G_f° of G_f since $\mathfrak{g}_f\subset\mathfrak{h}$. This condition will be essential in the following. We turn now to the problem of answering questions about whether certain obvious subgroups associated with $\mathfrak{h}$ are closed. We have the following [**9**, p. 68]:

PROPOSITION 1. *Let $\mathfrak{h}$ be a polarization for f, and let D° be the analytic subgroup corresponding to $\mathfrak{d}$. Then D° is closed, and if $\mathfrak{h}$ is stable under G_f, then $D=D^\circ\cdot G_f$ is closed and has D° for connected component.*

This settles the question of the closure of D; we come to closure of E presently – a problem that is somewhat more subtle.

We saw in the exponential solvable case that the "Pukanszky condition" played a crucial role in the irreducibility of representations. We introduce it here for complex polarizations as follows.

DEFINITION. Let the polarization $\mathfrak{h}$ for f be invariant under G_f. We say it satisfies the Pukanszky condition if $f+\mathfrak{e}^{\perp}\subset O$ where O is the orbit of f.

As before this is equivalent to the condition that $D^{\circ}\cdot f=f+\mathfrak{e}^{\perp}$ [**9**, p. 70]. Another by-product of this condition is the fact that if it is satisfied then $D^{\circ}\cap G_f=G_f^{\circ}$ [**9**, p. 72]. Notice now that we are dealing with a fundamentally new phenomenon, and that is the possibility that G_f may not be connected – as it always is for nilpotent and exponential solvable groups. This introduces totally new problems into the theory. Since the group G in question is always taken to be simply connected, we see by the exact sequence of homotopy groups that G_f modulo its connected component G_f° is precisely the fundamental group of the orbit O of f. In the exponential solvable or nilpotent case, the specified linear functional f gave rise to a unique character λ_f of $H=\exp(\mathfrak{h})$ for our real polarization $\mathfrak{h}$ which we used for forming induced representations. The group D is going to play the role of the group H for this purpose, and so we have to construct characters of this group. To do this we begin with the groups G_f and G_f° and we do the following:

DEFINITION. The form f is *integral* if there is a character σ_f of G_f such that $\sigma_f(\exp(x))=\exp(2\pi i f(x))$ for $x\in\mathfrak{g}_f$, the Lie algebra of G_f. (We will remark presently the equivalence of this with another condition of a cohomological nature.)

Note that the condition specifies σ_f on the connected component of G_f, and there could be many such σ_f. However, there may not be any in general, for, first of all, G_f° may not be simply connected and the formula above may be inconsistent. For simply connected solvable groups, which are our prime subject of interest, G_f° is simply connected so this problem does not arise and σ_f always exists on G_f°. However, the real problem is that of extending σ_f to all of G_f and that is what the condition of integrality asserts in this case. Once there is one extension there are many in general and all possible such characters may be very simply parameterized. If σ_f is one, then all others are of the form $\sigma_f'(x)=\sigma_f(x)\,s(\dot{x})$ where s is a character of $G_f/G_f^{\circ}\approx\pi_1(O)$ and where $\dot{x}$ is the image of x in this quotient. Thus if $L(f)$ is the set of all such characters, it is a principal homogeneous space for the dual group of $\pi_1(O)$ which we denote by $A(O)$; it is a compact abelian group. We now have the following [**9**, p. 72]:

THEOREM 18. *Let f be integral and let σ_f be any character of G_f as in the definition. Suppose that $\mathfrak{h}$ is a polarization for f invariant under G_f and satisfying the Pukanszky condition. Then there is a unique character λ_f of $D=D^{\circ}\cdot G_f$ extending σ_f and such that $\lambda_f(\exp(x))=\exp(2\pi i f(x))$ for $x\in\mathfrak{d}$. Hence the set of such characters $L(f,\mathfrak{h})$ is a principal homogeneous space for $A(O)$.*

The final piece of information which we need is something about the closure of the groups E and E°, where E° is the analytic subgroup corresponding to the subalgebra $\mathfrak{e}$ of the polarization and $E = E^\circ \cdot G_f$ where it is assumed that $\mathfrak{h}$ is invariant under G_f, so this makes sense. For this and for many other questions we introduce the notion of a polarization being admissible relative to an ideal $\mathfrak{a}$ of $\mathfrak{g}$.

DEFINITION. The polarization $\mathfrak{h}$ for f is admissible for an ideal $\mathfrak{a}$ if $\mathfrak{h} \cap \mathfrak{a}_C$ is a polarization for the restriction of f to $\mathfrak{a}$.

The idea of this condition is that we will select a nice $\mathfrak{a}$, say a nilpotent ideal such that $\mathfrak{g}/\mathfrak{a}$ is also nice, such as nilpotent or even abelian. This can of course be done when $\mathfrak{g}$ is solvable. As an example we have the following [**9**, p. 71]:

THEOREM 19. *Let $\mathfrak{a}$ be a nilpotent ideal with $\mathfrak{g}/\mathfrak{a}$ abelian. Let $f \in \mathfrak{g}^*$ and let $\mathfrak{h}$ be a polarization invariant under G_f and admissible for $\mathfrak{g}$. Then $\mathfrak{h}$ satisfies the Pukanszky condition and E is closed.*

At this point we have stacked up a number of conditions on a polarization for $f \in \mathfrak{g}^*$, the important ones being positivity, the Pukanszky condition, invariance by G_f, and admissibility for an ideal $\mathfrak{a}$. Before going further it would be wise to have an existence theorem – and a good one exists due to Auslander and Kostant [**5**]; see also [**9**, p. 82].

THEOREM 20. *Let $\mathfrak{g}$ be solvable and $f \in \mathfrak{g}^*$, and let $\mathfrak{a}$ be any ideal, and f' the restriction of f to $\mathfrak{a}$, and G_f and $G_{f'}$ the stability groups of f and f'. Then there exists a positive polarization $\mathfrak{h}$ satisfying the Pukanszky condition, admissible for $\mathfrak{a}$ and such that $\mathfrak{h}$ is invariant under G_f and $\mathfrak{h} \cap \mathfrak{a}_C$ is invariant under $G_{f'}$.*

The proof given in [**9**] is quite direct and conceptually simple, although a bit complicated technically. The idea of the proof is that one takes an appropriately chosen composition series of ideals in $\mathfrak{g}_C$, say $\mathfrak{g}_i$ with some $\mathfrak{g}_i = \mathfrak{a}_C$, and one lets f_i be the restriction of f to $\mathfrak{g}_i$, and let V_i be the singular subspace in $\mathfrak{g}_i$ of B_{f_i}. Then $\mathfrak{h} = \sum V_i$ essentially does the trick. As a by-product of the argument one gets some useful corollaries, namely that if $\mathfrak{n}$ is nilpotent, and a is any automorphism of $\mathfrak{n}$ fixing some f, then $\mathfrak{n}$ admits a positive polarization for f invariant under the automorphism a. The method of argument is also capable of yielding a number of other kinds of related results on the existence of polarizations with special properties, one of which is the existence of polarizations as described in Theorem 20 which are simultaneously admissible for two nilpotent ideals $\mathfrak{a}_1$ and $\mathfrak{a}_2$ with $\mathfrak{a}_1 \subset \mathfrak{a}_2$.

Now that we have the notion of a polarization for $f \in \mathfrak{g}^*$ together with special properties, and an existence theorem, let us consider how one obtains from such an object a unitary representation. This is done essentially by the device of holomorphic induction. Let f be an integral linear functional and let $\mathfrak{h}$ be a positive polarization satisfying the Pukanszky condition, invariant by G_f, and admissible for some nilpotent ideal $\mathfrak{a}$ with $\mathfrak{g}/\mathfrak{a}$ abelian. Finally let $\sigma_f \in L(f)$ and let

$\lambda_f \in L(f, \mathfrak{h})$ be its unique extension to D given by Theorem 18. The subgroup D is closed and let Δ_D denote its modular function and Δ_G that of G, and $m(d) = \Delta_D/\Delta_G(d)$. We note that $m(\exp(x)) = \exp(\mathrm{Tr}(\mathrm{ad}_{\mathfrak{g}/\mathfrak{d}}(x)))$ for $x \in \mathfrak{d}$ and that $\mathrm{Tr}(\mathrm{ad}_{\mathfrak{g}/\mathfrak{d}}(x)) = \mathrm{Tr}(\mathrm{ad}_{\mathfrak{g}/\mathfrak{e}}(x)) = \mathrm{Tr}(\mathrm{ad}_{\mathfrak{g}_c/\mathfrak{e}_c}(x))$ for $x \in \mathfrak{d}$, since $\mathrm{ad}(x)$ preserves a skew bilinear form on $\mathfrak{e}/\mathfrak{d}$ and is traceless there.

Now consider all C^∞ functions ϕ on G satisfying

(1) $$\phi(gd) = \lambda_f(d)^{-1} m(d)^{1/2} \phi(g),$$

(2) $$\phi * X = -2\pi i f(X) + \tfrac{1}{2}\,\mathrm{Tr}\,(\mathrm{ad}_{\mathfrak{g}/\mathfrak{e}}(X)), \qquad X \in \mathfrak{h},$$

and noting that (1) insures that

(*) $$|\phi(gd)|^2 = m(d)\,|\phi(g)|^2,$$

(3) $$|\phi|^2 = \int_{G/D} |\phi|^2\, d\mu < \infty,$$

where this integral is the usual invariant linear functional on functions satisfying (*). In (2) we are of course interpreting $X \in \mathfrak{h}$ as a left invariant vector field on G. The extra term involving the trace on $\mathfrak{g}/\mathfrak{d}$ is necessary to make (1) and (2) compatible. This space of functions has a norm given in (3) and we let $H(f, \sigma_f, \mathfrak{h}, G)$ denote its completion. Finally we define a representation $\pi(f, \sigma_f, \mathfrak{h}, G)$ on this Hilbert space by left translation $(\pi(f, \sigma_f, \mathfrak{h}, G)(g)\,\phi)(s) = \phi(g^{-1}s)$.

This unitary representation of course depends on f, and σ_f, but also *a priori* on $\mathfrak{h}$. If we are to have any analogue of the theory for nilpotent and exponential solvable groups we have to show independence of $\mathfrak{h}$ and this is one of the really hard problems. After resolving this to obtain a representation $\pi(f, \sigma_f)$, the dependence only on the orbit of f is quite easy. The proof of independence of $\mathfrak{h}$ involves finding a different construction for the representation $\pi(f, \sigma_f, \mathfrak{h}, G)$ in which it is more or less manifestly independent of $\mathfrak{h}$. This different construction involves crucially the Mackey little group method. We select a nilpotent ideal $\mathfrak{n}$ such that $\mathfrak{g}/\mathfrak{n}$ is abelian, and let $N = \exp(\mathfrak{n})$ be the corresponding closed normal subgroup which is type I. We let f' be the restriction of f to $\mathfrak{n}$ and let $\pi(f')$ be the irreducible unitary representation of N corresponding to f' where G operates on $\mathfrak{n}$ (and $\mathfrak{n}^*$) via the adjoint representation. The first fact that is evident is that $S = G_1 N = N G_1 \subset G$ is the stability group of $\pi(f')$, for if $g \in G_1$ and $g \cdot \pi(f') = \pi(f')$, then g carries f' into some linear functional on $\mathfrak{n}$ giving the same representation as f' and hence in the same N-orbit as f'. Thus $g \cdot f' = n \cdot f'$ for some $n \in N$ and $g = n(n^{-1}g)$ with $n^{-1}g \in G_1$ as desired. Mackey's little group method [**34**] tells us to look for irreducible representations of S which restrict on N to a multiple of $\pi(f')$ and then to induce these representations up to G. This yields a family of irreducible representations parameterized by the irreducible representations of S restricting to a multiple of $\pi(f')$ on N. This is achieved in two steps; we first form the semidirect product $N \times G_1$ where G_1 is viewed as operating via automorphisms on N. The group S of interest to us is a quotient group of this semidirect product by the normal subgroup $L=\{(g, g^{-1})$ with $g \in N \cap G_1 = N_{f'}$, the isotropy group of f' in $N\}$. The problem, recall, is to extend $\pi(f')$ to S; the following gives an extension to $N \times G_1$, instead, which we shall use as a basis for later work.

THEOREM 21. *There exists a "canonical" extension $\varrho(f')$ of $\pi(f')$ to $N \times G_1$. This representation satisfies $\varrho(f')(g) = \lambda_{f'}(g)^{-1}\pi(f')(g)$ whenever $g \in N_{f'} \subset G_1$ where $\lambda_{f'}$ is the usual character of $N_{f'}$ determined by f'.*

The proof of this theorem is rather involved, but it comes down to the selection of $\varrho(f')(g)$ for $g \in G_1$; these operators by their definition are intertwining operators for the equivalent representations $n \to \pi_{f'}(n)$ and $n \to \pi_{f'}(gng^{-1})$ of N. Such an intertwining operator is determined up to a scalar of absolute value one, and the point of the proof is to select these intertwining operators very carefully. We note that $\mathrm{ad}(g)$ is an automorphism of $\mathfrak{n}$ fixing f', and we make use very crucially of an earlier statement that f' admits an $\mathrm{ad}(g)$ invariant positive polarization. In the course of the proof one therefore has to essentially prove the main theorem on independence of polarization in the nilpotent case. This is reasonably easy and worth a short digression at this point.

Recall that we discussed such polarizations for Heisenberg groups but gave no other examples; the point is that there are no other examples; more precisely if there is a positive (nonreal) polarization in the nilpotent case, the group or at least that part of it associated with the nonreal part of the polarization is essentially a Heisenberg group.

THEOREM 22. *If $\mathfrak{n}$ is nilpotent, and $\mathfrak{h}$ a positive nonreal polarization for $f \in \mathfrak{n}^*$, let $\mathfrak{d}$ and $\mathfrak{e}$ be defined as before, and let $\mathfrak{b} = \mathfrak{d} \cap \ker(f)$. Then $\mathfrak{b}$ and $\mathfrak{d}$ are ideals in $\mathfrak{e}$ and $\mathfrak{e}/\mathfrak{b}$ is a Heisenberg algebra with center $\mathfrak{d}/\mathfrak{b}$.*

Since we have shown independence of polarization in the Heisenberg algebras by explicitly constructing the intertwining operators, Theorem 22 establishes it in general by reduction to the special case.

Let us return to the situation at hand in Theorem 21. If $N \cap G_1 = (e)$ so that $S \cong N \times G_1$, Mackey's method would provide us with a simple method for constructing all irreducible representations of S restricting on N to a multiple of $\pi(f')$ – namely we pick an irreducible representation of G_1, lift it back to a representation of $N \times G_1$, and tensor it with the representation $\varrho(f')$ provided by

Theorem 21. In general $N \cap G_1 \neq (e)$ but the same idea works – namely we look for irreducible representations μ of G_1 such that $\mu(g) = \lambda_{f'}(g) \cdot 1$, and then pull μ back to a representation of the semidirect product $N \times S$, call it μ and tensor it with $\varrho(f')$ of Theorem 21. The resulting tensor product vanishes on the kernel L of the map of $N \times G_1$ onto S and hence becomes a representation of S as desired. The point now is to construct such a representation μ – and indeed not any old such representation, but one which is determined by the data at hand, namely $f \in \mathfrak{g}^*$ and σ_f, a character of G_f.

Recall that G_1 is the isotropy group of f', and we let G_1° be its connected component and let $\mathfrak{g}_1$ be its Lie algebra and we let f_1 be the restriction of f to $\mathfrak{g}_1$, and let $G_1(f_1)$ be the stabilizer of f_1 in G_1. Recall that $N_{f'} = G_1 \cap N$ is connected and let Q be the connected component of the kernel of $\lambda_{f'}$ in $N_{f'}$. Then it may be shown [**48**, p. 125] that Q and $N_{f'}$ are normal in G_1 and that G_1/Q is nilpotent with connected component G_1°/Q. Since the representation μ we are seeking must vanish on Q and the linear functional f_1 vanishes on its Lie algebra $\mathfrak{q}$ anyway by its defining property, we see that we are in the nilpotent case, and the linear functional f_1 determines a unique irreducible representation μ° of G_1° by Kirillov theory. To get an irreducible representation of G_1 we apply the Mackey little group method again and look for the isotropy subgroup S° of μ° in G_1. Instead of S° we consider $G_1(f_1)$ as above and note that $S^\circ = G_1^\circ \cdot G_1(f_1) = G_1(f_1) \cdot G_1^\circ$. One shows [**48**, Proposition 4.5.4, p. 129] that there exists a positive polarization for f_1 on $\mathfrak{g}_1$, invariant under $G_1(f_1)$ using the fact that $G_1(f_1) = G_f \cdot N_f$. Just as before this allows us to construct a representation ν° of the semidirect product $G_1^\circ \times G_1(f_1)$ which extends μ° of G_1°. This representation as it stands does not factor through to a representation of $S^\circ = G_1^\circ \cdot G_1(f_1)$, but a minor modification of it will. More precisely, using the fact that $G_1(f_1) = G_f \cdot N_{f'}$ [**48**, p. 129] one shows that there is a unique character α_f of $G_1(f_1)$ which extends the given character σ_f of G_f and which satisfies $\alpha_f(\exp(x)) = \exp(2\pi i f(x))$ for $x \in \mathfrak{g}_1(f_1)$, the Lie algebra of $G_1(f_1)$. We then view α_f as a character of $G_1 \times G_1(f_1)$ via projection to the second factor, and tensor this character with the given representation ν° of this semidirect product. This tensor product may be verified now to drop down to a representation of $S^\circ = G_1^\circ \cdot G_1(f_1)$. Let us call this representation $\mu^1(f, \sigma_f)$ to denote the ingredients used in its formation; it is an extension of the representation μ° of G_1°.

The Mackey theory tells us that when we induce this representation from S° up to all of G_1, we get an irreducible representation whose restriction to $N_{f'}$ is a multiple of the character $\lambda_{f'}$ of this group. Let this induced representation be denoted by $\mu(f, \sigma_f)$ and this completes the second step of our construction.

Recall that we had started with the representation $\pi(f')$ of N determined by the restriction of the linear functional to its Lie algebra $\mathfrak{n}$ (any nilpotent ideal with $\mathfrak{g}/\mathfrak{n}$ abelian) and had constructed a representation $\varrho(f')$ of the semidirect product

$N \times G_1$. We now tensor $\varrho(f')$ with the representation $\mu(f, \sigma_f)$ of G_1 to obtain another representation $\varrho(f, \sigma_f)$ of $N \times G_1$. We recall that the stability group of $\pi(f')$ in G is $S = N \cdot G_1$, a quotient of the semidirect product $N \times G_1$, and the construction has been precisely carried out so that $\varrho(f, \sigma_f)$ drops down to a representation of S. This representation by construction is restricted on N to a multiple of $\pi(f')$. Finally we follow the Mackey little group method and induce $\varrho(f, \sigma_f)$ up to a representation of G which we denote by $\varrho(f, \sigma_f, N, G)$. This is an irreducible representation associated to the pair (f, σ_f), σ_f being a character of G_f, and, in our previous notation, a member of the set $L(f)$ which is a principal homogeneous space for the dual group $A(O)$ of the fundamental group of the orbit O of f. It has been constructed using the Mackey little group method (using it twice, in fact – which is what makes things so complicated) and using relatively easily established facts about polarizations only in the nilpotent case. It also appears to depend specifically on the choice of N which played such a key role. In any case it is an irreducible representation which one may reasonably say is associated to the pair (f, σ_f) by extrapolating from the nilpotent case and not using anything about polarizations in general.

Recall now that we had also attached to the pair (f, σ_f) another representation $\pi(f, \sigma_f, \mathfrak{h}, G)$ by means of holomorphic induction starting with a "*good*" polarization relative to some $\mathfrak{a}$. The following theorem ties all of these threads together and settles two nagging problems at once, namely the independence of polarization for π, and the independence of choice of N for ϱ.

THEOREM 23. *Let f, σ_f be given with f integral, and let $\mathfrak{n}$ be a nilpotent ideal with $\mathfrak{g}/\mathfrak{n}$ abelian. Finally let $\mathfrak{h}$ be a positive polarization, invariant by $G(f)$ satisfying the Pukanszky condition and admissible for $\mathfrak{n}$. Then $\pi(f, \sigma_f, \mathfrak{h}, G)$ and $\varrho(f, \sigma_f, N, G)$ are unitarily equivalent, and hence $\pi(f, \sigma_f, \mathfrak{h}, G) = \pi(f, \sigma_f)$ is independent of $\mathfrak{h}$ and $\varrho(f, \sigma_f, N, G) = \varrho(f, \sigma_f)$ is independent of N.*

This theorem was initially proved under the additional hypothesis that $\mathfrak{h} \cap \mathfrak{n}$ is invariant under $G(f')$ where f' is the restriction of f to $\mathfrak{n}$ and $G(f')$ is the isotropy group of f' in G. However, this condition can be deleted; see [**48**] and [**21**]. The proof of the theorem involves going back and looking closely at the construction of these representations and one uses very strongly the independence of polarization in the nilpotent case, to which the general case is finally reduced; see ([**5**], [**9**]).

To see that $\pi(f, \sigma_f, N, G)$ is independent of N choose two such N_1 and N_2 and let $N_3 = N_1 \cap N_2$ with Lie algebras $\mathfrak{n}_i$; it suffices to show say that the representations are the same for $\mathfrak{n}_1$ and $\mathfrak{n}_3$, with $\mathfrak{n}_3 \subset \mathfrak{n}_1$. Recall from the discussion following Theorem 20 that we can find one of our good polarizations which is simultaneously admissible for $\mathfrak{n}_1$ and $\mathfrak{n}_3$, and then our assertion follows. The independence of $\mathfrak{h}$ follows immediately – of course we have restricted $\mathfrak{h}$ very sharply and one might ask if the representation $\pi(f, \sigma_f, \mathfrak{h}, G)$ is independent of $\mathfrak{h}$ as $\mathfrak{h}$ ranges through a wider class.

We have now reproduced the Kirillov construction for a simply connected solvable group and have associated to each pair (f, σ_f) an irreducible representation. Notice that G operates on such pairs by conjugating f via the coadjoint representation and by moving σ_f in a corresponding way. The following is straightforward [**5**, p. 343].

THEOREM 24. *$\pi(f, \sigma_f)=\pi(f', \sigma_{f'})$ if and only if there is some $g \in G$ with $g \cdot (f, \sigma_f)=(f', \sigma_{f'})$.*

This last theorem tells us that we may associate to each integral orbit O, a set $L(O)$ which is a principal homogeneous space for $A(O)$, the dual of the fundamental group of O (say $L(O)=L(f)$ for $f \in O$) and then we have a map from $L(O)$ into $\hat{G}$ the irreducible representations. We then let $L=\bigcup (L(O))$, $O \in \mathfrak{g}^*/\mathrm{ad}^*(G)$, O integral, be the union of all of these spaces so that there is a one-one map k (the Kirillov map): $L \to \hat{G}$ given by Theorems 23 and 24.

The obvious question remaining now is a completeness theorem – that is, whether the Kirillov map k from L into $\hat{G}$ is surjective in analogy with the nilpotent and exponential solvable cases. One suspects on general principles that k could not possibly be surjective if G is not type I for then $\hat{G}$ is so bad that it would not admit such a nice description as the map k would provide. The amazing fact is that the condition that G be type I is sufficient to give the completeness theorem and this is one of the major results of Auslander and Kostant; another is a necessary and sufficient condition for G to be type I.

We first want to rephrase our condition of integrality for a linear functional f; recall that this was that the isotropy group admits a character σ_f with $\sigma_f(\exp(x))=\exp(2\pi i f(x))$ for $x \in \mathfrak{g}_f$. Let O be the orbit of f, and we saw before that O carries a canonically defined closed two form ω which is defined at each $f \in O$ by transferring the form B_f, a skew symmetric bilinear form on $\mathfrak{g}/\mathfrak{g}_f$ over to the tangent space to O at f via the natural isomorphism. We let $[\omega]$ denote the de Rham class of ω in $H^2(O, R)$, and we have the following ([**33**] or [**9**, p. 41 ff.]):

THEOREM 25. *The linear functional f is integral if and only if $[\omega]$ is an integral class, i.e., is in the image of $H^2(O, Z)$ into $H^2(O, R)$.*

Now let us turn to the question of the unitary type of G and the structure of $\hat{G}$; we recall that the Mackey little group method is potentially powerful enough to settle the unitary type of a group G, for the method gives a way at least, if certain conditions are satisfied, to construct all irreducible representations, and indeed all primary representations and to show that they are all type I. The method is also capable of showing that a group is not type I, especially when complemented by the analysis of Chapter I of [6]. We go back then to the construction via the little group method of the representations $\varrho(f, \sigma_f, N, G)$ which we based on the choice of a fixed nilpotent normal subgroup N which is used as the starting point for applying the Mackey theory. Intuitively G should be type I if and only if G operates smoothly (see Chapter I of [6]) on $\hat{N}$ and all the isotropy groups S_π for $\pi \in \hat{N}$ are such that S_π/N is type I for the Mackey obstruction cohomology class in $H^2(S_\pi/N, T)$; see [35], [6]. (When this latter condition is satisfied one says that G is isotropically type I relative to N.) This statement as it stands is false but with a minor modification it is true, and the modification is that we have to look at the action of G not on $\hat{N}$ but on the appropriate pieces of the projective dual space relative to the Mackey obstruction of S_π – that is at certain spaces of projective representations – and demand that that action of G be smooth there. In any case one shows first that if G is type I so that G is necessarily isotropically type I, then all orbits are integral [5, p. 346]. This is based on a very close analysis of the structure of the isotropy group S_π/N and its Mackey obstruction class, and the relation of this to the structure of G_f and the existence of a character σ_f as in the definition of integrality, and uses simple necessary and sufficient conditions given in Chapter IV of [6] that such a group be type I for a specified cohomology class. One then has to analyze the action of G on $\hat{N}$ and on the various spaces of projective representations mentioned above. The smoothness condition mentioned above turns out to be equivalent to the assertion that the coadjoint action of G on $\mathfrak{g}^*$ is smooth. Since this is a topological transformation group on a locally compact space $\mathfrak{g}^*$, Glimm's results [26] give simpler equivalent conditions for smoothness – one being that all the orbits are locally closed or equivalently are all G_δ's. In any case when this is all put together we get a necessary and sufficient condition for G to be type I.

THEOREM 26. *The simply connected solvable Lie group is type* I *if and only if every orbit is integral and is locally closed.*

In summary the integrality comes from the isotropic type I condition and the topological condition on the orbits from the smoothness of action conditions for type I. We note that the orbits for f and tf, $t \neq 0$, are homeomorphic, and, under this homeomorphism, the class $[\omega_t]$ on the orbit of tf is t times the class $[\omega]$ on the orbit of f. Thus the only way the integrality condition can be satisfied for all orbits is if $[\omega]=0$ for all orbits.

Finally as we have noted, the Mackey little group method in the presence of the required smoothness of actions gives an effective procedure for constructing all irreducible representations. This procedure applied to the case at hand shows that the representations $\varrho(f, \sigma_f, N, G)=\varrho(f, \sigma_f)$ as constructed in the discussion prior to Theorem 23 exhaust all the irreducible representations.

THEOREM 27. *If G is type* I, *the map k from* $L = \bigcup L(0)$ *into* $\hat{G}$ *is onto; or equivalently, every irreducible representation is of the form* $\varrho(f, \sigma_f) = \pi(f, \sigma_f)$.

We see now that Theorems 24, 26, and 27 provide essentially as complete a generalization of the Kirillov theory to general type I solvable groups as one could hope for. We note of course in the nilpotent and exponential solvable cases that the orbits are always locally closed ([**42**, p. 101] or [**9**, p. 4 ff.]) so that this condition is automatic. Moreover as G_f is connected, every orbit is diffeomorphic to Euclidean space and integrality is automatic.

Let us briefly turn now to the further properties of the irreducible representations of G and in particular whether they are CCR and whether they have distribution characters. In the study of exponential groups we were led to formulate a general principle to the effect that whenever orbits correspond to representations, closed orbits should correspond to CCR representations. For a type I solvable group this question can be made very precise by the conjecture that a representation $\pi \in \hat{G}$ which comes via k from an orbit O is CCR if and only if O is closed. We saw that that was true for exponential groups which are ad-algebraic. More generally it is true that any ad-algebraic solvable group is type I [**44**, p. 440] and moreover Pukanszky shows in [**44**] that all representations associated with a closed orbit for an ad-algebraic solvable group are CCR. A further piece of evidence comes from combining the results of Chapter V of [**6**] and the results in [**37**]. More precisely let G be type I; then, by [**6**], G is CCR if and only if G is type R, but, by [**37**] and Theorem 26, all orbits of G in $\mathfrak{g}^*$ are closed if and only if G is type R so that the condition for G to be CCR and the condition for all orbits in $\mathfrak{g}^*$ to be closed coincide. Finally one knows in general that if G is type I and unimodular, then almost all representations (with respect to Plancherel measure) are CCR. Now granting a point from the next section on the Plancherel measure this should mean that all orbits in $\mathfrak{g}^*$, except those constituting a null set for Lebesgue measure in $\mathfrak{g}^*$, are closed. This is an interesting problem that appears to be unanswered for general type I solvable groups having only to do with orbits in $\mathfrak{g}^*$.

Let us turn to the question of existence of global characters for representations of a type I, or perhaps even of a general, solvable Lie group. Pukanszky [**44**] obtained a variant of Theorem 17 valid for any ad-algebraic solvable group; however, Duflo [**9**], [**20**] has improved this result and we shall use his formulation. The first new twist in the general (nonexponential) case is that a reparameterization of representations in terms of orbits is in order for the discussion of their characters. What this amounts to is the exact analogue of the situation for compact semisimple groups where one has to translate highest weights by adding half the sum of the positive roots. We proceed to define the analogue of this in general – for the following see [**9**, p. 118].

PROPOSITION. *Let $\mathfrak{g}$ be any Lie algebra, $f \in \mathfrak{g}^*$ and $\mathfrak{h}$ a positive polarization for f invariant under $G(f)$, and let $\mathfrak{d}$ and $\mathfrak{e}$ be as before. Then for $y \in \mathfrak{d}$, $O(y) = \mathrm{Tr}(\mathrm{ad}_{\mathfrak{h}/\mathfrak{d}_c}(y)) = -\mathrm{Tr}(\mathrm{ad}_{\mathfrak{e}_c/\mathfrak{h}}(y))$ is purely imaginary, and the restriction of θ to $\mathfrak{g}_f$ is independent of $\mathfrak{h}$. Moreover, if $\mathfrak{g}$ is solvable, θ vanishes on $[\mathfrak{g}, \mathfrak{g}] \cap \mathfrak{d}$ and hence is the restriction to $\mathfrak{d}$ of a homomorphism $\bar{\theta}$ of $\mathfrak{g}$ into the complex numbers.*

For a compact semisimple algebra $\mathfrak{g}$ and a linear functional f in general position, it is evident that $\mathfrak{g}_f$ is a Cartan subalgebra, and that f restricted to $\mathfrak{g}_f$ is in the interior of some Weyl chamber, and that a positive polarization for f is a Borel subalgebra of $\mathfrak{g}_C$ containing $\mathfrak{g}_f$ and consisting of the positive roots relative to the Weyl chamber of f. Thus evidently the θ of the Proposition is the sum of the positive roots on $\mathfrak{g}_f$. In the solvable case which is of interest to us now, we consider $i\theta$, and note that it is the restriction to $\mathfrak{d}$ of some $i\bar{\theta} \in \mathfrak{g}^*$ so that $i\bar{\theta}$ vanishes on $[\mathfrak{g}, \mathfrak{g}]$. There is therefore a homomorphism ϱ of all of G into the circle group T such that the differential $d\varrho$ of ϱ on $\mathfrak{d}$ is $-(\theta/2)$. Now we have made a choice to get $\bar{\theta}$, but let us consider two linear functionals $i\bar{\theta}_1$ and $i\bar{\theta}_2$ which vanish on $[\mathfrak{g}, \mathfrak{g}]$ and which are extensions of $i\theta$ restricted to $\mathfrak{g}_f$; i.e., $\theta_1(y) = \theta_2(y) = \theta(y)$ for $y \in \mathfrak{g}_f$. Thus $i(\bar{\theta}_1 - \bar{\theta}_2)$ vanishes on $[\mathfrak{g}, \mathfrak{g}] + \mathfrak{g}_f$ and it is a routine exercise to see that $f - i(\bar{\theta}_1/2)$ and $f - i(\bar{\theta}_2/2)$ lie in the same G-orbit in $\mathfrak{g}^*$. Thus to the linear functional f with orbit $O \subset \mathfrak{g}^*$ we have associated in a canonical way another orbit which we denote by O_0, and this is the orbit containing $f - i(\bar{\theta}/2)$. Moreover, let O be integral so that there is a character σ_f of G_f of differential $2\pi i f$ and hence a character λ_f of D of differential $2\pi i f$. We now pick and fix some $f - i(\bar{\theta}/2)$ as above and let ϱ be the homomorphism of G to the wide T with differential $-\theta/2$. Then $\lambda_f(\varrho/D)$ is a homomorphism of D into T with differential $2\pi i f + (\theta/2)$. This implies that the orbit O_0 is integral and conversely one sees that if O_0 is integral, so is O.

This construction yields our new parameterization of the irreducible representations. We fix f and σ_f as above, and let $\pi_0(f, \sigma_f)$ be the representation $\pi(f - i(\bar{\theta}/2), \sigma_f(\varrho/G_f))$ where $\bar{\theta}$ and ϱ are as above. We have already verified that the orbit of $f - i(\bar{\theta}/2)$ is independent of the choice of $\bar{\theta}$, and using Theorem 24, one sees by the way ϱ is chosen that $\pi_0(f, \sigma_f)$ does not depend on the choice of $\bar{\theta}$ either. The point is, as Duflo first noted, that a formula tying characters to orbits must tie the character of $\pi_0(f, \sigma_f)$ to the orbit O and not the character of $\pi(f, \sigma_f)$ to the orbit O. We note of course that the collection of all representations $\pi(f, \sigma_f)$ coincides with the collection of all representations $\pi_0(f, \sigma_f)$ and that the analogue of Theorem 24 can be shown to hold for this new parameterization. Finally $\theta = 0$ and hence $\pi(f, \sigma_f) = \pi_0(f, \sigma_f)$ whenever f admits a real polarization, and so there is no change in the case of exponential groups.

We shall now state Duflo's results which say in some sense that Theorems 17 and 17′ carry over to the general case. Recall that condition (H) on an orbit O with canonical invariant measure ν_O is the condition that there is a norm $|\cdot|$ on $\mathfrak{g}$

and an integer m so that

$$\int_O (1+|x|)^{-m}\, dv_O < \infty .$$

The cleanest version of the general formula for characters involves integration over the Lie algebra rather than the group (these are equivalent in the exponential case); hence we introduce the following sets $V=\{X \mid \exp$ has maximal rank at $X\}=\{X \mid \lambda(X) \neq 2\pi i n$ for any root λ of $\mathfrak{g}$ and any nonzero integer $n\}$, and $V_0=\{X \mid |\mathrm{Im}\,\lambda(X)|<2\pi, \forall \text{ roots } \lambda\}$ which is the connected component of V containing O. Of course $V=V_0=\mathfrak{g}$ in the exponential case. Recall from the discussion leading up to Theorem 17 that we introduced for each $f\in\mathfrak{g}^*$, or equivalently to its orbit O, two sets of roots F and F_1 and functions L_O and K_O on $\mathfrak{g}$. It is an easy matter to see that all of these objects are the same for f or for $f-i(\bar{\theta}/2)$. As before we introduce in place of the character χ_π, the object $\tilde{\chi}_\pi$ defined as follows:

$$\tilde{\chi}_\pi(\phi)=\mathrm{Tr}\left(\int_{\mathfrak{g}} \phi(x)\,\pi(\exp(x))\,dx\right)$$

whenever the right-hand side exists; here ϕ is a C^∞ function of compact support on the Lie algebra.

THEOREM 28. *Let $\mathfrak{g}$ be solvable and let O be an integral orbit in $\mathfrak{g}^*$ satisfying condition* (H) *and let σ_f be a character of G_f as above, and let $\pi=\pi_0(f,\sigma_f)$ be the irreducible representation attached to this data as discussed above. Then π has a global distribution character χ_π, and $\tilde{\chi}_\pi$ is a distribution at least on V; more precisely if ϕ is supported in V, then*

$$\tilde{\chi}_\pi(\phi)=\int (\phi K_O^{-1})^\wedge\, d\mu_O .$$

In short we have the same formula as in the exponential case which says that, up to multiplication by the function K_O, $\tilde{\chi}_\pi$ is the Fourier transform of the orbit measure μ_O at least for functions ϕ supported in V. Note that this says that the restriction of $\tilde{\chi}_\pi$ $(\pi=\pi_0(f,\sigma_f))$ to the open dense set V depends only on O and not on σ_f. Finally, we also have an analogue of Theorem 17′ for generic orbits; however, its domain of validity is smaller, namely V_0 instead of V.

THEOREM 28′. *Let us maintain the hypotheses and notation of Theorem* 28, *and let us also assume that $\mathfrak{g}_f$ is nilpotent. Then for ϕ supported in V_0, we have*

$$\tilde{\chi}_\pi(\phi)=\int (\phi K^{-1})^\wedge\, d\mu_O$$

where K is as in Theorem 17′.

It is notable that all the representations associated to the same orbit O have the same "character" locally around e, i.e., their global characters, if they exist, agree in the neighborhood V_0 (which can be described explicitly).

5. Plancherel formula. We shall now turn to the Plancherel formula on a general solvable Lie group, that is, the decomposition of the regular representation, or more precisely the decomposition of the Dirac delta measure δ at e as an integral of characters (in the type I case). It turns out that one can obtain a decomposition of the regular representation of any solvable group (type I or not) into primary representations. This is the result of the lengthy paper [**45**] of Pukanszky and we will summarize the major points here. The idea is to modify the Kirillov construction in the non-type I case so that instead of associating to the pair (f, σ_f) an irreducible representation, one associates to something like this a primary representation of G which may not be type I, and this in a sense is the ultimate generalization of the Kirillov construction for solvable groups. One then obtains a family of primary representations which turn out to be sufficient to decompose the regular representation. The corresponding Plancherel formula becomes quite subtle, for, in addition, now we are dealing with groups that are possibly nonunimodular and the Plancherel formula must be understood in a somewhat modified sense (there has been work on this in general by several authors ([**32**], [**38**], [**56**]), in addition to [**45**]).

We begin therefore with a simply connected solvable Lie group G, with Lie algebra $\mathfrak{g}$ and $f \in \mathfrak{g}^*$. Then we have a character σ_f° on G_f°, the connected component of the isotropy group G_f given by $\sigma_f^\circ(\exp(x)) = \exp(2\pi i f(x))$, $x \in \mathfrak{g}_f$. The idea is that even though f is not integral we can maybe extend σ_f° to some larger part of G_f than G_f°. Indeed let $\tilde{G}_f$ be the kernel of σ_f° and we look at the quotient $G_f/\tilde{G}_f$ and view σ_f° as a character of $G_f^\circ/\tilde{G}_f$. Now if $G_f/\tilde{G}_f$ were abelian, σ_f° could be extended to $G_f/\tilde{G}_f$ (and hence to G_f), so that f is integral. Thus if f is not integral, we can still extend σ_f° to abelian subgroups of $G_f/\tilde{G}_f$ and in particular to its center which we denote by $\bar{G}_f/\tilde{G}_f$. Thus going back to G_f, we have an extension of σ_f to a character $\bar{\sigma}_f$ of $\bar{G}_f$ where $\bar{G}_f/\tilde{G}_f$ is the center of $G_f/\tilde{G}_f$. We have

PROPOSITION. *The set of such extensions $\bar{\sigma}_f$ is a principal homogeneous space $\bar{L}(f)$ for the dual group $K(f)$ to $\bar{G}_f/\tilde{G}_f$. The group $\bar{G}_f$ is called the reduced stabilizer and $\bar{G}_f = G_f$ if and only if f, or the corresponding orbit O, is integral. Evidently $K(f)$ and $\bar{L}(f)$ depend only on the orbit O and are denoted by $K(O)$ and $\bar{L}(O)$.*

The last statement follows from the fact that G_f/G_f° is abelian as can easily be seen. With the notion of the reduced stabilizer we can proceed to construct the analogues of the representations $\pi(f, \sigma_f)$ as before. We select a positive polarization $\mathfrak{h}$, invariant by G_f, satisfying the Pukanszky condition and admissible for some nilpotent ideal $\mathfrak{n}$ with $\mathfrak{g}/\mathfrak{n}$ abelian and $\mathfrak{h} \cap \mathfrak{n}_C$ $G(f')$-invariant. Instead of the groups $D = D^\circ \cdot H_f$ and $E = E^\circ \cdot G_f$ we form the groups $\bar{D} = D^\circ \cdot \bar{G}_f$ and $\bar{E} = E^\circ \cdot \bar{G}_f$ which are both closed, and as before there is a unique character $\bar{\lambda}_f$ extending $\bar{\sigma}_f$ and having the right behavior on D°. We then form the space of "holomorphic" functions $H(f, \bar{\sigma}_f, \mathfrak{h}, G)$ as before by replacing D by $\bar{D}$ and λ_f by $\bar{\lambda}_f$, and define the holomorphically induced representation $\pi(f, \bar{\sigma}_f, \mathfrak{h}, G)$. We note also that G operates on the set of pairs $(f, \bar{\sigma}_f)$.

THEOREM 29. *The representation $\pi(f, \bar{\sigma}_f, \mathfrak{h}, G)$ is a semifinite primary representation, and is independent of the polarization $\mathfrak{h}$, and is denoted $\pi(f, \bar{\sigma}_f)$. Two such are quasi-equivalent if and only if the data $(f, \bar{\sigma}_f)$ and $(f', \bar{\sigma}_{f'})$ are conjugate under G; and in this case they are equivalent. Finally $\pi(f, \bar{\sigma}_f)$ is type* I *if and only if $\bar{G}_f$ is of finite index in G_f.*

The proof [**45**] is an analogue of the proof of a corresponding theorem for integral f. This result tells us what happens when integrality fails, but the primary representations $\pi(f, \bar{\sigma}_f)$ are still not what we want for the regular representation for it still remains to take account of the possible lack of smoothness of the action of G on $\mathfrak{g}^*$. What we have to do is to group together the $\pi(f, \bar{\sigma}_f)$ for f in a number of different orbits which are all equivalent under a coarser equivalence relation than that of G-orbits. This coarser equivalence relation on $\mathfrak{g}^*$ can be obtained in a canonical way from the following general theorem:

THEOREM 30. *Let G be any connected Lie group operating linearly* (*or affinely*) *on a finite-dimensional vector space V. Then there exists a unique equivalence relation R on V such that the classes of R are G-invariant, locally closed, and for every $x \in V$, the G-orbit of x is dense in the R-class of x. For this equivalence relation, the G-orbit of a point is an R-class if and only if it is locally closed.*

We remark that in case the action of G is distal [**37**], the classes of R are simply the orbit closures. It would be interesting to know if Theorem 30 is valid in the context of a locally compact group operating on a general locally compact space. Theorem 30 can now be applied to the case at hand of the action of G on $\mathfrak{g}^*$ for a solvable Lie group G, and we obtain larger equivalence classes for the relation R; these classes are unions of G-orbits. We fix attention now on a single R-class P in $\mathfrak{g}^*$, and for $f \in P$, consider the principal homogeneous space $\bar{L}(f)$ of characters $\bar{\sigma}_f$ of the reduced stability group $\bar{G}_f$ as above, and we let $\bar{L}(P) = \bigcup \bar{L}(f), f \in P$. Then Pukanszky shows that $\bar{L}(f)$ $(f \in P)$ is independent of f (this is clearly the case as f varies over a G-orbit, but not so clearly the case as f varies over the entire R-class P). Let us denote by $\bar{K}(P)$ the common fiber isomorphic to $\bar{L}(f)$, for $f \in P$ – it is a principal homogeneous space for $\bar{A}(P)$ the dual group of $\bar{G}_f / G_f^\circ$ for any $f \in P$.

He also shows that it can be given a topology so that it is a fiber bundle, and that that bundle is trivial. We note that $\bar{L}(P)$ is a G-space and Pukanszky shows that $\bar{L}(P)$ (which covers P) admits an equivalence relation S with S-classes locally closed, G-invariant and with G-orbits dense in the S-classes. This is done making use of precise structural information on how G acts on $\bar{L}(P)$. In the case when P is a single G-orbit, one may verify at once that the S-classes are G-orbits – and in particular if G is type I, so that $P=O$ is an orbit, then the S-classes may be identified to points in $L(O)$ (cf. Theorem 27). Now let Q be an S-orbit in $\bar{L}(P)$, and one can then find a distinguished G-invariant measure μ_Q on Q (Q is a homogeneous space for a larger group than G and μ_Q is invariant under this larger group and hence uniquely specified).

The idea now is that we wish to attach to each Q a primary representation $\pi(Q)$ which is the analogue of the Kirillov construction in the type I case. Indeed we simply form the representations $\pi(f, \bar{\sigma}_f)=\pi(q)$ as the pair $(f, \bar{\sigma}_f)=q$ ranges over Q and form the direct integral

$$\pi(Q)=\int_Q \pi(q)\, d\mu_Q.$$

Of course there is the technicality of showing that the $\pi(q)$ are a measurable family of representations so that we may form this direct integral, but this point may be handled without difficulty [**45**].

THEOREM 31. *The representation $\pi(Q)$ is primary and semifinite and is type* I *if and only if Q is a G-orbit and if the reduced stabilizer $\bar{G}_f$ for $(f, \bar{\sigma}_f)\in Q$ is of finite index in G_f. In particular if G is type* I *then Q may be regarded as a point in $x\in L(O)$ where the image of Q in $\mathfrak{g}^*$ is O, and $\pi(Q)$ is a multiple of the Kirillov representation $k(x)$ specified in Theorem* 27.

Finally we have the following:

THEOREM 32. *The regular representation of G may be decomposed as a central direct integral of multiples of the representations $\pi(Q)$.*

This assertion gives the Plancherel theorem for G in a somewhat crude initial form. One has to form the space of S-classes Q corresponding to all different R-classes in $\mathfrak{g}^*$ into a Borel space, say $X(G)$, and then one has to specify the measure class on $X(G)$ to be used in Theorem 32. A more precise Plancherel formula would select an actual measure from this class and use this to write the Dirac delta measure as an integral of the characters of the representations $\pi(Q)$, to the extent that these exist (at least in the unimodular case).

We note that the space $X(G)$ has a natural projection onto $\mathfrak{g}^*/R$, where R is the equivalence relation given by Theorem 30. Let us call this projection b, and note that inverse image of a point $b^{-1}(P)$, P an R-class, consists of the set of all S-classes (see above) in the set $\bar{L}(P)$. In particular $\bar{L}(P)$ is homeomorphic to the product $P \times \bar{K}(P)$ in terms of our previous notation. Now P is an orbit of a group containing G and hence carries a unique measure class of measures quasi-invariant under that larger group. Let one such be μ_P. Now $\bar{K}(P)$ is a principal homogeneous space for a compact group and hence carries a unique invariant probability measure ν and $\mu_P \times \nu$ specifies a distinguished measure on the set $\bar{L}(P)$. The set $b^{-1}(P)$ is then a quotient of $\bar{L}(P)$ and we denote by ν_P^* the quotient measure in $b^{-1}(P)$ and give $b^{-1}(P)$ the quotient Borel structure from $\bar{L}(P)$. In case $P=O$ is a single G-orbit, we have seen that S-classes are single G-orbits and so in that case $b^{-1}(P)$ can be canonically identified to $\bar{K}(P)$ and under this identification ν_P^* is of course the invariant measure ν. Thus in general, we have measures (really only measure classes) ν_P^* determined in each fiber of the map b of $X(G)$ into $\mathfrak{g}^*/R$. The base $\mathfrak{g}^*/R$ admits a distinguished measure (or rather a measure class) by taking the quotient measure of Lebesgue measure in $\mathfrak{g}^*/R$ – let us call this measure v. We now have the ingredients to specify the Plancherel measure (class), call it μ, in Theorem 32. The following seems to be implicitly contained in [**45**], but is not made explicit there.

THEOREM 33. *The space $X(G)$ has a standard Borel structure so that $Q \to \pi(Q)$ is a Borel isomorphism into the set of primary representations of G, and so that $b: X(G) \to \mathfrak{g}^*/R$ is Borel and so that the fibers $b^{-1}(P)$ inherit the Borel structure given them above. Finally the Plancherel measure μ of Theorem 32 has an integral decomposition relative to b,*

$$\mu = \int (\nu_P^*)\, dv(P).$$

Thus in particular, $Y \subset X(G)$ is a Plancherel null set if and only if $Y \cap b^{-1}(P)$ is a ν_P^* null set for v-almost all classes $P \in \mathfrak{g}^*/R$.

For further information and discussion the reader might consult the paper of Kleppner and Lipsman [**32**] (see also, these Proceedings). Since all of the primary representations $\pi(Q)$ are semifinite, we can conclude [**45**] that the regular representation of G has no type III component. Indeed the same result is true for an arbitrary connected locally compact group ([**18**], [**46**]). Using structure theory for such groups, the argument comes down in the end by a series of reductions to the case of a solvable Lie group.

In the case of a unimodular type I group, we do not have much additional information about the specific Plancherel measure for which the Plancherel formula holds. Theorem 9 provides some information in the nilpotent case, but we do not seem to know very much more. In the nonunimodular case, we have hinted above that some basic modifications must be made in the Plancherel formula even in the meaning of characters. If π is an irreducible representation, then to the extent that it exists $\phi \to \mathrm{Tr}(\pi(\phi)) = \chi_\pi(\phi)$, for ϕ C^∞ on G, should be a linear functional of ϕ invariant under conjugation; however, in the nonunimodular case a simple calculation shows that it is not, but rather it transforms like the modular function under conjugation. Moreover, even in the simplest cases, χ_π is not defined for all ϕ which are C^∞ of compact support. One solves both of these problems simultaneously by trying to find an unbounded positive selfadjoint operator D_π associated to π such that $\pi(g)\, D_\pi \pi(g)^{-1} = \Delta(g)^{-1} D_\pi$; such an operator is unique up to scalars if it exists, and one can form what we shall call the reduced character $\tilde{\chi}_\pi(\phi) = \mathrm{Tr}(D_\pi \pi(\phi))$. These reduced characters are invariant under conjugation and for a type I group, the modified Plancherel formula would read

$$\delta(\psi) = \int_{\hat{G}} \tilde{\chi}_\pi(\psi)\, d\mu(\pi).$$

Since there is some question about the set of ψ for which this holds, it is perhaps easier to replace this with an L^2 version with $\psi = \phi * \phi^*$ so that the formula becomes

$$(*) \qquad |\phi|^2 = \int_{\hat{G}} \chi_\pi(\phi * \phi^*)\, d\mu(\pi) = \int_{\hat{G}} \mathrm{Tr}(D_\pi \pi(\phi)\, \pi(\phi)^* D_\pi)\, d\mu(\pi).$$

Indeed one sees that we do not even need the group to be type I for this to make sense – we take the central decomposition of the regular representation into primary representations and use these in place of $\hat{G}$, and Tr in the formula means a trace on the von Neumann algebra generated by the primary representation (which is unique up to a scalar multiple). As usual one has to establish the validity of a formula such as (*) for a dense set of ϕ in $L_1(G) \cap L_2(G)$ as the starting point and then extend it to all $\phi \in L_2(G)$ by continuity. The remaining question is to say something about the operator D_π; one hopes for a Lie group G that one could find a semi-invariant D in the universal enveloping algebra $U(\mathfrak{g})$ so that $\mathrm{ad}(g) \cdot D = \Delta(g)^{-1} D$ for $g \in G$ so that D_π may be obtained as $d\pi(D)$ where $d\pi$ is the infinitesimal representation of $U(\mathfrak{g})$ associated to π. A quick calculation with the $ax+b$ group, the simplest nonunimodular group, shows that the element B of $\mathfrak{g}$ will work. (Recall that $\mathfrak{g}$ has a basis of the form A, B with $[A, B] = B$.) For a general solvable group one can do almost as well, but there are still some open questions. We have the following [**45**]:

THEOREM 34. *Let G be a simply connected solvable group and $\mathfrak{g}$ its Lie algebra; then there exist two elements D_1 and D_2 in $U(\mathfrak{g})$, in fact in the center of $U(\mathfrak{n})$ where $\mathfrak{n}$*

is the nilradical of $\mathfrak{g}$ *such that* D_1/D_2 *transforms like the inverse of the modular function and such that the closure of* $d\pi(Q)(D_1)(d\pi(Q)(D_2))^{-1}$ *is an operator* $D_{\pi(Q)}$ *for almost all* $Q \in \chi(G)$, *i.e., for Plancherel almost all primary representations of* G. *Moreover there exists another element* C *in* $U(\mathfrak{n})$ *such that if* $\phi = C\psi$ *for* ψ C^∞ *of compact support, then*

$$|\phi|^2 = \int_{X(G)} \tilde{\chi}_{\pi(Q)}(\phi * \phi^*)\, d\mu(Q)$$

where it is understood that this means that $\tilde{\chi}_{\pi(Q)}(\phi * \phi^*) = \mathrm{Tr}(D_{\pi(Q)}\pi(Q)(\phi * \phi^*)D_{\pi(Q)})$ *is finite for almost all* Q, *where* Tr *is a trace on the von Neumann algebra of the representation* $\pi(Q)$. *Finally the set of such* ϕ*'s is dense in* $L_2(G)$.

This theorem, essentially Theorem 4 in [**45**], gives a rather reasonable explicit version of the Plancherel formula in this setting. Finally, we may ask if any version at all of the completeness theorem (Theorem 27) carries over to the non-type I case; and we shall find that the answer is indeed 'yes.' Several examples have made clear that in studying non-type I groups or algebras, one should simply ignore $\hat{G}$, the space of equivalence classes of irreducible representations of G, and replace it by $P(G)$ which we define to be the space of primitive ideals of the group C^*-algebra of G [**17**]. The space $P(G)$ carries a natural topology – the Jacobson topology – and a Borel structure so that it is a standard Borel space. Moreover, there is always a map ϕ from $\hat{G}$ to $P(G)$ which associates to each equivalence class the kernel of any representation in it, when it is viewed as a representation of the group C^*-algebra. It may be shown that ϕ is always surjective and finally one of the fundamental results of Glimm (see [**17**]) is that ϕ is injective if and only if G is type I. Thus in the type I case, $\hat{G} \simeq P(G)$ are the same, and our point is that in the non-type I case, we should really forget $\hat{G}$ and look at $P(G)$ instead. Moreover, one may also show that if π is any primary representation of G, then the corresponding representation of the group C^*-algebra has for its kernel a primitive ideal $\phi(\pi)$ [**17**]. Moreover, this ideal in $P(G)$ depends only on the quasi-equivalence of π, and so we immediately have a map ϕ from the space of quasi-equivalence classes of primary representations of G onto $P(G)$. It is very natural, therefore, to consider the kernels in $C^*(G)$ of the representations $\pi(Q)$ of Theorem 31, and indeed $Q \mapsto \phi(\pi(Q)) = k(Q)$ gives rise to a map k from the space $X(G)$ of Theorem 31 into $P(G)$. This may reasonably be called the Kirillov map in this context and the obvious analogue of Theorem 27 would be the assertion that k is a bijection of $X(G)$ on $P(G)$. Pukanszky in his lecture at this Institute has announced that he now has a proof of this fact [**46**].

There is one final point we would like to take up in this connection, and that is the question of discrete series representations. Let π be an irreducible representation of a locally compact group G which has center Z. One knows that π is scalar on Z so that there exists a character λ of Z so that $\pi(z)=\lambda(z)\cdot 1$ for $z\in Z$. One says that π is a discrete series representation if π is a summand of the induced representation $\mathrm{ind}_{Z\to G}(\lambda)$. (If Z is compact, this is the same as π being a summand of the regular representation of G.) Before we formulated a general principle about CCR representations to the effect that whenever orbits in $\mathfrak{g}^*$ correspond to irreducible representations, closed orbits correspond to CCR representations. One may formulate a similar principle for discrete series representations which should be the following: Let G be a general Lie group G, with center Z and Lie algebra $\mathfrak{g}$; to the extent that $f\in\mathfrak{g}^*$ or its orbit O corresponds to an irreducible representation π, π is discrete series if and only if G_f/Z is compact, where G_f as usual is the isotropy group of f. This statement has been verified precisely for nilpotent groups [**39**], and, moreover, if one puts the known representation theory of semisimple groups into the Kirillov framework as in [**33**], then one sees that the condition above precisely characterizes the discrete series representations. In view of the state of knowledge about representation theory of solvable groups, one ought to be able to formulate a theorem. The following seems reasonable as a first approximation.

CONJECTURE. *Let $Q\in X(G)$, then $\pi(Q)$ is type* I, *and the corresponding irreducible representation is in the discrete series if and only if Q is an orbit and, for any $(f, \bar{\sigma}_f)\in Q$, G_f/Z is compact.*

In this context one might remark that, for unimodular groups, an irreducible representation π is in the discrete series if and only if the absolute value of one (or equivalently all) of its matrix coefficients $(\pi(g)\, x, y)$ is square integrable on G/Z. Moreover one has Schur orthogonality relations,

$$\int_{G/Z} (\pi(g)\, x_1, y_1)\, \overline{(\pi(g)\, x_2, y_2)}\, dg = d_\pi^{-1} (x_1, x_2)\, (y_1, y_2)$$

where d_π is a number called the formal degree. For a nonunimodular group this is no longer true, but we do have a good replacement for it, using the operators D_π which we introduced above. If the integration on the left-hand side is with respect to left (right) Haar measure, the right-hand side has to be replaced by $(D_\pi^{-1} x_1, x_2)\, (y_1, y_2)$ (or $(x_1, x_2)\, (D_\pi^{-1} y_1, y_2)$) with suitable conventions about convergence of the integrals tied to the appropriate vectors being in the domain of the appropriate unbounded operator. It is quite revealing to carry out this computation in detail for the $ax+b$ group.

6. Nil and solvmanifolds. We turn now to the final topic of our discussion and that is the discussion of discrete uniform subgroups Γ of a solvable Lie group G and of harmonic analysis on the solvmanifold G/Γ. More precisely we note that

G/Γ has a finite invariant measure and we form $L_2(G/\Gamma)$ with the natural action of G by translations. This representation U is simply the induced representation $\mathrm{ind}_{\Gamma\to G}(1)$ where 1 denotes the trivial representation. The problem is to decompose U into irreducible pieces and to study the resulting harmonic analysis or generalized Fourier series. (We note that the case $G=R$ and $\Gamma=Z$ gives rise to ordinary Fourier series.) Such results are of importance in resolving questions about ergodicity of flows on G/Γ and in studying equidistribution questions – analogues on G of problems about equidistribution mod 1.

The representation U is CCR in the sense that $U(\phi)$ is a compact operator for any C^∞ function of compact support on G – a fact that can be seen by displaying $U(\phi)$ rather explicitly as an integral operator on G/Γ. The following results from general principles:

THEOREM 35. *The representation U is a discrete direct sum $U=\sum n_\pi\pi$ $(\pi\in\hat{G})$ of irreducible representations, each occurring with finite multiplicity n_π. All the π's which occur are CCR representations.*

Since we have a complete model of $\hat{G}$ at least if G is type I, given by Theorem 27, one asks immediately for any given Γ, which representations occur, and how often. There is now a complete answer for nilpotent groups ([**36**], [**28**], [**49**]), but the answer in general is open. Since $\hat{G}$ looks something like $\mathfrak{g}^*/\mathrm{ad}^*(G)$ in general, and since Γ gives a kind of integrality condition in G, and hence something like an integrality condition in $\mathfrak{g}$, and hence in $\mathfrak{g}^*$, one is led to suppose that those π's which occur are those whose corresponding orbit is integral in $\mathfrak{g}^*$ in the sense of containing an integral f. This is precisely the case for $G=R^n$ and $\Gamma=Z^n$, and it is almost the right theorem for nilpotent groups, but it requires some minor modification. Let N be simply connected nilpotent with Lie algebra $\mathfrak{n}$ and let Γ be a discrete uniform subgroup in N. We consider first a special case – namely when Γ is what is called a lattice subgroup [**36**]. This means that $\log(\Gamma)\subset\mathfrak{n}$ is a lattice in $\mathfrak{n}$ and one may show that any Γ' is commensurable with a lattice group Γ. In this case we have a very precise notion of integrality of linear functionals $f\in\mathfrak{n}^*$, namely that they take integral values on $\log(\Gamma)$. Let L^* be the set of such f. We call an orbit $O\subset\mathfrak{n}^*$ integral if O meets L^* and we note then that $O\cap L^*$ is a discrete set upon which Γ operates. Let m_O denote the number of Γ-orbits on $O\cap L^*$. Then we have the following [**36**]:

THEOREM 36. *Let $\pi\in\hat{N}$ and let O be the corresponding orbit, and let Γ be a lattice group and U the corresponding representation. Thus π occurs in U if and only if O is integral and the multiplicity n_π is no greater than m_O.*

Howe and Richardson ([**29**], [**49**]) independently discovered the necessary and sufficient condition for occurrence in general and the exact formula for the multiplicity; this goes as follows: We note that Γ induces the structure of a rational Lie algebra on $\mathfrak{n}$ [2] so we may speak of rational subalgebras or of rational $f \in \mathfrak{n}^*$. Thus, given a rational $f \in \mathfrak{n}^*$ (the rationality is easily seen to be a necessary condition for the corresponding representation to occur) one can always find a rational polarization for it.

Let us consider the set of all pairs $(\lambda, \mathfrak{h})$ consisting of a character λ of $H = \exp(\mathfrak{h})$ with $\lambda(\exp(x)) = \exp(2\pi i f(x))$, $x \in \mathfrak{h}$, where $\mathfrak{h}$ is a polarization for f. The group N acts by conjugation on such pairs in a natural way, and the isotropy group of $(\lambda, \mathfrak{h})$ is precisely $H = \exp(\mathfrak{h})$. We call a pair $(\lambda, \mathfrak{h})$ integral if $\mathfrak{h}$ is a rational subalgebra and if λ vanishes on $(\exp(H) \cap L)$; we note that Γ operates on the set of integral pairs. Finally by the Kirillov theory each pair $(\lambda, \mathfrak{h})$ gives rise to a $\pi \in \hat{N}$ by induction.

THEOREM 37. *If $(\lambda, \mathfrak{h})$ induces π, then π occurs in the representation U on $L_2(G/\Gamma)$ with multiplicity equal to the number of Γ-orbits on the set of integral points in the N-orbit of $(\lambda, \mathfrak{h})$ – with the understanding that π occurs at all if and only if the N-orbit of $(\lambda, \mathfrak{h})$ contains at least one integral point.*

This theorem completely answers the multiplicity and occurrence problem in the nilpotent case. We remark that in the case of discrete series representations for nilpotent groups one may give a very much simpler formula [39].

Of course much remains to be done even when one has the multiplicities, for one is interested in finer decompositions of the primary blocks into irreducible pieces and in more detailed problems of harmonic analysis. These investigations are under way and some of the results are being reported at this meeting.

REFERENCES

1. L. Auslander, *An exposition of the algebraic theory of solvmanifolds*, Bull. Amer. Math. Soc. **79** (1973), 227–285.

2. L. Auslander, et al., *Flows on homogeneous spaces*, Ann. of Math. Studies, no. 53, Princeton Univ. Press, Princeton, N.J., 1963. MR **29** #4841.

3. L. Auslander and J. Brezin, *Invariant subspace theory for three dimensional nilmanifolds*, Bull. Amer. Math. Soc. **78** (1972), 255–258. MR **44** #6903.

4. L. Auslander and B. Kostant, *Quantization and representations of solvable Lie groups*, Bull. Amer. Math. Soc. **73** (1967), 692–695. MR **39** #2910.

5. ———, *Polarization and unitary representations of solvable Lie groups*, Invent. Math. **14** (1971), 255–354.

6. L. Auslander and C. C. Moore, *Unitary representations of solvable Lie groups*, Mem. Amer. Math. Soc. No. 62 (1966). MR **33** #7723.

7. V. Bargmann, *On a Hilbert space of analytic functions and an associated integral transform*. I, Comm. Pure Appl. Math. **14** (1961), 187–214. MR **28** #486.

8. P. Bernat, *Sur les représentations unitaires des groupes de Lie résolubles*, Ann. Sci. École Norm. Sup. (3) **82** (1965), 37–99. MR **33** #2763.

9. P. Bernat, et al., *Représentations des groupes de Lie résolubles*, Dunod, Paris, 1972.

10. J. Brezin, *Unitary representation theory for solvable Lie groups*, Mem. Amer. Math. Soc. No. 79 (1968). MR **37** #2896.

11. R. Blattner, *On induced representations*, Amer. J. Math. **83** (1961), 79–98. MR **23** #A2757.

12. ———, *On induced representations*. II. *Infinitesimal induction*, Amer. J. Math. **83** (1961), 499–512. MR **26** #2885.

13. J. Dixmier, *Sur les représentations unitaires des groupes de Lie nilpotents*. I, II, III, IV, V, VI, Amer. J. Math. **81** (1959), 160–170; Bull. Soc. Math. France **85** (1957), 325–388; Canad. J. Math. **10** (1958), 321–348; Canad. J. Math. **11** (1959), 321–344; Bull. Soc. Math. France **87** (1959), 65–79; Canad. J. Math. **12** (1960), 324–352. MR **21** #2705; **20** #1928; #1929; **22** #5900a, b; **21** #5693.

14. ———, *Sur les représentations unitaires des groupes de Lie algèbriques*, Ann. Inst. Fourier (Grenoble) **7** (1957), 315–328. MR **20** #5820.

15. ———, *L'application exponentielle dans les groupes de Lie résolubles*, Bull. Soc. Math. France **85** (1957), 113–121. MR **19**, 1182.

16. ———, *Representations induites holomorphes des groupes resolubles algèbriques*, Bull. Soc. Math. France **94** (1966), 181–206. MR **34** #7724.

17. ———, *Les C*-algèbres et leurs représentations*, Cahiers Scientifiques, fasc. 29, Gauthier-Villars, Paris, 1969. MR **39** #7442.

18. ———, *Sur la représentation régulière d'un groupe localement compact connexe*, Ann. Sci. École Norm. Sup. (4) **2** (1969), 423–436. MR **41** #5553.

19. ———, *Polarizations dans les algèbres de Lie*, Ann. Sci. École Norm. Sup. **4** (1971), 321–355.

20. M. Duflo, *Caractères des groupes et des algèbres de Lie resolubles*, Ann. Sci. École Norm. Sup. (4) **3** (1970), 23–74. MR **42** #4672.

21. ———, *Sur les extensions des représentations irreductibles des groupes de Lie nilpotents*, Ann. Sci. École Norm. Sup. (4) **5** (1972), 71–120.

22. ———, *Construction of primitive ideals in an enveloping algebra* (Budapest Congress, 1971).

23. M. Duflo and M. Vergne, *Une propriété de la représentation coadjointe d'une algèbre de Lie*, C. R. Acad. Sci. Paris Sér. A–B **268** (1969), A583–A585. MR **39** #6935.

24. J. M. G. Fell, *The dual space of C*-algebras*, Trans. Amer. Math. Soc. **94** (1960), 365–403. MR **26** #4201.

25. ———, *A new proof that nilpotent groups are CCR*, Proc. Amer. Math. Soc. **13** (1962), 93–99. MR **24** #A3238.

26. J. Glimm, *Locally compact transformation groups*, Trans. Amer. Math. Soc. **101** (1961), 124–138. MR **25** #146.

27. R. Howe, *On the representations of discrete finitely generated torsion free nilpotent groups* (to appear).

28. ———, *On Frobenius reciprocity for unipotent algebraic groups over Q*, Amer J. Math. **93** (1971), 163–172. MR **43** #7556.

29. ———, *Kirillov theory for compact p-adic groups* (to appear).

30. A. A. Kirillov, *Unitary representations of nilpotent Lie groups*, Uspehi Mat. Nauk. **17** (1962), no.4 (106), 57–110 = Russian Math. Surveys **17** (1962), no. 4, 53–104. MR **25** #5396.

31. ———, *Plancherel's formula for nilpotent groups*, Funkcional. Anal. i Priložen. **1** (1967), no. 4, 84–85. (Russian) MR **37** #347.

32. A. Kleppner and R. Lipsman, *The Plancherel formula for group extensions* (to appear).

33. B. Kostant, *Quantization and unitary representations*, Lecture Notes in Math., no. 170, Springer-Verlag, Berlin and New York, 1970, pp. 87–207.

34. G. W. Mackey, *The theory of group representations*, Lecture Notes (Summer, 1955), Department of Mathematics, University of Chicago, Chicago, Ill., 1955. MR **19**, 117.

35. ———, *Unitary representations of group extensions*, Acta Math. **99** (1958), 265–311. MR **20** #4789.

36. C. C. Moore, *Decomposition of unitary representations defined by discrete subgroups of nilpotent groups*, Ann. of Math. (2) **82** (1965), 146–182. MR **31** #5928.

37. ———, *Distal affine transformation groups*, Amer. J. Math. **90** (1968), 733–751. MR **38** #1210.

38. ———, *The Plancherel formula for non-unimodular groups*, Abstract for Internat. Conf. on Functional Analysis, University of Maryland, College Park, Md., 1971.

39. C. C. Moore and J. A. Wolf, *Square integrable representations of nilpotent groups*, Trans. Amer. Math. Soc. **183** (1973).

40. E. Nelson, *Analytic vectors*, Ann. of Math. (2) **70** (1959), 572–615. MR **21** #5901.

41. L. Pukanszky, *On the theory of exponential groups*, Trans. Amer. Math. Soc. **126** (1967), 487–507. MR **35** #301.

42. ———, *On unitary representations of exponential groups*, J. Functional Analysis **2** (1968), 73–113. MR **37** #4205.

43. ———, *On characters and the Plancherel formula of nilpotent groups*, J. Functional Analysis **1** (1967), 255–280. MR **37** #4236.

44. ———, *Characters of algebraic solvable groups*, J. Functional Analysis **3** (1969), 435–494. MR **40** #1539.

45. ———, *Representations of solvable Lie groups*, Ann. Sci. École Norm. Sup. **4** (1971), 464–608.

46. ———, *The primitive ideal space of solvable Lie groups* (preprint).

47. ———, *Leçons sur les représentations des groupes*, Monographies de la Société Mathématique de France, no. 2, Dunod, Paris, 1967. MR **36** #311.

48. S. R. Quint, *Representations of solvable Lie groups*, Lecture Notes, University of California, Berkeley, Calif., 1972.

49. L. Richardson, *Decomposition of the L_2-space of a general compact nilmanifold*, Amer. J. Math. **93** (1971), 173–190. MR **44** #1771.

50. I. E. Segal, *An extension of Plancherel's formula to separable unimodular groups*, Ann. of Math. (2) **52** (1950), 272–292. MR **12**, 157.

51. ———, *Hypermaximality of certain operators on Lie groups*, Proc. Amer. Math. Soc. **3** (1952), 13–15. MR **14**, 448.

52. W. F. Stinespring, *Integration theorems for gages and duality for unimodular groups*, Trans. Amer. Math. Soc. **90** (1959), 15–56. MR **21** #1547.

53. W. F. Stinespring and E. Nelson, *Representations of elliptic operators in an enveloping algebra*, Amer. J. Math. **81** (1959), 547–560. MR **22** #907.

54. R. F. Streater, *The representations of the oscillator group*, Comm. Math. Phys. **4** (1967), 217–236. MR **34** #7721.

55. O. Takenouchi, *Sur la facteur représentation des groupes de Lie résoluble de type E*, Math. J. Okayama Univ. **7** (1957), 151–161. MR **20** #3933.

56. N. Tatsumma, *Plancherel formula for non unimodular locally compact groups*, J. Math. Kyoto Univ. **12** (1972), 176–261.

57. M. Vergne, *Construction de sous-algèbres subordonnes a un élément du dual d'une algèbre de Lie résolubles*, C. R. Acad. Sci. Paris Sér. A–B **270** (1970), A173–A175. MR **40** #7323.

58. ———, *Construction de sous-algèbres subordonnées à un élément du dual d'une algèbre de Lie résoluble*, C. R. Acad. Sci. Paris Sér. A–B **270** (1970), A704–A707. MR **41** #288.

59. ———, *Étude de certaines représentations induites d'un groupe de Lie résoluble exponentiel*, Ann. Sci. École Norm. Sup. (4) **3** (1970), 353–384. MR **43** #4966.

60. A. Weil, *Adeles and algebraic groups*, Notes by M. Demazure and T. Ono, Institute for Advanced Study, Princeton, N.J., 1961.

UNIVERSITY OF CALIFORNIA, BERKELEY

THE THEORY OF CHARACTERS AND THE DISCRETE SERIES FOR SEMISIMPLE LIE GROUPS

V. S. VARADARAJAN*

Contents

AMS (MOS) subject classifications (1970). Primary 22E30; Secondary 22E45.

* The work of this author was supported at various stages by NSF grant GP-18127. This support is gratefully acknowledged.

1. Introduction

1.1. Summary. The purpose of this survey is to describe, with occasional indications of proofs, some of the main results in harmonic analysis on real semisimple Lie groups. More precisely we shall discuss the theory of characters and some of its important consequences, especially the determination of the discrete series of representations, i.e., the irreducible unitary representations with square integrable matrix coefficients. The central results of this theory were obtained by Harish-Chandra in a series of papers [**4**], [**5**], [**6**] in the past two decades, and there are clear indications that the ideas and methods of these articles will continue to play a pivotal and definitive role in Fourier analysis on semisimple Lie groups.

Let G be a connected real semisimple Lie group with finite center. The guiding principle behind Harish-Chandra's approach to harmonic analysis may be formulated as follows: Each Cartan subgroup of G makes a separate contribution to the Plancherel formula on G, the compact Cartan subgroups (when they exist) contribute to the discrete part of the Plancherel formula, and the contribution from an arbitrary Cartan subgroup is determined by the discrete series of a suitable reductive subgroup of G ([**4g**], [**6i**]). More precisely, let $L=L_IL_R$ be a Cartan subgroup where L_R is a vector group and L_I the maximal compact subgroup of L, and let P be a parabolic subgroup with Langlands decomposition $P=ML_RN$. Then L_I is a (compact) Cartan subgroup of M, and the part of the Plancherel formula due to L is determined by those unitary representations of G that are induced by the representations $man\mapsto\zeta(a)\,\gamma(m)$ $(m\in M,\, a\in L_R,\, n\in N)$ of P, where ζ is a unitary character of L_R and γ is a discrete series representation of M (these representations of G are irreducible for almost all ζ and have a finite Jordan decomposition for all ζ). From this point of view the construction of the discrete series is an indispensable prerequisite for carrying out the L^2 Fourier theory on G.

However, the determination of the discrete series turned out to be an extremely difficult problem. The early work of Bargmann [**1**] succeeded in the case $G=SL(2, \boldsymbol{R})$; but its subsequent generalizations aimed at realizing these representations in Hilbert spaces of holomorphic functions on suitable homogeneous spaces of G were only partially successful ([**4d**], [**4e**], [**4f**]; see however [**9**], [**10**], [**11**]). It was only relatively recently that Harish-Chandra succeeded in determining all the characters of the discrete series through a profound study of the differential equations satisfied by them ([**6f**], [**6h**]). Our aim in this article is to sketch an outline of the main features of Harish-Chandra's work leading to this construction. Roughly speaking, there are four parts to this program.

The first deals with the local theory of characters, and more generally, invariant eigendistributions. We shall come to this a little bit later.

In the second part (§3) we assume that G has a compact Cartan subgroup B. The problem here is to construct, corresponding to each regular character ξ of B, an invariant eigendistribution on G which is given on B' by a formula similar to Weyl's formula in the compact case. Note that except in the case of compact G, G will have noncompact Cartan subgroups, and consequently it is not reasonable to expect an invariant eigendistribution to be determined by its values on B'; for example, the characters of the unitary principal series all vanish on B'. However, by imposing an extra *global* condition ((ii) of (3.1.1)) on the distribution, that restricts its behaviour at infinity on G, Harish-Chandra was able to prove the existence of a unique distribution Θ_ξ with the required properties[1] (the condition (ii) of (3.1.1) is nothing more than saying that Θ_ξ is tempered).

In the third part (§6) we examine the behaviour at infinity of the analytic functions which are the Fourier components of the Θ_ξ with respect to a maximal compact subgroup K. Clearly the main question to be settled here is the square integrability of these functions. We describe Harish-Chandra's solution to this problem [**6h**]; it is based on a systematic study of the asymptotic behaviour of K-finite tempered eigenfunctions on G, many ideas of which go back to his papers [**5e**], [**5f**].

In the fourth part (§§5, 7) we consider the closed subspace of $L^2(G)$ spanned by the translates of the Fourier components of the Θ_ξ. This subspace can be shown to be contained in the Hilbert space ${}^\circ L^2(G)$ spanned by the matrix coefficients of the discrete series of G, and the problem is to show that it is all of ${}^\circ L^2(G)$, i.e., a completeness question. This was accomplished by Harish-Chandra through the fundamental technique of integrating over the conjugacy classes. The main point here is that this technique reduces the harmonic analysis of the matrix coefficients

[1] In our later notation, $\Theta_\xi=\Theta_\lambda$ when $\xi=\xi_\lambda$.

of the discrete series to questions of Fourier analysis on B, and the fact that we have a distribution Θ_ξ corresponding to each regular character ξ of B is decisive here. This method is similar to that used by Weyl in the compact case. It must be pointed out, however, that this similarity is essentially formal. The conjugacy classes of noncompact G are unbounded, and in order to construct a satisfactory theory of integration over them, it is absolutely essential to overcome convergence problems at infinity (§§4, 5.3–5.5).

Fundamental to all this development and preceding it is the local theory of invariant eigendistributions. We describe briefly in §2 the main results of this aspect of analysis on G. The theorems dealing with this topic were established by Harish-Chandra in two stages – first by reducing them to analogous questions on the Lie algebra, and deducing the latter from general theorems on invariant eigendistributions on semisimple Lie algebras. This survey makes no attempt to discuss either of these in any detail. We wish to point out, however, that the above mentioned reduction to the Lie algebra is much more than a technical device. It actually establishes a remarkable connection between harmonic analysis on G and that on the vector space $\mathfrak{g}$. As a major illustration of this we mention the crucial role played by the theory of Fourier transforms on $\mathfrak{g}$ in the construction of the distributions Θ_ξ (§3.5).

1.2. Notation and preliminaries. We shall work with a real Lie group G, not necessarily connected, with Lie algebra $\mathfrak{g}$; $\mathfrak{g}_c$ is the complexification of $\mathfrak{g}$ and $\mathfrak{G}$, the universal enveloping algebra of $\mathfrak{g}_c$. We write $\mathfrak{Z}$ for the center of $\mathfrak{G}$. The important case is when G is connected, semisimple and has a finite center; but the technical necessities of many proofs make it often convenient to operate in the wider context of groups of class $\mathscr{H}$. We shall say that G is of class $\mathscr{H}$ if it has the following properties: (i) $\mathfrak{g}$ is reductive and $\mathrm{Ad}[G]$ is contained in the connected complex adjoint group of $\mathfrak{g}_c$; (ii) G, and the centralizer of $\mathfrak{g}$ in G, have both finitely many connected components; (iii) if ${}^\circ G = \bigcap_\chi \mathrm{kernel}(\chi)$ where the intersection is over all continuous homomorphisms of G into the positive reals, G splits as ${}^\circ G \times V$ where V is a vector group; and (iv) the analytic subgroup of G defined by $[\mathfrak{g}, \mathfrak{g}]$ is closed. As a rule the structure theory of semisimple groups carries over to this more general case. K will denote a fixed maximal compact subgroup of G, and θ, the corresponding involution of G with $\theta(x) = x^{-1}$, $\forall x \in V$. $\mathfrak{g} = \mathfrak{k} + \mathfrak{p}$ is the associated Cartan decomposition of $\mathfrak{g}$, $\mathfrak{k}$ being the Lie algebra of K; $G \simeq K \cdot \exp \mathfrak{p}$ as usual. $\langle \cdot, \cdot \rangle$ denotes a nonsingular G-invariant symmetric bilinear form on $\mathfrak{g} \times \mathfrak{g}$ such that (i) it is the Killing form on $\mathfrak{g}_1 \times \mathfrak{g}_1$ ($\mathfrak{g}_1 = [\mathfrak{g}, \mathfrak{g}]$), (ii) $\mathfrak{k}$ and $\mathfrak{p}$ are orthogonal, and (iii) $X \mapsto \|X\|^2 = -\langle X, \theta X \rangle$ is positive definite on $\mathfrak{g}$, thus converting $\mathfrak{g}$ into a Hilbert space.

For $x \in G$, $D(x)$ is the coefficient of t^l in $\det(\mathrm{Ad}(x) - 1 + t)$ where $l = \mathrm{rank}(G)$. $G' = \{x : x \in G, D(x) \neq 0\}$ is the set of regular points of G. For any Cartan subalgebra

(CSA) $\mathfrak{l}$, the corresponding Cartan subgroup (CSG) is defined as the centralizer L of $\mathfrak{l}$ in G; $L'=L\cap G'$. For any root α of $(\mathfrak{g}, \mathfrak{l})$ we write ξ_α for the corresponding global root on L; α is said to be *real* (*imaginary*) if its values on $\mathfrak{l}$ are all real (imaginary). Any CSG is conjugate to one which is θ-stable; if L above is θ-stable, $L=(L\cap K)\cdot\exp(\mathfrak{l}\cap\mathfrak{p})$. G may not admit compact CSG's, but if it does, they are all conjugate. Suppose $G\subseteq G_c$ where G_c is a complex semisimple group, and $\mathfrak{l}$ is as above. We write L_c for the centralizer of $\mathfrak{l}_c$ in G_c. If $\mu\in\mathfrak{l}_c^*$, we write ξ_μ for the complex character of L_c such that $\xi_\mu(\exp H)=e^{\mu(H)}$ $(H\in\mathfrak{l}_c)$, whenever this exists. If P is a positive system of roots of $(\mathfrak{g}, \mathfrak{l})$, $\delta_P=\frac{1}{2}\sum_{\alpha\in P}\alpha$, and ξ_{δ_P} exists, we write $\Delta_{L,P}=\xi_{-\delta_P}\prod_{\alpha\in P}(\xi_\alpha-1)$. W_L (W_{L_c}) is the normalizer of L in G (L_c in G_c). As usual we say $\lambda\in\mathfrak{l}_c^*$ is *regular* if $\langle\lambda, \alpha\rangle\neq 0$, $\forall\alpha\in P$, *integral* if $2\langle\lambda, \alpha\rangle/\langle\alpha, \alpha\rangle\in\boldsymbol{Z}$, $\forall\alpha\in P$.

$G=KAN$ and $\mathfrak{g}=\mathfrak{k}+\mathfrak{a}+\mathfrak{n}$ are fixed Iwasawa decompositions, with $\mathfrak{a}\subseteq\mathfrak{p}$. $\mathfrak{a}^+$ is the positive chamber in $\mathfrak{a}$, $A^+=\exp\mathfrak{a}^+$, and $\log: A\to\mathfrak{a}$ inverts $\exp:\mathfrak{a}\to A$. $\varrho(H)=\frac{1}{2}\operatorname{tr}(\operatorname{ad}H)_{\mathfrak{n}}$ $(H\in\mathfrak{a})$. Σ denotes the set of simple roots of $(\mathfrak{g}, \mathfrak{a})$.

A subalgebra $\mathfrak{q}$ of $\mathfrak{g}$ is *parabolic* if $\mathfrak{q}_c$ contains a Borel subalgebra of $\mathfrak{g}_c$; the normalizer Q of $\mathfrak{q}$ in G is the corresponding *parabolic subgroup* (*psgrp*). $\mathfrak{q}=\mathfrak{m}_1+\mathfrak{n}_1$, where $\mathfrak{m}_1=\mathfrak{q}\cap\theta(\mathfrak{q})$ is reductive and $\mathfrak{n}_1$ is the nil radical of $\mathfrak{q}\cap[\mathfrak{g}, \mathfrak{g}]$, the sum being direct. If $\mathfrak{c}=\operatorname{center}(\mathfrak{m}_1)\cap\mathfrak{p}$, and $\mathfrak{m}$ is the orthogonal complement of $\mathfrak{c}$ in $\mathfrak{m}_1$, we have the Langlands decomposition $\mathfrak{q}=\mathfrak{m}+\mathfrak{c}+\mathfrak{n}_1$. If $C=\exp\mathfrak{c}$, M_1, the centralizer of $\mathfrak{c}$ in G, $M={}^\circ M_1$ and $N_1=\exp\mathfrak{n}_1$, we have the Langlands decomposition $Q=MCN_1$; moreover $M_1=Q\cap\theta(Q)$ and $\mathfrak{m}$ is the Lie algebra of M. If $\mathfrak{l}$ is a θ-stable CSA we can always choose $\mathfrak{q}$ such that $\mathfrak{c}=\mathfrak{l}\cap\mathfrak{p}$. We denote by d_Q the character $m_1\mapsto|\det\operatorname{Ad}(m_1)_{\mathfrak{n}_1}|^{1/2}$ of M_1.

Let $F\subseteq\Sigma$ and let $\mathfrak{p}_F=\mathfrak{m}_{1F}+\mathfrak{n}_F=\mathfrak{m}_F+\mathfrak{a}_F+\mathfrak{n}_F$ where $\mathfrak{a}_F$ is the null space of F, $\mathfrak{m}_{1F}$ is the centralizer of $\mathfrak{a}_F$ in $\mathfrak{g}$, and $\mathfrak{n}_F$ is the span of the root spaces $\mathfrak{g}_\lambda$ for those positive roots λ of $(\mathfrak{g}, \mathfrak{a})$ which are not in $\boldsymbol{R}\cdot F$. Then $\mathfrak{p}_F$ is parabolic, and the above direct sums are its Langlands decompositions. We call these *standard* and denote their corresponding global counterparts by P_F, M_{1F}, A_F, N_F. Any psgrp is conjugate via K to a unique standard one. $\mathfrak{K}$, $\mathfrak{A}$, $\mathfrak{N}$, $\mathfrak{M}_{1F}$, $\mathfrak{M}_F$, $\mathfrak{A}_F$, $\mathfrak{N}_F$ are the subalgebras (containing 1) of $\mathfrak{G}$ generated respectively by $\mathfrak{k}$, $\mathfrak{a}$, $\mathfrak{n}$, $\mathfrak{m}_{1F}$, $\mathfrak{m}_F$, $\mathfrak{a}_F$, $\mathfrak{n}_F$. $\mathfrak{Z}_F$ denotes the center of $\mathfrak{M}_{1F}$. We put $d_F=d_{P_F}$.

The elements of $\mathfrak{G}$ act as differential operators in G from both left and right. We use Harish-Chandra's notation and put, for a smooth function f, a, $b\in\mathfrak{G}$, and $x\in G$, $(afb)(x)=f(b; x; a)$; and for $X\in\mathfrak{g}$, $f(X; x)=(d/dt)\,(f(\exp tXx))_{t=0}$, $f(x; X)=(d/dt)\,(f(x\exp tX))_{t=0}$ $(x\in G)$. We denote the adjoint of $a\in\mathfrak{G}$ by $a^\dagger$, so that $\int_G af\cdot g\,dx=\int_G f\cdot a^\dagger g\,dx$ for all $f, g\in C_c^\infty(G)$. $\mathfrak{G}$ also acts on distributions on G; if T is a distribution and $a\in\mathfrak{G}$, $(aT)(f)=T(a^\dagger f)$ $(f\in C_c^\infty(G))$. T is said to be invariant if it is invariant under the inner automorphisms of G; it is said to be an eigendistribution if for some homomorphism $\chi:\mathfrak{Z}\to\boldsymbol{C}$, $zT=\chi(z)\,T$, $\forall z\in\mathfrak{Z}$. If M is any Lie subgroup of G with Lie algebra $\mathfrak{m}$ and if χ is any character of M, then, for $a \in \mathfrak{M}$ (=subalgebra of $\mathfrak{G}$ generated by $(1, \mathfrak{m})$), $\chi \circ a \circ \chi^{-1} \in \mathfrak{M}$; if $\chi = e^{\mu}$ where $\mu \in \mathfrak{m}_c^*$, $a \mapsto \chi \circ a \circ \chi^{-1}$ is the unique automorphism of $\mathfrak{M}$ such that $\chi \circ X \circ \chi^{-1} = X - \mu(X)\,1\,(X \in \mathfrak{m})$.

Let $\mathfrak{l}$ be a CSA and P a positive system of roots of $(\mathfrak{g}, \mathfrak{l})$. If $\delta = \frac{1}{2}\sum_{\alpha \in P} \alpha$ and X_α are root vectors, then, for any $z \in \mathfrak{Z}$, there exists a unique element $\mu_{\mathfrak{g}/\mathfrak{l}}(z)$ in the subalgebra $\mathfrak{L}$ of $\mathfrak{G}$ generated by $(1, \mathfrak{l})$ such that $z - e^{-\delta} \circ \mu_{\mathfrak{g}/\mathfrak{l}}(z) \circ e^{\delta} \in \sum_{\alpha \in P} \mathfrak{G} X_\alpha$; $\mu_{\mathfrak{g}/\mathfrak{l}}(z)$ is independent of P and $\mu_{\mathfrak{g}/\mathfrak{l}}$ is an isomorphism of $\mathfrak{Z}$ onto the algebra of all elements of $\mathfrak{L}$ that are invariant under the Weyl group of $(\mathfrak{g}_c, \mathfrak{l}_c)$. If $\lambda \in \mathfrak{l}_c^*$, the map

$$\chi_\lambda^{\mathfrak{l}} : z \mapsto \mu_{\mathfrak{g}/\mathfrak{l}}(z)(\lambda) \qquad (z \in \mathfrak{Z})$$

is a homomorphism of $\mathfrak{Z}$ into $\boldsymbol{C}$; we often write χ_λ for $\chi_\lambda^{\mathfrak{l}}$. $\chi_\lambda = \chi_{\lambda'}$ if and only if λ and λ' are in the same orbit of the Weyl group of $(\mathfrak{g}_c, \mathfrak{l}_c)$; every homomorphism of $\mathfrak{Z}$ into $\boldsymbol{C}$ is of the form χ_λ for some $\lambda \in \mathfrak{l}_c^*$. If $\mathfrak{h}_c$ is another CSA of $\mathfrak{g}_c$ and y is an element of the complex adjoint group of $\mathfrak{g}_c$ such that $\mathfrak{l}_c^y = \mathfrak{h}_c$, then $\chi_\lambda^{\mathfrak{l}} = \chi_{\lambda \circ y^{-1}}^{\mathfrak{h}}$. χ_λ is called *regular* if λ is regular.

Suppose $\mathfrak{m}$ is a subalgebra of $\mathfrak{g}$ which is reductive in $\mathfrak{g}$ and has the same rank as $\mathfrak{g}$. Let $\mathfrak{M}$ be as above, and $\mathfrak{Z}_{\mathfrak{m}}$ the center of $\mathfrak{M}$. Then there exists a unique injection $\mu_{\mathfrak{g}/\mathfrak{m}}$ of $\mathfrak{Z}$ into $\mathfrak{Z}_{\mathfrak{m}}$ with the following property: For any CSA $\mathfrak{l} \subseteq \mathfrak{m}$, $\mu_{\mathfrak{g}/\mathfrak{l}} = \mu_{\mathfrak{m}/\mathfrak{l}} \circ \mu_{\mathfrak{g}/\mathfrak{m}}$. $\mathfrak{Z}_{\mathfrak{m}}$ is a free finite module over $\mu_{\mathfrak{g}/\mathfrak{m}}[\mathfrak{Z}]$, of rank equal to the index of the Weyl group of $(\mathfrak{m}_c, \mathfrak{l}_c)$ in that of $(\mathfrak{g}_c, \mathfrak{l}_c)$. On the other hand, with $\mathfrak{l}$ as above, we have a natural "restriction" isomorphism $p \mapsto p_{\mathfrak{l}}$, of the algebra $I(\mathfrak{g})$ of G-invariant elements of the symmetric algebra $S(\mathfrak{g}_c)$ onto the algebra of Weyl group invariants of $S(\mathfrak{l}_c)$. So we have a canonical isomorphism $\zeta \mapsto \tilde{\zeta}$ of $\mathfrak{Z}$ onto $I(\mathfrak{g})$ such that $\mu_{\mathfrak{g}/\mathfrak{l}}(\zeta) = (\tilde{\zeta})_{\mathfrak{l}}$. Suppose $\mathfrak{q} \subseteq \mathfrak{g}$ is a parabolic subalgebra and $\mathfrak{m}_1, \mathfrak{n}_1, Q$ are as defined earlier. Then, for any $z \in \mathfrak{Z}$, $z_1 = d_Q^{-1} \circ \mu_{\mathfrak{g}/\mathfrak{m}_1}(z) \circ d_Q$ is the unique element of the center of the enveloping algebra of $\mathfrak{m}_1$ such that $z - z_1 \in \theta(\mathfrak{n}_1)\,\mathfrak{G}\mathfrak{n}_1$. For $F \subseteq \Sigma$ we write μ_F for $\mu_{\mathfrak{g}/\mathfrak{m}_{1F}}$.

We assume the reader is familiar with the basic concepts and results of representation theory (cf. [**4a**], [**4b**], [**4c**]). Let $\mathscr{E}(G)$ denote the set of all equivalence classes of irreducible unitary representations of G. If $\pi \in \omega \in \mathscr{E}(G)$, the multiplicities $[\omega : \mathfrak{d}]$ with which the classes $\mathfrak{d} \in \mathscr{E}(K)$ enter in the restriction of π to K are all finite, and, in fact, there is a constant $c > 0$ such that $[\omega : \mathfrak{d}] \leq c \dim(\mathfrak{d})$, $\forall \omega \in \mathscr{E}(G)$ and $\mathfrak{d} \in \mathscr{E}(K)$. It follows from this that, for $\pi \in \omega \in \mathscr{E}(G)$ and any $f \in C_c^\infty(G)$, the operator $\pi(f) = \int_G f(x)\,\pi(x)\,dx$ is of trace class, and $\Theta_\omega : f \mapsto \operatorname{tr} \pi(f)$ is an invariant distribution on G that depends only on the class ω. Let χ_ω be the infinitesimal character of ω, so that, for any $z \in \mathfrak{Z}$, $\pi(z) = \chi_\omega(z) \cdot 1$ on the Gårding subspace of π; then $z\Theta_\omega = \chi_\omega(z)\,\Theta_\omega$, for all $z \in \mathfrak{Z}$. Thus Θ_ω, the *global character* of ω, is an invariant eigendistribution corresponding to the eigenhomomorphism χ_ω. In addition Θ_ω is of the positive definite type, i.e.,

$$(1) \qquad \Theta_\omega(f * \tilde{f}) \geqq 0 \qquad (f \in C_c^\infty(G));$$

here $*$ denotes convolution and $\tilde{f}(x)=f(x^{-1})^{\text{conj}}$ $(x\in G)$. Of course the most crucial property of Θ_ω is that it determines ω completely: $\omega_1=\omega_2\Leftrightarrow\Theta_{\omega_1}=\Theta_{\omega_2}$.

Let $X_1,\ldots, X_r$ be an orthonormal basis for $\mathfrak{k}$ and let

(2) $$\Omega=1-(X_1^2+\cdots+X_r^2).$$

Ω is independent of the choice of the basis and $\Omega^k=\Omega$, $\forall k\in K$. If $\mathfrak{d}\in\mathscr{E}(K)$, the members of $\mathfrak{d}$ map Ω into a real scalar $c(\mathfrak{d})\geqq 1$ and it is known that for some $q\geqq 0$, $\dim(\mathfrak{d})=O(c(\mathfrak{d})^q)$ and $\sum_{\mathfrak{d}} c(\mathfrak{d})^{-q}<\infty$. Let $\pi\in\omega\in\mathscr{E}(G)$; $\mathfrak{H}$, the Hilbert space of π; $\mathfrak{H}_{\mathfrak{d}}$ $(\mathfrak{d}\in\mathscr{E}(K))$ the isotypical subspaces of $\mathfrak{H}$ and $E_{\mathfrak{d}}:\mathfrak{H}\to\mathfrak{H}_{\mathfrak{d}}$ the corresponding orthogonal projections. Then, $\forall\varphi\in\mathfrak{H}_{\mathfrak{d}}$, $\varphi'\in\mathfrak{H}$, $f\in C_c^\infty(G)$, and any integer $s\geqq 0$,

(3) $$(\pi(f)\,\varphi,\varphi')=c(\mathfrak{d})^{-s}\int_G(\Omega^s f)(x)\,(\pi(x)\,\varphi,\varphi')\,dx,$$

and so $\|\pi(f)\,E_{\mathfrak{d}}\|\leqq c(\mathfrak{d})^{-s}\|\Omega^s f\|_1$ ($\|\cdot\|_p$ is the L^p-norm). These estimates, together with the bounds $[\omega:\mathfrak{d}]\leqq c\dim(\mathfrak{d})$, easily imply that $\pi(f)$ is of trace class and that for some $C>0$, $q\geqq 0$,

(4) $$|\Theta_\omega(f)|\leqq C\,\|\Omega^q f\|_1 \qquad (f\in C_c^\infty(G));$$

and further, that if the matrix coefficients of ω are in $L^2(G)$,

(5) $$|\Theta_\omega(f)|\leqq C\,\|\Omega^q f\|_2 \qquad (f\in C_c^\infty(G)).$$

The theory of representations and characters is intimately related to the theory of (matrix as well as scalar) spherical functions. To define the latter in sufficient generality we proceed as follows. Let U be a finite-dimensional Hilbert space and $\tau=(\tau_1,\tau_2)$ a unitary double representation of K: This means that τ_1 is a unitary representation and τ_2 a unitary antirepresentation of K in U, such that $\tau_1(k_1)$ and $\tau_2(k_2)$ commute $\forall k_1, k_2\in K$; in view of this we allow τ_1 to act from the left and τ_2 to do so from the right. A function $f:G\to U$ is said to be *τ-spherical* if $f(k_1xk_2)=\tau_1(k_1)\,f(x)\,\tau_2(k_2)$, $\forall x\in G$, $k_1, k_2\in K$. $C^\infty(G:\tau)$ is the space of all τ-spherical functions of class C^∞. Of special interest are those functions in $C^\infty(G:\tau)$ which are $\mathfrak{Z}$-finite. These are all analytic, and they arise in a natural fashion from irreducible representations of G. For example, let $\pi\in\omega\in\mathscr{E}(G)$ and the notation be as in the previous paragraph. Fix $\mathfrak{d}_1,\mathfrak{d}_2\in\mathscr{E}(K)$ and let U be the Hilbert space of linear maps $u:\mathfrak{H}_{\mathfrak{d}_2}\to\mathfrak{H}_{\mathfrak{d}_1}$ with the Hilbert-Schmidt norm. Let $\pi_{\mathfrak{d}}(k)$ $(k\in K$, $\mathfrak{d}\in\mathscr{E}(K))$ be the restriction of $\pi(k)$ to $\mathfrak{H}_{\mathfrak{d}}$ and let $\tau_1(k_1)\,u=\pi_{\mathfrak{d}_1}(k_1)\,u$, $u\tau_2(k_2)=u\pi_{\mathfrak{d}_2}(k_2)$ $(k_1, k_2\in K$, $u\in U)$. Then $f:x\mapsto E_{\mathfrak{d}_1}\pi(x)\,E_{\mathfrak{d}_2}$ is an element of $C^\infty(G:\tau)$, and $zf=\chi_\omega(z)f$, for all $z\in\mathfrak{Z}$. It is interesting to consider the special case when $\mathfrak{d}_1=\mathfrak{d}_2$ is the trivial class of K; the function f is then spherical in the usual sense $(f(k_1xk_2)=f(x)$, $\forall x\in G$, $k_1, k_2\in K)$ and is an eigenfunction for every element of the centralizer of K in $\mathfrak{G}$.

The importance and usefulness in harmonic analysis of the τ-spherical functions which are eigenfunctions for $\mathfrak{Z}$ lies of course in the fact that they can be studied directly on the group with the help of their differential equations. As an illustration of this remark we mention the following theorem, which can be used to prove the fundamental finite multiplicity theorems of representation theory.

THEOREM 1. *Let τ, U and the other notation be as above. Fix an ideal $\mathfrak{Z}_0$ in $\mathfrak{Z}$ such that $m=\dim(\mathfrak{Z}/\mathfrak{Z}_0)<\infty$. Let $F(\mathfrak{Z}_0:\tau)$ be the space of all $f\in C^\infty(G:\tau)$ such that $zf=0$, for all $z\in\mathfrak{Z}_0$. Then*

$$\dim(F(\mathfrak{Z}_0:\tau))\leqq mw\dim(U) \tag{6}$$

where w is the order of a Weyl group of $\mathfrak{g}_c$.

COROLLARY 2. *Let $\omega\in\mathcal{E}(G)$. Then $[\omega:\mathfrak{d}]$ is finite for all $\mathfrak{d}\in\mathcal{E}(K)$ and*

$$[\omega:\mathfrak{d}]\leqq w^{1/2}\dim(\mathfrak{d}) \qquad (\mathfrak{d}\in\mathcal{E}(K)). \tag{7}$$

COROLLARY 3. *Let f be a C^∞ function on G with values in a finite-dimensional vector space V such that* (i) $\dim(\mathfrak{Z}f)<\infty$, (ii) *the left and right translates of f by elements of K span a finite-dimensional space. Then there exist α, $\beta\in C_c^\infty(G)$ invariant under inner automorphisms by elements of K such that $f=\alpha*f*\beta$.*

For a representation-theoretic proof of this see [**6h**, Theorem 1].

2. Local behaviour of invariant eigendistributions

We now take up the description of the local behaviour of invariant eigendistributions on a semisimple group. Throughout this section G is a connected real form of a simply connected complex semisimple Lie group G_c. However, with suitable modifications, the main results may be shown to be valid for all connected reductive groups.

2.1. Formulation of the main theorems. Let V and W be open subsets of G with $W\subseteq V$; Θ, a distribution on V and $\Theta_W=\Theta \mid W$. Θ is said to be $\mathfrak{Z}$-*finite* on W if $\dim(\mathfrak{Z}\Theta_W)<\infty$.

An invariant open set $V\subseteq G$ is said to be *completely invariant* when it has the following property: If $x\in V$ and x_s is the semisimple part in the Jordan decomposition of x, then $x_s\in V$.[2] In Theorems 1–4, V is an arbitrary completely invariant open subset of G.

[2] This definition appears to be slightly weaker than Harish-Chandra's [**6e**, p. 461], but is actually equivalent to his.

THEOREM 1. *Let Θ be an invariant $\mathfrak{Z}$-finite distribution on V. Then Θ is a locally summable function that is analytic on $V \cap G'$. Let L be a CSG and P any positive system of roots of* $(\mathfrak{g}, \mathfrak{l})$. *Define $\Phi_{L,P}(a) = \Delta_{L,P}(a)\,\Theta(a)\ (a \in L' \cap V)$. Then*

$$(1) \qquad \mu_{\mathfrak{g}/\mathfrak{l}}(z)\,\Phi_{L,P} = 0, \quad \forall z \in \mathfrak{Z}_\Theta,$$

where $\mathfrak{l}$ is the CSA corresponding to L and $\mathfrak{Z}_\Theta = \{z : z \in \mathfrak{Z},\ z\Theta = 0\}$.

Let $\mathfrak{F}$ be the space of all $f \in C^\infty(\mathfrak{l})$ such that $\mu_{\mathfrak{g}/\mathfrak{l}}(z) f = 0$, $\forall z \in \mathfrak{Z}_\Theta$. Suppose that $a \in L \cap V$, $\mathfrak{v}$ is a sufficiently small connected neighborhood of 0 in $\mathfrak{l}$, and $\mathfrak{v}^\times = \{H : H \in \mathfrak{v},\ a \exp H \in L'\}$. Then, for each connected component $\mathfrak{v}^+$ of $\mathfrak{v}^\times$, there exists $f_{\mathfrak{v}^+} \in \mathfrak{F}$ such that $\Phi_{L,P}(a \exp H) = f_{\mathfrak{v}^+}(H)\ (H \in \mathfrak{v}^+)$. It must however be kept in mind that $f_{\mathfrak{v}^+}$ will in general vary with $\mathfrak{v}^+$. It is clearly a very important problem to elucidate the relations that obtain among the $f_{\mathfrak{v}^+}$ on the interfaces between the various $\mathfrak{v}^+$. Theorem 2 deals with this question.

Let L, P be as in Theorem 1. Put $\varpi_{L,P} = \prod_{\alpha \in P} H_\alpha$ and regard $\varpi_{L,P}$ as a differential operator on L. Define

$$(2) \qquad L'(R) = \{a : a \in L,\ \xi_\alpha(a) \neq 1 \text{ for each real root } \alpha\}.$$

THEOREM 2. *Let the notation be as in Theorem 1. Then $\Phi_{L,P}$ extends to an analytic function on $L'(R) \cap V$ while $\varpi_{L,P}\Phi_{L,P}$ extends to a continuous function Ψ_L on $L \cap V$; Ψ_L is independent of the choice of P. If L_1 and L_2 are two CSG's, $\Psi_{L_1} = \Psi_{L_2}$ on $L_1 \cap L_2 \cap V$.*

THEOREM 3. *Let $\chi : \mathfrak{Z} \to \mathbf{C}$ be a regular homomorphism and let Θ' be an invariant analytic function on $G' \cap V$ such that $z\Theta' = \chi(z)\,\Theta'$, for all $z \in \mathfrak{Z}$. Then Θ' is locally integrable around each point of V.*

Put $\Theta(f) = \int_{G' \cap V} \Theta'(x) f(x)\,dx\ (f \in C_c^\infty(V))$. Then, in order that the distribution Θ satisfy the differential equations $z\Theta = \chi(z)\,\Theta\ (z \in \mathfrak{Z})$ on V, it is sufficient that the functions $\Phi_{L,P}$ possess the properties described in Theorem 2.

Let $a \in G$ be a semisimple element. We shall say that a is *semiregular* if the derived algebra of the centralizer of a in $\mathfrak{g}$ has dimension 3.

THEOREM 4. *Let Θ be an invariant locally summable function on V that is analytic and $\mathfrak{Z}$-finite on $G' \cap V$. In order that Θ be $\mathfrak{Z}$-finite on V, it is necessary and sufficient that, for each (semisimple) semiregular $a \in V$, there should exist an open neighborhood $N_a \subseteq V$ of a such that Θ is $\mathfrak{Z}$-finite on N_a.*

Theorems 1 and 2 are due to Harish-Chandra [**6e**], while Theorems 3 and 4 are virtually implicit in his work ([**6c**], [**6e**]; see also [**8**]). Harish-Chandra's method of proving these theorems rests (mainly) on transferring the study of an invariant distribution in a neighborhood of a semisimple point $a \in G$ to the study of an associated invariant distribution in a neighborhood of 0 of the centralizer of a in $\mathfrak{g}$. We now wish to describe this procedure more precisely.

Let $a \in G$ be semisimple and let $\mathfrak{m}_a$ (resp. M_a) be its centralizer in $\mathfrak{g}$ (resp. G). A system $(\mathfrak{u}, V)$ is said to be *adapted to a* if the following conditions are satisfied:

(i) $\mathfrak{u}$ is an M_a-invariant open neighborhood of 0 in $\mathfrak{m}_a$, which is star-like[3] at 0, and which contains the semisimple parts of each of its elements; $V = U^G$ where $U = a \exp \mathfrak{u}$.

(ii) $X \mapsto a \exp X$ is an analytic diffeomorphism of $\mathfrak{u}$ on U.

(iii) For $y \in U$, $D_a(y) = \det(\mathrm{Ad}(y) - 1)_{\mathfrak{g}/\mathfrak{m}_a} \neq 0$.

(iv) If $x \in G$, $X, X' \in \mathfrak{u}$, and $(a \exp X)^x = a \exp X'$, then $x \in M_a$ and $X^x = X'$.

Under these circumstances V can be shown to be open and completely invariant. V and $\mathfrak{u}$ are both connected. We write

$$J_a(X) = \det\left(\frac{\exp(\mathrm{ad}_{\mathfrak{m}_a} X) - 1}{\mathrm{ad}_{\mathfrak{m}_a} X}\right) \qquad (X \in \mathfrak{u}); \tag{3}$$

$J_a(X) > 0$, $\forall X \in \mathfrak{u}$.

Let $I(\mathfrak{m}_a)$ be the algebra of all elements of the symmetric algebra over $(\mathfrak{m}_a)_c$ that are invariant under the adjoint group of $\mathfrak{m}_a$. Let $\zeta \mapsto \tilde{\zeta}$ be the canonical isomorphism (cf. §1.2) of the center of the enveloping algebra of $(\mathfrak{m}_a)_c$ onto $I(\mathfrak{m}_a)$.

THEOREM 5. (i) *Let V be a completely invariant open subset of G, and V_s the set of semisimple points of V. Then, for each $a \in V_s$, there exists $(\mathfrak{u}_a, V_a)$ adapted to a with $V_a \subseteq V$; moreover, for any such choices, $V = \bigcup_{a \in V_s} V_a$.*

(ii) *Let $a \in G$ be semisimple and let $(\mathfrak{u}, V)$ be adapted to a. Then there is a linear isomorphism $\Theta \mapsto \Theta_a$ of the space of invariant distributions on V onto the space of M_a-invariant distributions on $\mathfrak{u}$ with the following properties:* (a) *for any $z \in \mathfrak{Z}$,* $(z\Theta)_a = (\mu_{\mathfrak{g}/\mathfrak{m}_a}(z))^{\sim} \Theta_a$, (b) *$\Theta$ is a locally summable function on V if and only if Θ_a is a locally summable function on $\mathfrak{u}$; moreover, in this case, $\forall X \in \mathfrak{u}$,*

$$\Theta_a(X) = \Theta(a \exp X)\, |D_a(a \exp X)|^{1/2}\, J_a(X)^{1/2}. \tag{4}$$

2.2. Some remarks on the proofs. These theorems are quite difficult to prove and we do not propose to go into their proofs in any detail. We shall restrict ourselves to a few comments on the main lines of argument.

We begin with Theorem 5. Concerning (i), let $a \in G$ be semisimple, and let $V_a(\varepsilon) = (a \exp \mathfrak{u}_a(\varepsilon))^G$ where $\varepsilon > 0$ and

$$\mathfrak{u}_a(\varepsilon) = \{X : X \in \mathfrak{m}_a,\ |\lambda| < \varepsilon, \text{ for all eigenvalues } \lambda \text{ of } \mathrm{ad}\, X\}.$$

One can then prove that $(\mathfrak{u}_a(\varepsilon), V_a(\varepsilon))$ is adapted to a for all sufficiently small

[3] This means that if $X \in \mathfrak{u}$ and $|t| \leq 1$, $tX \in \mathfrak{u}$.

$\varepsilon>0$, and that, given any invariant open set W containing a, $V_a(\varepsilon)\subseteq W$ for some $\varepsilon>0$.

Part (ii) is incomparably more difficult to establish. There are three main stages in its proof. For the first we need the following two lemmas.

LEMMA 1. *Let M (resp. N) be an analytic orientable manifold of dimension m (resp. n), and let ω_M (resp. ω_N) be an analytic m-form (resp. n-form) on M (resp. N) that is everywhere >0. Let ψ be an analytic submersion of M onto N. Then, for each $\alpha\in C_c^\infty(M)$, there exists $f_\alpha\in C_c^\infty(N)$ such that*

$$\int_N gf_\alpha\omega_N=\int_M (g\circ\psi)\,\alpha\omega_M \qquad (g\in C^\infty(N)). \tag{1}$$

The map $\alpha\mapsto f_\alpha$ is linear, maps $C_c^\infty(M)$ onto $C_c^\infty(N)$, and is continuous in the Schwartz topologies; moreover, $\operatorname{supp}(f_\alpha)\subseteq\psi[\operatorname{supp}\alpha]$. *If D_M (resp. D_N) is a C^∞ differential operator on M (resp. N) and if D_M and D_N are ψ-related,[4] then $f_{D^\dagger_M\alpha}=D^\dagger_N f_\alpha$, for all $\alpha\in C_c^\infty(M)$, the adjoints being taken with respect to ω_M and ω_N.*

This is essentially a local result [**6a**, Theorem 1].

For any distribution Θ on N let τ_Θ be the distribution on M such that $\tau_\Theta(\alpha)=\Theta(f_\alpha)$; $\Theta\mapsto\tau_\Theta$ is linear and injective.

LEMMA 2. *For the map $\Theta\mapsto\tau_\Theta$ we have* (i) $\operatorname{supp}(\tau_\Theta)\subseteq\psi^{-1}[\operatorname{supp}\Theta]$, (ii) *if D_M and D_N are as in Lemma* 1, $D_M\tau_\Theta=\tau_{D_N\Theta}$, (iii) *if S is a measurable function on N, S is locally summable on N if and only if $S\circ\psi$ is locally summable on M; in this case, $\tau_S=S\circ\psi$.*

In the context of Theorem 5 we take $M=G\times\mathfrak{u}$, $N=V$, $\psi(x,X)=(a\exp X)^x$, $\omega_M=dxdX$, $\omega_N=dX$. If Θ is invariant we can write $\tau_\Theta=1\otimes\sigma_\Theta$ for a unique distribution σ_Θ on $\mathfrak{u}$, and σ_Θ is M_a-invariant. The map $\Theta\mapsto\sigma_\Theta$ is a linear isomorphism of the space of invariant distributions on V onto the space of M_a-invariant distributions on $\mathfrak{u}$.

The second stage consists in establishing

LEMMA 3. *Given any analytic invariant differential operator E on V, there is an analytic M_a-invariant differential operator $R(E)$ on $\mathfrak{u}$ such that $\sigma_{E\Theta}=R(E)\,\sigma_\Theta$ for all invariant distributions Θ on V.*

For this see [**6e**, §§5–7].

Let $\lambda_a(X)=|D_a(a\exp X)|^{1/2}\,J_a(X)^{1/2}$ $(X\in\mathfrak{u})$. The third step consists in proving that, for any $z\in\mathfrak{Z}$,

[4] This means that $(D_Ng)\circ\psi=D_M(g\circ\psi)$ for all $g\in C^\infty(N)$.

(2) $$R(z)\,S=(\lambda_a^{-1}\circ(\mu_{\mathfrak{g}/\mathfrak{m}_a}(z))^{\sim}\circ\lambda_a)\,S$$

for all M_a-invariant distributions S on $\mathfrak{u}$. Once this is done we obtain at once the relation

(3) $$\sigma_{z\Theta}=(\lambda_a^{-1}\circ(\mu_{\mathfrak{g}/\mathfrak{m}_a}(z))^{\sim}\circ\lambda_a)\,\sigma_\Theta \qquad (z\in\mathfrak{Z})$$

for all invariant distributions Θ on V, so that we may take $\Theta_a=\lambda_a\sigma_\Theta$. The proof of (2) is simple if S is any M_a-invariant C^∞ function on $\mathfrak{u}$; for distributions, (2) follows from the following lemma.

LEMMA 4. *Let F be an analytic M_a-invariant differential operator on $\mathfrak{u}$. Suppose $FS=0$ for every M_a-invariant C^∞ function S on $\mathfrak{u}$. Then $FS=0$ for every M_a-invariant distribution S on $\mathfrak{u}$.*

Theorem 5 follows from these results. However Lemma 4 is difficult to establish and its proof is based on some of the deeper aspects of invariant analysis on reductive Lie algebras [**6d**, Theorems 4 and 5].

Theorem 5 enables one to reduce the proofs of Theorems 1–4 to those of analogous results on reductive Lie algebras. We shall show how this is done for Theorems 1 and 2 while referring the reader to Harish-Chandra's papers ([**6a**], [**6b**], [**6c**], [**6d**]) for the theory of invariant distributions on reductive Lie algebras.

We first consider Theorem 1. By Theorem 5, Θ_a is M_a-invariant and $I(\mathfrak{m}_a)$-finite on $\mathfrak{u}_a$, and hence is a locally summable function on $\mathfrak{u}_a$ by Theorem 1 of [**6d**]. This shows that Θ is a locally summable function around a. As $a\in V_s$ is arbitrary, Θ is a locally summable function on V.

The reduction of Theorem 2 is based on the following lemma; here $a\in G$ is semisimple and $(\mathfrak{u}, V)$ is adapted to a.

LEMMA 5. *Let $\mathfrak{l}\subseteq\mathfrak{m}_a$ be a CSA and let P (resp. P_a) be a positive system of roots of $(\mathfrak{g}, \mathfrak{l})$ (resp. $(\mathfrak{m}_a, \mathfrak{l})$). Write $\pi_{P_a}=\prod_{\alpha\in P_a}\alpha$, $\varpi_{P_a}=\prod_{\alpha\in P_a}H_\alpha$, $\varpi_{P/P_a}=\prod_{\alpha\in P/P_a}H_\alpha$. Then there exists a constant $c(\mathfrak{l}, P, P_a)\neq 0$ such that, for all $X\in\mathfrak{l}\cap\mathfrak{u}$,*

(4) $$\Delta_{L,P}(a\exp X)=c(\mathfrak{l}, P, P_a)\,\pi_{P_a}(X)\,|D_a(a\exp X)|^{1/2}\,J_a(X)^{1/2}.$$

Moreover, if ζ_a is the element of $I(\mathfrak{m}_a)$ whose restriction to $\mathfrak{l}_c$ is $c(\mathfrak{l}, P, P_a)\,\varpi_{P/P_a}$, then ζ_a is independent of the choices of $\mathfrak{l}$, P and P_a.

Suppose Θ is an invariant $\mathfrak{Z}$-finite distribution on V. We note that for $X\in\mathfrak{l}\cap\mathfrak{u}$, $a\exp X\in L'$ (resp. $L'(R)$) if and only if no root (resp. no real root) of $(\mathfrak{m}_a, \mathfrak{l})$ vanishes at X. Let $\tilde\Theta_a=\zeta_a\Theta_a$. From Lemma 5 we then obtain, for all $X\in\mathfrak{l}\cap\mathfrak{u}$ with $\pi_{P_a}(X)\neq 0$,

$$\Phi_{L,P}(a\exp X)=c(\mathfrak{l}, P, P_a)\,\pi_{P_a}(X)\,\Theta_a(X), \qquad (\varpi_{L,P}\Phi_{L,P})\,(a\exp X)=\tilde\Theta_a(X;\,\varpi_{P_a}\circ\pi_{P_a}).$$

Theorem 2 now follows from Theorems 2 and 3 and Lemma 19 of [**6d**].

2.3. Examples and remarks. Let $\chi: \mathfrak{Z} \to C$ be a homomorphism and let

$$(1) \qquad \mathfrak{J}(\chi) = \{\Theta : \Theta \text{ an invariant distribution on } G, z\Theta = \chi(z)\,\Theta, \forall z \in \mathfrak{Z}\}.$$

It follows from Theorems 1 and 2 of §2.1 and the theory of differential equations invariant with respect to a finite reflexion group [**13**] that

$$(2) \qquad \dim(\mathfrak{J}(\chi)) \leqq Nw$$

where w is the order of a Weyl group of $\mathfrak{g}_c$ and N is the total number of connected components of the various $L_i'(R)$, $L_1, \dots, L_r$ being a complete system of mutually nonconjugate CSG's of G. In particular there cannot exist more than Nw mutually inequivalent irreducible unitary representations with the same infinitesimal character. In fact we have the following more general result as a direct consequence of (2):[5] Let π be a representation of G in a Banach space V such that (a) all the K-multiplicities of π are finite; (b) π has both global and infinitesimal characters; then there is an integer $r \geqq 1$ and closed π-invariant subspaces $V_0 = V \supseteq V_1 \supseteq \cdots \supseteq V_r = \{0\}$ such that the representations induced in V_i/V_{i+1} are irreducible for all $i = 0, 1, \dots, r-1$.

We should also note that an invariant $\mathfrak{Z}$-finite distribution on G is an analytic function on an open set that is somewhat larger than G' [**4i**, Theorem 6]. Let $'G$ be the set of all semisimple points $a \in G$ whose centralizers in $\mathfrak{g}$ have compact adjoint groups. $'G$ is easily seen to be an invariant open subset of G. Any invariant $\mathfrak{Z}$-finite distribution Θ on a completely invariant open set V is actually an analytic function on $'G \cap V$. In fact, let $a \in {'G} \cap V$ and let $(\mathfrak{u}_a, V_a)$ be adapted to a with $V_a \subseteq V$; then $I(\mathfrak{m}_a)$ contains an elliptic element $\Box$, and Θ_a is annihilated on $\mathfrak{u}_a$ by $\Box^k + c_1 \Box^{k-1} + \cdots + c_k$ for suitable constants $c_1, \dots, c_k$.

It is an interesting problem to determine $\mathfrak{J}(\chi)$ as explicitly as possible for an arbitrary χ (see [**8**] for many explicit calculations involving $SU(p, q)$). We merely limit ourselves to a consideration of some examples. We write $r(G)$ for the maximum number of mutually nonconjugate CSG's of G.

EXAMPLE 1. $r(G) = 1$. Let L be a θ-stable CSG, $\mathfrak{l}$ its Lie algebra. Then $\mathfrak{l} = \mathfrak{c} + \mathfrak{a}$ where $\mathfrak{c} = \mathfrak{l} \cap \mathfrak{k}$, and $\mathfrak{a} = \mathfrak{l} \cap \mathfrak{p}$ is maximal abelian in $\mathfrak{p}$; also $L = CA$ where $C = L \cap K$ and $A = \exp \mathfrak{a}$. For $a \in L$ we write a_I and a_R for the components of a in C and A respectively. Write $\Delta = \Delta_{L,P}$. Let $\mathscr{C}$ be the set of all $\mu \in \mathfrak{l}_c^*$ for which $\exp H \mapsto \exp\{\mu(H)\}$ is well defined on $\exp \mathfrak{c}$.

For any $\mu \in \mathfrak{l}_c^*$ let $\mathfrak{F}(\mu)$ be the space of all analytic functions φ on L such that (i) $\varphi^s = \varepsilon(s)\,\varphi$, $\forall s \in W_L$, and (ii) $v\varphi = v(\mu)\,\varphi$, $\forall v \in \mu_{\mathfrak{g}/\mathfrak{l}}$ [3]. A simple argument shows that $\mathfrak{F}(\mu) = \{0\}$ if $W_{L_c} \cdot \mu \cap \mathscr{C} = \emptyset$. Now $(\mathfrak{g}, \mathfrak{l})$ has no real roots while every semiregular point of G is already in $'G$. From these facts and Theorems 2, 4 and 5 of

[5] This was pointed out to me by Harish-Chandra in 1968.

§2.1 we then obtain the following result: Fix $\lambda\in\mathfrak{l}_c^*$, and for any $\Theta\in\mathfrak{J}(\chi_\lambda)$ let φ_Θ be the analytic function on L such that $\varphi_\Theta(a)=\Delta(a)\,\Theta(a)\;(a\in L')$; then $\Theta\mapsto\varphi_\Theta$ is a linear isomorphism of $\mathfrak{J}(\chi_\lambda)$ onto $\mathfrak{J}(\lambda)$. In particular, $\mathfrak{J}(\chi_\lambda)=\{0\}$ when $W_{L_c}\cdot\lambda\cap\mathscr{C}=\emptyset$.

For any character (not necessarily unitary) ξ of L let $\mu_\xi\in\mathfrak{l}_c^*$ be defined by $\xi(\exp H)=\exp\{\mu_\xi(H)\}\;(H\in\mathfrak{l})$. For any $\mu\in\mathfrak{l}_c^*$ let W_μ be the stabilizer of μ in W_{L_c} and let $P(\mu)$ be the space of all polynomials p on $\mathfrak{l}_c^*$ such that (a) p is harmonic[6] with respect to W_μ, and (b) $p(H_1+H_2)=p(H_2)$, $\forall H_1\in\mathfrak{c}$, $H_2\in\mathfrak{a}$. Given any W_L-orbit $\mathfrak{o}$ in $\mathscr{C}$ let $A'(\mathfrak{o})$ be the space spanned by functions of the form $a\mapsto\xi(a)\,p(\log a_R)$ where ξ is a character of L with $\mu_\xi\in\mathfrak{o}$ and $p\in P(\mu_\xi)$; $A'(\mathfrak{o})$ is stable under W_L and we write $A(\mathfrak{o})$ for the subspace of all $\varphi\in A'(\mathfrak{o})$ such that $\varphi^s=\varepsilon(s)\,\varphi$, $\forall s\in W_L$. It is then not difficult to prove that for any $\lambda\in\mathfrak{l}_c^*$, $\mathfrak{F}(\lambda)$ is precisely the direct sum of the spaces $A(\mathfrak{o}_i)\;(1\leqq i\leqq r)$ where $\mathfrak{o}_1,\dots,\mathfrak{o}_r$ are the distinct W_L-orbits in $W_{L_c}\cdot\lambda\cap\mathscr{C}$.

Let ξ be a character of L. Then the function $\varphi_\xi=\sum_{s\in W_L}\varepsilon(s)\,\xi^s$ lies in $\mathfrak{F}(\mu_\xi)$, and so there exists an invariant eigendistribution Θ_ξ such that $\Delta\cdot(\Theta_\xi\mid L')$ is a nonzero multiple of φ_ξ. It can be verified that for a suitable choice of this constant, Θ_ξ is the character of a principal series representation of G and that all such characters are obtained in this manner (see [**4h**, Theorem 2], which gives the explicit formula for the principal series characters). Thus, if $\xi_1,\dots,\xi_N$ is a maximal set of characters of L with $\mu_{\xi_i}\in W_{L_c}\cdot\lambda\;(1\leqq i\leqq N)$ such that no two of the ξ_i are conjugate under W_L, the $\Theta_{\xi_i}\;(1\leqq i\leqq N)$ are linearly independent members of $\mathfrak{J}(\chi_\lambda)$ and are precisely all the principal series characters in $\mathfrak{J}(\chi_\lambda)$. It is possible that $N<\dim(\mathfrak{J}(\chi_\lambda))$.

Suppose now that λ is regular. Then $\mathfrak{F}(\lambda)$ is spanned by the functions φ_ξ corresponding to the characters ξ with $\mu_\xi\in W_{L_c}\cdot\lambda\cap\mathscr{C}$, and so $\mathfrak{J}(\chi_\lambda)$ is spanned by the principal series characters that belong to it. If $\mathscr{C}$ is connected and λ has the additional property that $W_{L_c}\cdot\lambda\cap\mathscr{C}$ is a single W_L-orbit, $\dim(\mathfrak{J}(\chi_\lambda))=1$; for a *complex* G, C is connected and the condition on λ implies that the associated principal series representation is irreducible.

EXAMPLE 2. *$r(G)=2$ and the symmetric space G/K has rank* 1. We now have two θ-stable CSG's L and B with Lie algebras $\mathfrak{l}$ and $\mathfrak{b}$. Concerning L we use the same notation as in Example 1. C has at most two connected components and we may assume that $C\subseteq B\subseteq K$. We can select $y\in G_c$ such that y fixes C pointwise and $\mathfrak{l}_c^y=\mathfrak{b}_c$. Let P_L be a positive system of roots of $(\mathfrak{g},\mathfrak{l})$, $\Delta_L=\Delta_{L,P_L}$, and $\Delta_B=\Delta_L\circ y^{-1}$; P_L contains a single real root α. Let $\beta=\alpha\circ y^{-1}$; then $\xi_\alpha=\xi_\beta=1$ on C. The components of $L'(R)$ are of the form $C^+A^\pm$ where C^+ is a component of C and $A^\pm$ are the subsets of A where $\xi_\alpha\gtrless 1$. If $W_{L,I}$ is the group generated by the Weyl reflexions corresponding to the imaginary roots of $(\mathfrak{g},\mathfrak{l})$, then $W_L=W_{L,I}\cup s_\alpha W_{L,I}$; it can further be shown that $y\circ W_{L,I}\circ y^{-1}\subseteq W_B\cup s_\beta W_B$. If C^+ is

[6] This means that $Dp=0$ for all homogeneous W_μ-invariant differential operators (on $\mathfrak{l}_c$) with constant coefficients and positive order.

a connected component of C, there is an $x^+ \in C^+$ that is fixed by W_L (cf. [**6g**, §24] for this structure theory).

Fix a regular $\lambda \in \mathfrak{l}_c^*$ and let $\mu = \lambda \circ y^{-1}$. As in the previous example $\mathfrak{J}(\chi_\lambda) = \{0\}$ if $W_{L_c} \cdot \lambda \cap \mathscr{C} = \emptyset$. Let $\mathfrak{J}^\circ(\chi_\lambda)$ be the subspace of all Θ in $\mathfrak{J}(\chi_\lambda)$ that are 0 on B'. For $\Theta \in \mathfrak{J}^\circ(\chi_\lambda)$ let φ_Θ be the analytic function on CA^+ such that $\varphi_\Theta(a) = \Delta_L(a)\,\Theta(a)$ $(a \in CA^+ \cap L')$. Then Theorems 1–3 of §2.1 imply that $\Theta \mapsto \varphi_\Theta$ is a linear isomorphism of $\mathfrak{J}^\circ(\chi_\lambda)$ onto the subspace $\mathfrak{F}^\circ(\lambda)$ of all analytic functions φ on CA^+ such that (i) $\varphi^s = \varepsilon(s)\,\varphi$, $\forall s \in W_{L,I}$, (ii) $v\varphi = v(\lambda)\,\varphi$, $\forall v \in \mu_{\mathfrak{g}/\mathfrak{l}}[\mathfrak{Z}]$, and (iii) $\varpi_L \varphi$ has boundary values 0 on C. Let $\xi_1, \ldots, \xi_N$ be a maximal set of characters of L with $\mu_{\xi_i} \in W_{L_c} \cdot \lambda$, for all i, no two of which are conjugate under W_L; define $\varphi_i = \sum_{s \in W_{L,I}} \varepsilon(s)\,(\xi_i^s + \xi_i^{s_\alpha s})$. Then $\{\varphi_1, \ldots, \varphi_N\}$ is a basis for $\mathfrak{F}^\circ(\lambda)$. One may conclude from this that $\mathfrak{J}^\circ(\chi_\lambda)$ is spanned by the principal series characters it contains.

By Theorem 2 of §2.1, $\mathfrak{J}(\chi_\lambda) = \mathfrak{J}^\circ(\chi_\lambda)$ if λ is nonintegral. Now let λ be assumed to be integral. Given $\Theta \in \mathfrak{J}(\chi_\lambda)$, there are constants c_s with $c_{ts} = c_s$ $(s \in W_{B_c}, t \in W_B)$ such that $\Phi_{B,P_B} = \sum_{s \in W_{B_c}} \varepsilon(s)\, c_s \xi_{s\mu}$. We shall now prove the existence of a unique $\Theta_\mu \in \mathfrak{J}(\chi_\lambda)$ such that

$$(3) \qquad \Delta(b)\,\Theta_\mu(b) = \sum_{s \in W_B} \varepsilon(s)\,\xi_{s\mu}(b) \quad (b \in B'), \qquad \sup_{x \in G'} |D(x)|^{1/2} |\Theta_\mu(x)| < \infty .$$

Assume this for a moment; then, if $\{s_1, \ldots, s_p\}$ is a system of representatives for $W_B \backslash W_{B_c}$, $\mathfrak{J}(\chi_\lambda)$ is the direct sum of $\mathfrak{J}^\circ(\chi_\lambda)$ and the $\Theta_{s_i\mu}$ $(1 \leq i \leq p)$.

For the existence and uniqueness of Θ_μ it is enough to fix a connected component C^+ of C and show that there is exactly one choice of the constants c_s^+ $(s \in W_{B_c})$ for which the function $\varphi = \sum_{s \in W_{B_c}} \varepsilon(s)\, c_s^+\, \xi_{(s\mu)\circ y}$ on C^+A^+ has the following properties: (i) φ is bounded on C^+A^+, (ii) $\varphi^s = \varepsilon(s)\,\varphi$, $\forall s \in W_{L,I}$, (iii) $\varpi_L \varphi = \sum_{s \in W_B} \xi_{s\mu}$ on C^+. The conditions (i) and (iii) already imply that $c_s^+ = 1$ if $s \in W_B \cup s_\beta W_B$ and $\langle s\mu, \beta \rangle > 0$, while $c_s^+ = 0$ for all other s. The relation $y \circ W_{L,I} \circ y^{-1} \subseteq W_B \cup s_\beta W_B$ shows that for these choices of the c_s^+, (ii) is automatic. In addition to the existence and uniqueness of Θ_μ the above discussion leads to the following formula valid for all $a = b \exp H \in CA^+ \cap L'$ $(b \in C, H \in \mathfrak{a}^+)$ [**6g**, §24]:

$$(4) \qquad \Delta_L(a)\,\Theta_\mu(a) = \sum_{s \in W_B} \varepsilon(s)\,\xi_{s\mu}(b) \exp\{-|((s\mu)\circ y)(H)|\}.$$

3. The distributions Θ_λ

3.1. Formulation of the main theorem. Let G be as in §2, and in addition let $\mathrm{rk}(G) = \mathrm{rk}(K)$. We propose to discuss the construction of the invariant eigendistributions on G that will eventually turn out to be the characters of the discrete series of G. Since $\mathrm{rk}(G) = \mathrm{rk}(K)$, G has compact CSG's, all such being connected and mutually conjugate. We fix one of them, say B, $\subseteq K$, and denote its Lie algebra by $\mathfrak{b}$. All roots of $(\mathfrak{g}, \mathfrak{b})$ are imaginary. Let $\mathscr{L}$ be the additive group of all integral elements of $\mathfrak{b}_c^*$, and $\mathscr{L}'$, the subset of all regular elements of $\mathscr{L}$. The basic result of the theory is the following theorem of Harish-Chandra [**6f**, Theorem 3].

THEOREM 1. *Let P be a positive system of roots of* $(\mathfrak{g}, \mathfrak{b})$ *and let* $\Delta = \Delta_{B,P}$. *Let* $\lambda \in \mathscr{L}'$. *Then there exists a unique invariant eigendistribution* Θ_λ *on* G *such that*

$$(1)\qquad \begin{aligned} &\text{(i)}\quad \Theta_\lambda(b)\,\Delta(b) = \textstyle\sum_{s \in W_B} \varepsilon(s)\,\xi_{s\lambda}(b) \qquad (b \in B'),\\ &\text{(ii)}\quad \sup_{x \in G'} |D(x)|^{1/2}\,|\Theta_\lambda(x)| < \infty. \end{aligned}$$

In the following sections we shall examine the main steps in Harish-Chandra's proof of this theorem.

3.2. The class $\mathscr{E}$ of invariant open sets. For any $x \in G$ let $\mathfrak{m}_x$ (resp. M_x) be the centralizer of x in $\mathfrak{g}$ (resp. G). x is called *elliptic* if it is in B^G $(=K^G)$, or, equivalently, if x is semisimple and all eigenvalues of $\mathrm{Ad}(x)$ are of modulus unity. Any $x \in G$ can be uniquely written as $a \exp Y$, where a is elliptic, $Y \in \mathfrak{m}_a$ and all eigenvalues of ad Y are real; a is called the *elliptic component* of x. $\mathscr{E}$ is the class of all invariant open subsets V of G with the following property: If $x \in V$ and a is the elliptic component of x, then $a \exp X \in V$ for all $X \in \mathfrak{m}_a$ for which ad X has only real eigenvalues. Members of $\mathscr{E}$ are clearly completely invariant. Given any elliptic $a \in G$ and $\varepsilon > 0$, we put

$$(1)\qquad \begin{gathered} \mathfrak{g}[\varepsilon] = \{X : X \in \mathfrak{g},\ |\mathrm{Im}\,\lambda| < \varepsilon,\ \forall \text{ eigenvalues } \lambda \text{ of } \mathrm{ad}\,X\},\\ \mathfrak{u}_a[\varepsilon] = \mathfrak{m}_a \cap \mathfrak{g}[\varepsilon], \qquad U_a[\varepsilon] = a \exp \mathfrak{u}_a[\varepsilon], \qquad V_a[\varepsilon] = U_a[\varepsilon]^G. \end{gathered}$$

LEMMA 1. *If $a \in G$ is elliptic, there exists ε_a with $0 < \varepsilon_a < \pi$ such that the system* $(\mathfrak{u}_a[\varepsilon], V_a[\varepsilon])$ *is adapted to a for any ε with* $0 < \varepsilon \leqq \varepsilon_a$.

LEMMA 2. *Let $a \in B$ and $0 < \varepsilon \leqq \varepsilon_a$. Then*
(i) $U_a[\varepsilon] \cap B = a \exp(\mathfrak{b} \cap \mathfrak{u}_a[\varepsilon])$, *and*
(ii) $V_a[\varepsilon] \cap B = \bigcup_{s \in W_B} (U_a[\varepsilon] \cap B)^s$.

LEMMA 3. *$\mathscr{E}$ is closed under finite intersections. If $a \in G$ is elliptic, then $V_a[\varepsilon] \in \mathscr{E}$ for $0 < \varepsilon \leqq \varepsilon_a$. If $V \in \mathscr{E}$ and $a \in V$ is elliptic, there exists ε with $0 < \varepsilon \leqq \varepsilon_a$ such that $V_a[\varepsilon] \subseteq V$.*

LEMMA 4. *Let $V \in \mathscr{E}$. For each $a \in V \cap B$ let δ_a be chosen so that $0 < \delta_a \leqq \varepsilon_a$ and $V_a[\delta_a] \subseteq V$. Then $V = \bigcup_{a \in V \cap B} V_a[\delta_a]$. If $V = G$, there exists a finite subset $F \subseteq B$ such that $G = \bigcup_{a \in F} V_a[\delta_a]$.*

These lemmas are not particularly difficult to prove.

3.3. Reduction to the Lie algebra. Let $\mathfrak{m}$ be a reductive subalgebra of $\mathfrak{g}$ containing $\mathfrak{b}$. We now wish to formulate two lemmas dealing with certain spaces of distributions on $\mathfrak{m}$ and deduce Theorem 3.1.1 from them.

Let M be the analytic subgroup of G defined by $\mathfrak{m}$. For any CSA $\mathfrak{l}\subseteq\mathfrak{m}$, let $P_{\mathfrak{m},\mathfrak{l}}$ be a positive system of roots of $(\mathfrak{m},\mathfrak{l})$, and let $\pi_{\mathfrak{m},\mathfrak{l}}=\prod_{\alpha\in P_{\mathfrak{m},\mathfrak{l}}}\alpha$, $\varpi_{\mathfrak{m},\mathfrak{l}}=\prod_{\alpha\in P_{\mathfrak{m},\mathfrak{l}}}H_\alpha$. Put $'\mathfrak{l}=\{H:H\in\mathfrak{l},\ \pi_{\mathfrak{m},\mathfrak{l}}(H)\neq 0\}$. Denote by $W_{M,B}$ the subgroup of W_B that comes from M and write

$$p(\mathfrak{m})=\text{index of } W_{M,B} \text{ in } W_{B_c}. \tag{1}$$

Let $\nu\in\mathfrak{b}_c^*$ be regular and take only imaginary values on $\mathfrak{b}$. For any $\varepsilon>0$ let $\mathfrak{u}[\varepsilon]=\mathfrak{m}\cap\mathfrak{g}[\varepsilon]$, and let $\mathfrak{J}_{\nu,\varepsilon}(\mathfrak{m})$ be the vector space of all M-invariant distributions T on $\mathfrak{u}[\varepsilon]$ having the following properties:

$$\begin{aligned}&\text{(i)}\quad (\mu_{\mathfrak{g}/\mathfrak{m}}(z))^{\sim}\,T=\chi_\nu(z)\,T, \text{ for all } z\in\mathfrak{Z}.\\ &\text{(ii)}\quad \text{For each CSA } \mathfrak{l}\subseteq\mathfrak{m},\ \sup_{H\in{}'\mathfrak{l}\cap\mathfrak{u}[\varepsilon]}|\pi_{\mathfrak{m},\mathfrak{l}}(H)\,T(H)|<\infty.\end{aligned} \tag{2}$$

LEMMA 1. *Let $T\in\mathfrak{J}_{\nu,\varepsilon}(\mathfrak{m})$. Then $T=0$ if and only if $T(H)=0$, $\forall H\in{}'\mathfrak{b}\cap\mathfrak{u}[\varepsilon]$.*

LEMMA 2. $\dim(\mathfrak{J}_{\nu,\varepsilon}(\mathfrak{m}))=p(\mathfrak{m})$.

We shall now indicate how Theorem 3.1.1 may be deduced from these two lemmas. We begin with the uniqueness. It is convenient to prove it in the following form:

LEMMA 3. *Let $\lambda\in\mathscr{L}'$, $V\in\mathscr{E}$, and let Θ be an invariant distribution on V such that $z\Theta=\chi_\lambda(z)\,\Theta$, $\forall z\in\mathfrak{Z}$. Suppose that*

$$\begin{aligned}&\text{(i)}\quad \sup_{x\in V\cap G'}|D(x)|^{1/2}|\Theta(x)|<\infty,\\ &\text{(ii)}\quad \Theta(b)=0,\quad \forall b\in V\cap B'.\end{aligned} \tag{3}$$

Then $\Theta=0$.

Let $a\in V\cap B$, and let ε be such that $0<\varepsilon\leqq\varepsilon_a$ and $V_a[\varepsilon]\subseteq V$. It is enough to prove that $\Theta\mid V_a[\varepsilon]=0$ (Lemma 3.2.4), or $\Theta_a=0$ on $\mathfrak{u}_a[\varepsilon]$, in view of Lemma 3.2.1 and Theorem 2.1.5. A simple calculation based on (2.2.4) shows that $\Theta_a\in\mathfrak{J}_{\lambda,\varepsilon}(\mathfrak{m}_a)$. Further, by (ii) of (3), $\Theta_a(H)=0$, $\forall H\in{}'\mathfrak{b}\cap\mathfrak{u}_a[\varepsilon]$. So $\Theta_a=0$ on $\mathfrak{u}_a[\varepsilon]$ by Lemma 1.

To accomplish the construction of Θ_λ we proceed as follows: Let $a\in B$. For each $T\in\mathfrak{J}_{\lambda,\varepsilon}(\mathfrak{m}_a)$, let φ_T be the analytic function on $\mathfrak{b}[\varepsilon]=\mathfrak{b}\cap\mathfrak{u}_a[\varepsilon]$ such that $\varphi_T(H)=\pi_{\mathfrak{m}_a,\mathfrak{b}}(H)\,T(H)$ for all $H\in{}'\mathfrak{b}\cap\mathfrak{u}_a[\varepsilon]$. By Lemma 1, $T\mapsto\varphi_T$ is a linear injection of $\mathfrak{J}_{\lambda,\varepsilon}(\mathfrak{m}_a)$ into the vector space $\mathfrak{U}_{\lambda,\varepsilon}$ of all analytic functions φ on $\mathfrak{b}[\varepsilon]$ such that (i) $\varphi^s=\varepsilon(s)\,\varphi$, $\forall s\in W_{M,B}$, and (ii) $\varphi=\sum_{s\in W_{Bc}}\varepsilon(s)\,c_s e^{s\lambda}$ for suitable constants c_s. But $\dim(\mathfrak{U}_{\lambda,\varepsilon})=p(\mathfrak{m}_a)$, and so, by Lemma 2, $T\mapsto\varphi_T$ is an isomorphism onto $\mathfrak{U}_{\lambda,\varepsilon}$. In particular we can find $T^{(a)}\in\mathfrak{J}_{\lambda,\varepsilon}(\mathfrak{m}_a)$ such that $\varphi_{T^{(a)}}=\sum_{s\in W_B}\varepsilon(s)\,\xi_{s\lambda}(a)\,e^{s\lambda}$.

The expression for $\varphi_{T^{(a)}}$ shows that $T^{(a)} \mid {}'\mathfrak{b}\cap\mathfrak{b}[\varepsilon]$ is invariant under the subgroup of W_B that fixes a. Lemma 3 then implies that $T^{(a)}$ is invariant under M_a. Theorem 2.1.5 and Lemma 2.2.5 now lead to the existence of an invariant eigendistribution $\Theta_\lambda^{(a)}$ on $V_a[\varepsilon]$ such that

$$(4)\qquad \begin{aligned} &\text{(i)}\quad \Theta_\lambda^{(a)}(b)\,\Delta(b)=\textstyle\sum_{s\in W_B}\varepsilon(s)\,\xi_{s\lambda}(b),\ b\in a\exp(\mathfrak{b}[\varepsilon]\cap{}'\mathfrak{b}),\\ &\text{(ii)}\quad \sup_{x\in V_a[\varepsilon]\cap G'}|D(x)|^{1/2}|\Theta_\lambda^{(a)}(x)|<\infty.\end{aligned}$$

From (ii) of Lemma 3.2.2 it follows that (i) of (4) is valid for all $b\in B'\cap V_a[\varepsilon]$.

For each $a\in B$, select δ_a with $0<\delta_a\leqq\varepsilon_a$ and let $\Theta_\lambda^{(a)}$ be the distribution constructed as above on $V_a[\delta_a]$. If $a,a'\in B$, then $\Theta_\lambda^{(a)}$ and $\Theta_\lambda^{(a')}$ coincide on $B'\cap V_a[\delta_a]\cap V_{a'}[\delta_{a'}]$, and hence on $V_a[\delta_a]\cap V_{a'}[\delta_{a'}]$ by Lemma 3, as the latter belongs to $\mathscr{E}$ in view of Lemma 3.2.3. The existence of Θ_λ on G now follows from Lemma 3.2.4.

3.4. Proof of Lemma 3.3.1. It remains to indicate how the proofs of Lemmas 1 and 2 of § 3.3 may be carried out. We consider first Lemma 3.3.1. Let $T\in\mathfrak{J}_{\nu,\varepsilon}(\mathfrak{m})$ be such that $T(H)=0$, for all $H\in{}'\mathfrak{b}\cap\mathfrak{u}[\varepsilon]$. It must be shown that for each CSA $\mathfrak{l}\subseteq\mathfrak{m}$, $T=0$ on ${}'\mathfrak{l}\cap\mathfrak{u}[\varepsilon]$. Let $\mathfrak{l}_I$ (resp. $\mathfrak{l}_R$) be the subspace of all elements of $\mathfrak{l}$ where all the roots of $(\mathfrak{g},\mathfrak{l})$ take imaginary (resp. real) values. Then the proof uses induction on $\dim(\mathfrak{l}_R)$. We may assume $\dim(\mathfrak{l}_R)>0$; for otherwise, $\mathfrak{l}$ and $\mathfrak{b}$ are conjugate under M and there is nothing to prove. Let $\mu\in\mathfrak{l}_c^*$ be such that $\chi_\mu=\chi_\nu$. Define

$$(1)\qquad {}'\mathfrak{l}(R)=\{H: H\in\mathfrak{l},\ \text{no real root of } (\mathfrak{m},\mathfrak{l}) \text{ vanishes at } H\}.$$

If Γ is any connected component of ${}'\mathfrak{l}(R)\cap\mathfrak{u}[\varepsilon]$, there are constants $c_s(\Gamma)$ $(s\in W_{L_c})$ such that

$$(2)\qquad \pi_{\mathfrak{m},\mathfrak{l}}(H)\,T(H)=\sum_{s\in W_{L_c}}\varepsilon(s)\,c_s(\Gamma)\exp\{s\mu(H)\}\qquad (H\in{}'\mathfrak{l}\cap\Gamma).$$

One has to prove that $c_s(\Gamma)=0$, for all s, Γ.

LEMMA 1. (i) *There exists $m\in M$ such that $\mathfrak{l}_I^m\subseteq\mathfrak{b}$.* (ii) *Let $\mathfrak{m}_1$ be the centralizer of $\mathfrak{l}_I$ in $\mathfrak{m}$. Then $\mathfrak{l}_I$ is the center of $\mathfrak{m}_1$ and $\mathfrak{l}_R$ is a CSA of $[\mathfrak{m}_1,\mathfrak{m}_1]$.*

For proving (i) let $\mathfrak{c}=\text{center}(\mathfrak{m})$, $\bar{\mathfrak{l}}=\mathfrak{l}\cap[\mathfrak{m},\mathfrak{m}]$ and let $\bar{\mathfrak{l}}_I(\mathfrak{m})$ (resp. $\bar{\mathfrak{l}}_R(\mathfrak{m})$) be the set of points of $\bar{\mathfrak{l}}$ where all roots of $(\mathfrak{m},\mathfrak{l})$ take only imaginary (resp. real) values. As both $\mathfrak{b}$ and $\mathfrak{l}$ are CSA's of $\mathfrak{m}$, $\mathfrak{c}\subseteq\mathfrak{b}\cap\mathfrak{l}\subseteq\mathfrak{l}_I$; further, $\bar{\mathfrak{l}}_I(\mathfrak{m})\subseteq\mathfrak{l}_I$, $\bar{\mathfrak{l}}_R(\mathfrak{m})\subseteq\mathfrak{l}_R$.[7] A dimension argument then gives $\mathfrak{l}_I=\mathfrak{c}+\bar{\mathfrak{l}}_I(\mathfrak{m})$, $\mathfrak{l}_R=\bar{\mathfrak{l}}_R(\mathfrak{m})$. It is easy to see that for a suitable $m\in M$, $(\bar{\mathfrak{l}}_I(\mathfrak{m}))^m\subseteq\mathfrak{b}\cap[\mathfrak{m},\mathfrak{m}]$; then $\mathfrak{l}_I^m\subseteq\mathfrak{b}$. For proving (ii) we may, in view of (i), assume that $\mathfrak{l}_I\subseteq\mathfrak{b}$. Then both $\mathfrak{b}$ and $\mathfrak{l}$ are CSA's of $\mathfrak{m}_1$, so that $\mathfrak{l}_I\supseteq\mathfrak{b}\cap\mathfrak{l}\supseteq\text{center}(\mathfrak{m}_1)\supseteq\mathfrak{l}_I$. The roots of $(\mathfrak{m}_1,\mathfrak{l})$ are all real and $\mathfrak{l}\cap[\mathfrak{m}_1,\mathfrak{m}_1]$ is spanned

[7] It follows from representation theory that if $\mathfrak{g}_1$ is a *semisimple* subalgebra of $\mathfrak{g}$ and $X\in\mathfrak{g}_1$, then all eigenvalues of $\operatorname{ad}X$ are real (resp. imaginary) if and only if $\operatorname{ad}_{\mathfrak{g}_1}X=\operatorname{ad}X\mid\mathfrak{g}_1$ has this property.

by the H_α's corresponding to them. So $\mathfrak{l}\cap[\mathfrak{m}_1, \mathfrak{m}_1]\subseteq\mathfrak{l}_R$; by dimensionality, $\mathfrak{l}\cap[\mathfrak{m}_1, \mathfrak{m}_1]=\mathfrak{l}_R$.

Lemma 1 leads at once to

LEMMA 2. *Let $\mathfrak{l}_I[\varepsilon]=\{H:H\in\mathfrak{l}_I, |\beta(H)|<\varepsilon$, for all roots β of $(\mathfrak{g}, \mathfrak{l})\}$; and for any simple system S of roots of $(\mathfrak{m}_1, \mathfrak{l})$, let $\mathfrak{l}_R^+(S)=\{H:H\in\mathfrak{l}_R, \beta(H)>0, \forall\beta\in S\}$. Then the connected components of $'\mathfrak{l}(R)\cap\mathfrak{u}[\varepsilon]$ are precisely all the sets of the form $\mathfrak{l}_I[\varepsilon]+\mathfrak{l}_R^+(S)$ (as S varies). If $W_{\mathfrak{m},\mathfrak{l}}(R)$ is the group generated by the reflexions s_β ($\beta\in S$), then $W_{\mathfrak{m},\mathfrak{l}}(R)$ fixes each element of $\mathfrak{l}_I$ and acts simply transitively on the collection $\{\mathfrak{l}_R^+(S)\}$. If S is fixed and $\tau\in\mathfrak{l}_c^*$ is real valued on $\mathfrak{l}_R$, there exists $s\in W_{\mathfrak{m},\mathfrak{l}}(R)$ such that $(s\tau)(H)\geqq 0$, for all $H\in\mathfrak{l}_R^+(S)$.*

LEMMA 3. *Let S be as above and $\beta\in S$. Define $\mathfrak{l}_{R,\beta}^+=\{H: H\in\mathfrak{l}_R, \beta(H)=0, \alpha(H)>0, \forall\alpha\in S\backslash\{\beta\}\}$. Then there is a CSA $\tilde{\mathfrak{l}}\subseteq\mathfrak{m}$ such that $\dim(\tilde{\mathfrak{l}}_R)=\dim(\mathfrak{l}_R)-1$ and $\mathfrak{l}_I[\varepsilon]+\mathfrak{l}_{R,\beta}^+\subseteq(\mathfrak{u}[\varepsilon]\cap\tilde{\mathfrak{l}})\cap\mathrm{Cl}(\mathfrak{l}_I[\varepsilon]+\mathfrak{l}_R^+(S))$.*

Let $X_{\pm\beta}\in\mathfrak{m}$ be the root vectors corresponding to $\pm\beta$ such that $\beta([X_\beta, X_{-\beta}])=2$. Let $\mathfrak{l}_\beta$ be the null space of β in $\mathfrak{l}$. It suffices to take $\tilde{\mathfrak{l}}=\mathfrak{l}_\beta+\boldsymbol{R}\cdot(X_\beta-X_{-\beta})$.

The proof of Lemma 3.3.1 may now be completed as follows. Take $\Gamma=\mathfrak{l}_I[\varepsilon]+\mathfrak{l}_R^+(S)$ in (2). Then by Lemma 3 and the induction hypothesis, the continuous function on $\mathfrak{u}[\varepsilon]\cap\tilde{\mathfrak{l}}$ that extends $\varpi_{\mathfrak{m},\tilde{\mathfrak{l}}}(\pi_{\mathfrak{m},\tilde{\mathfrak{l}}}(T\mid{}'\tilde{\mathfrak{l}}\cap\mathfrak{u}[\varepsilon]))$ must vanish on $\mathfrak{l}_I[\varepsilon]+\mathfrak{l}_{R,\beta}^+$. So $\sum_{s\in W_{L_c}} c_s(\Gamma)\,e^{s\mu}\equiv 0$ on $\mathfrak{l}_\beta$, leading to the relations

$$c_s(\Gamma)+c_{s_\beta s}(\Gamma)=0 \qquad (s\in W_{L_c}, \beta\in S). \tag{3}$$

Suppose $c_t(\Gamma)\neq 0$ for some $t\in W_{L_c}$. Select $s'\in W_{\mathfrak{m},\mathfrak{l}}(R)$ such that $(s't\mu)(H)\geqq 0$, $\forall H\in\mathfrak{l}_R^+(S)$. Then $c_{s't}(\Gamma)\neq 0$ by (3). On the other hand, it follows easily from the boundedness of $\sum_{s\in W_{L_c}}\varepsilon(s)\,c_s(\Gamma)\,e^{s\mu}$ on Γ that $s\mu(H)\leqq 0$, $\forall H\in\mathfrak{l}_R^+(S)$, if $c_s(\Gamma)\neq 0$. So $s't\mu\mid\mathfrak{l}_R=0$. In particular, $\langle s't\mu, \beta\rangle=0$, $\forall\beta\in S$, contradicting the regularity of μ.

3.5. Tempered invariant distributions on $\mathfrak{m}$. Proof of Lemma 3.3.2. We now take up Lemma 3.3.2. The main step in its proof is to show that if an invariant distribution T on $\mathfrak{u}[\varepsilon]$ satisfying (i) of (3.3.2) is tempered,[8] then it also satisfies (ii) therein, and hence belongs to $\mathfrak{J}_{v,\varepsilon}(\mathfrak{m})$. It follows from this that $\mathfrak{J}_{v,\varepsilon}(\mathfrak{m})$ contains the restrictions to $\mathfrak{u}[\varepsilon]$ of the Fourier transforms of the invariant measures on suitably chosen M-orbits in $\mathfrak{m}$, enabling us to verify that $\dim(\mathfrak{J}_{v,\varepsilon}(\mathfrak{m}))=p(\mathfrak{m})$.

LEMMA 1. *Let Γ_1 be an M-invariant open subset of $\mathfrak{m}$; let $\mathfrak{l}\subseteq\mathfrak{m}$ be a CSA and let $\Gamma=(\Gamma_1\cap{}'\mathfrak{l})^M$. Suppose f is any M-invariant continuous function on Γ such that the distribution defined by f is tempered. Let $f_{\mathfrak{l}}=f\mid\Gamma\cap{}'\mathfrak{l}$. Then, for some integer $r\geqq 0$, the distribution defined by $\pi_{\mathfrak{m},\mathfrak{l}}^r f_{\mathfrak{l}}$ on $\Gamma\cap{}'\mathfrak{l}$ is tempered.*

---

[8] A distribution defined on an open subset U of a real vector space V is said to be *tempered* if it is continuous in the topology induced by the seminorms $f\mapsto\sup_U|Ef|$ ($f\in C_c^\infty(U)$) where E is a differential operator on V with polynomial coefficients.

For this lemma, see [**6f**, Lemma 17]. The proof of this lemma goes in three stages. In what follows, for any real vector space V, we write $D(V)$ for the algebra of differential operators on V with polynomial coefficients.

Let L be the CSG of G corresponding to $\mathfrak{l}$ and $\bar{M}=M/L\cap M$. For $m\in M$ let $\bar{m}$ be its image in $\bar{M}$ and let $d\bar{m}$ be the invariant measure in $\bar{M}$. For any $g\in C_c^\infty(\Gamma)$ let $\varphi(g:H)=\int_{\bar{M}} g(H^m)\,d\bar{m}\,(H\in\Gamma\cap'\mathfrak{l},\ H^{\bar{m}}=H^m)$; $\varphi(g:\cdot)\in C_c^\infty(\Gamma\cap'\mathfrak{l})$. Using the well-known formula for integration on $\mathfrak{m}$, we find the following consequence of the tempered nature of f: $\exists D_i\in D(\mathfrak{m})$ such that, for all $g\in C_c^\infty(\Gamma)$,

$$(1)\qquad \left|\int_{\Gamma\cap'\mathfrak{l}} \pi_{\mathfrak{m},\mathfrak{l}}(H)^2\varphi(g:H)\,f(H)\,dH\right|\leqq \sum_{1\leqq i\leqq k}\sup_{\Gamma}|D_i g|.$$

The second stage consists in "inverting" the map $g\mapsto\varphi(g:\cdot)$. Let $W_{M,L}$ be the normalizer of $\mathfrak{l}$ in M modulo $L\cap M$. Then $W_{M,L}$ is a finite group that acts naturally on $\bar{M}$ and preserves $d\bar{m}$. We select $\bar{\gamma}\in C_c^\infty(\bar{M})$ invariant under this action such that $\int_{\bar{M}}\bar{\gamma}(\bar{m})\,d\bar{m}=1$. Let $\gamma(m)=\bar{\gamma}(\bar{m})$ $(m\in M)$ and let C be a compact set in M whose image in $\bar{M}$ contains $\operatorname{supp}\bar{\gamma}$. It is then easy to show that there is a unique $g_\beta\in C_c^\infty(\Gamma)$ such that $g_\beta(H^m)=\bar{\gamma}(\bar{m})\,\bar{\beta}(H)$ $(H\in\Gamma\cap'\mathfrak{l},\ m\in M,\ \bar{\beta}=\sum_{s\in W_{M,L}}\beta^s)$; we have $\varphi(g_\beta:H)=\bar{\beta}(H)$, $\forall H\in\Gamma\cap'\mathfrak{l}$. From (1) we then obtain, for all $\beta\in C_c^\infty(\Gamma\cap'\mathfrak{l})$, with w denoting the order of $W_{M,L}$,

$$(2)\qquad \left|\int_{\Gamma\cap'\mathfrak{l}} \pi_{\mathfrak{m},\mathfrak{l}}(H)^2\beta(H)\,f(H)\,dH\right|\leqq w^{-1}\sum_{1\leqq i\leqq k}\ \sup_{m\in C,\,H\in\Gamma\cap'\mathfrak{l}}|(D_i g_\beta)(H^m)|.$$

For the third step, let $\psi(m:H)=H^m$. Then an elementary analysis of the differential of ψ leads to the following result: If $E\in D(\mathfrak{m})$, there exists an integer $l\geqq 0$, $E_j\in D(\mathfrak{l})$, $\xi_j\in\mathfrak{M}$ (= subalgebra of $\mathfrak{G}$ generated by $(1,\mathfrak{m})$) and analytic functions h_j on M such that, for all $g\in C^\infty(\Gamma)$, $H\in\Gamma\cap'\mathfrak{l}$, $m\in M$,

$$(3)\qquad (Ef)(H^m)=\pi_{\mathfrak{m},\mathfrak{l}}(H)^{-l}\sum_{1\leqq j\leqq q} h_j(m)\,(g\circ\psi)(m;\xi:H;E_j)$$

(this follows from Lemmas 3–5 of §3 of [**5a**]).

If we now use (3) in (2) with $E=D_j$, $g=g_\beta$, the required conclusion about $f_{\mathfrak{l}}$ follows without difficulty.

LEMMA 2. *Let T be a tempered invariant distribution on $\mathfrak{u}[\varepsilon]$ satisfying* (i) *of* (3.3.2). *Then $T\in\mathfrak{I}_{v,\varepsilon}(\mathfrak{m})$.*

Let $\mathfrak{l}\subseteq\mathfrak{m}$ be a CSA. Then T is given by (3.4.2) with $\Gamma=\mathfrak{l}_I[\varepsilon]+\mathfrak{l}_R^+(S)$ (cf. §3.4). By Lemma 1, for some integer $r\geqq 0$, the function

$$H\mapsto\pi_{\mathfrak{m},\mathfrak{l}}(H)^r\sum_{s\in W_{L_c}}\varepsilon(s)\,c_s(\Gamma)\,e^{s\mu(H)}$$

defines a tempered distribution on Γ. It is not difficult to deduce from this (cf. [**6f**, Lemma 15]) that $s\mu(H)\leqq 0$, $\forall H\in \mathfrak{l}_R^+(S)$, whenever $c_s(\Gamma)\neq 0$. But then T must satisfy (ii) of (3.3.2).

Let $\mathscr{C}(\mathfrak{m})$ be the Schwartz space of rapidly decreasing functions on $\mathfrak{m}$. For $g\in\mathscr{C}(\mathfrak{m})$, $\hat{g}\in\mathscr{C}(\mathfrak{m})$ is defined by $\hat{g}(X)=\int_{\mathfrak{m}} g(Y)\, e^{i\langle X, Y\rangle}\, dY$. For any tempered distribution T on $\mathfrak{m}$, its Fourier transform $\hat{T}(g)=T(\hat{g})$ $(g\in\mathscr{C}(\mathfrak{m}))$.

Let ν be as in Lemma 3.3.2. Define $H_\nu\in i\mathfrak{b}$ by $\langle H_\nu, H\rangle=\nu(H)$ $(H\in\mathfrak{b})$.

LEMMA 3. *For any $s\in W_{B_c}$, the orbit $(-iH_\nu^s)^M$ is closed and admits an M-invariant measure $\sigma_{\nu,s}$. $\sigma_{\nu,s}$, regarded as a distribution on $\mathfrak{m}$, is tempered, and $\hat{\sigma}_{\nu,s}$ satisfies the differential equations $(\mu_{\mathfrak{g}/\mathfrak{m}}(z))^{\sim}\,\hat{\sigma}_{\nu,s}=\chi_\nu(z)\,\hat{\sigma}_{\nu,s}$, for all $z\in\mathfrak{Z}$.*

That the orbit X^M is closed and admits an invariant measure for any regular $X\in\mathfrak{m}$ is well known. That it is tempered when regarded as a distribution in $\mathfrak{m}$ is proved in [**5b**] (cf. §5.9). The last assertion is proved by a straightforward calculation.

Let $s_1,\ldots, s_p$ $(p=p(\mathfrak{m}))$ be representatives of $W_{M,B}\backslash W_{B_c}$. By Lemmas 2 and 3, $\hat{\sigma}_{\nu,s_j}\mid \mathfrak{u}[\varepsilon]\in\mathfrak{J}_{\nu,\varepsilon}(\mathfrak{m})$, $1\leqq j\leqq p$. Since the orbits $(-iH_\nu^{s_j})^M$ $(1\leqq j\leqq p)$ are disjoint, the σ_{ν,s_j} are linearly independent. So the $\hat{\sigma}_{\nu,s_j}$ are also linearly independent. A simple argument based on Lemma 3.3.1 now implies the linear independence of $\hat{\sigma}_{\nu,s_j}\mid\mathfrak{u}[\varepsilon]$, $1\leqq j\leqq p$. Thus $\dim(\mathfrak{J}_{\nu,\varepsilon}(\mathfrak{m}))\geqq p(\mathfrak{m})$. On the other hand, if $T\in\mathfrak{J}_{\nu,\varepsilon}(\mathfrak{m})$, there are constants $c_s(T)$ $(s\in W_{B_c}$, $c_{ts}=c_t$, $\forall t\in W_{M,B})$ such that $\pi_{\mathfrak{m},\mathfrak{b}}(H)\,T(H)$ $=\sum_{s\in W_{B_c}}\varepsilon(s)\,c_s(T)\,e^{s\nu(H)}$, $\forall H\in{}'\mathfrak{b}\cap\mathfrak{u}[\varepsilon]$. Lemma 3.3.1 then yields the estimate $\dim(\mathfrak{J}_{\nu,\varepsilon}(\mathfrak{m}))\leqq p(\mathfrak{m})$.

REMARK. A simple Fubini argument applied to the integration formula on $\mathfrak{m}$ shows that for almost all ν the invariant measures $\sigma_{\nu,s}$ are tempered $\forall s\in W_{B_c}$. For such ν, the above discussion is certainly valid. The proof of Lemma 3.4.2 for the exceptional (but still regular) ν may then be completed by a limiting process. We may thus avoid using the highly nontrivial Theorem 3 of [**5b**].

3.6. The distribution Θ_λ^*. Let $\lambda\in\mathscr{L}'$. It follows from the preceding theory that there is a unique invariant eigendistribution Θ_λ^* on G such that

$$(1)\qquad \begin{aligned} &\text{(i)}\quad \Theta_\lambda^*(b)\,\Delta(b)=\textstyle\sum_{s\in W_{Bc}}\varepsilon(s)\,\xi_{s\lambda}(b)\ (b\in B'),\\ &\text{(ii)}\quad \sup_{x\in G'}|D(x)|^{1/2}\,|\Theta_\lambda^*(x)|<\infty.\end{aligned}$$

Clearly $z\Theta_\lambda^*=\chi_\lambda(z)\,\Theta_\lambda^*$, $\forall z\in\mathfrak{Z}$. The distribution Θ_λ^* is somewhat less singular than Θ_λ. For instance, on B, Θ_λ^* is a finite Fourier series, and is therefore bounded. In this section we shall formulate analogous results for the other CSG's [**6f**, §24]. These play an important role in invariant analysis on G.

THEOREM 1. *Let L be a CSG and $\mathfrak{l}$ its Lie algebra. Let $W_{\mathfrak{l}}(l)$ be the group* *generated by the Weyl reflexions corresponding to the imaginary roots of* $(\mathfrak{g}, \mathfrak{l})$. *Then* $W_{\mathrm{I}}(I)$ *leaves* L *(and* L'*) invariant and* $\Theta^*_\lambda(a^s) = \Theta^*_\lambda(a)$ $(a \in L', s \in W_{\mathrm{I}}(I))$. *Moreover, if* $\mathfrak{z}$ *is the centralizer of* $\mathfrak{l}_R$ *in* $\mathfrak{g}$, *there is a constant* $C > 0$ *such that*

$$|\Theta^*_\lambda(a)| \leqq C|\det(\mathrm{Ad}(a) - 1)_{\mathfrak{g}/\mathfrak{z}}|^{-1/2} \qquad (a \in L'). \tag{2}$$

We use induction on $\dim(\mathfrak{l}_R)$. We may assume that $\mathfrak{l}$ is θ-stable, $\mathfrak{l}_I \subseteq \mathfrak{b}$, and $\dim(\mathfrak{l}_R) > 0$. If $L_I = L \cap K$ and $F = K \cap \exp(-1)^{1/2}\mathfrak{l}_R$, then $L = L_I \exp \mathfrak{l}_R$ and $L_I = F \exp \mathfrak{l}_I$. So $W_{\mathrm{I}}(I)$ leaves fixed each component of L_I and each element of $\exp \mathfrak{l}_R$. Let $\mu \in \mathfrak{l}^*_c$ be such that $\chi_\mu = \chi_\lambda$.

Let L_I^+ be a component of L_I, and let $\mathfrak{m}_1$ be the centralizer of L_I^+ in $\mathfrak{g}$. Clearly $\mathrm{rk}(\mathfrak{g}) = \mathrm{rk}(\mathfrak{m}_1) = \mathrm{rk}(\mathfrak{m}_1 \cap \mathfrak{k})$. Proceeding as in §3.4 (and using the same notation) we find that $\mathfrak{l}_R$ is a CSA of $[\mathfrak{m}_1, \mathfrak{m}_1]$, and that the connected components of $L'(R)$ are precisely all sets of the form $L_I^+ \exp \mathfrak{l}_R^+(S)$, where L_I^+ is as above and S is a simple system of roots of $(\mathfrak{m}_1, \mathfrak{l})$. Fix L_I^+, S and let c_s be constants such that

$$\Delta_L(a)\, \Theta^*_\lambda(a) = \sum_{s \in W_{L_c}} \varepsilon(s)\, c_s \xi_{s\mu}(a) \qquad (a \in L' \cap L_I^+ \exp \mathfrak{l}_R^+(S));$$

here $\Delta_L = \Delta_{L, P_L}$ for some positive system P_L of roots of $(\mathfrak{g}, \mathfrak{l})$.

LEMMA 2. *Fix* $\beta \in S$. *Then we can find a CSA* $\tilde{\mathfrak{l}}$ *with CSG* $\tilde{L}$ *and an element* $y \in G_c$ *having the following properties:* (i) $\tilde{\mathfrak{l}}$ *is* θ*-stable and* $\dim(\tilde{\mathfrak{l}}_R) = \dim(\mathfrak{l}_R) - 1$, (ii) $\mathfrak{l}_c^y = \tilde{\mathfrak{l}}_c$, (iii) y *fixes each element of* $L_I^+ \exp(\mathfrak{l}_R \cap \mathfrak{l}_\beta)$, (iv) *there is a connected component* $\tilde{L}^+$ *of* $\tilde{L}'(R)$ *such that* $\mathrm{Cl}(\tilde{L}^+) \cap \mathrm{Cl}(L_I^+ \exp \mathfrak{l}_R^+(S))$ *contains* $L_I^+ \exp \mathfrak{l}_{R,\beta}^+$, *and* (v) $y \circ W_{\mathrm{I}}(I) \circ y^{-1} \subseteq W_{\tilde{\mathrm{I}}}(I)$.

Select root vectors $X_{\pm\beta} \in \mathfrak{m}_1$ corresponding to $\pm\beta$ such that $\beta([X_\beta, X_{-\beta}]) = 2$ and $X_{-\beta} = -\theta(X_\beta)$ (this is possible). Take $\tilde{\mathfrak{l}} = \mathfrak{l}_\beta + \boldsymbol{R} \cdot (X_\beta - X_{-\beta})$ and $y = \exp(-(-1)^{1/2}\pi(X_\beta + X_{-\beta})/4)$.

Let $\Delta^L = \Delta_L \circ y^{-1}$ and let d_s be constants such that

$$\Delta_{\tilde{L}}(h)\, \Theta^*_\lambda(h) = \sum_{s \in W_{L_c}} \varepsilon(s) \quad d_s \xi_{s\mu \circ y^{-1}}(h) \qquad (h \in \tilde{L}^+ \cap \tilde{L}').$$

From Lemma 2 and Theorem 2.1.2 it follows that $\sum_{s \in W_{L_c}} (c_s - d_s)\, \xi_{s\mu}$ vanishes on $L_I^+ \exp(\mathfrak{l}_R \cap \mathfrak{l}_\beta)$. This gives the relations $c_s + c_{s_\beta s} = d_s + d_{s_\beta s}$, $\forall s$. Fix $t \in W_{\mathrm{I}}(I)$ and write $\tilde{c}_s = c_{ts} - c_s$. From Lemma 2 (v), the invariance of $\Theta^*_\lambda \mid \tilde{L}^+$ with respect to $W_{\tilde{\mathrm{I}}}(I)$, and the fact that $W_{\tilde{\mathrm{I}}}(I)$ commutes with s_β, we obtain

$$\tilde{c}_s + \tilde{c}_{s_\beta s} = 0 \qquad (s \in W_{L_c}, \beta \in S). \tag{3}$$

Further, if $\tilde{c}_s \neq 0$, then either c_s or c_{ts} is $\neq 0$, and, in either case, $s\mu(H) \leqq 0$, $\forall H \in \mathfrak{l}_R^+(S)$.

We now argue as in §3.4 to conclude that $\tilde{c}_s=0$, $\forall s$. Thus $\Theta_\lambda^*(a^s)=\Theta_\lambda^*(a)$ $(a\in L'$, $s\in W_{\mathrm{I}}(I))$.

For proving (2), let $P_{L,I}$ be the set of imaginary roots in P_L. For $a\in L$, write a_I and a_R for its components in L_I and $\exp\mathfrak{l}_R(a=a_Ia_R)$. Define, with $\delta=\frac{1}{2}\sum_{\alpha\in P_L}\alpha$,

$$\Delta_L^+(a)=\xi_\delta(a_R)\prod_{\alpha\in P_L\backslash P_{L,I}}(\xi_\alpha(a)-1),\qquad \Delta_{L,I}(a)=\xi_\delta(a_I)\prod_{\alpha\in P_{L,I}}(\xi_\alpha(a)-1).$$

It is then an immediate consequence of the invariance under $W_{\mathrm{I}}(I)$ that, for some constant $C>0$ and all $a\in L'\cap(L_I^+\exp\mathfrak{l}_R^+(S))$,

$$|\Delta_L^+(a)\,\Theta_\lambda^*(a)|\leqq C\left|\Delta_{L,I}(a)^{-1}\sum_{s\in W_{\mathrm{I}}(I)}\varepsilon(s)\,\xi_{s\mu}(a_I)\right|.$$

It is not difficult to show that the expression on the right is bounded over L_I^+ and that $|\Delta_L^+(a)|^2=|\det(\mathrm{Ad}(a)-1)_{\mathfrak{g}/\mathfrak{z}}|$, $\forall a\in L$. The estimate (2) is now immediate.

4. The Schwartz space

Throughout this section G is a group of class $\mathscr{H}$.

4.1. The functions Ξ and σ. For $x\in G$, $\sigma(x)=\sigma(x^{-1})$ is the distance between the cosets K and xK in the Riemannian space G/K. σ is a spherical function, $\sigma(\exp X)=\|X\|$ $(X\in\mathfrak{p})$, and

(1) $$\sigma(xy)\leqq\sigma(x)+\sigma(y)\qquad(x,y\in G).$$

Let π be the unitary representation of G induced by the trivial representation of a minimal psgrp P. The trivial representation of K occurs exactly once in $\pi\mid K$. We define $\Xi(x)=\Xi_G(x)=(\pi(x)\,\psi,\psi)$ $(x\in G)$ where ψ is a unit vector fixed by $\pi[K]$. Ξ does not depend on the choice of P. For $x\in G$ let $x=k\exp H(x)\,n$ where $k\in K$, $H(x)\in\mathfrak{a}$, $n\in N$; $\varrho(X)=\frac{1}{2}\mathrm{tr}(\mathrm{ad}\,X\mid \mathfrak{n})$ $(X\in\mathfrak{a})$. Then it follows from the explicit form of π that [**4b**, p. 43]

(2) $$\Xi(x)=\Xi(x^{-1})=\int_K\exp\{-\varrho(H(xk))\}\,dk\qquad(x\in G).$$

Ξ is an analytic spherical function, $\Xi(1)=1$, and $0<\Xi(x)\leqq 1$, for all x. Further, for any $b\in\mathfrak{G}$ let $a_b\in\mathfrak{A}$ be the unique element such that $b-a_b\in\mathfrak{k}\mathfrak{G}+\mathfrak{G}\mathfrak{n}$ [**5e**, Lemma 3]; then, denoting by $\mathfrak{Q}$ the centralizer of K in $\mathfrak{G}$,

(3) $$q\Xi=a_q(-\varrho)\,\Xi\qquad(q\in\mathfrak{Q}).$$

It is well known [7, §3 of Chapter X] that Ξ is uniquely determined by the differential equations (2) and the condition $\Xi(1)=1$. This observation leads to the relation

$$\int_K \Xi(xky)\,dk=\Xi(x)\,\Xi(y) \qquad (x, y\in G). \tag{4}$$

In the following results we collect together a few estimates involving Ξ and σ. Corollary 2 and Theorem 3 make clear the importance of the function Ξ in Fourier analysis.

THEOREM 1. (i) *Given $a, b\in\mathfrak{G}$, there exists $C=C(a, b)>0$ such that $|\Xi(a; x; b)|\leqq C\Xi(x)$, $\forall x\in G$.*

(ii) *If E is any compact subset of G, there exists $C=C(E)>0$ such that $\Xi(y_1xy_2)\leqq C\Xi(x)$, $\forall y_1, y_2\in E$, $x\in G$.*

(iii) *There exists $C>0$ and $d\geqq 0$ such that, for all $h\in A^+$,*

$$\exp\{-\varrho(\log h)\}\leqq\Xi(h)\leqq C\exp\{-\varrho(\log h)\}\,(1+\sigma(h))^d. \tag{5}$$

The estimates (i) and (ii) may be derived from the explicit form of the representation π. The proof of (iii) requires a study of the differential equations (3) [**5e**, Theorem 3].

COROLLARY 2. *There exists $r>0$ such that $\Xi^2(1+\sigma)^{-r}\in L^1(G)$.*

For some constant $c>0$, $c\,dx=J(h)\,dk_1dhdk_2$ $(x=k_1hk_2, k_1, k_2\in K, h\in A^+)$ where

$$J(h)=\prod_{\lambda>0}\{e^{\lambda(\log h)}-e^{-\lambda(\log h)}\}^{m(\lambda)} \tag{6}$$

(the product is over the positive roots of $(\mathfrak{g}, \mathfrak{a})$ and $m(\lambda)=$dimension of the root space of λ; cf. [7, p. 382]). The corollary follows from this and (5).

THEOREM 3.[9] *Fix p, $1\leqq p\leqq 2$. Then, given $a, b\in\mathfrak{G}$, there exist $a_i, b_i\in\mathfrak{G}$ $(1\leqq i\leqq m)$ with the following property: For any $f\in C^\infty(G)$ with $ufv\in L^p(G)$, $\forall u, v\in\mathfrak{G}$,*

$$\|\Xi^{-2/p}(afb)\|_\infty\leqq\sum_{1\leqq i\leqq m}\|a_ifb_i\|_p. \tag{7}$$

In view of the closed graph theorem it is enough to prove that $\|\Xi^{-2/p}(ufv)\|_\infty<\infty$, $\forall u, v\in\mathfrak{G}$. Let $g=afb$. By the work of §5.8, there exist $\zeta_j\in\mathfrak{A}$ $(1\leqq j\leqq r)$ such

[9] For $1\leqq p\leqq\infty$, $\|\cdot\|_p$ is the L^p-norm.

that for all $h \in A^+$ and $F \in H^p(A^+, J)$, $|F(h)|^p \leqq \sum_j \|\zeta_j F\|_{p,J}^p$. We apply this estimate to

$$F(h) = \exp\ (2\varrho(\log h)/p)\, g_h(\eta;\, k_1 : k_2;\, \xi)$$

where $\xi, \eta \in \mathfrak{K}$, $g_h(k_1 : k_2) = g(k_1 h k_2)$ $(k_1, k_2 \in K)$, and integrate over $K \times K$. Thus we find that

$$\sup_{h \in A^+} \exp\ (2\varrho(\log h)/p)\ \|\zeta g_h\|_p < \infty, \quad \forall \zeta \in \mathfrak{K} \otimes \mathfrak{K}.$$

By (5) and Sobolev's lemma, $\sup_{h \in A^+} \Xi(h)^{-2/p} \|g_h\|_\infty = \|\Xi^{-2/p} g\|_\infty < \infty$. For details, see [**14b**, §3].

4.2. Definition of $\mathscr{C}(G)$. Following Harish-Chandra [**6h**, §9], we define the Schwartz space $\mathscr{C}(G)$ to be the space of all $f \in C^\infty(G)$ such that for all $m \geqq 0$, $a, b \in \mathfrak{G}$,

(1) $$\mu_{a,b:m}(f) = \|\Xi^{-1}(1+\sigma)^m (afb)\|_\infty < \infty.$$

The seminorms $\mu_{a,b:m}$ convert $\mathscr{C}(G)$ into a Fréchet space (cf. also [**2**] for $G = SL(2, \boldsymbol{R})$). In this definition, $1+\sigma$ may be replaced by any other function having the same growth. It is not difficult to construct C^∞ spherical functions τ, for example, such that (i) $0 < \alpha \leqq (1+\sigma)^{-1}\tau \leqq \beta < \infty$, and (ii) the derivatives $X_1 \cdots X_r \tau Y_1 \cdots Y_s$ $(r+s \geqq 1,\ X_1, Y_j \in \mathfrak{g})$ are all bounded; for instance, if $G = {}^\circ G$, we may take $\tau = 1 - \log \Xi$. That $\mathscr{C}(G)$ is a natural object of study is made clear by the following result which is a consequence of Theorem 4.1.3.

THEOREM 1. *$\mathscr{C}(G)$ is precisely the space of all $f \in C^\infty(G)$ such that $(1+\sigma)^m (afb) \in L^2(G)$, $\forall m \geqq 0$, $a, b \in \mathfrak{G}$, and its topology coincides with the one induced by the seminorms $f \mapsto \|(1+\sigma)^m (afb)\|_2$.*

THEOREM 2. *$C_c^\infty(G)$ is a dense subspace of $\mathscr{C}(G)$, and the natural inclusion map is continuous. $\mathscr{C}(G)$ is a Fréchet algebra under convolution.*

Let χ_t $(t > 0)$ be the characteristic function of the set B_t where $\sigma \leqq t$, and $\beta_t = \beta * \chi_t * \beta$, β being spherical and $\in C_c^\infty(G)$. Using the facts that (i) $\sup_{t>0} \|a\beta_t b\|_\infty < \infty$ $(a, b \in \mathfrak{G})$ and (ii) for some $h > 0$, $\beta_t = 1$ on B_{t-h}, $\forall t \geqq h$ (cf. (4.1.1)), we find that for any $f \in \mathscr{C}(G)$, $\beta_t f \in C_c^\infty(G)$ and $\beta_t f \to f$ in $\mathscr{C}(G)$ as $t \to +\infty$. For proving the second assertion let $\Xi_s = \Xi(1+\sigma)^{-s}$ $(s > 0)$. Then, by (4.1.1),

(2) $$\Xi_s(y^{-1}kx) \leqq \Xi(y^{-1}kx)\,(1+\sigma(y))^s\,(1+\sigma(x))^{-s} \qquad (x, y \in G,\ k \in K).$$

So, if r is as in Corollary 4.1.2 and $s_1 \geqq s+r$, (2) and (4.1.4) give the estimate (with $c = \int_G \Xi^2 (1+\sigma)^{-r}\, dx$)

(3) $$\Xi_{s_1} * \Xi_s(x) \leqq c\,\Xi_s(x) \qquad (x \in G).$$

The estimate (3) shows that, for $f, g\in\mathscr{C}(G)$, $f*g\in\mathscr{C}(G)$, and that the map $f, g\mapsto f*g$ is continuous.

4.3. Tempered distributions on G. A distribution T on G is *tempered* if T has an extension (which is unique by Theorem 4.2.2) to a continuous linear functional on $\mathscr{C}(G)$.

If ψ is a locally summable function on G such that

$$|\psi(x)|\leqq C\Xi(x)(1+\sigma(x))^p \qquad \text{(for almost all } x\in G) \tag{1}$$

for some $C>0$, $p\geqq 0$, then ψ defines a tempered distribution since $|\int_G \psi f\,dx|\leqq \text{const}\,\mu_{1,1:p+r}(f)$, $\forall f\in C_c^\infty(G)$, r being as in Corollary 4.1.2. We shall discuss in this section two special cases where growth properties analogous to (1) are consequences of the property of being tempered.

The first result is an analogue of the classical result on measures of slow growth [**12**, p. 97]; it is substantially Theorem 4 of [**6h**].

THEOREM 1. *Let $\mathscr{M}$ be a set of nonnegative Borel measures on G. Then the following statements are equivalent:* (i) *there is a continuous seminorm γ on $\mathscr{C}(G)$ such that $|\int f\,dm|\leqq\gamma(f)$, for all $m\in\mathscr{M}$ and all spherical $f\in C_c^\infty(G)$, and* (ii) *there exists $C>0$ and $q\geqq 0$ such that $\int \Xi(1+\sigma)^{-q}\,dm\leqq C$, for all $m\in\mathscr{M}$. In this case, the convergence of the integral is uniform for $m\in\mathscr{M}$.*

Assume (i). Let $G={}^\circ G\cdot V$, $x={}^\circ x^1x$ ($x\in G$, ${}^\circ x\in{}^\circ G$, ${}^1x\in V$). For $t_1, t_2>0$, let $C_{t_1,t_2}=\{x:x\in G, \Xi({}^\circ x)\geqq e^{-t_1}, \sigma({}^1x)\leqq t_2\}$. Select spherical $\beta\in C_c^\infty(G)$ with $\beta(x)=\beta(x^{-1})\geqq 0$ for all x and $\int\beta\,dx=1$, and let $f_{t_1,t_2}=\beta*\chi_{t_1,t_2}*\beta$ where χ_{t_1,t_2} is the characteristic function of C_{t_1,t_2}. Let $a>0$ be such that $e^{-a}\Xi(x)\leqq\Xi(y_1xy_2)\leqq e^a\Xi(x)$ and $\sigma({}^1y)\leqq\frac{1}{2}a$, $\forall x\in G$, $y, y_1, y_2\in\operatorname{supp}\beta$. Then $0\leqq f_{t_1,t_2}\leqq 1$, $f_{t_1,t_2}=1$ on C_{t_1-a,t_2-a} and $=0$ outside C_{t_1+a,t_2+a}. Taking $\gamma=\sum_{1\leqq i\leqq n}\mu_{a_i,b_i:s}$ we find, $\forall m\in\mathscr{M}$, $t_1, t_2>0$,

$$\begin{aligned} m(C_{t_1,t_2})&\leqq\sum\mu_{a_i,b_i:s}(f_{t_1+a,t_2+a})\\ &\leqq\sum\|a_i\beta\|_1\,\|\beta b_i\|_1\sup\{\Xi(x)^{-1}(1+\sigma(x))^s:x\in C_{t_1+2a,t_2+2a}\}.\end{aligned}$$

So there exist $B>0$ and $l\geqq 0$ such that, for all $m\in\mathscr{M}$, $t_1, t_2>0$,

$$m(C_{t_1,t_2})\leqq Be^{t_1}(1+t_1)^l(1+t_2)^l. \tag{2}$$

The estimate (2) implies (ii). Actually (2) is equivalent to (ii).

Our second result deals with the $\mathfrak{Z}$-finite K-finite functions. It is essentially Theorem 9 of [**6h**].

THEOREM 2. *Let $\varphi\in C^\infty(G)$ be $\mathfrak{Z}$-finite and K-finite. If the distribution defined by φ is tempered, then φ satisfies* (1) *for some $C>0$, $p\geqq 0$.*

Write $T_\varphi(f)=\int \varphi f\,dx$ ($f\in C_c^\infty(G)$). Then there exist $a_i, b_i\in\mathfrak{G}$ and $p\geqq 0$ such that, for all $f\in C_c^\infty(G)$),

(3) $$|T_\varphi(f)|\geqq \sum_{1\leqq i\leqq n} \mu_{a_i,b_i:p}(f).$$

Now, by Corollary 1.2.3, there exist $\beta_1, \beta_2\in C_c^\infty(G)$ such that $\varphi=\beta_1*\varphi*\beta_2$ so that $T_\varphi(f)=T_\varphi(\beta_1'*f*\beta_2')$ ($\beta_i'(x)=\beta_i(x^{-1})$). Transferring the differentiations in (3) to the β_i' we find that, for some $C>0$, $|T_\varphi(f)|\leqq C\mu_{1,1:p}(f)$, $\forall f\in C_c^\infty(G)$. An elementary measure-theoretic argument then gives $|\varphi|\ \Xi(1+\sigma)^{-p}\in L^1(G)$. Now, if $a, b\in\mathfrak{G}$, $a\varphi b$ is both $\mathfrak{Z}$-finite and K-finite and $T_{a\varphi b}$ satisfies an estimate of the form (3) with the *same* p. Hence $|a\varphi b|\ \Xi(1+\sigma)^{-p}\in L^1(G)$. Replacing σ by a function τ as described in §4.2 we find $a(\Xi(1+\tau)^{-p}\varphi)\,b\in L^1(G)$, $\forall a, b\in\mathfrak{G}$. Theorem 4.1.3 now implies that $\|\Xi^{-1}(1+\tau)^{-p}a\varphi b\|_\infty<\infty$ for all $a, b\in\mathfrak{G}$.

As one may infer from Harish-Chandra's work **[6i]**, only those irreducible unitary representations (the so-called tempered representations) whose characters and matrix coefficients are tempered distributions play a role in the L^2 Fourier theory on G, and therein lies the real importance of the tempered distributions. For instance, the representations of the discrete series are tempered, as one may conclude from the L^2 estimates (1.2.5), as are the representations that are associated with the various CSG's (cf. §1.1). If π is an irreducible unitary representation whose character Θ_π is tempered, the matrix coefficients of π defined by K-finite vectors are of the form $a(\Theta_\pi*f)\,b$ where $a, b\in\mathfrak{G}$, f is a matrix coefficient of K, and the convolution is over K; they are therefore tempered. The converse is also true, though less trivial to establish: If the matrix coefficients defined by K-finite vectors of an irreducible unitary π are tempered, then Θ_π is tempered.

It is an important problem to determine the conditions for an invariant $\mathfrak{Z}$-finite distribution to be tempered. We shall come to this later.

5. Invariant analysis on G

In this section we shall assume that G is of class $\mathscr{H}$ and discuss some aspects of the theory of integration over the conjugacy classes of G. This is one of the most important techniques of harmonic analysis; it enables us to reduce many problems on G to (presumably easier) questions on the Cartan subgroups of G **[6h]**.

We set up the map $f\mapsto {}'F_{f,L}$ for a CSG L, and formulate the main results in §5.1. In §5.2 we examine how these can be reduced to the case of compact L. The key estimate used in the proofs is discussed in §5.4. These are then applied in §§5.5 and 5.7 to study various questions on harmonic analysis. There are two appendices: The first (§5.8) deals with some estimates of a classical nature; the second (§5.9) examines some aspects of the theory of tempered invariant eigendistributions on a real semisimple Lie algebra.

5.1. The map $f \mapsto {'F_f}$. Let $\mathfrak{l}$ be a θ-stable CSA, $\mathfrak{l}_R = \mathfrak{l} \cap \mathfrak{p}$, $\mathfrak{l}_I = \mathfrak{l} \cap \mathfrak{k}$, and $\mathfrak{L}$, the subalgebra of $\mathfrak{G}$ generated by $(1, \mathfrak{l})$; L is the corresponding CSG, $L_I = L \cap K$, $L_R = \exp \mathfrak{l}_R$. M_1 (resp. $\mathfrak{m}_1$) is the centralizer of $\mathfrak{l}_R$ in G (resp. $\mathfrak{g}$); $M = {}^\circ M_1$, and $\mathfrak{m}$, the Lie algebra of M. $D_\mathfrak{l}$ is the invariant function on G vanishing outside $(L')^G$ such that $D_\mathfrak{l}(b) = \det(1 - \mathrm{Ad}(b))_{\mathfrak{g}/\mathfrak{m}_1}$ $(b \in L')$; if L is compact, $D_\mathfrak{l}$ is the characteristic function of the regular elliptic set $(L')^G$, and is denoted also by φ_L. P_I is a positive system of roots of $(\mathfrak{m}_1, \mathfrak{l})$; ${'\Delta_I} = \prod_{\alpha \in P_I} (\xi_\alpha - 1)$; $\delta_I = \frac{1}{2} \sum_{\alpha \in P_I} \alpha$. $\bar{G} = G/L_R$, $x \mapsto \bar{x}$ is the natural map of G on $\bar{G}$, and $d\bar{x}$ is an invariant measure on $\bar{G}$; if $b \in L$, $x \in G$, we write $b^{\bar{x}}$ for b^x. $L'(I)$ is the set of all $b \in L$ such that $\xi_\beta(b) \neq 1$ for any singular[10] $\beta \in P_I$. For any open set $U \subseteq L$, let $\mathscr{C}(U)$ be the Schwartz space of U. For any continuous function f on G we put

$$(1) \qquad {'F_f}(b) = {'F_{f,L}}(b) = {'\Delta_I}(b)\, |D_\mathfrak{l}(b)|^{1/2} \int_{\bar{G}} f(b^{\bar{x}})\, d\bar{x} \qquad (b \in L')$$

whenever this integral is absolutely convergent for all $b \in L'$. Put

$$(2) \qquad {'\zeta} = e^{\delta_I} \circ \zeta \circ e^{-\delta_I} \qquad (\zeta \in \mathfrak{L}).$$

THEOREM 1. *For each $f \in \mathscr{C}(G)$, ${'F_f}$ is well defined and lies in $\mathscr{C}(L')$. The map $f \mapsto {'F_f}$ is continuous from $\mathscr{C}(G)$ to $\mathscr{C}(L')$. Moreover, for all $z \in \mathfrak{Z}$, $f \in \mathscr{C}(G)$,*

$$(3) \qquad {'F_{zf}} = {'\mu_{\mathfrak{g}/\mathfrak{l}}(z)}\, {'F_f}.$$

THEOREM 2. *Fix $f \in \mathscr{C}(G)$ and $b \in L$.* (i) *If $b \in L'(I)$, then ${'F_f}$ extends as a C^∞ function around b.* (ii) *Let $b \notin L'(I)$ and let $S_I(b)$ be the set of all singular $\beta \in P_I$ with $\xi_\beta(b) = 1$. Then, for any $\zeta \in \mathfrak{L}$ for which $\zeta^{s_\beta} = -\zeta$, $\forall \beta \in S_I(b)$, ${'\zeta}\, {'F_f}$ extends as a continuous function around b. In particular, if $\varpi_I = \prod_{\beta \in P_I} H_\beta$, ${'\varpi_I}\, {'F_f}$ extends to a continuous function on L.*

Let $\beta \in P_I$ be singular, $L_\beta = \{b : b \in L,\ \xi_\beta(b) = 1\}$, and let L'_β be the set of all $b \in L_\beta$ such that $\xi_{\pm\beta}$ are the only global roots that are equal to 1 at b. Let $\mathfrak{z}$ be the centralizer of L_β in $\mathfrak{g}$ and select a CSA $\mathfrak{l}_1 \subseteq \mathfrak{z}$ that is not conjugate to $\mathfrak{l}$ in G; let L_1 be the corresponding CSG. Note that $b \in L'_1(I)$ and so, for any $f \in \mathscr{C}(G)$, ${'F_{f,L_1}}$ extends (by Theorem 2) to a C^∞ function in a neighborhood of b in L_1. For any function $g \in \mathscr{C}(L'(I))$ and any $b \in L'_\beta$ we write

$$g(b\pm) = \lim_{t \to 0\pm} g(b \exp(-1)^{1/2} t H_\beta).$$

Let y be an element in the adjoint group of $\mathfrak{z}_c$ such that $\mathfrak{l}_c^y = (\mathfrak{l}_1)_c$.

---

[10] Given $\beta \in P$ which is either real or imaginary, let $\mathfrak{z} = \mathfrak{g} \cap (C \cdot H_\beta + C \cdot X_\beta + C \cdot X_{-\beta} + C \cdot X_{-\beta})$. β is called singular if $\mathfrak{z}$ is not of compact type.

THEOREM 3. *Let the notation be as above. Then there is an automorphism $\zeta \to \mathring{\zeta}$ of $\mathfrak{L}$ and a C^∞ function c on L'_β such that, for all $\sigma \in \mathfrak{L}$, $b \in L'_\beta$,*

$$(4) \qquad (\zeta\, 'F_{f,L})(b+) - (\zeta\, 'F_{f,L})(b-) = c(b)\,(\mathring{\zeta}^y\, 'F_{f,L_1})(b).$$

Suppose G is as in §3. The above results then suggest a simple modification of the definition of $'F_f$ to simplify some of its formal properties. Let L be a CSG; $\Delta = \Delta_{L,P}(P = P_L)$ is a positive system of roots of $(\mathfrak{g}, \mathfrak{l})$. Let

$$\varepsilon_R(b) = \operatorname{sign} \prod_{\alpha \in P,\, \alpha \text{ real}} (\xi_\alpha(b) - 1) \qquad (b \in L').$$

Then we define, for all $f \in \mathscr{C}(G)$,

$$(5) \qquad F_f(b) = F_{f,L}(b) = \varepsilon_R(b)\,\Delta(b) \int_{\bar{G}} f(b^{\bar{x}})\, d\bar{x} \qquad (b \in L').$$

Changing P to another positive system P' results in merely multiplying the RHS of (5) by a constant $C(P, P') = \pm 1$. Let us now choose P so that it contains the complex conjugate of each of its nonimaginary roots, and let P_I be the set of imaginary roots in P. Then a simple calculation shows that

$$(6) \qquad F_f(b) = \xi_{-\delta}(b_I)\, 'F_f(b) \qquad (b \in L',\ f \in \mathscr{C}(G))$$

where $\delta = \frac{1}{2} \sum_{\alpha \in P} \alpha$. So

$$(7) \qquad F_{zf} = \mu_{\mathfrak{g}/\mathfrak{l}}(z)\, F_f \qquad (f \in \mathscr{C}(G),\ z \in \mathfrak{Z}).$$

Moreover, if ζ is as in Theorem 2, ζF_f extends continuously around b. In particular, $\varpi_I F_f$ extends to a continuous function on L. Finally, Theorem 3 may be reformulated in the following manner: There is a nowhere vanishing locally constant function c on L'_β such that, for all $\zeta \in \mathfrak{L}$, $b \in L'_\beta$,

$$(8) \qquad (\zeta F_{f,L})(b+) - (\zeta F_{f,L})(b-) = c(b)\,(\zeta^y F_{f,L_1})(b).$$

5.2. Reduction to a compact CSG. The first step in proving these theorems is to come down to the case when L is compact. Let $Q = MCN_1$ be a psgrp with $C \subseteq L_R$. For any continuous function f on G we write

$$(1) \qquad f_Q(m_1) = d_Q(m_1) \int_{N_1} f(m_1 n)\, dn \qquad (m_1 \in M_1 = MC)$$

whenever this integral is absolutely convergent for all m_1 (certainly for $f \in C_c(G)$). If $f \in C_c^\infty(G)$, then $f_Q \in C_c^\infty(M_1)$, and for all $z \in \mathfrak{Z}$, $a, b \in \mathfrak{m}_1$,

$$bf_Q a = ('bfa')_Q \qquad (a' = d_Q^{-1} \circ a \circ d_Q,\ 'b = d_Q \circ b \circ d_Q^{-1}) \tag{2}$$

$$(zf)_Q = \mu_{\mathfrak{g}/\mathfrak{m}_1}(z)\, f_Q.$$

If $f \in C_c(G)$ is spherical, $f_Q \in C_c(M_1)$ and is spherical on M_1; moreover,

$$\int_{M_1} \Xi_{M_1} f_Q \, dm_1 = \int_G \Xi f \, dx \tag{3}$$

with dx and dm_1 suitably normalized (independently of f). These results are not difficult to establish.

LEMMA 1. *There exist $q \geqq 0$, and for each $l \geqq 0$ a constant $C_l > 0$, such that for all $m_1 \in M_1$,*

$$d_Q(m_1) \int_{N_1} \Xi(m_1 n)\,(1+\sigma(m_1 n))^{-(q+l)}\, dn \leqq C_l \Xi_{M_1}(m_1)\,(1+\sigma(m_1))^{-l}, \tag{4}$$

the integrals converging uniformly when m_1 varies over compact subsets of M_1.

First assume $l=0$. Let r be such that $c = \int_G \Xi^2 (1+\sigma)^{-r}\, dx < \infty$. Then a direct calculation shows[11] that

$$\int_{M_1 \times N_1} \Xi(m_1 n)\,(1+\sigma(m_1 n))^{-r} \Xi_{M_1}(m_1)\, d_Q(m_1)\, dm_1\, dn = c.$$

From this and (2) we conclude that given $u, v \in \mathfrak{M}_1$, there exist $u_i \in \mathfrak{M}_1$ $(1 \leqq i \leqq s)$ such that, for all $f \in C_c^\infty(G)$,

$$\int_{M_1} |u(\Xi_{M_1} f_Q)\, v|\, dm_1 \leqq \sum_{1 \leqq i \leqq s} \int_{M_1} \Xi_{M_1} |u_i f v_i|_Q \, dm_1 \leqq c \sum_{1 \leqq i \leqq s} \mu_{u_i, v_i : r}(f).$$

Theorems 4.1.3 and 4.3.1 now give us what we want. The result for $l > 0$ follows from the following estimate [**6h**, §§42–43]: There exists $c > 0$ such that

$$1 + \sigma(m_1 n) \geqq c(1 + \sigma(m_1)) \qquad (m_1 \in M_1,\ n \in N_1). \tag{5}$$

Lemma 1 yields at once

[11] In proving this we use the following generalization of (4.1.2): If τ is the function $km_1 n \mapsto \Xi_{M_1}(m_1)\, d_Q(m_1)^{-1}$ on G, then (cf. [**6h**, p. 101])

$$\Xi(x) = \int_{K_1} \tau(xk_1)\, dk_1 \qquad (x \in G,\ K_1 = K \cap M_1). \tag{$*$}$$

One proves $(*)$ by showing that the RHS is a spherical function satisfying (4.1.3).

LEMMA 2. *If $f \in \mathscr{C}(G)$, f_Q is well defined and lies in $\mathscr{C}(M_1)$; $f \mapsto f_Q$ is continuous from $\mathscr{C}(G)$ to $\mathscr{C}(M_1)$; and (2) is valid for all $f \in \mathscr{C}(G)$.*

LEMMA 3. *There exists a constant $c > 0$ such that, for all $f \in C_c(G)$,*

$$(6) \qquad 'F_f(b) = c' F_{\bar{f}_Q}(b) \qquad (b \in L')$$

where $\bar{f}$ is defined by $\bar{f}(x) = \int_K f(kxk^{-1})\, dk$ $(x \in G)$.

Observe that L is a CSG of M_1. If $C = L_R$, these lemmas give us the required reduction, as L_I is a compact CSG of M. For Lemma 3 note that $(KN_1M)^- = \bar{G}$ and that $dk\, dn\, dm \sim d\bar{x}$. So after a simple calculation we find that there is a constant $c_1 > 0$ such that for all $f \in C_c(G)$, $b \in L'$,

$$\int_{\bar{G}} f(b^{\bar{x}})\, d\bar{x} = c_1 \cdot \int_{N_1 \times M} \bar{f}(m^* n^*)\, dn\, dm$$

where $m^* = mbm^{-1}$ and $n^* = m^* n m^{*-1} n^{-1}$. For fixed m, $n \mapsto n^*$ is an analytic diffeomorphism of N_1 onto itself, and $dn^* = |\det(\mathrm{Ad}(m^*) - 1)_{\mathfrak{n}_1}| \cdot dn = d_Q(m^*)^{-1} \cdot |D_I(b)|^{1/2}\, dn$. This gives (6).

5.3. Proofs of the main theorems. The second and key step in the proof of Theorem 5.1.1 is the following lemma which we shall discuss in §5.4.

LEMMA 1. *Let L be compact and φ_L, the characteristic function of $(L')^G$. Then there exists $q \geqq 0$ such that $\varphi_L \Xi (1+\sigma)^{-q} \in L^1(G)$. In particular, for any $a, b \in \mathfrak{G}$, $f \mapsto \|\varphi_L(afb)\|_1$ is a well-defined continuous seminorm on $\mathscr{C}(G)$.*

We shall indicate briefly how Theorem 5.1.1 (for compact L) follows from this. First we show that, for any $f \in C_c^\infty(G)$, $'F_f \in C^\infty(L')$ and satisfies (5.1.2) (cf. [**5d**]). Write $'\Delta = '\Delta_I$, $\delta = \delta_I$, $\alpha(z) = e^\delta \circ \mu_{\mathfrak{g}/\mathfrak{l}}(z) \circ e^{-\delta}$ $(z \in \mathfrak{Z})$. We then obtain the following estimate: There exists $c > 0$ such that, for all $f \in C_c^\infty(G)$,

$$(1) \qquad \int_{L'} |\alpha(z)' F_f|\, |'\Delta|\, db \leqq c\, \|\varphi_L(zf)\|_1 .$$

We now apply[12] Theorem 5.8.4 and Lemma 1 to conclude that $'F_f \in H^\infty(L')$, and that there is a continuous seminorm ν on $\mathscr{C}(G)$ for which $|'F_f(b)| \leqq \nu(f)$, $\forall b \in L'$, $f \in C_c^\infty(G)$. Theorem 4.3.1 now gives the following estimate for some $q \geqq 0$; this estimate leads easily to Theorem 5.1.1:

[12] If $f \in C_c^\infty(G)$, we may already conclude from Theorem 5.8.4 and Lemma 5.2.3 that for any CSG L_1, $'F_{f,L_1}$, vanishes outside a compact subset of L_1 and each of its derivatives is bounded on L_1'; Lemma 1 is needed only when $f \in \mathscr{C}(G)$.

(a) $$\sup_{b \in L'} |D(b)|^{1/2} \int_G \Xi(b^x)(1+\sigma(b^x))^{-q}\, dx < \infty .$$

It is interesting to observe that for general L (not necessarily compact or θ-stable) one can deduce the analogue of (2) from Theorem 5.1.1 with the help of Theorem 4.3.1. In fact, given L, there exists $c>0$ such that

(3) $$1+\sigma(b^x) \geqq c(1+\sigma(b)) \qquad (b \in L,\ x \in G)$$

as may be deduced from (5.2.5). Consequently we have the following estimate [**6h**, §17]: there exists $q=q(L) \geqq 0$ such that for all $l \geqq 0$,

(4) $$\sup_{b \in L'} |D(b)|^{1/2}(1+\sigma(b))^l \int_{\bar{G}} \Xi(b^{\bar{x}})(1+\sigma(b^{\bar{x}}))^{-(q+l)}\, d\bar{x} < \infty .$$

We now consider Theorem 2. One may assume L to be compact and establish the results concerning $'F_f$ around each semiregular point of L; Theorem 2 would then follow from the fact that $'F_f$ and its derivatives are bounded in L'.[13] In other words, both Theorems 2 and 3 of §5.1 would follow from a study of $'F_f$ in the neighborhood of an arbitrary semiregular point of L. Furthermore, in view of the continuity of the map $f \mapsto {'F_f}$, it is enough to do this for f lying in $C_c^\infty(G)$ or a dense subspace thereof. This observation enables us to come down to the case when $\mathfrak{g}$ is semisimple.

Fix a semiregular $b \in L$. We select a system $(\mathfrak{u}, V)$ adapted to b (we shall use the notation of §§2, 3). Now there are invariant functions $g \in C^\infty(G)$ such that $g=1$ in a neighborhood of b in G and supp $g \subseteq V$. Consequently it is enough to study $'F_f$ in a neighborhood of b in L for $f \in C_c^\infty(V)$.

Let $\tilde{G} = G/M_b$, $M_b^* = M_b/M_b \cap L$ and let $\tilde{x}, d\tilde{x}, y^*, dy^*$ have their obvious meanings. Given $f \in C_c^\infty(V)$, $x \in G$, and $H \in {'\mathfrak{l}} \cap \mathfrak{u}$, we define $\tilde{f}(x:H) = f((b \exp H)^x)$. Then there exists a constant $c>0$ such that, for all H, f as above,

(5) $$\int_{\bar{G}} f((b \exp H)^{\bar{x}})\, d\bar{x} = c \int_{\tilde{G}} \left(\int_{M^*_b} \tilde{f}(x:H^{y^*})\, dy^* \right) d\tilde{x};$$

[13] The principle we are appealing to can be formulated as follows. Let X be a real Hilbert space of finite dimension d; B, an open ball with center at 0; $W_1, \ldots, W_n$ distinct linear subspaces of dimension $d-1$; and f, a C^∞ function on $B' = B \backslash \bigcup_i W_i$, such that each derivative of f is bounded on B'. Suppose $0 \leqq k \leqq \infty$, and that for each i $(1 \leqq i \leqq n)$ and each $x \in B \cap (W_i \backslash \bigcup_{j \neq i} W_j)$, there is a neighborhood N_x of x and a function ψ_x of class C^k on N_x such that $\psi_x = f$ on $N_x \cap B'$. Then there is a function ψ of class C^k on B such that $\psi = f$ on B'.

here the inner integral on the RHS depends only on $\tilde{x}$ and defines a function belonging to $C_c^\infty(\tilde{G})$, and it is this function that is integrated over $\tilde{G}$. Since $\dim[\mathfrak{m}_b, \mathfrak{m}_b]=3$, the orbital integrals over M_b^* in (5) are accessible through explicit calculation. Theorems 2 and 3 follow without much difficulty from these calculations [**6c**, §§7–10].

5.4. Continuity over $\mathscr{C}(G)$ of L^1-norms on the elliptic set. Proof of Lemma 5.3.1. Lemma 5.3.1 would follow if we establish the following: there exists $c>0$ such that, for almost all $x \in G$,

$$(1) \qquad \int_K \varphi_L(xk)\, dk \leqq c\Xi(x) \qquad (L \text{ compact}).$$

We shall obtain (1) as a consequence of the following more general results [**6g**, Theorems 4 and 5]:

Let L be any θ-stable CSG; then the function $\varphi_L : x \mapsto |D_\mathfrak{l}(x)|^{-1/2}$ is locally summable on G, and there is a constant $c>0$ such that, for almost all $x \in G$,

$$(2) \qquad \int_K \varphi_L(xk)\, dk \leqq c\Xi(x).$$

The function $|D|^{-1/2}$ is locally summable[14] on G [**6e**, §28]; φ_L is locally summable, as $\sup_{x\in G'}(|D(x)|^{1/2}\, \varphi_L(x))<\infty$. The LHS of (2) is thus finite for almost all $x \in G$. We prove (2) by induction on $\dim(G)$. Let I_c^+ be the set of all $f \in C_c(G)$ that are spherical and $\geqq 0$. For any invariant locally summable function Θ on G write $\Theta_0(x)=\int_K \Theta(xk)\, dk$; Θ_0 is locally summable, spherical, and $\int_G \Theta f\, dx=\int_G \Theta_0 f\, dx$, $\forall f \in I_c^+$; if Θ is $\mathfrak{Z}$-finite, $\Theta_0 \in C^\infty(G)$. We may also assume $G={}^\circ G$.

Let $\dim(L_R)>0$ and $\varphi_{L_I}^M$ be the characteristic function of $(L_I')^M$, M being as in §5.2 with $C=L_R$. Then, by Lemma 5.2.3, there exists $c_1>0$ such that, for all $f \in I_c^+$,

$$(3) \qquad \int_G \varphi_L f\, dx \leqq c_1 \int_{M\times L_R} \varphi_{L_I}^M(m)\, f_Q(mb)\, dm\, db.$$

Estimating $\varphi_{L_I}^M$ by the induction hypothesis and using (5.2.3), we find that for some constant $c_2>0$ and all $f \in I_c^+$, the RHS of (3) is $\leqq c_2\int_G \Xi f\, dx$. We are thus left with the case of compact L. Since both sides of (2) depend only on $\mathrm{Ad}(x)$ we may assume that $G \subseteq G_c$ where G_c is complex, semisimple, and simply connected. As L is now connected, $(L')^G$ is contained in the component of 1 of G, and so we may

[14] This follows from the fact that $\int_{L_1} |'F_{f,L_1}|\, db<\infty$ for all $f \in C_c^\infty(G)$ and all CSG's L_1 (cf. footnote 12).

suppose that G is connected. We write $L=B$ and use notation of §3.

The proof of (1) is difficult because for a given $x\in G$ it is not a simple matter to determine the set of $k\in K$ for which xk is elliptic. If Θ_π is the character of a finite-dimensional representation π of G, $|\Theta_\pi(y)|\leqq \dim(\pi)$ for all elliptic y; and by considering all possible π it is not difficult to establish the following weaker result as a first step towards (1) [**6g**, Lemma 42];

(4)
$$\lim_{x\to\infty} (\varphi_B)_0\,(x)=0.$$

The main idea in Harish-Chandra's proof of (1) is to use the distribution $\Theta^*=\Theta^*_\delta$ in place of Θ_π; here $\delta=\frac{1}{2}\sum_{\alpha\in P}\alpha$ and Θ^*_δ is defined by (3.6.1). $\Theta^*=1$ on $(B')^G$; *and, unlike* Θ_π, *it is of slow growth on all the CSG's of G.*

A simple calculation shows that if $\{X_1,\ldots,X_r\}$ and $\{X_{r+1},\ldots,X_n\}$ are orthonormal bases of $\mathfrak{k}$ and $\mathfrak{p}$ respectively, and if $q=X_1^2+\cdots+X_n^2$, then $q\Theta^*=0$, whence $q\Theta^*_0=0$ also. Further, the estimates of Theorem 3.6.1 imply that, for some constant $c>0$ and almost all x,

(5)
$$|\Theta^*_0(x)-(\varphi_B)_0(x)|\leqq c\sum_{1\leqq i\leqq s}(\varphi_{L_i})_0\,(x),$$

$L_1,\ldots,L_s$ being a complete system of noncompact CSG's of G. From (4), (5) and the induction hypothesis we find that $\lim_{x\to\infty}\Theta^*_0(x)=0$. The maximum principle for the (second degree) elliptic operator q implies that $\Theta^*_0=0$. But then (1) follows from (5) and the induction hypothesis.

It remains to sketch a proof of (4). Let Σ be the simple system of roots of $(\mathfrak{g},\mathfrak{a})$. Extend $\mathfrak{a}$ to a CSA $\mathfrak{h}$ and let S be the simple system corresponding to an ordering of the roots of $(\mathfrak{g},\mathfrak{h})$ that is compatible with $\mathfrak{a}^+$. If (4) were false, we can find $\gamma>0$, $F\subsetneqq\Sigma$, and $\{H_n\}_{n\geqq 1}$ from $\mathrm{Cl}(\mathfrak{a}^+)$ such that (i) $\lambda(H_n)=O(1)$ or $\to+\infty$ according as $\lambda\in F$ or $\in\Sigma\backslash F$, and (ii) $(\varphi_B)_0\,(a_n)\geqq\gamma>0$ for all n, where $a_n=\exp H_n$. Let S' be the set of all $\beta\in S$ whose restrictions to $\mathfrak{a}$ are in $\boldsymbol{R}\cdot F$, and let π be the irreducible representation of G whose highest weight Λ is such that $\langle\Lambda,\beta\rangle$ is $=0$ or >0 according as $\beta\in S'$ or $\in S\backslash S'$. If Γ is the set of $\mu\in\mathfrak{h}^*$ of the form $\sum_{\beta\in S}c(\beta)\,\beta$ with $c(\beta)\geqq 0$ for all β and $c(\beta)>0$ for some $\beta\in S\backslash S'$, then it can be shown [**6g**, §22] that $\Lambda\in\Gamma$, and that, for any weight $\Lambda'\neq\Lambda$ of π, $\Lambda-\Lambda'\in\Gamma$. Let $u_1,\ldots,u_p$ be a basis of weight vectors of π, with Λ_i as the weight of u_i and $\Lambda_1=\Lambda$; and let $(a_{ij}(x))$ be the matrix of $\pi(x)$ $(x\in G)$ in this basis. If $K_n=\{k:k\in K,\ a_nk$ is elliptic$\}$, the inequality $|\Theta_\pi(a_nk)|\leqq p$ $(k\in K_n)$ then implies that

$$|a_{11}(k)|\leqq p\exp\{-\Lambda(H_n)\}+\sum_{2\leqq i\leqq p}|a_{ii}(k)|\exp\{-(\Lambda-\Lambda_i)(H_n)\}\qquad(k\in K_n).$$

So $K_n\subseteq\{k:|a_{11}(k)|\leqq\tau_n\}$ where $\tau_n\to 0$. Since $\{k:k\in K,\ a_{11}(k)=0\}$ has measure zero, we find that $(\varphi_B)_0\,(a_n)=\int_{K_n}dk\to 0$, a contradiction.

It is clear from our discussion so far that the map $f \mapsto \bar{f}_Q$ establishes an intimate connexion between problems of invariant analysis on G and those on M_1. This connexion actually goes much deeper than the above results would seem to suggest, and it may be of some interest to look into this a little more closely. We have $G = K \exp(\mathfrak{m} \cap \mathfrak{p})\, CN_1$, and for $x \in G$ let $\mu(x)$, $c(x)$ and $n(x)$ be the components of x in $\exp(\mathfrak{m} \cap \mathfrak{p})$, C and N_1 respectively. Then we have the following formula, valid for all continuous functions f on G with $\|\Xi^{-1}(1+\sigma)^r f\|_\infty < \infty$ for all $r \geqq 0$:

$$(6) \qquad (\bar{f}^x)_Q = \int_K [(f^{k^{-1}})_Q]^{\mu(xk)}\, dk \qquad (x \in G).$$

Let us now write T_Q for the map which is dual to the map $f \mapsto \bar{f}_Q$. Then, (6) shows that for any tempered invariant distribution τ on M_1, $T_Q(\tau)$ is a tempered invariant distribution on G. Moreover, for such τ,

$$(7) \qquad z T_Q(\tau) = T_Q(\mu_{\mathfrak{g}/\mathfrak{m}_1}(z)\, \tau) \qquad (z \in \mathfrak{Z}).$$

In particular, if τ is an eigendistribution on M_1, $T_Q(\tau)$ has the same property on G. Explicit calculations show that if τ is the character of an irreducible unitary representation π_Q of Q that is trivial on N_1, $T_Q(\tau)$ is the character of the representation of G that is induced by π_Q.

5.5. Tempered invariant eigendistributions. In this and the next two paragraphs we shall use the foregoing theory to study various questions of harmonic analysis on G. For simplicity we assume in this paragraph that G is as in §§2 and 3 and use the notation therein.

THEOREM 1. *Let Θ be an invariant $\mathfrak{Z}$-finite distribution on G. Then the following statements are equivalent:*

(i) *Given any CSG L, there exist $C = C_L > 0$, $q = q_L \geqq 0$ such that $|\Phi_{L,P}(a)| \leqq C(1+\sigma(a))^q$ for all $a \in L'$.*

(ii) *There exist $C > 0$, $q \geqq 0$ such that $|\Theta(x)| \leqq C|D(x)|^{-1/2}(1+\sigma(x))^q$ for all $x \in G'$.*

(iii) *Θ is tempered.*

If these conditions are satisfied, then

$$(1) \qquad \Theta(f) = \int_G \Theta(x)\, f(x)\, dx \qquad (f \in \mathscr{C}(G))$$

the integral converging absolutely.

From (5.5.3) we see that (i)$\Rightarrow$(ii). Further, (5.3.4) shows that, for some $q \geqq 0$,

(2) $$\int_G |D(x)|^{-1/2}\Xi(x)\,(1+\sigma(x))^{-q}\,dx<\infty.$$

The implication (ii)⇒(iii) as well as the last assertion follows at once from (2). For the proof that (iii)⇒(i) see [**6h**, § 19]; the main argument here is similar to but more delicate than Lemma 3.5.1.

COROLLARY 2. *The distributions* Θ_λ $(\lambda\in\mathscr{L}')$ *are tempered.*

COROLLARY 3. *Let* Θ *be an invariant eigendistribution such that the corresponding eigenhomomorphism is regular. Then* Θ *is tempered if and only if*

(3) $$\sup_{x\in G'}|D(x)|^{1/2}\,|\Theta(x)|<\infty.$$

5.6. The relation $(\varpi F_f)(1)=cf(1)$**.** If G is compact and B is a maximal torus, then $F_{f,B}$ is a C^∞ function on B and its harmonic analysis leads at once to the Plancherel formula. It is tempting to suppose that the same method would yield significant results even in the noncompact case. While this turns out to be substantially the case, complications arise because the functions F_f are no longer C^∞, and the jumps of these functions and their derivates cannot be ignored. Following Harish-Chandra we shall formulate Theorem 1 below as an important step towards the Plancherel formula ([**5d**], [**6h**]; see also [**3**]). In §7 we shall see that it leads to a complete determination of the discrete series.

A CSG L will be called *fundamental* if L_I has the maximum possible dimension. It is known that any two fundamental CSG's are conjugate in G and that L is fundamental if and only if $(\mathfrak{g},\mathfrak{l})$ has no real roots [**5b**, §8].

THEOREM 1. *Let G be of class* $\mathscr{H}$. *Fix a CSG L and let* $\varpi_L=\prod_{\alpha\in P_L}H_\alpha$. *Then, if L is not fundamental,* $('\varpi_L{}'F_{f,L})(1)=0$ *for all* $f\in\mathscr{C}(G)$.

Suppose L is fundamental. Let $q=\frac{1}{2}(\dim G/K-\operatorname{rk}G+\operatorname{rk}K)$.[15] *Then there is a constant* $c>0$ *such that for all* $f\in\mathscr{C}(G)$,

(1) $$f(1)=(-1)^q\,c('\varpi_L'F_{f,L})(1).$$

If G is as in §3, $(\varpi F_{f,L})(1)=0$ *for all* $f\in\mathscr{C}(G)$ *for nonfundamental L, while for fundamental L,*

(2) $$f(1)=(-1)^q\,c(\varpi_L F_{f,L})(1)\qquad (f\in\mathscr{C}(G)).$$

We discuss this when G is as in §3. If L is not fundamental and α is any real root of $(\mathfrak{g},\mathfrak{l})$, we observe that $F_{f,L}$ is invariant under s_α on $L'\cap\exp\mathfrak{l}$. So $(\varpi_L F_{f,L})^{s_\alpha}=-(\varpi_L F_{f,L})$ on $\exp\mathfrak{l}$, implying $(\varpi_L F_{f,L})(1)=0$. Let L be fundamental and let

[15] q is an integer $\geqq 0$.

(3) $$\Gamma(f)=(\varpi_L F_{f,L})(1) \qquad (f\in\mathscr{C}(G)).$$

Γ is an invariant tempered distribution. A simple argument shows that the support of Γ is contained in the set of unipotent elements of G. We may thus transfer Γ to the Lie algebra. The theorem then follows from the work of § 5.9.

5.7. Cusp forms. Let G be a group of class $\mathscr{H}$. Following Harish-Chandra [**6i**, p. 538] we shall define a *cusp form* on G to be any $f\in\mathscr{C}(G)$ such that, for any psgrp $Q=MCN_1\neq G$,

(1) $$\int_{N_1} f(xn)\,dn=0 \qquad (x\in G).$$

${}^\circ\mathscr{C}(G)$ denotes the set of all cusp forms on G. It is a closed subspace of $\mathscr{C}(G)$ invariant under all translations; the estimates (4.2.3) and (5.2.4) imply easily that it is a two-sided ideal in $\mathscr{C}(G)$. The importance of ${}^\circ\mathscr{C}(G)$ for harmonic analysis on G lies in the fact that its closure in $L^2(G)$ is precisely the closed linear span of all the subspaces of $L^2(G)$ that are irreducibly invariant under the regular representation. We shall prove this in § 7 (heuristically, (1) expresses the condition that f be orthogonal to the principal series of representations associated with Q). In this paragraph we shall formulate some of the properties of cusp forms. Theorem 1 below is especially noteworthy; it reveals the real reason why the harmonic analysis of a cusp form involves only the discrete series.

THEOREM 1. *Let $f\in{}^\circ\mathscr{C}(G)$. Then $'F_{f,L}=0$ for any noncompact CSG. If L is compact, $'F_{f,L}$ extends to a C^∞ function on L.*

The relation (5.2.6) (extended to $\mathscr{C}(G)$) gives the first assertion. If L is compact, Theorem 5.1.3 implies that $'F_{f,L}$ is C^∞ around each regular or semiregular point; this implies the second assertion (cf. footnote 13).

THEOREM 2. *Let f be a $\mathfrak{Z}$-finite function in $\mathscr{C}(G)$. Then $f\in{}^\circ\mathscr{C}(G)$. Moreover, ${}^\circ\mathscr{C}(G)$ is the closure of the space of $\mathfrak{Z}$-finite K-finite functions in $\mathscr{C}(G)$.*

For the second part see § 7. For the first let $f\in\mathscr{C}(G)$ be $\mathfrak{Z}$-finite and $Q=MCN_1\neq G$ a psgrp. Then for all $m\in M$, $a\mapsto f_Q(ma)$ is $\mathfrak{C}$-finite on C, hence $\equiv 0$. So $f_Q\equiv 0$. This implies easily that $f\in{}^\circ\mathscr{C}(G)$.

THEOREM 3. *${}^\circ\mathscr{C}(G)\neq\{0\}\Leftrightarrow G$ has a compact CSG.*

Suppose there exists $f\in{}^\circ\mathscr{C}(G)$, $f\neq 0$. If no CSG is compact, $'F_{f,L}=0$ for all L. Hence $f(1)=0$ by the work of § 5.6. Replacing f by its translates, $f(x)=0$ for all x. The converse follows from Theorems 6.1.2 and 6.1.3.

5.8. Appendix. Some estimates in a classical setting. The results described in this paragraph assert that in certain situations one can estimate sup norms by L^p-norms ($1 \leqq p < \infty$). These estimates are closely related to those obtained by Harish-Chandra in the context of Theorem 3 of [**6e**].

We fix a real Hilbert space V of finite dimension d; S is the symmetric algebra over V_c; $S_0 \subseteq S$ a subalgebra such that $1 \in S_0$ and S is a finite module over S_0. If $U \subseteq V$ is open and $w \in C^\infty(U)$ is >0, $H^p(U, w, S_0)$ denotes the space of all $f \in C^\infty(U)$ for which $\|\eta f\|_{p,w} = (\int_U |\eta f|^p w\, dx)^{1/p} < \infty$, $\forall \eta \in S_0$ (we write $H^p(U, w)$ when $S_0 = S$); $H^\infty(U)$ is the space of all $f \in C^\infty(U)$ for which $\|\eta f\|_\infty < \infty$, $\forall \eta \in S$. We topologize these spaces by the corresponding seminorms. $B(x, a)$ is the closed ball of center x and radius $a > 0$.

LEMMA 1. *Fix U, w as above and a real function ε on U such that $0 < \varepsilon(x) \leqq 1$ and $B(x, \varepsilon(x)) \subseteq U$ for all $x \in U$. Put $\omega(x) = \inf_{y \in B(x, \varepsilon(x))} w(y)$. Then there exists an integer $b \geqq 0$, and, for each $\xi \in S$, a continuous seminorm ν_ξ on $H^p(U, w, S_0)$ such that*

$$(1) \qquad |f(x;\xi)| \leqq \varepsilon(x)^{-b}\, \omega(x)^{-1/p}\, \nu_\xi(f) \qquad (x \in U, f \in H^p(U, w, S_0)).$$

In particular $H^p(U, w, S_0)$ is Fréchet.

As S is a finite module over S_0 we may prove (1) with some $b = b(\xi)$. Let Δ be the Laplacian of V; $D = 1 - \Delta$; and k_r $(2r > d)$, the tempered fundamental solution of D^r. $k_r \in C^{(2r-d-1)}(V)$ is C^∞ on $V \backslash \{0\}$, and, for $\eta \in S$ of degree n, $(\eta k_r)(x) = O(\|x\|^{-2n})$ for $x \to 0$. Fix $\xi \in S$, let $s = d + 1 + \deg(\xi)$, and choose $\eta_1, \ldots, \eta_m \in S_0$ such that $(D^s)^m = \sum_{1 \leqq j \leqq m} \eta_j (D^s)^{m-j}$. Then, for all $g \in C_c^\infty(V)$ and all $x \in V$,

$$(*) \qquad g(x;\xi) = \sum_{1 \leqq j \leqq m} \int_V k_{js}(x-y;\xi)\, g(y;\eta_j)\, dy.$$

We now select "localizing" functions $\psi_x \in C_c^\infty(U)$ for $x \in U$ such that (i) $0 \leqq \psi_x \leqq 1$ and $\psi_x(y) = 0$ or 1 according as $\|y - x\| > \frac{3}{4}\varepsilon(x)$ or $< \frac{1}{4}\varepsilon(x)$, and (ii) if $\zeta \in S$, there exists $c_\zeta > 0$ with $|(\zeta\psi_x)(y)| \leqq c_\zeta \varepsilon(x)^{-\deg(\zeta)}$ for all $y \in V$, $x \in U$ (cf. [**14b**, §3]). On the other hand there exist $\zeta_q, \sigma_{jq} \in S$ with the σ_{jq} having zero constant terms such that for all $h \in C^\infty(V)$, $\eta_j \circ h = h\eta_j + \sum_{1 \leqq q \leqq M} (\sigma_{jq} h)\, \zeta_q$ $(1 \leqq j \leqq m)$. Taking $g = f\psi_x$ in $(*)$ we obtain, for all $f \in C^\infty(U)$, $x \in U$,

$$f(x;\xi) = \sum_{1 \leqq j \leqq m} \int_{B(x, \varepsilon(x))} k_{js}(x-y;\xi)\, \psi_x(y)\, f(y;\eta_j)\, dy$$

$$+ \sum_{1 \leqq j \leqq m} \sum_{1 \leqq q \leqq M} \int_{\frac{1}{4}\varepsilon(x) \leqq \|y-x\| \leqq \frac{3}{4}\varepsilon(x)} F_{j,q,x}(y;\zeta_q^\dagger)\, f(y)\, dy;$$

here $F_{j,q,x}(y)=k_{js}(x-y;\xi)\,\psi_x(y;\sigma_{jq})$ and $\zeta_q^\dagger$ is the adjoint of ζ_q. The estimates for k_r and $\zeta\psi_x$ now lead to (1).

THEOREM 2. *Let $\lambda_j \in V_c^*$ be nonzero $(1 \leqq j \leqq q)$; $V'=\{x: x\in V,\ \lambda_j(x)\neq 0,\ \forall j\}$; U, a union of components of V'; and suppose that, for some $c>0$, $r\geqq 0$,*

$$w(x)\geqq c\left(1+\max_j |\lambda_j(x)|^{-1}\right)^{-r} \qquad (x\in U). \tag{2}$$

Then $H^p(U, w, S_0)\subseteq H^\infty(U)$, the inclusion map being continuous.

It is easy to come down to the case where the λ_j are real on V and U is the set where they are all >0; this can then be handled with the help of Lemma 1 (cf. [**14b**, §3]). We mention two situations where the conditions of Theorem 2 are satisfied:

(i) $w(x)=\prod_j |\lambda_j(x)|^{a_j}$ ($a_j>0$ are constants), and
(ii) the λ_j are real on V, U is the set where they are all >0, and $w=\prod_j(1-e^{-\lambda_j})$.

Let $\mathscr{C}(U)$ be the Schwartz space of U.

COROLLARY 3. *Let $w(x)=\prod_j |\lambda_j(x)|^{a_j}$ ($a_j>0$ constant). Fix a subalgebra P_0 of the algebra P of all polynomials on V_c such that $1\in P_0$ and P is a finite module over P_0. Then $\mathscr{C}(U)=\{f: gf\in H^p(U, w, S_0),\ \forall g\in P_0\}$, and the topology of $\mathscr{C}(U)$ coincides with that induced by the seminorms $f\mapsto \|\xi(gf)\|_{p,w}$ $(\xi\in S_0,\ g\in P_0)$.*

Let T be a compact Lie group whose identity component is abelian and has V as Lie algebra. Let $\Omega=\prod_{1\leqq j\leqq q}|\chi_j-1|$ where the χ_j are one-dimensional characters of T that are $\not\equiv 1$ on any component of T. Put $T'=\{b: b\in T,\ \Omega(b)\neq 0\}$ and define $H^p(T', \Omega, S_0)$ and $H^\infty(T')$ in the obvious way. These are Fréchet spaces.

THEOREM 4. *$H^p(T', \Omega, S_0)=H^\infty(T')$ as Fréchet spaces.*

We fix $f\in H^p(T', \Omega, S_0)$, $b\in T$, $\xi\in S$, and show that for some neighborhood N of b, $\|\xi f\|_{\infty, N\cap T'}<\infty$. We may assume that $\chi_j(b)=1$ for all j. Let $\lambda_j\in V^*$ be such that $\chi_j(b\exp x)=\exp\{(-1)^{1/2}\lambda_j(x)\}$ $(x\in V)$; write $w(x)=\prod_j|\lambda_j(x)|$ and V' as before. Clearly there exist $\alpha, \beta>0$ sufficiently small such that $|\Omega(b\exp x)|\geqq \beta w(x)$ if $\|x\|\leqq\alpha$, and $\varphi(x\mapsto f(b\exp x))$ lies in $H^p(B(0,2\alpha)\cap V', w, S_0)$. We now fix a component V^+ of V' and apply Lemma 1 with U as the (convex) open set $V^+\cap B(0,\alpha)$ and $\varepsilon(x)=\frac{1}{2}\min(2\alpha-\|x\|, \min\|\lambda_j\|^{-1}|\lambda_j(x)|)$. As $\omega(x)\geqq(\varepsilon(x)/2)^q$, there exists an integer $l\geqq 0$ such that, for all $\eta\in S$, $\sup_{x\in U}\varepsilon(x)^l\,|(\eta\varphi)(x)|<\infty$. This implies that $\|\xi\varphi\|_{\infty,U}<\infty$. In fact, fix $x_0\in U$ and write $\varphi_x(t)=\varphi(x_t;\xi)$ where $0\leqq t\leqq 1$, $x_t=(1-t)\,x+tx_0$ $(x\in U)$. Then there exists $\gamma>0$ such that $\varepsilon(x_t)\geqq\gamma t$ for all $x\in U$, $0\leqq t\leqq 1$, and so we can find constants $L_m>0$ $(m=0,1,\ldots)$ such that $|\varphi_x^{(m)}(t)|\leqq L_m t^{-l}$, $\forall m\geqq 0$, $x\in U$, $0<t\leqq 1$. The arguments of [**14b**, §3] now lead to the desired conclusion.

5.9. Appendix. Tempered invariant eigendistributions on a semisimple Lie algebra. In this appendix we shall describe some important results in the theory of tempered invariant eigendistributions on a semisimple Lie algebra $\mathfrak{g}$. The main references are **[5b]**, **[5c]**, **[6c]**.

Let $\mathfrak{g}$ be as above, and G, the adjoint group of $\mathfrak{g}$. Fix a CSA $\mathfrak{l}$ with CSG L. Let $G^*=G/L$; $x^*=xL$ $(x\in G)$; and dx^*, the invariant measure on G^*. Let P be a positive system of roots of $(\mathfrak{g}, \mathfrak{l})$; $\pi=\prod_{\alpha\in P}\alpha$; $\varpi=\prod_{\alpha\in P}H_\alpha$. For $p\in I(\mathfrak{g})$, $p_\mathfrak{l}$ is its restriction to $\mathfrak{l}$. We often write $\partial(q)$ for the differential operator corresponding to $q\in S(\mathfrak{g}_c)$. Write $\varepsilon_R(H)=\operatorname{sgn}\prod_{\alpha\in P,\,\alpha\text{ real}}\alpha(H)$ $(H\in\mathfrak{l}')$. $\mathscr{C}(\mathfrak{g})$ is the Schwartz space of $\mathfrak{g}$. We define the Fourier transform map on $\mathscr{C}(\mathfrak{g})$ and its dual in the usual way (using $\langle\cdot,\cdot\rangle$).

THEOREM 1. *For any $f\in\mathscr{C}(\mathfrak{g})$ and $H\in\mathfrak{l}'$,*

$$\psi_f(H)=\varepsilon_R(H)\,\pi(H)\int_{G^*} f(H^{x^*})\,dx^* \tag{1}$$

is well defined, the integral converging absolutely. $\psi_f\in\mathscr{C}(\mathfrak{l}')$ and $f\mapsto\psi_f$ is a continuous map of $\mathscr{C}(\mathfrak{g})$ into $\mathscr{C}(\mathfrak{l}')$. Moreover

$$\psi_{\partial(p)f}=\partial(p_\mathfrak{l})\,\psi_f \qquad (p\in I(\mathfrak{g}),\ f\in\mathscr{C}(\mathfrak{g})). \tag{2}$$

THEOREM 2. *Let $\mathfrak{l}'(I)$ be the set of all $H\in\mathfrak{l}$ where no singular imaginary root is 0. Fix $f\in\mathscr{C}(\mathfrak{g})$. Then ψ_f extends to a C^∞ function on $\mathfrak{l}'(I)$. Suppose $H\in\mathfrak{l}$ and $S_I(H)$ is the set of all singular imaginary roots in P vanishing at H. If $\zeta\in S(\mathfrak{l}_c)$ is such that $\zeta^{s_\beta}=-\zeta$ for all $\beta\in S_I(H)$, then $\partial(\zeta)\,\psi_f$ extends as a continuous function around H. In particular, $\partial(\varpi)\,\psi_f$ extends to a continuous function on $\mathfrak{l}$.*

The proofs are similar to those of the corresponding results on the group. In Theorem 1, the estimates furnished by Theorem 2 and Corollary 3 of §5.8 enable one to handle the convergence problems. It follows from these results that for any $H\in\mathfrak{l}'$ the invariant measure on H^G is tempered. We write

$$\sigma_H(f)=\int_{G^*} f(H^{x^*})\,dx^* \qquad (f\in\mathscr{C}(\mathfrak{g})). \tag{3}$$

THEOREM 3. *The distribution $\hat\sigma_H$ is invariant and $\partial(p)\,\hat\sigma_H=p((-1)^{1/2}H)\,\hat\sigma_H$ $(p\in I(\mathfrak{g}),\ H\in\mathfrak{l}')$. Suppose $\mathfrak{l}_1\subseteq\mathfrak{g}$ is a CSA, and y is an element of the adjoint group of $\mathfrak{g}_c$ such that $\mathfrak{l}_c^y=(\mathfrak{l}_1)_c$. Then there are uniquely defined locally constant functions $c_s(\cdot:\cdot)$ $(s\in W_{L_c})$ on $\mathfrak{l}'\times\mathfrak{l}_1'$ such that (writing $\pi_1=\pi\circ y^{-1}$)*

$$\pi_1(H_1)\,\pi(H)\,\hat\sigma_H(H_1)=\sum_{s\in W_{L_c}}\varepsilon(s)\,c_s(H:H_1)\,e^{i\langle H_1,(sH)^y\rangle} \qquad (H\in\mathfrak{l}',\ H_1\in\mathfrak{l}_1'). \tag{4}$$

Since $\hat{\sigma}_H$ is an invariant eigendistribution one obtains (4) with $c_s(H:\cdot)$ locally constant on $\mathfrak{l}'_1$ for all $H\in\mathfrak{l}'$. Also it is easy to show that $c_s(\cdot:H_1)$ is C^∞ for all $H_1\in\mathfrak{l}'_1$. The relations (2) are then used to conclude that $c_s(\cdot:H_1)$ is locally constant on $\mathfrak{l}'$ for all $H_1\in\mathfrak{l}'_1$.

THEOREM 4. *If* $\mathfrak{l}$ *is not fundamental,* $(\partial(\varpi)\,\psi_f)(0)=0$ *for all* $f\in\mathscr{C}(\mathfrak{g})$. *For fundamental* $\mathfrak{l}$, *there exists a constant* $c>0$ *such that, with* q *as in Theorem* 5.6.1,

$$(5)\qquad cf(0)=(-1)^q(\partial(\varpi)\psi_f)(0)\qquad (f\in\mathscr{C}(\mathfrak{g})).$$

The first assertion is proved as in Theorem 5.6.1. Let $\mathfrak{l}$ be fundamental and $\gamma(f)=(\partial(\varpi)\,\psi_f)(0)\ (f\in\mathscr{C}(\mathfrak{g}))$. It follows from (2) that $\hat{\gamma}$ is $I(\mathfrak{g})$-finite, and from (4) that $\hat{\gamma}$ is locally constant on $\mathfrak{g}'$ must be a constant [**6d**, §10]. So, for some constant c_1, $\gamma(f)=c_1f(0)$, $\forall f\in\mathscr{C}(\mathfrak{g})$. The proof that $(-1)^qc_1$ is real and >0 is, however, delicate. It depends on the construction (based on some work of de Rham) and properties of invariant fundamental solutions to the differential operators $\partial(\omega)^m$ $(m\geqq 1)$ where ω is the Casimir element in $I(\mathfrak{g})$ [**6c**, §§11–13].

6. Behaviour at infinity of eigenfunctions

6.1. Outline of the main results. We shall now take up the problem of showing that the Fourier components with respect to K of the distributions Θ_λ lie in $L^2(G)$. As these are tempered (Corollary 5.5.2), we may subsume this under the general problem of determining the behaviour, at infinity on G, of tempered, K-finite, $\mathfrak{Z}$-finite functions.

Let G be a group of class $\mathscr{H}$. For any finite-dimensional double representation $\tau=(\tau_1,\tau_2)$ of K in U, let $\mathscr{A}(G:\tau)$ be the space of all tempered $\mathfrak{Z}$-finite functions $f\in C^\infty(G:\tau)$. As $G=K\,\mathrm{Cl}(A^+)\,K$ our problem is that of determining, for $f\in\mathscr{A}(G:\tau)$, the behaviour of $f(a)$ as $a\to\infty$ in $\mathrm{Cl}(A^+)$. Since $\lambda(\log a)$ may not tend to ∞ for all $\lambda\in\Sigma$ (the set of simple roots of $(\mathfrak{g},\mathfrak{a})$), we put for each $F\subsetneqq\Sigma$, $\beta_F(H)=\min_{\lambda\in\Sigma\setminus F}\lambda(H)$ $(H\in\mathfrak{a})$, and study for arbitrary F how $f(a)$ behaves when $\beta_F(\log a)\to\infty$ ($\mathrm{Cl}(A^+)\ni a\underset{F}{\to}\infty$ in symbols).

Fix F and let $P_F=M_{1F}=M_FA_FN_F$ be the corresponding standard psgrp; let $M^+_{1F}=\bigcup_{\mathfrak{a}\ni H:\beta_F(H)>0}K_F(\exp H)\,K_F$. Then for any $z\in\mathfrak{Z}$, there is a differential operator E_z on M^+_{1F} such that (i) $zg=(d_F^{-1}\circ\mu_F(z)\circ d_F)\,g+E_zg$ on M^+_{1F} for all $g\in C^\infty(G:\tau)$, and (ii) as functions of (m,a) $(m\in M_F,\ a\in A_F)$, the coefficients of E_z go to zero when $a\underset{F}{\to}\infty$. Thus, if $f\in C^\infty(G:\tau)$ is any tempered eigenfunction $(zf=\chi(z)\,f$, $\forall z\in\mathfrak{Z})$, the function $m_1\mapsto d_F(m_1)\,f(m_1)$ on M^+_{1F} satisfies certain differential equations which are *perturbations* of the equations $\mu_F(z)\,h=\chi(z)\,h\,(z\in\mathfrak{Z})$. It follows from this that for a suitable tempered solution f_F of the unperturbed equations one can approximate $d_F(ma)\,f(ma)$ by $f_F(ma)$ when $A_F\ni a\underset{F}{\to}\infty$. The knowledge of these f_F, together with estimates for $|d_F(ma)\,f(ma)-f_F(ma)|$, then yield a complete picture of the asymptotic behaviour of f. When generalized so as to take care of $\mathfrak{Z}$-finite functions, this method leads to the following theorems which are the main results of the theory. In essence this is Harish-Chandra's method (cf. [**6h**, §§27–31], [**5f**, §§2–8]; cf. also [**14**]); f_F is the so-called *constant term of f along the psgrp P_F.*

THEOREM 1. *Let G be as above and fix $f\in\mathscr{A}(G:\tau)$. Then, for each $F\subsetneqq\Sigma$, there is a unique $f_F\in\mathscr{A}(M_{1F},\tau_F)$ $(\tau_F=\tau\mid K_F)$ such that, for all $m\in M_{1F}$,*

(1) $$|d_F(m\exp tH)f(m\exp tH)-f_F(m\exp tH)|\to 0 \qquad (t\to+\infty,\ H\in\mathfrak{a}_F^+);$$

$\mu_F(z)f_F=0$ for all $z\in\mathfrak{Z}$ for which $zf=0$. For any $\kappa>0$ let

(2) $$A^+(F:\kappa)=\{a:a\in\mathrm{Cl}(A^+),\ \beta_F(\log a)\geqq\kappa\varrho(\log a)\}.$$

Then there exist $\gamma>0$, $q\geqq 0$ and for each $\kappa>0$ a constant $C_\kappa>0$ such that

(3) $$|f(a)-d_F(a)^{-1}f_F(a)|\leqq C_\kappa\Xi(a)^{1+\gamma\kappa}(1+\sigma(a))^q \qquad (a\in A^+(F:\kappa)).$$

Finally, $F'\subseteqq F\subsetneqq\Sigma$, we have the transitivity relation

(4) $$(f_F)_{F'}=f_{F'}.$$

THEOREM 2. *Let $G, f\neq 0$ be as in Theorem 1. Then the following statements are equivalent:* (i) *$f\in L^2(G)\otimes U$,* (ii) *$f_F=0$ for all $F\subsetneqq\Sigma$, and* (iii) *$f\in\mathscr{C}(G)\otimes U$. If these are satisfied, then $G={}^\circ G$, and there exists $\gamma>0$ such that $|afb|=O(\Xi^{1+\gamma})$ for all $a, b\in\mathfrak{G}$.*

THEOREM 3. *Let* $\mathrm{rk}(G)=\mathrm{rk}(K)$. *Let $\mathfrak{b}\subset\mathfrak{k}$ be a CSA and λ, a regular element of $\mathfrak{b}_c^*$ that is real valued on $(-1)^{1/2}\mathfrak{b}$. Then any $f\in C^\infty(G:\tau)$, which is tempered and satisfies the differential equations $zf=\chi_\lambda(z)f$ for all $z\in\mathfrak{Z}$, lies in $\mathscr{C}(G)\otimes U$. In particular, if G and Θ_λ are as in §3 ($\lambda\in\mathscr{L}'$), the Fourier components of Θ_λ are all in $\mathscr{C}(G)$.*

Our aim now is to sketch the main lines of arguments in the proofs of these theorems. We may assume $G={}^\circ G$.

6.2. The differential equations on M_{1F}^+. Initial estimates. Fix F. For $m\in M_{1F}$ let $\gamma_F(m)=\|\mathrm{Ad}(m^{-1})_{\mathfrak{n}_F}\|$; $m\in M_{1F}^+\Leftrightarrow\gamma_F(m)<1$. Let $\mathscr{S}_F$ be the algebra of functions on M_{1F}^+ generated (without 1) by the derivatives of the matrix coefficients of the mappings

$$b_F(m\mapsto(\mathrm{Ad}(m^{-1})-\mathrm{Ad}(\theta(m^{-1})))_{\mathfrak{n}_F}^{-1}) \quad\text{and}\quad c_F(m\mapsto\mathrm{Ad}(m^{-1})_{\mathfrak{n}_F}b_F(m)).$$

It is known [**14b**, §4] that for each $g\in\mathscr{S}_F$, there is $c=c(g)>0$ and $r=r(g)\geqq 0$ such that

(1) $$|g(m)| \leqq c\gamma_F(m)\,(1-\gamma_F(m))^{-r} \qquad (m \in M_{1F}^+).$$

We note that $\mathfrak{G}$ is the direct sum of $\theta(\mathfrak{n}_F)\,\mathfrak{G}$ and $\mathfrak{M}_{1F}\mathfrak{K}$. Let $v_F: \mathfrak{G} \to \mathfrak{M}_{1F}\mathfrak{K}$ be the corresponding projection. It can be shown that

(2) $$v_F(z) = d_F^{-1} \circ \mu_F(z) \circ d_F \qquad (z \in \mathfrak{Z}).$$

The transfer of the differential equations satisfied by members of $\mathscr{A}(G:\tau)$, from G to M_{1F}^+, is based on the following [**14b**, §4]:

LEMMA 1. *Let $b \in \mathfrak{G}$. Then there exist $\eta_i \in \mathfrak{M}_{1F}$, ξ_i, $\zeta_i \in \mathfrak{K}$, and $g_i \in \mathscr{S}_F$ $(1 \leqq i \leqq q)$ such that, for all $g \in C^\infty(G:\tau)$ and $m \in M_{1F}^+$,*

(3) $$g(m;b) = g(m; v_F(b)) + \sum_{1 \leqq i \leqq q} g_i(m)\,\tau_1(\xi_i)\,g(m;\eta_i)\,\tau_2(\zeta_i).$$

Let $f \in \mathscr{A}(G:\tau)$ be nonzero and $\mathfrak{Z}_f = \{z : z \in \mathfrak{Z},\ zf = 0\}$. Define $\mathfrak{Z}_{f,F} = \mathfrak{Z}_F \mu_F[\mathfrak{Z}_f]$; then $l = \dim(\mathfrak{Z}_F/\mathfrak{Z}_{f,F}) < \infty$. We select $u_1 = 1, u_2, \ldots, u_l$ in $\mathfrak{Z}_F$ to be linearly independent modulo $\mathfrak{Z}_{f,F}$. Clearly there is a unique $l \times l$ matrix representation $\Gamma : \xi \mapsto \Gamma(\xi) = (c_{ij}(\xi))$ of $\mathfrak{Z}_F$ such that, for each $\xi \in \mathfrak{Z}_F$ and $1 \leqq j \leqq l$,

(4) $$\xi u_j = \sum_{1 \leqq i \leqq l} c_{ji}(\xi)\,u_i + \zeta_{j,\xi} \qquad (\zeta_{j,\xi} \in \mathfrak{Z}_{f,F}).$$

We define $\hat{U} = U \otimes \boldsymbol{C}^l$, $\hat{\tau} = \tau \otimes 1$, and choose an orthonormal basis $\{e_1, \ldots, e_l\}$ for $\boldsymbol{C}^l$. For $m \in M_{1F}$, $\xi \in \mathfrak{Z}_F$, put

(5) $$\Phi(m) = \sum_{1 \leqq j \leqq l} f(m; u_j \circ d_F) \otimes e_j,$$
$$\Psi(m:\xi) = \sum_{1 \leqq j \leqq l} f(m; \zeta_{j,\xi} \circ d_F) \otimes e_j.$$

We then have the differential equations

(6) $$\Phi(m;\xi) = (1 \otimes \Gamma(\xi))\,\Phi(m) + \Psi(m:\xi).$$

It is convenient to rewrite these in the following form.

LEMMA 2. *Let $\Sigma = \{\lambda_1, \ldots, \lambda_p\}$, $F = \{\lambda_{d+1}, \ldots, \lambda_p\}$ and let $\{H_1, \ldots, H_p\}$ be the basis[16] of $\mathfrak{a}$ dual to $\{\lambda_1, \ldots, \lambda_p\}$. For $\mathbf{t} = (t_1, \ldots, t_d) \in \boldsymbol{R}^d$, put*

$$a(\mathbf{t}) = \exp(t_1 H_1 + \cdots + t_d H_d).$$

Then, for all $m \in M_{1F}^+$, $\eta \in \mathfrak{M}_{1F}$, $1 \leqq j \leqq d$,

(7) $$\frac{\partial}{\partial t_j}\Phi(ma(\mathbf{t});\eta) = (1 \otimes \Gamma(H_j))\,\Phi(ma(\mathbf{t});\eta) + \Psi(ma(\mathbf{t});\eta : H_j).$$

[16] Note that $H_j \in \mathfrak{Z}_F$ for $1 \leqq j \leqq d$, and they span $\mathfrak{a}_F$.

For $\boldsymbol{t}=(t_1,\ldots,t_d)$, let $|\boldsymbol{t}|=(t_1^2+\cdots+t_d^2)^{1/2}$, $\min(\boldsymbol{t})=\min(t_1,\ldots,t_d)$. Define $\boldsymbol{R}_+^d=\{\boldsymbol{t}:\boldsymbol{t}\in\boldsymbol{R}^d, \min(\boldsymbol{t})>0\}$.

LEMMA 3. *Let $\Xi_F=\Xi_{M_{1F}}$. Then for each $\eta\in\mathfrak{M}_{1F}$, $\xi\in\mathfrak{Z}_F$ we can find $B=B_{\xi,\eta}>0$ and $q=q_{\xi,\eta}\geqq 0$ such that for all $m\in M_{1F}^+$, $\boldsymbol{t}\in\boldsymbol{R}_+^d$,*

$$
\begin{aligned}
&|\Phi(ma(\boldsymbol{t});\eta)|\leqq B\Xi_F(m)\,(1+\sigma(m))^q\,(1+|\boldsymbol{t}|)^q,\\
&|\Psi(ma(\boldsymbol{t});\eta:\xi)|\leqq B\Xi_F(m)\,(1+\sigma(m))^q\gamma_F(m)\,(1-\gamma_F(m))^{-q}(1+|\boldsymbol{t}|)^q e^{-\min(\boldsymbol{t})}.
\end{aligned}
\tag{8}
$$

For proving these we need to use the following consequence of (4.1.5): there exist $c_0>0$, $r_0\geqq 0$ such that

$$d_F(m)\,\Xi(m)\leqq c_0\Xi_F(m)\,(1+\sigma(m))^{r_0}\qquad (m\in M_{1F}^+). \tag{9}$$

In addition, while deriving the estimates for Ψ we use: (i) the following inequality which follows from (1)–(3): given $\zeta\in\mathfrak{M}_{1F}$ and $z\in\mathfrak{Z}_f$, we can find $c_1>0$, $r_1\geqq 0$ such that, for all $m\in M_{1F}^+$,

$$|f(m;\zeta\mu_F(z)\circ d_F)|\leqq c_1 d_F(m)\,\Xi(m)\,(1+\sigma(m))^{r_1}\gamma_F(m)\,(1-\gamma_F(m))^{-r_1};$$

and (ii) the inequality $\gamma_F(ma(\boldsymbol{t}))\leqq\gamma_F(m)\exp\{-\min(\boldsymbol{t})\}$ for all $m\in M_{1F}^+$, $\boldsymbol{t}\in\boldsymbol{R}_+^d$.

6.3. On some differential equations of first order. We shall now describe the technique which enables us to determine the asymptotic behaviour of Φ from the first order differential equations (6.2.7) and the estimates (6.2.8).

Let W be a Hilbert space of dimension $n<\infty$; $\Gamma_1,\ldots,\Gamma_d$, mutually commuting endomorphisms of W. For $1\leqq j\leqq d$ and $\mu\in\boldsymbol{C}$, $W_{j,\mu}=\{w:w\in W, (\Gamma_j-\mu 1)^s w=0$ for some $s\geqq 0\}$. We put

$$^\circ W=\bigcap_{1\leqq j\leqq d}\left(\sum_{\mu:\,\mathrm{Re}\,\mu=0} W_{j,\mu}\right),\qquad {}^\circ\Gamma_j=\Gamma_j\,|\,{}^\circ W; \tag{1}$$

of course, $^\circ W$ is invariant under all the Γ_j and the $^\circ\Gamma_j$ have only pure imaginary eigenvalues. We now consider functions F and G_j $(1\leqq j\leqq d)$, defined and C^1 in a neighborhood of $\mathrm{Cl}(\boldsymbol{R}_+^d)$ with values in W, and having the following properties: (i) there exist $c>0$, $\beta>0$, $r\geqq 0$ such that, for all $\boldsymbol{t}\in\boldsymbol{R}_+^d$, $1\leqq j\leqq d$,

$$
\begin{aligned}
&|F(\boldsymbol{t})|\leqq c(1+|\boldsymbol{t}|)^r,\qquad |G_j(\boldsymbol{t})|\leqq c(1+|\boldsymbol{t}|)^r\,e^{-\beta\min(\boldsymbol{t})},\\
&(\partial F/\partial t_j)\;(\boldsymbol{t})=\Gamma_j F(\boldsymbol{t})+G_j(\boldsymbol{t}).
\end{aligned}
\tag{2}
$$

LEMMA 1. *Let the notation be as above. Then there is a unique $w\in{}^\circ W$ such that, with $F_\infty(\boldsymbol{t})=\exp(t_1{}^\circ\Gamma_1+\cdots+t_d{}^\circ\Gamma_d)\,w$, $|F(t\tau)-F_\infty(t\tau)|\to 0$ as $t\to+\infty$, for each $\tau\in\boldsymbol{R}_+^d$. Moreover, there are constants $C>0$, $\alpha>0$, depending only on r, β, n and the Γ_j such that, for all $\boldsymbol{t}\in\boldsymbol{R}_+^d$,*

$$(3)\qquad \begin{aligned} &|F(\boldsymbol{t})-F_\infty(\boldsymbol{t})| \leqq Cc(1+|\boldsymbol{t}|)^{r+2n}\exp\{-\alpha\min(\boldsymbol{t})\},\\ &|F_\infty(\boldsymbol{t})| \leqq Cc(1+|\boldsymbol{t}|)^{n}. \end{aligned}$$

Any endomorphism T of W can be uniquely written as $T'+T''$, where T', $T''\in \boldsymbol{C}[T]$, T' is semisimple, and the spectrum of T' (resp. T'') is real (resp. pure imaginary); moreover (see [**5e**, Lemma 60]) there is a constant $k(n)>0$ independent of T such that $\|e^{T''}\|\leqq k(n)\,(1+\|T''\|)^{n-1}$. We may thus replace Γ_j by Γ'_j and come down to the case when $\Gamma_j=c_j 1$ for all j, with $\boldsymbol{c}=(c_1,\dots,c_d)\in \boldsymbol{R}^d$. If some $c_i>0$, we select $\gamma_1,\dots,\gamma_d>0$ with[17] $(\gamma,\boldsymbol{c})>0$ and find that, for all $\boldsymbol{t}\in\boldsymbol{R}^d_+$,

$$F(\boldsymbol{t})=-\sum_j \gamma_j\int_0^\infty \exp\{-x(\gamma,\boldsymbol{c})\}G_j(\boldsymbol{t}+x\gamma)\,dx.$$

If $\boldsymbol{c}\neq\boldsymbol{0}$ but $c_i\leqq 0$ for all i, we find that, for all $\boldsymbol{t}\in\boldsymbol{R}^d_+$,

$$F(\boldsymbol{t})=\exp\{(\boldsymbol{c},\boldsymbol{t})\}\left\{F(\boldsymbol{0})+\sum_j t_j\int_0^1 \exp\{-x(\boldsymbol{c},\boldsymbol{t})\}\,G_j(x\boldsymbol{t})\,dx\right\}.$$

Finally, suppose $\boldsymbol{c}=\boldsymbol{0}$. Then $F(\tau,\dots,\tau)\to$ a limit w as $\tau\to+\infty$, and for all $\tau>0$, writing $\boldsymbol{\tau}=(\tau,\dots,\tau)$, $F(\boldsymbol{\tau})=w-\sum_j\int_\tau^\infty G_j(x,\dots,x)\,dx$. We then have, for all $\boldsymbol{t}\in\boldsymbol{R}^d_+$, with $\tau=\min(\boldsymbol{t})$,

$$F(\boldsymbol{t})-w=(F(\boldsymbol{\tau})-w)+\sum_j(t_j-\tau)\int_0^1 G_j(\boldsymbol{\tau}+x(\boldsymbol{t}-\boldsymbol{\tau}))\,dx.$$

The required estimates follow from these formulae.

6.4. Proofs of the theorems. We now apply the results of §6.3 to (6.2.7) and (6.2.8) to deduce the existence of a function $\Phi_F\in C^\infty(M_{1F}:\hat{\tau}_F)$ with values in ${}^\circ\hat{U}$, having the following properties:

(i) $\Phi_F(m\exp H)=e^{1\otimes\Gamma(H)}\,\Phi_F(m)$ for all $m\in M_{1F}$, $H\in\mathfrak{a}_F$,

(ii) $\Phi_F(m;\xi)=(1\otimes\Gamma(\xi))\,\Phi_F(m)$ for all $m\in M_{1F}$, $\xi\in\mathfrak{Z}_F$, and

(iii) there are constants $\alpha>0$, $B_1>0$, $q_1\geqq 0$ such that

$$(1)\qquad |\Phi_F(m)|\leqq B_1\Xi_F(m)\,(1+\sigma(m))^{q_1}\qquad (m\in M_{1F}),$$

$$(2)\qquad \begin{aligned} &|\Phi(m\exp H)-e^{1\otimes\Gamma(H)}\Phi_F(m)|\\ &\quad\leqq B_1\Xi_F(m)\,(1+\sigma(m))^{q_1}(1-\gamma_F(m))^{-q_1}\\ &\quad\cdot(1+\|H\|)^{q_1}\exp\{-\alpha\beta_F(H)\}\qquad (m\in M^+_{1F},\ H\in\mathfrak{a}^+_F). \end{aligned}$$

[17] $(\boldsymbol{a},\boldsymbol{b})=\sum_j a_j b_j$ for $\boldsymbol{a},\boldsymbol{b}\in\boldsymbol{R}^d$.

Actually, the existence of Φ_F with these properties on M_{1F}^+ is more or less immediate; the extension of Φ_F to M_{1F} so that (i), (ii) and (1) of (iii) above are valid for all $m \in M_{1F}$ is made possible by the following result (cf. [**6h**, Lemma 54]): Given $\bar{H} \in \mathfrak{a}_F^+$, there exists $c(\bar{H}) > 0$ such that $m \exp t\bar{H} \in M_{1F}^+$ for all $m \in M_{1F}$ and $t \geqq c(\bar{H}) \cdot \sigma(m)$.

Write $\Phi_F = \sum_{1 \leqq j \leqq l} \Phi_{F,j} \otimes e_j$ and define $f_F = \Phi_{F,1}$. Then $f_F \in \mathscr{A}(M_{1F}, \tau_F)$, and it is clear from (2) that (6.1.1) is valid. If $z \in \mathfrak{Z}_f$, $\Gamma(\mu_F(z)) = 0$, and so $\mu_F(z)\, f_F = 0$. For any $a \in A^+$ let us write $\log a = H_1 + H_2$ where $H_1 \in \mathfrak{a}_F$ and $\lambda(H_2) = 0$ for all $\lambda \in \Sigma \backslash F$. If we take $m = \exp(H_2 + \frac{1}{2}H_1)$, $H = \frac{1}{2}H_1$ in (2), and observe that for suitable constants $c' > 0$, $r' \geqq 0$, $d_F^{-1}\Xi_F \leqq c'\Xi(1+\sigma)^{r'}$ on A^+ (cf. (4.1.5)), we obtain the following estimate: There exist $B_2 > 0$, $q_2 \geqq 0$, $\gamma > 0$ such that for all $a \in \mathrm{Cl}(A^+)$ with $\beta_F(\log a) \geqq 1$,

$$(3) \qquad |f(a) - (d_F^{-1} f_F)(a)| \leqq B_2 (1 + \sigma(a))^{q_2} \exp\{-\varrho(\log a) - \gamma \beta_F(\log a)\}.$$

From this we get (6.1.3) and hence (6.1.4), without difficulty.

Now we can find $\kappa_0 > 0$ such that $\mathrm{Cl}(A^+) \subseteq \bigcup_{F \subsetneqq \Sigma} A^+(F:\kappa_0)$. So, if $f_F = 0$ for all F, (6.1.3) implies that $|f| = O(\Xi^{1+\gamma\kappa_0})$, thus proving the implication (ii)$\Rightarrow$(iii) of Theorem 6.1.2. Suppose now that $f \in L^2(G) \otimes U$. Let J be as in (4.1.6), and J_F, the corresponding function for M_{1F}. Clearly, given any $y > 0$, we can find $c(y) > 0$ such that for all $a \in A^+$ with $\beta_F(\log a) \geqq y$, $d_F(a)^2 J_F(a) \leqq c(y)\, J(a)$. So, writing $A_y^+(F:\kappa) = \{a : a \in A^+, \beta_F(\log a) \geqq \max(y, \kappa\varrho(\log a))\}$, we find from (6.1.3) that

$$(4) \qquad \int_{A^+{}_y(F:\kappa)} J_F(a)\, |f_F(a)|^2 \, da < \infty .$$

If we remember that $J_F(aa') = J_F(a')$ for all $a' \in A_F$ and that $a' \mapsto f_F(ma')$ is a tempered $\mathfrak{A}_F$-finite function on A_F, we can deduce from (4) that $f_F = 0$.

We now consider Theorem 3. We extend $\mathfrak{a}$ to a θ-stable CSA $\mathfrak{l}$, and assume that $zf = \chi_\Lambda(z) f$, $\forall z \in \mathfrak{Z}$, $\Lambda \in \mathfrak{l}_c^*$ being regular and real-valued on $(-1)^{1/2}(\mathfrak{l} \cap \mathfrak{l}) + (\mathfrak{l} \cap \mathfrak{p})$. We need the following lemma.

LEMMA 1. *Given $F \subsetneqq \Sigma$, there exists a unique ${}^\circ f_F \in \mathscr{A}(M_F : \tau_F)$ such that $f_F(ma) = {}^\circ f_F(m)$ for all $m \in M_F$, $a \in A_F$; moreover, $f_F = 0$ unless $s\Lambda \mid \mathfrak{a}_F = 0$ for some $s \in W_{L_c}$.*

To prove the lemma, one first uses the differential equation $\mu_F(z)\, f_F = \chi_\Lambda(z)\, f_F$ ($z \in \mathfrak{Z}$) and the regularity of Λ (cf. also Lemma 6.5.6) to conclude the following: There exist unique functions $f_{F,j}$ on M_F such that $f_F(ma) = \sum_{1 \leqq j \leqq N} \exp\{\Lambda_j(\log a)\} \cdot f_{F,j}(m)$ ($m \in M_F$, $a \in A_F$), $\Lambda_1, \ldots, \Lambda_N$ being all the distinct ones among the restrictions $s\Lambda \mid \mathfrak{a}_F$ ($s \in W_{L_c}$). As the Λ_j are real, the tempered nature of $a \mapsto f_F(ma)$ implies that $f_{F,j} = 0$ unless $\Lambda_j = 0$. The assertions of the lemma follow from this.

Suppose now that the conditions of Theorem 3 are satisfied but $f \notin \mathscr{C}(G) \otimes U$.

We select $F \subsetneq \Sigma$ of the smallest cardinality such that $f_F \neq 0$. Then $({}^\circ f_F)_{F'} = 0$ for all $F' \subsetneq F$, by (6.1.4). So ${}^\circ f_F \in \mathscr{C}(M_F) \otimes U$, implying that $\mathrm{rk}(M_F) = \mathrm{rk}(K_F)$ (§5.7). This means that, for some θ-stable CSA $\mathfrak{h}$, $\mathfrak{h} \cap \mathfrak{p} = \mathfrak{a}_F$. But $\mathfrak{h}$ is not conjugate to $\mathfrak{b}$ and so $(\mathfrak{g}, \mathfrak{h})$ has a real root α. This implies that $H_\beta \in \mathfrak{a}_F$ for some root β of $(\mathfrak{g}, \mathfrak{l})$. But then, by Lemma 1, $\langle s\Lambda, \beta \rangle = 0$ for some $s \in W_{L_c}$, contradicting the regularity of Λ.

6.5. Estimates uniform over the discrete spectrum. We have not made full use of the techniques of §§6.2 and 6.3 in proving the main theorems of this section. It turns out that one can obtain estimates describing the asymptotic behaviour of an eigenfunction that are actually uniform over the spectrum as well as over the set of parameters of the representations of K according to which the eigenfunction transforms when subjected to translations from K. Obviously, such estimates will play an important role in harmonic analysis on G (**[5e]**, **[5f]**, **[14a]**). In this section we shall illustrate what is involved by discussing the case when the eigenfunctions come from the discrete spectrum.

Let G be as in §3. $\mathfrak{l}$ is a θ-stable CSA containing $\mathfrak{a}$; $\mathscr{L}'(\mathfrak{l})$ the set of integral regular elements in $\mathfrak{l}_c^*$. Given $\Lambda \in \mathscr{L}'(\mathfrak{l})$ and τ, U as before, we define

$$(1) \qquad \mathscr{A}(G:\tau:\Lambda) = \{f : f \in \mathscr{A}(G:\tau), zf = \chi_\Lambda(z) f, \forall z \in \mathfrak{Z}\}.$$

Then $\mathscr{A}(G:\tau:\Lambda) \subseteq \mathscr{C}(G) \otimes U$. We define Ω as in §1, select a norm $\|\cdot\|$ in $\mathfrak{l}_c^*$, and put

$$(2) \qquad |\tau| = (1 + \|\tau_1(\Omega)\|)(1 + \|\tau_2(\Omega)\|), \qquad |\tau, \Lambda| = |\tau|(1 + \|\Lambda\|).$$

For any measurable function $\varphi: G \to U$, $\|\varphi\|_2$ is the L^2-norm of the function $x \to |\varphi(x)|$ whenever this is finite. Our main result in this section is then the following:

THEOREM 1. *We can find a constant $\alpha > 0$, and, corresponding to any $a, b \in \mathfrak{G}$, constants $C = C_{a,b} > 0$, $r = r_{a,b} \geqq 0$ such that, for all $\Lambda \in \mathscr{L}'(\mathfrak{l})$, all τ, and all $f \in \mathscr{A}(G:\tau:\Lambda)$,*

$$(3) \qquad |(afb)(x)| \leqq C|\tau, \Lambda|^r \|f\|_2 \Xi(x)^{1+\alpha} \qquad (x \in G).$$

We remark that it is enough to prove this theorem for $a = b = 1$; the general case can then be deduced through an elementary device. Further, there exists $\kappa > 0$ such that

$$\mathrm{Cl}(A^+) \subseteq \bigcup_{\lambda \in \Sigma} A^+(\Sigma \backslash \{\lambda\} : \kappa).$$

It is therefore sufficient to establish the following:

LEMMA 2. *Fix $\lambda \in \Sigma$ and let $F = \Sigma \backslash \{\lambda\}$. Then there exist $\alpha > 0$, $C > 0$, $q \geqq 0$ such* *that for all Λ, τ, $\kappa>0$, as above, $f\in\mathscr{A}(G:\tau:\Lambda)$,*

$$(4)\qquad |f(a)|\leqq C|\tau,\Lambda|^q\,\Xi(a)^{1+\alpha\kappa}\,\|f\|_2 \qquad (a\in A^+(F:\kappa)).$$

For the rest of this section we fix F as above, and $H\in\mathfrak{a}_F^+$; note that $\dim(\mathfrak{a}_F)=1$. First we have the following a priori estimates:

LEMMA 3. *Let $u, v\in\mathfrak{G}$. Then there exist $C=C_{u,v}>0$ and $s=s_{u,v}\geqq 0$ such that, for all Λ, τ as above and $f\in\mathscr{A}(G:\tau:\Lambda)$,*

$$(5)\qquad |(ufv)(x)|\leqq C|\tau,\Lambda|^s\,\Xi(x)\|f\|_2 \qquad (x\in G).$$

Given $a, b\in\mathfrak{G}$, we first prove the existence of a constant $C_1=C_{1,a,b}>0$ such that $\|afb\|_2\leqq C_1|\tau,\Lambda|^d\,\|f\|_2$ for all τ, Λ, f as above, where $d=\deg(a)+\deg(b)$ (see [**14b**, Lemma 5.5]); (5) now follows from Theorem 4.1.3.

We next introduce Φ and its differential equations. We select $v_1=1, v_2,\ldots, v_l\in\mathfrak{Z}_F$ such that $\mathfrak{Z}_F$ is the direct sum of $\mu_F[\mathfrak{Z}]\,v_i$ $(1\leqq i\leqq l)$. Define $\hat{U}=U\otimes\boldsymbol{C}^l$ and select an orthonormal basis $\{e_i\}$ for $\boldsymbol{C}^l$. If $v\in\mathfrak{Z}_F$, $vv_j=\sum_{1\leqq i\leqq l}\mu_F(z_{v:ij})\,v_i$ $(1\leqq j\leqq l)$ for unique $z_{v:ij}\in\mathfrak{Z}$ and we write $\Gamma(\Lambda:v)$ for the $l\times l$ matrix whose ijth element is $\mu_{\mathfrak{g}/\mathfrak{l}}(z_{v:ji})(\Lambda)$. $\Gamma(\Lambda:\cdot)$ is a representation of $\mathfrak{Z}_F$ in $\boldsymbol{C}^r$. Given $f\in\mathscr{A}(G:\tau:\Lambda)$ we write

$$(6)\qquad \Phi(m)=\sum_{1\leqq j\leqq l} f(m;v_j)\otimes e_j \qquad (m\in M_{1F}).$$

Proceeding as in §6.2, but taking into account the variability of τ, we get the following result [**14b**, Lemmas 5.2 and 5.4]:

LEMMA 4. *Let $v\in\mathfrak{Z}_F$. Then for each τ we can find a differential operator D_v^τ acting on $C^\infty(M_{1F}^+:\hat{U})$ such that*

(i) *for all τ, Λ as above and $f\in\mathscr{A}(G:\tau:\Lambda)$,*

$$(7)\qquad \Phi(m;v)=(1\otimes\Gamma(\Lambda:v))\,\Phi(m;v)+\Phi(m;D_v^\tau) \qquad (m\in M_{1F}^+);$$

(ii) *there exist $r\geqq 0$, $\omega_k\in\mathfrak{M}_{1F}$ $(1\leqq k\leqq k_0)$ such that, for all τ as above and all $g\in C^\infty(M_{1F}^+:\hat{U})$,*

$$(8)\qquad |g(m;D_v^\tau)|\leqq\gamma_F(m)\,(1-\gamma_F(m))^{-r}\,|\tau|^r\sum_{1\leqq k\leqq k_0}|f(m;\omega_k)|.$$

We now have the differential equations

$$(9)\qquad \frac{d}{dt}\,\Phi(m\exp tH)=(1\otimes\Gamma(H))\,\Phi(m\exp tH)+\Phi(m\exp tH;D_H^\tau)$$

valid for all $m\in M_{1F}^+$, $t\geqq 0$. Furthermore, as our eigenfunctions f are in $\mathscr{C}(G)\otimes U$,

$$\lim_{t \to +\infty} \Phi(m \exp tH) = 0. \tag{10}$$

We are therefore in a position to proceed as in §6.4. However, Lemma 6.3.1 cannot be used as it is, because the estimates given by it are not uniform over $\Gamma_1, \ldots, \Gamma_d$. We therefore use the following variant:

LEMMA 5. *Let W be as in* §6.3, *and Γ a semisimple endomorphism of W whose spectrum $S = S(\Gamma)$ is real. Let E_c ($c \in S$) be the spectral projections and define*

$$v(\Gamma) = \max_{c \in S} \|E_c\|, \qquad \sigma(\Gamma) = \min_{c \in S \setminus \{0\}} |c|. \tag{11}$$

Let F, H be functions of class C^1 on $[0, \infty)$ with values in W such that (i) *$dF/dt = \Gamma F + H$ on $[0, \infty)$,* (ii) *there exist $C > 0$, $r \geqq 0$, $\beta > 0$ such that $|F(t)| \leqq C(1+t)^r$, $|H(t)| \leqq C(1+t)^r e^{-\beta t}$ for all $t \geqq 0$, and* (iii) *the limit $\lim_{t \to +\infty} F(t)$ exists and is* 0. *Then, with $[S]$ denoting the number of elements of S and $A(r, \beta) > 0$ a constant depending on r and β but not on Γ or C,*[18] *we have*

$$|F(t)| \leqq A(r, \beta)\, C v(\Gamma)\, [S(\Gamma)]\, (1+t)^r \exp\{-\min(\beta, \sigma(\Gamma))\, t\} \qquad (t \geqq 0). \tag{12}$$

On the other hand, the spectral structure of the matrices $\Gamma(\Lambda : v)$ is known in great detail and one has the following lemma ([**5e**, §3], [**5f**, Lemma 19], [**14b**, Lemmas 5.1 and 7.2]):

LEMMA 6. *For each $v \in \mathfrak{Z}_F$, $\Lambda \in \mathscr{L}'(\mathfrak{l})$, $\Gamma(\Lambda : v)$ is semisimple and its eigenvalues are $\mu_{\mathfrak{m}_{1F}/\mathfrak{l}}(v)$ $(s\Lambda)$ $(s \in W_{L_c})$. In particular, the eigenvalues λ of $\Gamma(\Lambda : H)$ are real, and there exists $\alpha > 0$ such that $|\lambda| \geqq \alpha$ for all nonzero λ and all $\Lambda \in \mathscr{L}'(\mathfrak{l})$. Moreover, we can choose a basis $e_j(\Lambda)$ for $\mathbf{C}^l$ $(1 \leqq j \leqq l$, $\Lambda \in \mathscr{L}'(\mathfrak{l}))$ such that* (i) *all the $\Gamma(\Lambda : v)$ are diagonal in this basis, and* (ii) *if $E_j(\Lambda)$ are the projections $\mathbf{C}^l \to \mathbf{C} \cdot e_j(\Lambda)$, then there exist $C_0 > 0$, $r_0 \geqq 0$ such that, for all $\Lambda \in \mathscr{L}'(\mathfrak{l})$,*

$$\sum_{1 \leqq j \leqq l} \|E_j(\Lambda)\| \leqq C_0 (1 + \|\Lambda\|)^{r_0}. \tag{13}$$

Lemma 2 and thence Theorem 1 follow from these estimates more or less in the same way as in §6.4. For details see [**14b**].

The perturbation method is central in the entire theory of asymptotic behaviour of eigenfunctions, and the results obtained through its application go far beyond what we have indicated above. As further examples we mention the theorems that suitably formed "wave packets" (of eigenfunctions) over the spectrum belong to $\mathscr{C}(G)$ and even $\mathscr{C}^1(G)$ (cf. [**6i**], [**14a**]), as well as the results on the theory of integrable eigenfunctions (cf. [**14b**]).

---

[18] We can take $A(r, \beta) = \max(2, \int_0^\infty (1+u)^r e^{-\beta u}\, du)$.

7. The discrete series

We shall now describe briefly how the results of the preceding chapters lead to the determination of the discrete series. Exploiting the fact that the K-finite matrix coefficients of representations of the discrete series are cusp forms, one uses Theorem 5.7.1 to reduce their harmonic analysis to that on the compact CSG; the procedure is of course similar to that used by Weyl for determining the characters of compact groups.

7.1. The discrete series and the discrete part of the Plancherel formula. For simplicity we restrict ourselves to the case when G is a connected real form of a complex simply connected semisimple group G_c. We choose once and for all a Haar measure dx and fix it throughout this section. We write $\mathscr{E}_2(G)$ for the discrete series of G. For $\omega \in \mathscr{E}_2(G)$, $d(\omega)>0$ is its formal degree so that, for any $\pi \in \omega$ and unit vectors ϕ, ψ in the space of π,

$$\int_G |(\pi(x)\,\varphi, \psi)|^2\,dx = d(\omega)^{-1}. \tag{1}$$

$L^2_\omega(G)$ is the closed linear span of the matrix coefficients of ω; it coincides with the closed linear span of all subspaces of $L^2(G)$ that are irreducibly invariant under the right regular representation r and define a subrepresentation in ω. ${}^\circ L^2(G)$ is the (orthogonal) direct sum of the $L^2_\omega(G)$. ${}^\circ E$ and E_ω are the orthogonal projections of $L^2(G)$ on ${}^\circ L^2(G)$ and $L^2_\omega(G)$ respectively. ω^* is the class contragredient to ω.

THEOREM 1. *G has a discrete series if and only if* $\mathrm{rk}(G)=\mathrm{rk}(K)$, *i.e., G has a compact CSG.*

Assume now that $\mathrm{rk}(G)=\mathrm{rk}(K)$ and use the notation of §3. In particular, let P be a positive system of roots of $(\mathfrak{g}, \mathfrak{b})$; $\delta=\frac{1}{2}\sum_{\alpha\in P}\alpha$, $\Delta=\xi_{-\delta}\prod_{\alpha\in P}(\xi_\alpha-1)$, $\varpi=\prod_{\alpha\in P}H_\alpha$, $\varepsilon(\lambda)=\mathrm{sign}\,\varpi(\lambda)$ $(\lambda\in\mathscr{L}')$, $q=\frac{1}{2}\dim(G/K)$.

THEOREM 2. *For each $\lambda\in\mathscr{L}'$, there exists $\omega[\lambda]\in\mathscr{E}_2(G)$ such that $\Theta_{\omega[\lambda]}=(-1)^q\varepsilon(\lambda)\Theta_\lambda$, and every $\omega\in\mathscr{E}_2(G)$ is of the form $\omega[\lambda]$ for some $\lambda\in\mathscr{L}'$. $\omega[\lambda_1]=\omega[\lambda_2]$ if and only if λ_1 and λ_2 are in the same W_B-orbit. Moreover, there exists a constant $c(G)>0$ such that $d(\omega[\lambda])=c(G)\,[W_B]\,|\varpi(\lambda)|$ for all $\lambda\in\mathscr{L}'$. Finally, $\omega[-\lambda]=\omega[\lambda]^*$.*

Harish-Chandra has determined the value of $c(G)$ explicitly when the Haar measure dx is normalized in a canonical way (cf. [**6i**, pp. 537 and 540]).

For $\lambda\in\mathscr{L}'$ and $f\in\mathscr{C}(G)$, let

$$f_\lambda(x) = (-1)^q c(G)\,[W_B]\,\varpi(-\lambda)\,\Theta_{-\lambda}(r(x)\,f) \qquad (x\in G). \tag{2}$$

THEOREM 3. *For any $f \in \mathscr{C}(G)$ and $\lambda \in \mathscr{L}'$, $f_\lambda = E_{\omega[\lambda]} f$ and lies in $\mathscr{C}(G)$. Moreover, the series*

$$\sum_{\lambda \in \mathscr{L}'/W_B} f_\lambda \tag{3}$$

converges absolutely[19] *in $\mathscr{C}(G)$, and its sum ${}^\circ f$ is precisely ${}^\circ Ef$. Finally, ${}^\circ\mathscr{C}(G) = \mathscr{C}(G) \cap {}^\circ L^2(G)$, and $f \to {}^\circ f$ is a continuous map of $\mathscr{C}(G)$ onto ${}^\circ\mathscr{C}(G)$.*

7.2. Theorems 1 and 2. Standard arguments from representation theory [**6h**, Lemma 77] show that $\mathscr{E}_2(G) \neq \emptyset$ if and only if there are nonzero K-finite eigenfunctions for $\mathfrak{Z}$ in $L^2(G)$. Theorem 1 is then immediate from Theorems 5.7.3, 6.1.2 and 6.1.3.

We now come to Theorem 2 (for complete details, see [**6h**, §§40, 41]). Let $\mathrm{rk}(G) = \mathrm{rk}(K)$. For any $g \in C^\infty(B)$ let $\hat{g}: \lambda \mapsto \int g\xi_\lambda \, db$ ($\lambda \in \mathscr{L}$) be its Fourier transform ($\int_B db = 1$). Define

$$F_f(b) = \Delta(b) \int_G f(xbx^{-1}) \, dx \qquad (b \in B', f \in \mathscr{C}(G)). \tag{1}$$

Then Theorems 5.5.1 and 5.7.1 imply easily that, for all cusp forms f and all tempered invariant eigendistributions Θ,

$$\Theta(f) = (-1)^m [W_B]^{-1} \int_B \Phi F_f \, db \qquad (m = \tfrac{1}{2} \dim(G/B)), \tag{2}$$

Φ being the analytic function on B that extends $\Delta(\Theta \mid B')$; in particular, for all $\lambda \in \mathfrak{L}'$,

$$\Theta_\lambda(f) = (-1)^m \hat{F}_f(\lambda). \tag{3}$$

If we now take Fourier transforms in the relation (5.6.2) and remember that $F_f \in C^\infty(B)$, we obtain the following result: there is a constant $c(G) > 0$ ($c(G)$ is the constant c of (5.6.2)) such that

$$f(x) = (-1)^q c(G) \sum_{\lambda \in \mathscr{L}'} \varpi(\lambda) \, \Theta_\lambda(r(x) f) \qquad (x \in G, f \in {}^\circ\mathscr{C}(G)). \tag{4}$$

It is clear from (4) that the harmonic analysis of the cusp forms is completely controlled by the distributions Θ_λ.

We shall now indicate how the transition from (4) to Theorem 2 is carried out. Fix a homomorphism χ of $\mathfrak{Z}$ into $\boldsymbol{C}$ and let $\mathscr{E}_{2,\chi}$ be the set of all $\omega \in \mathscr{E}_2(G)$ with $\chi_\omega = \chi$. If $\omega \in \mathscr{E}_{2,\chi}$ and g is a matrix coefficient of ω, then $\Theta_\mu(r(x) g) = 0$ unless

[19] This means that for any continuous seminorm ν on $\mathscr{C}(G)$, $\sum_\lambda \nu(f_\lambda) < \infty$.

$\chi=\chi_{-\mu}$. So, taking $f=g$ in (4) we find that $\mathscr{E}_{2,\chi}=\emptyset$ unless $\chi=\chi_\lambda$ for some $\lambda\in\mathfrak{L}'$. For such a χ, we obtain from Corollary 5.5.3 and the results of §3 the following: Let $s_1=1, s_2,\dots, s_r$ be a complete system of representatives for $W_B\backslash W_{B_c}$: iet Φ_λ (resp. Φ_ω) be the analytic function on B extending $\Delta(\Theta_\lambda \mid B')$ (resp. $\Delta(\Theta_\omega \mid B')$); then, there is a unique $c_{\omega i}\in\boldsymbol{C}\,(\omega\in\mathscr{E}_{2,\chi}, 1\leqq i\leqq r)$ such that

$$(5) \qquad \Theta_\omega=\sum_{1\leqq i\leqq r} c_{\omega i}\Theta_{s_i\lambda}, \qquad \Phi_\omega=\sum_{1\leqq i\leqq r} c_{\omega i}\Phi_{s_i\lambda}.$$

We now use the orthogonality relations satisfied by the matrix coefficients of the discrete series to obtain the following relations, valid for all $\omega\in\mathscr{E}_2(G)$:

$$(6) \qquad \begin{aligned} F_f&=d(\omega)^{-1}f(1)\,\Phi_\omega && (f\in\mathscr{C}(G)\cap L^2_\omega(G)),\\ \Theta_\omega(f)&=d(\omega)^{-1}f(1)\,\delta_{\omega'\omega^*} && (f\in\mathscr{C}(G)\cap L^2_{\omega'}(G)).\end{aligned}$$

It follows from these relations that the functions $[W_B]^{-1/2}\,\Phi_\omega$ $(\omega\in\mathscr{E}_{2,\chi})$ are orthonormal in $L^2(B)$. Moreover, they have the same span as the $\Phi_{s_i\lambda}$ $(1\leqq i\leqq r)$. For, if this were not so, we could find a nonzero linear combination Θ of the $\Theta_{s_i\lambda}$ such that $\Theta(f)=0$ for all K-finite eigenfunctions f for $\mathfrak{Z}$ in $\mathscr{C}(G)$; taking f to be an arbitrary Fourier component of Θ^{conj} we find that $\Theta=0$. It follows at this stage that $\mathscr{E}_{2,\chi}$ has r elements and that the matrix $(c_{\omega i})$ $(\omega\in\mathscr{E}_{2,\chi}, 1\leqq i\leqq r)$ is unitary.

One now argues that the $c_{\omega i}$ are integers. To see this, let $\omega\in\mathscr{E}_{2,\chi}$ and let $n(\mathfrak{d})=[\omega:\mathfrak{d}]$, $\psi_\mathfrak{d}=$character of $\mathfrak{d}$ $(\mathfrak{d}\in\mathscr{E}(K))$. Then $\sum_{\mathfrak{d}\in\mathscr{E}(K)} n(\mathfrak{d})\,\psi_\mathfrak{d}$ is a well-defined distribution on K, and one can show that it coincides on $K\cap G'$ with the distribution defined thereon by the function Θ_ω. It follows without difficulty from this that all the $c_{\omega i}\in\boldsymbol{Z}$. This completes the proof of Theorem 2 except for the formula for the formal degree and the sign factors in $\Theta_{\omega[\lambda]}$.

For determining the signs we argue as follows: We have $\Theta_{\omega[\lambda]}=\zeta(\lambda)\,\Theta_\lambda$ where ζ is a W_B-skew function on $\mathscr{L}'$ with values ±1. Fix $\lambda\in\mathscr{L}'$ and use (4) with $x=1$ and $f=g*\tilde{g}$ where $g\neq0$ is a K-finite matrix coefficient of $\omega[\lambda]^*$. Then $f\in{}^\circ\mathscr{C}(G)\cap L^2_{\omega[\lambda]^*}(G)$, and

$$\|g\|^2=(-1)^q\,c(G)\,\varpi(\lambda)\,\zeta(\lambda)\,\Theta_{\omega[\lambda]}(g*\tilde{g}).$$

This determines $\zeta(\lambda)$ and $d(\omega[\lambda])$, and yields the *discrete part* of the Plancherel formula:

$$(7) \qquad \|f\|^2=c(G)\,[W_B]\sum_{\lambda\in\mathscr{L}'/W_B}|\varpi(\lambda)|\Theta_{\omega[\lambda]}(f*\tilde{f}) \qquad (f\in{}^\circ\mathscr{C}(G)).$$

7.3. Outline of the proof of Theorem 3. For any $\lambda\in\mathscr{L}'$ and $f\in\mathscr{C}(G)$, the relation $f_\lambda=E_{\omega[\lambda]}f$ follows from a "real variables" argument. Also, $f\in\mathscr{C}(G)$ is a differentiable vector for both the left and right regular representations, as can be easily deduced from the estimates of §4.1. So ${}^\circ Ef$ and $E_\omega f$ are differentiable likewise. Consequently they are C^∞ functions all of whose derivatives lie in $L^2(G)$.

Write $\mathscr{L}^+$ for the set of all $\lambda \in \mathscr{L}'$ such that $\langle \lambda, \alpha \rangle > 0$ for all compact roots $\alpha \in P$.[20] For each $\lambda \in \mathscr{L}^+$ we select a Hilbert space $\mathfrak{H}_\lambda$, a $\pi_\lambda \in \omega[\lambda]$ acting in $\mathfrak{H}_\lambda$, and an orthonormal basis $\{e_{\lambda,i} : i \in N_\lambda\}$ for $\mathfrak{H}_\lambda$ such that each $e_{\lambda,i}$ belongs to a subspace irreducibly invariant under $\pi_\lambda[K]$. Let

$$a_{\lambda,i,j}(x) = d(\omega[\lambda])^{1/2} (\pi_\lambda(x) e_{\lambda,j}, e_{\lambda,i}) \qquad (x \in G).$$

Clearly, the $a_{\lambda,i,j}$ form an orthonormal basis of ${}^\circ L^2(G)$, and, for all $f \in \mathscr{C}(G)$,

$$(1) \qquad {}^\circ Ef = \sum_{\lambda \in \mathscr{L}^+} f_\lambda, \qquad f_\lambda = \sum_{i,j \in N_\lambda} f_{\lambda,i,j}, \qquad f_{\lambda,i,j} = (f, a_{\lambda,i,j})\, a_{\lambda,i,j}.$$

Let Ω be as in §§1 and 6.5, $z = \omega + \langle \delta \cdot \delta \rangle + 1$ where ω is the Casimir of G. Then $z a_{\lambda,i,j} = (1 + \|\lambda\|^2)\, a_{\lambda,i,j}$, $\Omega^r a_{\lambda,i,j} \Omega^r = c^r_{\lambda,i} c^r_{\lambda,j} a_{\lambda,i,j}$; moreover, the $c_{\lambda,i}$ are constants $\geqq 1$ and have the property that, for some $q \geqq 0$,

$$(2) \qquad c = \sup_{\lambda \in \mathscr{L}^+} \sum_{i \in N_\lambda} c_{\lambda,i}^{-q} < \infty.$$

We now use the uniform estimates of §6.5 to establish the following: there exist $\alpha > 0$, $C > 0$, $p \geqq 0$ such that for all $\lambda \in \mathscr{L}^+$, $i, j \in N_\lambda$, $x \in G$,

$$(3) \qquad |a_{\lambda,i,j}(x)| \leqq C[c_{\lambda,i} c_{\lambda,j} (1 + \|\lambda\|^2)]^p\, \Xi(x)^{1+\alpha}.$$

From (2) and (3) it is easy to deduce the following: There exist $C > 0$ and $m \geqq 0$ such that, for all $f \in \mathscr{C}(G)$,

$$(4) \qquad \sum_{\lambda \in \mathscr{L}^+} \sum_{i,j \in N_\lambda} \|\Xi^{-(1+\alpha)} f_{\lambda,i,j}\|_\infty \leqq C \|\Omega^m z^m f \Omega^m\|_2.$$

The estimate (4), together with those obtained by replacing f with ufv ($u, v \in \mathfrak{G}$), led to Theorem 3 (cf. **[6i]**, **[14b]**).

The above discussion also shows that the topology of ${}^\circ\mathscr{C}(G)$ is precisely the one induced by the seminorms $f \mapsto \|ufv\|_2$ ($u, v \in \mathfrak{G}$). It is even possible to restrict ourselves only to the seminorms $f \mapsto \|\xi^r f \xi^s\|_2$ ($r, s \geqq 0$, $\xi = 1 - (X_1^2 + \cdots + X_n^2)$ where the X_i are an orthonormal basis of $\mathfrak{g}$); this however needs some more work which we do not go into here.

I wish to acknowledge my very great indebtedness to the many long conversations with Professor Harish-Chandra on problems of harmonic analysis on semisimple Lie groups. In addition I have profited a great deal from my discussions with friends and colleagues here as well as at other institutions, and I am grateful to all of them.

[20] $\mathscr{L}^+$ is a system of representatives for $\mathscr{L}'/W_B$.

References

1. V. Bargmann, *Irreducible unitary representations of the Lorentz group*, Ann. of Math. (2) **48** (1947), 568–640. MR **9**, 133.

2. L. Ehrenpreis and F. I. Mautner

a) *Some properties of the Fourier transform on semisimple Lie groups*. I, Ann. of Math. (2) **61** (1955), 406–439. MR **16**, 1017.

b) *Some properties of the Fourier transform on semisimple Lie groups*. II, Trans. Amer. Math. Soc. **84** (1957), 1–55. MR **18**, 745.

c) *Some properties of the Fourier transform on semisimple Lie groups*. III, Trans. Amer. Math. Soc. **90** (1959), 431–484. MR **21** #1545.

3. I. M. Gel'fand and M. I. Graev

a) *On a general method of decomposing the regular representation of a Lie group into irreducible representation*, Dokl. Akad. Nauk. SSSR **92** (1953), 221–224. (Russian) MR **15**, 601.

b) *Analogue of the Plancherel formula for the classical groups*, Trudy Moskov. Mat. Obšč. **4** (1955), 375–404; English Transl., Amer. Math. Soc. Transl. (2) **9** (1958), 123–154. MR, **17**, 173; **19**, 1181.

4. Harish-Chandra

a) *Representations of a semisimple Lie group on a Banach space*. I, Trans. Amer. Math. Soc. **75** (1953), 185–243. MR **15**, 100.

b) *Representations of semisimple Lie groups*. II, Trans. Amer. Math. Soc. **76** (1954), 26–65. MR **15**, 398.

c) *Representations of semisimple Lie groups*. III, Trans. Amer. Math. Soc. **76** (1954), 234–253. MR **16**, 11.

d) *Representations of semisimple Lie groups*. IV, Amer. J. Math. **77** (1955), 743–777. MR **17**, 282.

e) *Representations of semisimple Lie groups*. V, Amer. J. Math. **78** (1956), 1–41. MR **18**, 490.

f) *Representations of semisimple Lie groups*. VI. *Integrable and square-integrable representations*, Amer. J. Math. **78** (1956), 564–628. MR **18**, 490.

g) *Representations of semisimple Lie groups*, Proc. Internat. Congr. Math. (Amsterdam, 1954), vol. 1, Noordhoff, Groningen; North-Holland, Amsterdam, 1957, pp. 299–304. MR **20** #1926.

h) *The Plancherel formula for complex semisimple Lie groups*, Trans. Amer. Math. Soc. **76** (1954), 485–528. MR **16**, 111.

i) *The characters of semisimple Lie groups*, Trans. Amer. Math. Soc. **83** (1956), 98–163. MR **18**, 318.

5. Harish-Chandra

a) *Differential operators on a semisimple Lie algebra*, Amer. J. Math. **79** (1957), 87–120. MR **18**, 809.

b) *Fourier transforms on a semisimple Lie algebra*. I, Amer. J. Math. **79** (1957). MR **19**, 293.

c) *Fourier transforms on a semisimple Lie algebra*. II, Amer. J. Math. **79** (1957), 653–686. MR **20** #2396.

d) *A formula for semisimple Lie groups*, Amer. J. Math. **79** (1957), 733–760. MR **20** #2633.

e) *Spherical functions on a semisimple Lie group*. I, Amer. J. Math. **80** (1958), 241–310. MR **20** #925.

f) *Spherical functions on a semisimple Lie group*. II, Amer. J. Math. **80** (1958), 553–613. MR **21** #92.

6. Harish-Chandra

a) *Invariant distributions on Lie algebras*, Amer. J. Math. **86** (1964), 271–309. MR **28** #5144.

b) *Invariant differential operators and distributions on a semisimple Lie algebra*, Amer. J. Math. **86** (1964), 534–564. MR **31** #4862a.

c) *Some results on an invariant integral on a semisimple Lie algebra*, Ann. of Math. (2) **80** (1964), 551–593. MR **31** #4862b.

d) *Invariant eigendistributions on a semisimple Lie algebra*, Inst. Hautes Études Sci. Publ. Math. No. 27 (1965), 5–54. MR **31** #4862c.

e) *Invariant eigendistributions on a semisimple Lie group*, Trans. Amer. Math. Soc. **119** (1965), 457–508. MR **31** #4862d.

f) *Discrete series for semisimple Lie groups.* I, Acta Math. **113** (1965), 241–318. MR **36** #2744.

g) *Two theorems on semisimple Lie groups*, Ann. of Math. (2) **83** (1966), 74–128. MR **33** #2766.

h) *Discrete series for semisimple Lie groups.* II. *Explicit determination of the characters*, Acta Math. **116** (1966), 1–111. MR **36** #2745.

i) *Harmonic analysis on semisimple Lie groups*, Bull. Amer. Math. Soc. **76** (1970), 529–551. MR **41** #1933.

7. S. Helgason, *Differential geometry and symmetric spaces*, Pure and Appl. Math., vol. 12, Academic Press, New York, 1962. MR **26** #2986.

8. T. Hirai

a) *Invariant eigendistributions of Laplace operators on real semisimple Lie groups*, Japan J. Math. **39** (1970), 1–68.

b) *The Plancherel formula for $SU(p, q)$*, J. Math. Soc. Japan **22** (1970), 134–179. MR **42** #3230.

9. M. S. Narasimhan and K. Okamoto, *An analogue of the Borel-Weil-Bott theorem for hermitian symmetric pairs of noncompact type*, Ann. of Math. (2) **91** (1970), 486–511. MR **43** #419.

10. R. Parthasarathy,

a) *Dirac operator and the discrete series*, Ann. Math. **96** (1972), 1–30.

b) *A note on the vanishing of certain 'L^2-cohomologies'*, J. Math. Soc. Japan **23** (1971), 676–691.

11. W. Schmid, *On a conjecture of Langlands*, Ann. of Math. (2) **93** (1971), 1–42. MR **44** #4149.

12. L. Schwartz, *Théorie des distributions*. Tome II, Actualités Sci. Indust. no. 1122, Hermann, Paris, 1951. MR **12**, 833.

13. R. Steinberg, *Differential equations invariant under finite reflexion groups*, Trans. Amer. Math. Soc. **112** (1964), 392–400. MR **29** #4807.

14. P. C. Trombi and V. S. Varadarajan

a) *Spherical transforms on semisimple Lie groups*, Ann. of Math. (2) **94** (1971), 246–303. MR **44** #6913.

b) *Asymptotic behaviour of eigenfunctions on a semisimple Lie group: the discrete spectrum*, Acta Math. **129** (1972), 237–280.

UNIVERSITY OF CALIFORNIA, LOS ANGELES

FUNCTIONS ON SYMMETRIC SPACES*

SIGURDUR HELGASON

Contents

AMS (MOS) subject classifications (1970). Primary 22E45, 43A80, 43A85, 22E30, 43A90, 53C35, 58G99, 31B25, 31B10, 32M15.

* Research supported by NSF contract no. GP-22928.

I. Introduction

1. Preliminary comments and motivation. The aim of these lectures is to give a survey of various topics in function theory on symmetric spaces. If taken literally this would for example include analysis in Euclidean space and on the unit disk, so some limitation is clearly necessary. Thus we concentrate on basic results, which hold for symmetric spaces in general, and when we invoke special cases like the sphere or the non-Euclidean disk it is usually in order to illustrate a more general result or to motivate a conjecture. But even with this limitation our survey is quite incomplete; thus we leave out all discussion of automorphic forms and other function-theoretic topics related to discrete groups.

A simply connected Riemannian globally symmetric space can be written as a product manifold

$$X_+ \times X_0 \times X_-,$$

where X_+ is compact, of positive sectional curvature, X_0 is a Euclidean space and X_- is noncompact without a compact factor and of negative sectional curvature.

We write $X_+ = G_+/K_+$, $X_0 = G_0/K_0$, $X_- = G_-/K_-$, where G_+, G_0, G_- are the identity components of the groups of isometries of X_+, X_0, X_-, respectively. Although X_- is our main concern it will be useful for the sake of motivation to state a few facts concerning X_+ and X_0.

In his paper [**4**], É. Cartan studied harmonic analysis on the space X_+, refining the Peter-Weyl theory on G_+. Cartan proved the following result:

THEOREM 1.1. *In the sense of Hilbert space decomposition*

$$L^2(X_+) = \sum_{\delta} V_\delta, \tag{1.1}$$

where δ runs over the set of irreducible representations of G_+ for which $\delta(K_+)$ has a fixed vector, the subspace $V_\delta \subset L^2(X_+)$ is invariant under G_+, and the natural representation on it is equivalent to δ. The K_+-invariant vector φ in V_δ is unique up to a scalar multiple and if it is normalized by $\varphi(\{K_+\})=1$ then

$$\varphi(gK_+) = \langle \varphi, \delta(g)\varphi\rangle / \langle \varphi, \varphi\rangle, \tag{1.2}$$

where $\langle\ ,\ \rangle$ is the scalar product.

The decomposition (1.1) shows that each δ occurs with multiplicity 1. Cartan also showed that if X_+ is a sphere the spaces V_δ are just the eigenspaces of the Laplacian on X_+ – hence his term *zonal spherical function* for the function φ.

Next we turn to the space X_0 which is just the Euclidean space $\boldsymbol{R}^n$. Here $K_0 = \boldsymbol{SO}(n)$ and G_0 is the semidirect product of $\boldsymbol{R}^n$ and $\boldsymbol{SO}(n)$. Let $(\ ,\ )$ denote the inner product on $\boldsymbol{R}^n$. For each $\lambda \neq 0$ we consider the Hilbert space $\mathscr{H}_\lambda$ of functions

$$f(x) = \int_{S^{n-1}} \exp\{i\lambda(x, \omega)\} F(\omega)\, d\omega, \qquad F \in L^2(\boldsymbol{S}^{n-1}),$$

the norm of f being (legitimately) defined as $(\int |F(\omega)|^2\, d\omega)^{1/2}$. If $\lambda \in \boldsymbol{R} - \{0\}$ the natural representation T_λ of G_0 on $\mathscr{H}_\lambda$ is unitary and (as follows from Mackey [**36**, §14]) irreducible. (One can actually prove irreducibility for each $\lambda \in \boldsymbol{C} - \{0\}$.) If we now write out the Fourier transform $f \to \tilde{f}$ in polar coordinates, we have, for $\lambda \geqq 0$, $\omega \in \boldsymbol{S}^{n-1}$,

$$\tilde{f}(\lambda\omega) = \int_{\boldsymbol{R}^n} f(x) \exp\{-i\lambda(x, \omega)\}\, dx, \qquad f \in L^1(\boldsymbol{R}^n), \tag{1.3}$$

and, for $f \in C_c^\infty(\boldsymbol{R}^n)$,

$$f(x) = (2\pi)^{-n} \int_{\boldsymbol{R}^+} \left(\int_{S^{n-1}} \exp\{i\lambda(x, \omega)\} \tilde{f}(\lambda\omega)\, d\omega \right) \lambda^{n-1}\, d\lambda. \tag{1.4}$$

This and the corresponding Plancherel formula give a continuous analog of (1.1),

$$L^2(X_0) = \int_{\lambda > 0} \mathscr{H}_\lambda\, d\lambda^*, \qquad T_{X_0} = \int_{\lambda > 0} T_\lambda\, d\lambda^*, \tag{1.5}$$

in the sense of direct integrals of Hilbert spaces. Here T_{X_0} is the natural representation of G_0 on $L^2(X_0)$ and $d\lambda^* = (2\pi)^{-n} \lambda^{n-1}\, d\lambda$. As for X_+ there is up to a constant factor a unique vector in $\mathscr{H}_\lambda$ invariant under $T_\lambda(K_0)$, namely $\int_{S^{n-1}} \exp\{i\lambda(x, \omega)\}\, d\omega$ and the analog of (1.2) still holds.

These results indicate the type of questions we will be discussing for X_-. However, many features will appear which would not be meaningful for X_+ or X_-. Except for occasional motivational digressions we will therefore deal exclusively with the space X_-.

2. Notation.

(a) *General notation.* We shall use the standard notation $\boldsymbol{Z}$, $\boldsymbol{R}$, $\boldsymbol{C}$ for the ring of integers, the field of real numbers and the field of complex numbers, respectively; $\boldsymbol{Z}^+$ is the set of nonnegative integers, $\boldsymbol{R}^+$ the set of nonnegative real numbers. If S is a set, T a subset and f a function on S, the restriction of f to T is denoted $f \mid T$. If S is a topological space $\mathrm{Cl}(T)$ denotes the closure of T in S. The space of continuous functions on S is denoted by $C(S)$, $C_c(S)$ the set of those of compact support. Composition of functions and operators will often be denoted by $\circ$.

(b) *Manifolds*. If M is a manifold (satisfying the second countability axiom) and $m \in M$, the tangent space to M at m is denoted M_m. Following Schwartz [**43**] we write $\mathcal{D}(M)$ for the space of complex-valued C^∞ functions on M of compact support, topologized by means of uniform convergence of functions along with their derivatives; $\mathcal{D}'(M)$ denotes the dual space of all distributions on M. The space $\mathcal{E}(M)$ denotes the space of all complex-valued C^∞ functions on M topologized in a similar way as $\mathcal{D}(M)$, and $\mathcal{E}(M)$ denotes the dual space of distributions on M of compact support. If V is a vector space over $\boldsymbol{R}$, $\mathcal{S}(V)$ denotes the space of rapidly decreasing functions on V (Schwartz [**43**]). Let τ be a diffeomorphism of M onto itself, and let $f \in \mathcal{E}(M)$, $T \in \mathcal{D}'(M)$ and D be a differential operator on M. We put

$$f^\tau(m) = f(\tau^{-1}(m)), \qquad m \in M,$$
$$T^\tau(f) = T(f^{\tau^{-1}}), \qquad f \in \mathcal{D}(M),$$
$$D^\tau(f) = (Df^{\tau^{-1}})^\tau, \qquad f \in \mathcal{E}(M).$$

Then $f^\tau \in \mathcal{E}(M)$, $T^\tau \in \mathcal{D}'(M)$ and D^τ is another differential operator on M. The value of Df at a point m will usually be denoted by $(Df)(m)$ but sometimes it is convenient to write $D_m(f(m))$. If Φ is a differentiable mapping from a manifold M into another manifold, $d\Phi_m$ (and sometimes $d\Phi$) denotes the differential of Φ at m.

(c) *Lie groups*. If A is a group and $a \in A$, $L(a)$ denotes the left translation $x \to ax$ and $R(a)$ denotes the right translation $x \to xa$ on A. If $B \subset A$ is a subset we write $B^a = aBa^{-1}$; if B is a subgroup, A/B denotes the set of left cosets aB, $a \in A$. The transformation $xB \to axB$ of A/B will always be denoted by $\tau(a)$.

Lie groups will be denoted by italic capital letters and their Lie algebras by corresponding lower case German letters. If G is a Lie group and $\mathfrak{g}$ its Lie algebra the adjoint representation of G is denoted by Ad (or Ad_G) and the adjoint representation of $\mathfrak{g}$ by ad (or $\mathrm{ad}_{\mathfrak{g}}$).

We shall now list some standard notation concerning semisimple Lie groups which will be utilized throughout these notes. Let G be a connected semisimple Lie group with finite center, $\mathfrak{g}$ the Lie algebra of G, and $\langle\ ,\ \rangle$ (sometimes B) the Killing form of $\mathfrak{g}$. Let θ be a *Cartan involution* of $\mathfrak{g}$, that is an involutive automorphism such that the form $(X, Y) \to -\langle X, \theta Y\rangle$ is strictly positive definite on $\mathfrak{g} \times \mathfrak{g}$. Let $\mathfrak{g} = \mathfrak{k} + \mathfrak{p}$ be the decomposition of $\mathfrak{g}$ into eigenspaces of θ (a *Cartan decomposition*) and K the analytic subgroup of G with Lie algebra $\mathfrak{k}$. Let $\mathfrak{a} \subset \mathfrak{p}$ be a maximal abelian subspace, $\mathfrak{a}^*$ its dual, $\mathfrak{a}_c^*$ the complexification of $\mathfrak{a}^*$, i.e., the space of $\boldsymbol{R}$-linear maps of $\mathfrak{a}$ into $\boldsymbol{C}$. Let $A = \exp \mathfrak{a}$ and log the inverse of the map $\exp: \mathfrak{a} \to A$. For $\lambda \in \mathfrak{a}^*$ put

$$\mathfrak{g}_\lambda = \{X \in \mathfrak{g} \mid [H, X] = \lambda(H)\, X, \text{ for all } H \in \mathfrak{a}\}.$$

If $\lambda \neq 0$ and $\mathfrak{g}_\lambda \neq \{0\}$ then λ is called a (*restricted*) *root* and $m_\lambda = \dim(\mathfrak{g}_\lambda)$ is called its *multiplicity*. Let $\mathfrak{g}_c$ denote the complexification of $\mathfrak{g}$ and if $\mathfrak{s}$ is any subspace of $\mathfrak{g}$ let $\mathfrak{s}_c$ denote the complex subspace of $\mathfrak{g}_c$ spanned by $\mathfrak{s}$. If λ, $\mu \in \mathfrak{a}_c^*$, let $H_\lambda \in \mathfrak{a}_c$ be determined by $\lambda(H) = \langle H_\lambda, H \rangle$ for $H \in \mathfrak{a}$ and put $\langle \lambda, \mu \rangle = \langle H_\lambda, H_\mu \rangle$. Since $\langle \ , \ \rangle$ is positive definite on $\mathfrak{p}$ we put $|\lambda| = \langle \lambda, \lambda \rangle^{1/2}$ for $\lambda \in \mathfrak{a}^*$ and $|X| = \langle X, X \rangle^{1/2}$ for $X \in \mathfrak{p}$. Let $\mathfrak{a}'$ be the open subset of $\mathfrak{a}$ where all restricted roots are $\neq 0$. The components of $\mathfrak{a}'$ are called *Weyl chambers*. Fix a Weyl chamber $\mathfrak{a}^+$ and call a (restricted) root α *positive* if it is positive on $\mathfrak{a}^+$. Let $\mathfrak{a}_+^*$ denote the corresponding Weyl chamber in $\mathfrak{a}^*$, that is the pre-image of $\mathfrak{a}^+$ under the mapping $\lambda \to H_\lambda$. Let Σ denote the set of restricted roots, Σ^+ the set of positive roots and Σ^- the set of negatives of the members in Σ^+. Let $\Sigma_0 = \{\alpha \in \Sigma \mid \frac{1}{2}\alpha \notin \Sigma\}$, and put $\Sigma_0^+ = \Sigma^+ \cap \Sigma_0$, $\Sigma_0^- = \Sigma^- \cap \Sigma_0$. A root $\alpha \in \Sigma^+$ is called *simple* if it is not a sum of two positive roots. The walls of the Weyl chamber $\mathfrak{a}^+$ lie on the hyperplanes $\alpha_1 = 0, \ldots, \alpha_l = 0$, $\alpha_1, \ldots, \alpha_l$ being the simple roots. Let ϱ denote half the sum of the positive roots with multiplicity, i.e., $\varrho = \frac{1}{2} \sum_{\alpha \in \Sigma^+} m_\alpha \alpha$. Let $\mathfrak{n} = \sum_{\alpha > 0} \mathfrak{g}_\alpha$, $\bar{\mathfrak{n}} = \theta \mathfrak{n}$ and let N and $\bar{N}$ denote the corresponding analytic subgroups of G. Let M denote the centralizer of A in K, M' the normalizer of A in K, W the (finite) factor group M'/M, the *Weyl group*. The group W acts as a group of linear transformations of $\mathfrak{a}$ and also on $\mathfrak{a}_c^*$ by $(s\lambda)(H) = \lambda(s^{-1}H)$ for $H \in \mathfrak{a}$, $\lambda \in \mathfrak{a}_c^*$ and $s \in W$. Let w denote the order of W, and let $m'_1, \ldots, m'_w$ be a complete set of representatives in M' (mod M). Let $A^+ = \exp \mathfrak{a}^+$, $B = K/M$, $P = MAN$; then we have the decompositions

(2.1) $$G = K \operatorname{Cl}(A^+) K \qquad \text{(Cartan decomposition)},$$

(2.2) $$G = KAN \qquad \text{(Iwasawa decomposition)},$$

(2.3) $$G = \bigcup_{i=1}^{w} Pm'_i P \qquad \text{(Bruhat decomposition)}.$$

Here (2.1) means that each $g \in G$ can be written $g = k_1 A(g) k_2$ where $k_1, k_2 \in K$ and $A(g) \in \operatorname{Cl}(A^+)$; here $A(g)$ is actually unique. In (2.2) each $g \in G$ can be uniquely written $g = k(g) \exp H(g)\, n(g)$, $k(g) \in K$, $H(g) \in \mathfrak{a}$, $n(g) \in N$. In (2.3) the union is a disjoint union; exactly one of the summands, namely Pm^*P (where $m^* \in M'$ satisfies $\operatorname{Ad}(m^*)\, \mathfrak{a}^+ = -\mathfrak{a}^+$) is open in G. Thus the set $\bar{N}MAN$ is open in G; if $g \in \bar{N}MAN$ is written $g = \bar{n}man$, $\bar{n} \in \bar{N}$ is uniquely determined by g and will be denoted $\bar{n} = \bar{n}(g)$. Let s^* denote the coset m^*M in W.

The number l which equals $\dim \mathfrak{a}$ is called the *real rank* of G and the *rank* of the symmetric space $X = G/K$.

(d) *Normalization of measures.* It is convenient to make some conventions concerning the normalization of certain invariant measures. Let $l = \dim \mathfrak{a}$. The Killing form induces Euclidean measures on A, $\mathfrak{a}$ and $\mathfrak{a}^*$; multiplying these by the factor $(2\pi)^{-(1/2)l}$ we obtain invariant measures da, dH and $d\lambda$, and the inversion formula for the Fourier transform

$$f^*(\lambda)=\int_A f(a)\exp\{-i\lambda(\log a)\}\,da, \qquad \lambda\in\mathfrak{a}^*, \tag{2.4}$$

holds without any multiplicative constant,

$$f(a)=\int_{\mathfrak{a}^*} f^*(\lambda)\exp\{i\lambda(\log a)\}\,d\lambda, \qquad f\in\mathscr{S}(A). \tag{2.5}$$

We normalize the Haar measures dk and dm on the compact groups K and M, respectively, such that the total measure is 1. The Haar measures of the nilpotent groups N, $\bar{N}$ are normalized such that

$$\theta(dn)=d\bar{n}, \qquad \int_{\bar{N}} \exp\{-2\varrho(H(\bar{n}))\}\,d\bar{n}=1. \tag{2.6}$$

The Haar measure dg on G can be normalized such that

$$\int_G f(g)\,dg=\int_{KAN} f(kan)\exp\{2\varrho(\log a)\}\,dk\,da\,dn, \qquad f\in\mathscr{D}(G). \tag{2.7}$$

Since each automorphism $n\to mnm^{-1}$ of N by an element $m\in M$ preserves the measure dn, the measure $dm\,dn$ is a bi-invariant measure $d(mn)$ on MN.

For the Cartan decomposition we have the following integral formula

$$\int_G f(g)\,dg=c\int_K\int_K\int_A f(k_1ak_2)\,\delta(a)\,dk_1\,da\,dk_2 \tag{2.8}$$

where

$$\delta(\exp H)=\prod_{\alpha\in\Sigma^+}(e^{\alpha(H)}-e^{-\alpha(H)})^{m_\alpha}, \qquad H\in\mathfrak{a}, \tag{2.9}$$

and c is a constant.

Suppose U is a locally compact group and P a closed subgroup, both unimodular. Suppose du and dp are Haar measures on U and P, respectively. Let du_P denote the U-invariant measure on U/P given by

$$\int_U f(u)\,du=\int_{U/P}\left(\int_P f(up)\,dp\right)du_P, \qquad f\in C_c(U). \tag{2.10}$$

In particular we have a K-invariant measure dk_M on $B=K/M$ of total measure one. For the coset spaces $X=G/K$, $B=K/M$ we write for simplicity

(2.11) $$dx = dg_K, \qquad db = dk_M.$$

If S is a locally compact space with a measure μ and $p \geqq 1$, then $L^p(S)$ denotes the set of measurable functions f such that $|f|^p$ is μ-integrable.

(e) *Representations.* Let Q be a locally compact group, V a locally convex topological vector space, $\mathrm{Aut}(V)$ the group of linear homeomorphisms of V onto itself. A *representation* π of Q on V is a homomorphism of Q into $\mathrm{Aut}(V)$ such that the map $(q, v) \to \pi(q)\, v$ of $Q \times V$ into V is continuous; π is called *irreducible* if the only closed subspaces of V which are invariant under each $q \in Q$ are $\{0\}$ and V. If π' is a representation of Q on another locally convex topological vector space V' then π and π' are said to be *equivalent* if there exists a linear homeomorphism A of V onto V' such that $A\pi(q) = \pi'(q)\, A$ for all $q \in Q$. The operator A is called an *intertwining operator.*

(f) *Differential operators.* If A is a Lie group $\boldsymbol{D}(A)$ denotes the algebra of all left invariant differential operators on A with complex coefficients. If $B \subset A$ is a closed subgroup, $\boldsymbol{D}(A/B)$ denotes the algebra of A-invariant differential operators on A/B. The notation being as in (c) let $\boldsymbol{D}_0(G)$ denote the set of $D \in \boldsymbol{D}(G)$ which are invariant under all right translations from K. There is a homomorphism μ of $\boldsymbol{D}_0(G)$ onto $\boldsymbol{D}(G/K)$ such that

$$(\mu(D)\, f) \circ \pi = D(f \circ \pi) \quad \text{for } D \in \boldsymbol{D}_0(G),\ f \in \mathscr{E}(X),$$

π denoting the natural mapping of G onto G/K.

3. Summary. With most of the notation already explained we can now give a brief summary of the content of this article.

In Chapter II we discuss the spherical functions on G and their relationship to the spherical representations of G. We state Harish-Chandra's parametrization of the set of spherical functions by means of the set $\mathfrak{a}_c^*/W$ and describe various properties of the spherical functions (boundedness, positivity, positive definiteness) in terms of this parametrization.

If f is an integrable function on G, bi-invariant under K, its spherical transform is obtained by integrating it against a spherical function. Since these functions f form a commutative Banach algebra under convolution, abstract functional analysis gives an inversion formula and a Plancherel formula for the spherical transform. Since the group $G = KAK$ is determined up to local isomorphism by the space $\mathfrak{a}$ and the set Σ of roots (with multiplicities), the Plancherel measure for the spherical transform should, in principle, be expressible in terms of these data. This is accomplished primarily by means of Harish-Chandra's detailed study of the asymptotic behavior of the spherical functions at infinity.

The Plancherel formula gives a characterization of the spherical transforms of the square-integrable functions f on G, bi-invariant under K. In §4 of Chapter II, the spherical transforms of several other function spaces on G are similarly characterized.

The Radon transform on the symmetric space $X=G/K$ is defined in Chapter III, §3, after some preliminaries in §§1–2. It has applications to the representation theory of G, because the principal series is most naturally defined on function spaces over the space Ξ of horocycles in X. Here we are more concerned, however, with the applications of the Radon transform to differential equations on X, which are based on the following circumstances: When a G-invariant differential operator D on G/K is applied to a K-invariant function it becomes a singular differential operator on A. The Radon transform transforms this singular operator into a differential operator with constant coefficients. This fact, combined with the inversion formula for the spherical transform, proves that D has a fundamental solution.

In Chapter IV the spherical transform on G is generalized to a Fourier transform of arbitrary functions on X. The corresponding inversion formula, Plancherel formula, and Paley-Wiener-Schwartz correspondence theorems are described. The analogy with Euclidean space goes quite far; in particular, X is in a certain sense self-dual under the Fourier transform.

In Chapter V, §1, we outline a proof of the surjectivity of each G-invariant differential operator D on X. The proof rests on the inversion formula for the Fourier transform on X, and on some identities for the intertwining operators of the principal series, but primarily on the Paley-Wiener theorem for the Radon transform on X, which in turn uses the series expansion for Eisenstein integrals stated in Chapter II, §5.

Finally, in Chapter V, §§2–3, we describe the principal results about harmonic functions on symmetric spaces, the mean-value characterization, Poisson integral formulas, and various Fatou-type limit theorems.

II. Spherical functions on G

1. Definition and parametrization of classes of spherical functions. Cartan's formula (2) in I, §1 makes sense for unitary representations in a Hilbert space, and in their early papers on the representations of the complex classical groups Gel'fand and Naĭmark investigated these functions (see e.g. [**16a**]). In Gel'fand [**13a**] they are considered independently of representation theory. Following this and Godement's paper [**18a**] a *spherical function* on G is defined as a continuous function $\phi\not\equiv 0$ satisfying the functional equation

$$\int_K \phi(g_1 k g_2)\, dk \equiv \phi(g_1)\,\phi(g_2). \tag{1.1}$$

If π is a unitary irreducible representation of G on a Hilbert space $\mathscr{H}$ such that

$\pi(K)$ has a fixed unit vector $v \in \mathscr{H}$, $\phi(g) = \langle v, \pi(g)\, v\rangle$ is a positive definite spherical function and all positive definite spherical functions are obtained in this way (Godement [**18c**], Gel'fand-Naĭmark [**16b**]). We give later a generalization to all spherical functions.

Harish-Chandra ([**19b**], [**19d**]) found the following solution to (1.1):

THEOREM 1.1. *The spherical functions on G are precisely the functions*

$$\phi_\lambda(g) = \int_K \exp\{(i\lambda - \varrho)\,(H(gk))\}\, dk, \tag{1.2}$$

where $\lambda \in \mathfrak{a}_c^*$; *moreover* $\phi_\lambda \equiv \phi_\mu$ *if and only if* $\lambda = s\mu$ *for some* $s \in W$.

The spherical functions can also be characterized as the eigenfunctions ϕ of each $D \in \boldsymbol{D}_0(G)$ satisfying $\phi(e) = 1$ and the bi-invariance $\phi(kgk') \equiv \phi(g)$ ([**13a**], [**18a**]). In terms of formula (1.2),

$$D\phi_\lambda = \gamma(D)\,(i\lambda)\,\phi_\lambda \tag{1.3}$$

where γ is a homomorphism of $\boldsymbol{D}_0(G)$ onto $I(\mathfrak{a}_c)$, the set of Weyl group invariants in the symmetric algebra $S(\mathfrak{a}_c)$ [**19d**]. In particular, if $\Omega \in \boldsymbol{D}(G)$ is the Casimir operator,

$$\Omega\phi_\lambda = (-\langle\lambda, \lambda\rangle - \langle\varrho, \varrho\rangle)\,\phi_\lambda. \tag{1.4}$$

Consider now the convolution algebra $L^\natural(G)$ of integrable functions on G, bi-invariant under K. As noted by Gel'fand [**13a**] the symmetry of the space G/K implies that $L^\natural(G)$ is a commutative Banach algebra. Moreover, it is semisimple [**20b**, p. 453]. The continuous homomorphisms of $L^\natural$ into $\boldsymbol{C}$ are precisely the maps

$$f \to \int_G f(g)\,\phi(g)\,dg,$$

where ϕ is a *bounded* spherical function on G. The following theorem (Helgason-Johnson [**21**]) therefore gives a parametrization of the maximal ideal space of $L^\natural(G)$ in the spirit of Theorem 1.1.

THEOREM 1.2. *Let* C_ϱ *denote the convex hull of the points* $s\varrho\,(s \in W)$ *in* $\mathfrak{a}^*$. *Then the spherical function* ϕ_λ *is bounded if and only if* $\lambda \in \mathfrak{a}^* + iC_\varrho$.

Although the positivity of the spherical function is not of significance for us we mention the following result contained in Furstenberg [**11b**, Theorem 5.1] or Karpelevič [**26**, Theorem 17.11.1].

THEOREM 1.3. *The spherical function* ϕ_λ *is everywhere positive if and only if* $i\lambda\in\mathfrak{a}^*$.

Because of the above-mentioned connection with unitary representations one would like to know precisely for which $\lambda\in\mathfrak{a}_c^*$ the function ϕ_λ is positive definite. This is always the case if $\lambda\in\mathfrak{a}^*$ but the precise answer seems to be known only for rank $X=1$ (cf. Takahashi [**46**, p. 346] for the hyperbolic spaces, Kostant [**31**, Theorem 10] in general).

THEOREM 1.4. *Assume* rank $X=1$ *and let* α *(and possibly* 2α*) denote the positive restricted roots.*

(a) *Assume* 2α *is not a root. Then* ϕ_λ *is positive definite if and only if either* (i) $\lambda\in\mathfrak{a}^*$; (ii) $i\lambda\in\mathfrak{a}^*$ *and* $|\langle i\lambda,\alpha\rangle|\leqq\langle\varrho,\alpha\rangle$.

(b) *Assume* 2α *is a root. Then apart from the constant* $\phi_{\pm i\varrho}\equiv 1$, ϕ_λ *is positive definite if and only if either* (i) $\lambda\in\mathfrak{a}^*$; (ii) $i\lambda\in\mathfrak{a}^*$ *and* $|\langle i\lambda,\alpha\rangle|\leqq(\frac{1}{2}m_\alpha+1)\langle\alpha,\alpha\rangle$.

For X of any rank we have $\phi_{-\lambda}(g)\equiv\phi_\lambda(g^{-1})$ [**19d**, p. 294] which if ϕ_λ is positive definite equals the complex conjugate $(\phi_\lambda(g))^-$, which in the rank one case equals $(\phi_{-\lambda}(g))^-$. Thus if ϕ_λ is positive definite, it is real-valued and reaches its maximum at $g=e$. Hence (1.4) implies

$$\langle\lambda,\lambda\rangle+\langle\varrho,\varrho\rangle\geqq 0, \tag{1.5}$$

i.e., $\lambda\in\mathfrak{a}^*$ or $\lambda\in i\mathfrak{a}^*$ and $|\langle i\lambda,\alpha\rangle|\leqq\langle\varrho,\alpha\rangle$. This method is used in Faraut-Harzallah [**9**] to give a proof of Theorem 1.4 for the cases $m_{2\alpha}\leqq 1$.

Because of the correspondence mentioned above between the positive definite spherical functions and certain unitary representations we call a representation T of G on a vector space V a *spherical representation* if there exists a vector $v\neq 0$ in V fixed under $T(K)$. Then the correspondence above can be extended to all spherical functions.

THEOREM 1.5. *Let* $\lambda\in\mathfrak{a}_c^*$ *and* $\mathscr{E}_{(\lambda)}$ *the closed subspace of* $\mathscr{E}(G)$ *generated by the left translates of* ϕ_λ *under* G, T_λ *the natural representation of* G *on* $\mathscr{E}_{(\lambda)}$, T_λ^* *the contragredient representation of* G *on the dual space* $\mathscr{E}_{(\lambda)}^*$ *(strong topology). Then* T_λ *and* T_λ^* *are both irreducible,* $T_\lambda(K)$ *and* $T_\lambda^*(K)$ *admit fixed vectors* ϕ_λ *and* δ_λ, *respectively (unique up to scalar multiples) and, with a suitable normalization of* δ_λ,

$$\phi_\lambda(g)=\langle T_\lambda(g^{-1})\,\phi_\lambda,\delta_\lambda\rangle.$$

On the other hand, if T is an irreducible, quasisimple, spherical representation of G on a complete semireflexive locally convex space V, then there exists a $\lambda \in \mathfrak{a}_c^*$ such that T is weakly equivalent to T_λ.

Here "quasisimple" means that the center of G is mapped into scalars by T and that the representation dT of the universal enveloping algebra $\boldsymbol{D}(G)$ on the space V^∞ of differentiable vectors in V maps the center of $\boldsymbol{D}(G)$ into scalars. "Weak equivalence" means algebraic equivalence of the representations restricted to appropriate dense invariant subspaces. A proof of this theorem is given in [**20f**, Chapter III, §5] or [**20g**, Chapter II, §2]. For a related discussion of spherical induced representations see [**31**] and [**45**].

2. The abstract Plancherel formula. Let $C^\natural$ denote the space of continuous complex-valued functions on G, compactly supported and bi-invariant under K and $\mathscr{P}$ the set of positive definite spherical functions on G. If $f \in C^\natural(G)$, the *spherical transform* is defined by

$$\tilde{f}(\phi) = \int_G f(g)\,(\phi(g))^-\,dg.$$

The set $\mathscr{P}$ is given the weakest topology for which all the transforms are continuous. In [**18c**], Godement proved the following result (cf. also a related one by Mautner [**38**] and Harish-Chandra [**19a**]).

THEOREM 2.1. *The space $\mathscr{P}$ is locally compact and there exists a unique positive measure μ on $\mathscr{P}$ such that*

$$\int_G |f(g)|^2\,dg = \int_{\mathscr{P}} |\tilde{f}(\phi)|^2\,d\mu(\phi), \qquad f \in C^\natural(G). \tag{2.1}$$

The map $f \to \tilde{f}$ extends to an isometry of $L^2(K\backslash G/K)$ onto $L^2(\mathscr{P}, \mu)$.

Here $L^2(K\backslash G/K)$ denotes the set of functions in $L^2(G)$ bi-invariant under K.

The proof holds in much greater generality; in fact, it suffices for G to be a locally compact unimodular group, K a compact subgroup such that $C^\natural(G)$ is commutative under convolution.

Since $\phi_{-\lambda}(g) = \phi_\lambda(g^{-1})$ which equals $(\phi_\lambda(g))^-$ in case ϕ_λ is positive definite, we extend the definition of the spherical transform by

$$\tilde{f}(\lambda) = \int_G f(g)\,\phi_{-\lambda}(g)\,dg \tag{2.2}$$

for all $\lambda \in \mathfrak{a}_c^*$ and all functions f on G bi-invariant under K, for which (2.2) is absolutely convergent. Since the structure of G/K is known in great detail, it is a natural problem to express $\mathscr{P}$ and μ in terms of the parameter λ. Since the subalgebra $\mathfrak{a}$, together with the root system Σ (including the multiplicities m_α) determine the space G/K, one would expect that $\mathscr{P}$ and μ could, at least in principle, be expressed in terms of these data. We shall later see that for the applications to differential equations the explicit expression of μ is quite essential.

3. Harish-Chandra's expansion and the *c*-function. For the determination of the Plancherel measure μ formula (1.2) is insufficient and we shall now describe Harish-Chandra's series expansion for ϕ_λ into exponentials which was devised for this purpose. For the proof see [**19d**, pp. 263–283]; an exposition with some minor simplifications is in [**20g**, pp. 25–29].

Because of the Cartan decomposition for G (Chapter I, §2), ϕ_λ is completely determined by its values on A^+. If $D \in \boldsymbol{D}_0(G)$ and $f \in \mathscr{E}(G)$ bi-invariant under K, the mapping $\bar{f} \to (Df)^-$ (bar denoting restriction to A^+) can be realized by a differential operator $\Delta(D)$ on A^+, the "radial part" of D. Thus $\bar{\phi}_\lambda$ satisfies the differential equations

$$\Delta(D)\, u = \gamma(D)\,(i\lambda)\, u, \qquad D \in \boldsymbol{D}_0(G), \tag{3.1}$$

on A^+. Now the polynomial $\gamma(D)$ gives the leading term in $\Delta(D)$, and since there exist elements $u_1, \ldots, u_w$ ($w =$ order of W) such that $S(\mathfrak{a}) = \sum_i u_i I(\mathfrak{a})$ it can be proved that (3.1) has at most w linearly independent solutions. Now if Ω is the Casimir operator its radial part can be computed to be

$$\Delta(\Omega) = L_A + \sum_{\alpha \in \Sigma^+} m_\alpha (\coth \alpha)\, H_\alpha, \tag{3.2}$$

where L_A is the Laplacian on A and the vector H_α is viewed as a first-order differential operator on A^+. Let L denote the set of all linear combinations $\sum_{i=1}^{l} n_i \alpha_i$ ($n_i \in \boldsymbol{Z}^+$) where the α_i are the simple restricted roots. Since, by (1.4),

$$\Delta(\Omega)\, \bar{\phi}_\lambda = (-\langle \lambda, \lambda \rangle - \langle \varrho, \varrho \rangle)\, \bar{\phi}_\lambda, \tag{3.3}$$

we look for a solution Φ_λ of the equation

$$\Delta(\Omega)\, \Phi_\lambda = (-\langle \lambda, \lambda \rangle - \langle \varrho, \varrho \rangle)\, \Phi_\lambda \tag{3.4}$$

of the form

$$\Phi_\lambda(\exp H) = \sum_{\mu \in L} \Gamma_\mu(\lambda) \exp\{(i\lambda - \varrho - \mu)(H)\}, \qquad H \in \mathfrak{a}^+. \tag{3.5}$$

Using the expansion

$$\coth \alpha = 1 + 2 \sum_{k \geq 1} \exp\{-2k\alpha\}$$

the expression (3.2) gives a recursion formula for $\Gamma_\mu(\lambda)$, namely,

$$(3.6) \qquad \{\langle \mu, \mu \rangle - 2i\mu\} \Gamma_\mu = 2 \sum_{\alpha \in \Sigma^+} m_\alpha \sum_{k \geq 1} \Gamma_{\mu - 2k\alpha} \{\langle \mu + \varrho - 2k\alpha, \alpha \rangle - i\alpha\},$$

where k runs over the integers ≥ 1 for which $\mu - 2k\alpha \in L$. Putting $\Gamma_0 \equiv 1$, (3.6) defines Γ_μ recursively as a rational function on $\mathfrak{a}_c^*$. Rough estimates via (3.6) insure the convergence of (3.5) for all λ for which the denominators of all the $\Gamma_\mu(\lambda)$ are $\neq 0$.

Now it turns out that the solution Φ_λ to (3.4) is actually a solution of the entire system (3.1). In [**19d**, p. 273], Harish-Chandra derived this from the fact that this had to be so for those λ which correspond to the *finite-dimensional* spherical representations of G because then the finite-dimensional weight theory gives an expansion of ϕ_λ. That Φ_λ satisfies (3.1) can also be derived from the fact that each $\Delta(D)$ commutes with $\Delta(\Omega)$ and its coefficients have a rather special structure [**20g**, Chapter II, §1].

It is clear that, for each $s \in W$, $\Phi_{s\lambda}$ gives another solution of (3.1). For a generic λ these solutions are clearly linearly independent so by the statement about the solution space to (3.1), ϕ_λ is a linear combination

$$\phi_\lambda(a) = \sum_{s \in W} \boldsymbol{c}^s(\lambda)\, \Phi_{s\lambda}(a), \qquad a \in A^+,$$

for all such λ. Replacing λ here by $\sigma\lambda$ ($\sigma \in W$) we see that $\boldsymbol{c}^s(\lambda) = \boldsymbol{c}(s\lambda)$ where $\boldsymbol{c} = \boldsymbol{c}^e$. This then gives the following central result which marks the entrance of the remarkable $\boldsymbol{c}$-function.

THEOREM 3.1. *Suppose $\lambda \in \mathfrak{a}_c^*$ is such that $i(s\lambda - \sigma\lambda) \notin L$ for $s \neq \sigma$ in W and that $\langle \mu, \mu \rangle \neq 2i \langle \mu, s\lambda \rangle$ for all $\mu \in L - \{0\}$, $s \in W$. Then*

$$(3.7) \quad \phi_\lambda(\exp H) = \sum_{s \in W} \boldsymbol{c}(s\lambda) \exp\{(is\lambda - \varrho)(H)\} \sum_{\mu \in L} \Gamma_\mu(s\lambda) \exp\{-\mu(H)\} \quad (H \in \mathfrak{a}^+),$$

where Γ_μ is given by (3.6), *and* $\Gamma_0 \equiv 1$.

Let us consider the case rank $G/K = 1$. Then (3.3) is an ordinary differential equation which by a change of variables can be transformed into the hypergeometric equation. Using classical information about the behavior of the hypergeometric function at ∞ one finds [**19d**, p. 303] $\boldsymbol{c}(\lambda) = I(i\lambda)/I(\varrho)$, where

$$I(\nu)=\prod_{\alpha\in\Sigma^+} B(\tfrac{1}{2}m_\alpha, \tfrac{1}{4}m_{2\alpha}+\langle\nu, \alpha_0\rangle), \tag{3.8}$$

and B denotes the beta function and $\alpha_0=\alpha/\langle\alpha, \alpha\rangle$. The spectral theory of ordinary second-order differential operators (which motivated Harish-Chandra's work) shows because of (3.7) that the measure $d\mu$ in (2.1), which is the "spectral function" of the operator $\Delta(\Omega)$, is given by $|\boldsymbol{c}(\lambda)|^{-2}\, d\lambda$. (The details of this can be found in Flensted-Jensen [**10**] where the rank one case of the spherical transform is generalized such that the multiplicities m_α are allowed to be arbitrary positive real numbers.) Remarkably enough this is true even for rank $G/K>1$ and Harish-Chandra proved that $d\mu$ is supported on the subset $\mathfrak{a}^*/W$ and is given there by $|\boldsymbol{c}(\lambda)|^{-2}\, d\lambda$.

Considering now the case of arbitrary rank we want to combine Theorems 1.1 and 3.1 to get an integral formula for $\boldsymbol{c}(\lambda)$. The mapping $\bar{n}\to k(\bar{n})\, M$ is a diffeomorphism of $\bar{N}$ onto an open subset of K/M whose complement has lower dimension. The invariant measures are related by

$$\int_{\bar{N}} f(k(\bar{n})\, M) \exp\{-2\varrho(H(\bar{n}))\}\, d\bar{n}=\int_{K/M} f(kM)\, dk_M, \qquad f\in\mathscr{D}(K/M) \tag{3.9}$$

[**19d**, p. 287]. Since the integrand in (1.2) is right invariant under M, (1.2) and (3.9) give an expression for $\phi_\lambda(a)$ as an integral over $\bar{N}$. Combining this with (3.7) we obtain by letting $a\to\infty$ in A^+, provided $-H_\lambda\in\mathfrak{a}+i\mathfrak{a}^+$,

$$\boldsymbol{c}(\lambda)=\int_{\bar{N}} \exp\{-(i\lambda+\varrho)\,(H(\bar{n}))\}\, d\bar{n}. \tag{3.10}$$

This integral was evaluated by Gindikin-Karpelevič [**17**] as follows: Let $\alpha\in\Sigma_0^+$ and let $\mathfrak{g}^\alpha$ be the subalgebra of $\mathfrak{g}$ generated by the root spaces $\mathfrak{g}_\alpha$ and $\mathfrak{g}_{-\alpha}$. Then $\mathfrak{g}^\alpha$ is semisimple and has a Cartan decomposition

$$\mathfrak{g}^\alpha=\mathfrak{k}^\alpha+\mathfrak{p}^\alpha, \qquad \mathfrak{k}^\alpha=\mathfrak{g}^\alpha\cap\mathfrak{k}, \qquad \mathfrak{p}^\alpha=\mathfrak{g}^\alpha\cap\mathfrak{p}.$$

Let G^α and K^α denote the analytic subgroups of G corresponding to $\mathfrak{g}^\alpha$ and $\mathfrak{k}^\alpha$, respectively. The subspace $\mathfrak{a}^\alpha=\boldsymbol{R}H_\alpha$ is a maximal abelian subspace of $\mathfrak{p}^\alpha$; and if $A^\alpha=\exp\mathfrak{a}^\alpha$, $N^\alpha=G^\alpha\cap N$ then $G^\alpha=K^\alpha A^\alpha N^\alpha$ is an Iwasawa decomposition. The $\boldsymbol{c}$-function for G^α, denoted $\boldsymbol{c}^\alpha$, is given by an integral of the form (3.10) over the group $\bar{N}^\alpha=\bar{N}\cap G^\alpha$. The main result of [**17**] is

$$\boldsymbol{c}(\lambda)=a\prod_{\alpha\in\Sigma_0^+} \boldsymbol{c}^\alpha(\lambda^\alpha) \qquad (a=\text{constant}) \tag{3.11}$$

where λ^α is the restriction of λ to $\mathfrak{a}^\alpha$. Thus (3.8) holds for any rank. This work of

Gindikin-Karpelevič was motivated by Bhanu-Murthy ([**2a**], [**2b**]) where the $\boldsymbol{c}$-function, and even the vector $H(\bar{n})$, is explicitly determined for the cases $G=\boldsymbol{SL}(n, \boldsymbol{R})$, $G=\boldsymbol{Sp}(n, \boldsymbol{R})$.

4. The Plancherel measure and correspondence theorems for the spherical transform. Let $\mathscr{I}_c(G)$ be the set of functions in $\mathscr{D}(G)$ bi-invariant under K. We shall now define analogs for bi-invariant functions on G of the Schwartz space $\mathscr{S}(\boldsymbol{R})$. If $g\in G$ is written $g=k_1ak_2$ we put $|g|=|\log a|$ where $|\ |$ is the norm on A given by the Killing form. The spherical function ϕ_0 satisfies the inequalities

$$\exp\{-\varrho(\log a)\}\leqq\phi_0(a)\leqq c\exp\{-\varrho(\log a)\}\,(1+|\log a|^d),\qquad a\in A^+, \tag{4.1}$$

where c and d are positive constants [**19d**, p. 279]. For $p\geqq 0$ let $\mathscr{I}^p(G)$ denote the set of all $F\in\mathscr{E}(G)$ satisfying the following two conditions: (a) F is bi-invariant under K; (b) for each $D\in\boldsymbol{D}(G)$, $q\in\boldsymbol{Z}^+$,

$$\sup_g |(1+|g|)^q\,\phi_0(g)^{-2/p}(DF)(g)|<\infty. \tag{4.2}$$

Because of (4.1), $\mathscr{I}^p(G)\subset L^p(G)$.

The goal is now to prove the following result:

THEOREM 4.1. *For each $\lambda\in\mathfrak{a}^*$, $f\in\mathscr{I}^2(G)$, the integral*

$$\tilde{f}(\lambda)=\int_G f(g)\,\phi_{-\lambda}(g)\,dg \tag{4.3}$$

converges absolutely and

$$f(g)=w^{-1}\int_{\mathfrak{a}^*}\tilde{f}(\lambda)\,\phi_\lambda(g)\,|\boldsymbol{c}(\lambda)|^{-2}\,d\lambda, \tag{4.4}$$

$$\int_G |f(g)|^2\,dg=w^{-1}\int_{\mathfrak{a}^*}|\tilde{f}(\lambda)|^2\,|\boldsymbol{c}(\lambda)|^{-2}\,d\lambda. \tag{4.5}$$

While the expansion in Theorem 3.1 gives useful information about the behavior of ϕ_λ in the open Weyl chamber the determination of the Plancherel measure requires precise estimates also at the singular points.

THEOREM 4.2. *Let $D\in\boldsymbol{D}(G)$. Then there exist a $k\in\boldsymbol{Z}^+$ and a constant C such that*

$$(4.6) \qquad |(1+|\lambda|)^{-k}\pi(\lambda)\,(D\phi_\lambda)\,(a)\exp\{\varrho(\log a)\}| \leqq C$$

for $\lambda \in \mathfrak{a}^*$, $a \in A^+$. *Here* π *is the product of the positive roots.*

In [**19d**, p. 583], Harish-Chandra proves this estimate by induction on $\dim G$. For this, fix a point $a \in \mathrm{Cl}(A^+) - A^+$ and let G_1 be the connected semisimple part of the centralizer of a in G. Then G_1 has a Cartan decomposition $G_1 = K_1\,\mathrm{Cl}(A_1^+)\,K_1$ compatible with that of G. Harish-Chandra now obtains a precise relationship between certain limit values of ϕ_λ and its derivatives on the one hand and the spherical functions on G_1 on the other hand [**19d**, p. 573]. This leads to estimates of essentially the difference between the two spherical functions which then can be used to prove (4.6) by induction. We refer to the first part of [**19d**, second paper] for details.

The strategy for Theorem 4.1 is to define f by (4.4) and prove that it has the right Fourier transform. Let $\mathscr{S}(\mathfrak{a}^*)$ be the Schwartz space on $\mathfrak{a}^*$ and $\mathscr{I}(\mathfrak{a}^*)$ the space of W-invariants in $\mathscr{S}(\mathfrak{a}^*)$ with the induced topology. The continuous linear functionals on $\mathscr{I}(\mathfrak{a}^*)$ can be identified with the W-invariant tempered distributions on $\mathfrak{a}^*$.

THEOREM 4.3. *Let* $\mu \in \mathfrak{a}^*$. *Then the mapping*

$$(4.7) \qquad S_\mu : b \to \int_G \phi_{-\mu}(g) \left(\int_{\mathfrak{a}^*} b(\lambda)\,\phi_\lambda(g)\,|\mathbf{c}(\lambda)|^{-2}\,d\lambda \right) dg, \qquad b \in \mathscr{I}(\mathfrak{a}^*),$$

is a W-invariant tempered distribution on $\mathfrak{a}^*$.

The existence of the integral over $\mathfrak{a}^*$ follows easily from (3.8) and (3.11) and then the method of proof of Theorem 4.2 can be used to show that this integral gives a function in $\mathscr{I}^2(G)$. For the integral over G, the estimate (4.1) is also needed.

Since the repeated integral in (4.7) is the inverse spherical transform followed by the spherical transform itself, we want to prove $S_\mu(b) = wb(\mu)$. To begin with we prove

LEMMA 4.4. $pS_\mu = p(\mu)\,S_\mu$ *for each W-invariant polynomial p on* $\mathfrak{a}^*$.

By (1.3) there exists a $D \in \boldsymbol{D}_0(G)$ such that $D\phi_\lambda = p(\lambda)\,\phi_\lambda$. Then

$$pS_\mu(b) = S_\mu(bp) = \int_G \phi_{-\mu}(g) \left(\int_{\mathfrak{a}^*} b(\lambda)\,p(\lambda)\,\phi_\lambda(g)\,|\mathbf{c}(\lambda)|^{-2}\,d\lambda \right) dg.$$

Here we replace $p(\lambda)\,\phi_\lambda(g)$ by $(D\phi_\lambda)\,(g)$ and carry D over on $\phi_{-\mu} = \mathrm{conj}\,\phi_\mu$ by replacing it with its adjoint. The result is found to be $p(\mu)\,S_\mu(b)$ as desired.

From this lemma one can now conclude

(4.8) $$S_\mu(b)=\gamma(\mu)\, b(\mu), \qquad \mu\in\mathfrak{a}^*,\ b\in\mathscr{I}(\mathfrak{a}^*),$$

where γ is a function on $\mathfrak{a}^*$. For this fix $\mu\in\mathfrak{a}^*$ and suppose v is not on the orbit $W\cdot\mu$. Select a W-invariant polynomial p_0 such that $p_0(v)\neq p_0(\mu)$ and put $p=p_0-p_0(\mu)$. Then $pS_\mu=0$ so since $p(v)\neq 0$, v is not in the support of S_μ. Hence $\operatorname{supp} S_\mu\subset W\cdot\mu$ so, by the general distribution theory, S_μ is the sum of derivatives of the δ-functions at the points $s\mu$ $(s\in W)$. But then Lemma 4.4 implies that only zero order derivatives can occur, proving (4.8).

Now γ would be known if we could compute the integrals (4.7) for just one b. We wish to prove that γ actually is a constant. For $f\in\mathscr{I}^2(G)$ consider the function

(4.9) $$F_f(a)=\exp\{\varrho(\log a)\}\int_{\bar N} f(\bar n a)\, d\bar n, \qquad a\in A$$

(the integral actually converges absolutely). From (2.7) in Chapter I, it is readily seen that, at least formally,

(4.10) $$\tilde f(\lambda)=\int_G f(g)\,\phi_{-\lambda}(g)\,dg=\int_A F_f(a)\exp\{-i\lambda(\log a)\}\, da,$$

but both integrals are actually absolutely convergent and $\tilde f\in\mathscr{I}(\mathfrak{a}^*)$. We use this on the inner integral $\phi_b(g)$ in (4.7), and combine with the Euclidean inversion formula on A. Then we obtain

$$F_{\phi_b}(a)=\exp\{\varrho(\log a)\}\int_{\bar N}\phi_b(\bar n a)\,d\bar n=\int_{\mathfrak{a}^*}\tilde\phi_b(\lambda)\exp\{i\lambda(\log a)\}\,d\lambda$$

$$=\int_{\mathfrak{a}^*}S_\lambda(b)\exp\{i\lambda(\log a)\}\,d\lambda=\int_{\mathfrak{a}^*}\gamma(\lambda)\,b(\lambda)\exp\{i\lambda(\log a)\}\,d\lambda$$

$$=w^{-1}\int_{\mathfrak{a}^*}\gamma(\lambda)\,b(\lambda)\sum_{s\in W}\exp\{is\lambda(\log a)\}\,d\lambda.$$

The desired relation $\gamma(\lambda)\equiv w$ would therefore result from the following statement: The relation

(4.11) $$|c(\lambda)|^{-2}\exp\{\varrho(\log a)\}\int_{\bar N}\phi_\lambda(\bar n a)\,d\bar n=\sum_{s\in W}\exp\{is\lambda(\log a)\}$$

holds weakly in λ, that is, it gives the right result when integrated against any $b \in \mathscr{I}(\mathfrak{a}^*)$. Harish-Chandra's proof of this in [**19d**, §15] is too long to summarize here. Instead we give a vague heuristic argument, assuming that the integral in (4.11) converges (which actually is not the case). The function

$$g \to \int_{\bar{N}} \phi_\lambda(\bar{n}g)\, d\bar{n}, \qquad g \in G,$$

is an eigenfunction of each $D \in \boldsymbol{D}_0(G)$ right invariant under K, left invariant under $\bar{N}$. Viewing it as a function on A (since $G = \bar{N}AK$) the operators D give certain constant coefficient differential operators on A. The resulting differential equations on A can be solved explicitly and one obtains, if $\lambda \in \mathfrak{a}^*$ is regular,

$$\exp\{\varrho(\log a)\} \int_{\bar{N}} \phi_\lambda(\bar{n}a)\, d\bar{n} = \sum_{s \in W} a_s(\lambda) \exp\{is\lambda(\log a)\}, \tag{4.12}$$

where $a_s(\lambda) \in \boldsymbol{C}$. Since the left-hand side is W-invariant in λ, $a_s(\lambda)$ is independent of s. To see that it equals $|\boldsymbol{c}(\lambda)|^{-2}$ we use the expansion in Theorem 3.1 which implies that

$$\exp\{\varrho(\log a)\}\, \phi_\lambda(a) - \sum_{s \in W} \boldsymbol{c}(s\lambda) \exp\{is\lambda(\log a)\} \to 0 \tag{4.13}$$

as $a \to \infty$ in A^+. Now writing $\bar{n}a = k_1 a' k_2$ ($k_1, k_2 \in k$, $a, a' \in \mathrm{Cl}(A^+)$) one has the fairly elementary result [**19d**, p. 604] that

$$|\log a' - \log a - H(\bar{n})| \to 0$$

as $a \to \infty$ in A^+. Combining this with (4.13) we see that the left-hand side of (4.12) behaves for large a in A^+ as

$$\exp\{\varrho(\log a)\} \int_{\bar{N}} \exp\{-\varrho(\log a + H(\bar{n}))\} \sum_s \boldsymbol{c}(s\lambda) \exp\{is\lambda(\log a + H(\bar{n}))\}\, d\bar{n},$$

which, due to (3.10) and the relation $\boldsymbol{c}(s\lambda)\, \boldsymbol{c}(-s\lambda) = |\boldsymbol{c}(\lambda)|^2$ ($\lambda \in \mathfrak{a}^*$), reduces to

$$\sum_s |\boldsymbol{c}(\lambda)|^2 \exp\{is\lambda(\log a)\}.$$

This completes the heuristic justification of (4.11). This would prove $\gamma(\lambda) \equiv w$ so

$$b(\mu) = \int_G \phi_{-\mu}(g) \left(w^{-1} \int_{\mathfrak{a}^*} b(\lambda)\, \phi_\lambda(g)\, |\boldsymbol{c}(\lambda)|^{-2}\, d\lambda \right) dg, \qquad b \in \mathscr{I}(\mathfrak{a}^*). \tag{4.14}$$

In order for this to imply Theorem 4.1 it still remains to prove that each $f\in\mathscr{I}_c(G)$ can be written in the form

$$f(g)=\int_{\mathfrak{a}^*} b(\lambda)\,\phi_\lambda(g)\,|\mathbf{c}(\lambda)|^{-2}\,d\lambda,\qquad b\in\mathscr{I}(\mathfrak{a}^*), \tag{4.15}$$

because then (4.14) implies (4.4) for $f\in\mathscr{I}_c(G)$ and, by a density argument, for $f\in\mathscr{I}^2(G)$.

This was reduced to two conjectures in [**19d**, p. 612]. The first (a property of the c-function) turned out [**20j**, p. 506] to be a simple consequence of the product formula (3.11) and the second (the injectivity of $f\to F_f$ on $\mathscr{I}^2(G)$) was proved by Harish-Chandra in [**19g**, §21], as a by-product of his work on the discrete series. We shall now indicate a simpler proof of (4.15) in a sharper form, namely an intrinsic characterization of $\mathscr{I}_c(G)^{\sim}$. We call a holomorphic function F on $\mathfrak{a}_c^*$ a *rapidly decreasing holomorphic function of exponential type* if there exists a constant $R\geqq 0$ such that, for each $N\in\mathbf{Z}^+$,

$$\sup_{\xi,\eta\in\mathfrak{a}^*}\exp\{-R|\eta|\}\,(1+|\xi+i\eta|)^N\,|F(\xi+i\eta)|<\infty. \tag{4.16}$$

Here $|\xi+i\eta|=(|\xi|^2+|\eta|^2)^{1/2}$.

THEOREM 4.5. *The spherical transform $f\to\tilde f$ is a bijection of $\mathscr{I}_c(G)$ onto the space of W-invariant rapidly decreasing holomorphic functions on $\mathfrak{a}_c^*$ of exponential type.*

If $f\in\mathscr{I}_c(G)$ then F_f in (4.9) is easily seen to have compact support so by (4.10) and the Paley-Wiener theorem on A, $\tilde f$ satisfies an inequality (4.15). On the other hand, suppose F is W-invariant on $\mathfrak{a}_c^*$ and satisfies (4.15). Defining f in G by

$$f(g)=w^{-1}\int_{\mathfrak{a}^*} F(\lambda)\,\phi_\lambda(g)\,|\mathbf{c}(\lambda)|^{-2}\,d\lambda$$

it suffices, by (4.14) and the fact that the spherical transform is one-to-one on $\mathscr{I}_c(G)$, to prove that

$$f(\exp H)=0\quad\text{if } H\in\mathfrak{a}^+,\ |H|>R. \tag{4.17}$$

For this we use the expansion in Theorem 3.1, which we write in the form

$$\phi_\lambda(\exp H)=\sum_{\mu\in L}\psi_\mu(\lambda,H),\qquad H\in\mathfrak{a}^+, \tag{4.18}$$

where

$$\psi_\mu(\lambda, H) = \sum_{s \in W} c(s\lambda)\, \Gamma_\mu(s\lambda) \exp\{(is\lambda - \varrho - \mu)(H)\}.$$

We first prove an analog of (4.17) for each term in the expansion.

LEMMA 4.6. *Let $\mu \in L$. Then*

$$\int_{\mathfrak{a}^*} F(\lambda)\, \psi_\mu(\lambda, H)\, |c(\lambda)|^{-2}\, d\lambda = 0$$

for $H \in \mathfrak{a}^+$ and $|H| > R$.

We sketch the proof. For $\lambda \in \mathfrak{a}^*$ we have $|c(\lambda)|^2 = c(\lambda)\, c(-\lambda) = c(s\lambda)\, c(-s\lambda)$ $(s \in W)$. Since F is also W-invariant, it suffices to prove

$$\int_{\mathfrak{a}^*} F(-\xi)\, c(\xi)^{-1} \exp\{-i\xi(H)\}\, \Gamma_\mu(-\xi)\, d\xi = 0 \tag{4.19}$$

provided $H \in \mathfrak{a}^+$, $|H| > R$. To do this we shift the integration in (4.19) into the complexification $\mathfrak{a}_c^*$ in such a way that the singularities of the functions $\lambda \to c(\lambda)^{-1}$ and $\lambda \to \Gamma_\mu(-\lambda)$ are simultaneously avoided, Cauchy's theorem is applicable, and such that effective estimates can be obtained for the new integrand.

The singularities of $c(\lambda)^{-1}$ can be read off from the Gindikin-Karpelevič formula (3.8), the poles of the Γ function being at the nonpositive integers. The singularities of $\Gamma_\mu(-\lambda)$ can be read off from the recursion formula (3.6). It is therefore easy to see that the singularities of both functions are contained in the set

$$\{\lambda \in \mathfrak{a}_c^* \mid i\langle \nu, \lambda \rangle < 0 \text{ for some } \nu \in L\}.$$

Thus the function

$$F(-\lambda)\, c(\lambda)^{-1} \exp\{-i\lambda(H)\}\, \Gamma_\mu(-\lambda) \tag{4.20}$$

of the variable $\lambda = \xi + i\eta$ $(\xi, \eta \in \mathfrak{a}^*)$ is holomorphic in the tube

$$\{\xi + i\eta \mid \xi, \eta \in \mathfrak{a}^*, -H_\eta \in \mathfrak{a}^+\}.$$

Well-known estimates for the gamma function show that the behavior of the function (4.20) at ∞ is good enough for the Cauchy theorem to be applicable and give

$$\int_{\mathfrak{a}^*} F(-\xi)\, c(\xi)^{-1} \exp\{-i\xi(H)\}\, \Gamma_\mu(-\xi)\, d\xi$$
$$= \int_{\mathfrak{a}^*} F(-\xi-i\eta)\, c(\xi+i\eta)^{-1} \exp\{-i\xi(H)+\eta(H)\}\, \Gamma_\mu(-\xi-i\eta)\, d\xi$$

provided $-H_\eta \in \mathfrak{a}^+$. But this last integral, say $Q(H)$, is easily estimated by

$$|Q(H)| \leqq C \exp\{R|\eta|\} \exp\{\eta(H)\} \qquad \text{for all } H,\ -H_\eta \in \mathfrak{a}^+,$$

C being a constant. Put $H_\eta = -tH$ $(t>0)$. Then

$$|Q(H)| \leqq C \exp\{t|H|\,(R-|H|)\},$$

so, letting $t \to +\infty$, (4.19) follows. In order to conclude (4.17) we substitute the series expansion (4.18) for ϕ_λ into the integral defining $f(g)$. Integrating term-by-term, we would conclude, from the lemma,

$$f(\exp H) = \int_{\mathfrak{a}^*} F(\lambda)\, \phi_\lambda(\exp H)\, |c(\lambda)|^{-2}\, d\lambda$$
$$= \sum_{\mu \in L} \int_{\mathfrak{a}^*} F(\lambda)\, \psi_\mu(\lambda, H)\, |c(\lambda)|^{-2}\, d\lambda = 0$$

if $H \in \mathfrak{a}^+$, $|H| > R$. The justification of the interchange of summation and integration requires improved estimates of the coefficients $\Gamma_\mu(\lambda)$. The recursion formula (3.6) seems too unwieldy for this purpose but if instead of formula (3.2) for the radial part $\Delta(\Omega)$ we use the equivalent formula

$$\Delta(\Omega) = \delta^{-1/2} L_A \circ \delta^{1/2} - \delta^{-1/2} L_A(\delta^{1/2})$$

(cf. (2.9) in the Introduction) the expansion for $\delta^{1/2}(a)\, \phi_\lambda(a)$ has its coefficients given by a simpler recursion formula and can be readily estimated. The simple connection between the two expansions then gives the required estimates for $\Gamma_\mu(\lambda)$.

Gangolli's paper [**12a**] gave the justification of this interchange (cf. also [**20f**, p. 37]); the rest of the proof of Theorem 4.5 is in [**20e**].

The outlined proof of Theorem 4.1 clearly gives the following result as a by-product.

THEOREM 4.7. *The spherical transform $f \to \tilde{f}$ is a bijection of $\mathscr{I}^2(G)$ onto $\mathscr{I}(\mathfrak{a}^*)$.*

By a refinement of the induction technique used in the proof of Theorem 4.2, Trombi and Varadarajan [**47**] extended Theorem 4.7 to a characterization of $\mathscr{I}^p(G)^\sim$ as the space of W-invariant, rapidly decreasing holomorphic functions in a certain tube $\mathfrak{a}^* + i\mathfrak{a}_p^*$ in $\mathfrak{a}_c^*$ $(0<p<2)$. The width of the tube varies with p as $2/p-1$. For $p=1$ the tube is the tube $\mathfrak{a}^*+iC_\varrho$ of boundedness for the spherical function (Theorem 1.2). Here the result had been given in [**8a**, p. 417] for $G=\boldsymbol{SL}(2, \boldsymbol{R})$ and in [**20f**, p. 28] for G complex or of real rank one.

5. Eisenstein integrals. Harish-Chandra has in [**19f**] (cf. [**19e**]) extended the expansion in Theorem 3.1 to "Eisenstein integrals" which are functions on G generalizing the integral in Theorem 1.1. See [**51**, Chapter 9] for the proof.

Let V be a finite-dimensional vector space over $\boldsymbol{C}$ and $\sigma=(\sigma_1, \sigma_2)$ a double unitary representation of K on V, σ_1 operating on the left, σ_2 on the right. Let $\lambda \in \mathfrak{a}_c^*$ and consider the function

$$\phi(g:v)=\int_K \sigma_1(k(gk))\, v\sigma_2(k^{-1}) \exp\{(i\lambda-\varrho)(H(gk))\}\, dk \tag{5.1}$$

for $g\in G$, $v\in V$. This generalization of (1.2) satisfies $\phi(k_1gk_2:v)\equiv\sigma_1(k_1)\,\phi(g:v)\,\sigma_2(k_2)$, and we now state the generalization of Theorem 3.1.

THEOREM 5.1. *Let*

$$V_\sigma^M=\{v\in V \mid \sigma_1(m)\, v=v\sigma_2(m) \text{ for } m\in M\}.$$

There exist certain meromorphic functions $\boldsymbol{C}_s$ $(s\in W)$ on $\mathfrak{a}_c^*$ and rational functions Γ_μ $(\mu\in L)$ on $\mathfrak{a}_c^*$ all with values in $\operatorname{Hom}(V_\sigma^M, V_\sigma^M)$ *such that, for $a\in A^+$ and $v\in V_\sigma^M$,*

$$\exp\{\varrho(\log a)\}\int_K \sigma_1(k(ak))\, v\sigma_2(k^{-1}) \exp\{(i\lambda-\varrho)(H(ak))\}\, dk = \sum_{s\in W} \Phi(s\lambda:a)\, \boldsymbol{C}_s(\lambda)\, v, \tag{5.2}$$

where

$$\Phi(\lambda:a)=\exp\{i\lambda(\log a)\}\sum_{\mu\in L}\Gamma_\mu(\lambda)\exp\{-\mu(\log a)\}. \tag{5.3}$$

Here λ varies in a certain open dense subset ${}^*\mathfrak{a}_c'$ of $\mathfrak{a}_c^*$, and the functions Γ_μ are given by certain recursion formulas, depending on σ (cf. (5.4) below for $\sigma_2=1$).

We shall need this theorem just for the case when σ_2 is the identity representation. Therefore, in order to avoid unnecessary notation, we state the recursion formulas and describe the set $^*\mathfrak{a}_c'$ only for this case.

Extend $\mathfrak{a}_c$ to a Cartan subalgebra $\mathfrak{h}_c$ of $\mathfrak{g}_c$ and let P_+ be the set of roots β of $\mathfrak{g}_c$ with respect to $\mathfrak{h}_c$ such that the restriction $\bar{\beta}=\beta \mid \mathfrak{a}$ belongs to Σ^+. For each $\beta\in P_+$ select $X_\beta\in\mathfrak{g}_c$ such that $[H, X_\beta]=\beta(H)\, X_\beta (H\in\mathfrak{h}_c)$ and $\langle X_\beta, X_{-\beta}\rangle=1$ and put $Z_\beta=\frac{1}{2}(X_\beta+\theta X_\beta)$. Extending σ to a representation of the universal enveloping algebra of $\mathfrak{k}_c$ the functions Γ_μ are given by the recursion formula

$$
\begin{aligned}
&(\langle\mu, \mu\rangle-2i\langle\mu, \lambda\rangle)\, \Gamma_\mu(\lambda)\\
&\quad=2\sum_{\beta\in P_+}\sum_{k\geqq 1}\{\langle\bar{\beta}, \mu-2k\bar{\beta}+\varrho\rangle-i\langle\bar{\beta}, \lambda\rangle\}\, \Gamma_{\mu-2k\bar{\beta}}(\lambda)\\
&\quad\quad+8\sum_{\beta\in P_+}\sum_{k\geqq 1}k\sigma(Z_\beta Z_{-\beta})\, \Gamma_{\mu-2k\bar{\beta}}(\lambda), \qquad \Gamma_0\equiv 1,
\end{aligned}
\tag{5.4}
$$

where, just as in (3.6), k runs over the integers $\geqq 1$ for which $\mu-2k\bar{\beta}\in L$. The set $^*\mathfrak{a}_c'$ is the set of regular elements λ in $\mathfrak{a}_c^*$ for which all $\Gamma_\mu(s\lambda)$ are defined.

III. The Radon transform on X

1. The Euclidean case and the compact case. It was proved by Radon [**41**] and John [**25**] that a function $f\in\mathcal{D}(\boldsymbol{R}^n)$ can be determined explicitly from the integrals of f over the various hyperplanes in $\boldsymbol{R}^n$. If $\omega\in\boldsymbol{R}^n$ is a unit vector, $p\in\boldsymbol{R}$ and dm the Euclidean measure on the hyperplane $(x, \omega)=p$ then the function

$$
\hat{f}(\omega, p)=\int_{(x,\omega)=p} f(x)\, dm(x)
\tag{1.1}
$$

is called the *Radon transform* of f (for all f for which (1.1) is defined). Then if Δ is the Laplacian on $\boldsymbol{R}^n$, $d\omega$ the surface element on $\boldsymbol{S}^n$, and $c=\frac{1}{2}(2\pi i)^{1-n}$,

$$
f(x)=c\Delta_x^{(n-1)/2}\left(\int_{S^{n-1}}\hat{f}(\omega, (x, \omega))\, d\omega\right), \qquad f\in\mathcal{D}(\boldsymbol{R}^n),
\tag{1.2}
$$

where in case n is even the operator $\Delta^{(n-1)/2}$ is the customary fractional power of Δ. (In the papers of Radon and John the inversion formula looks different for odd n and even n. The unified version is verified in [**20d**, §4].) This formula has some remarkable features. First note that for a fixed ω the function $x\to\hat{f}(\omega, \langle\omega, x\rangle)$ is a *plane wave*, that is a function which is constant on each member of a family of parallel hyperplanes. Apart from the Laplacian, formula (1.2) gives a continuous decomposition of f into plane waves, in other words, a function of n variables is decomposed into functions of a single variable. This fact has direct applications to partial differential equations. Secondly, we note that (1.2) contains two integrations, dual to each other: First one integrates over the sets of points in a given hyperplane; then one integrates over the set of hyperplanes through a given point. This feature is important for generalizations of the Radon transform. Thus if ϕ is a continuous function on the space $\boldsymbol{H}^n$ of oriented hyperplanes in $\boldsymbol{R}^n$ we define

$$\check{\phi}(x)=\int_{x\in\xi}\phi(\xi)\,d\mu(\xi), \tag{1.3}$$

where $d\mu$ is the measure $d\omega$ carried over on the set of oriented hyperplanes through x. Then (1.2) can be written

$$f=c\Delta^{(n-1)/2}((\hat{f})^{\vee}). \tag{1.4}$$

The transforms $f\to\hat{f}$, $\varphi\to\check{\varphi}$ and their inversion formulas can also be expressed by means of Fourier integral operators in the sense of [**23b**]. In fact, if $d\xi$ is a suitably normalized invariant measure on $\boldsymbol{H}^n$ and $\langle x,\xi\rangle=p-(x,\omega)$ the (signed) distance from x to ξ: $(x,\omega)=p$, then

$$\hat{f}(\xi)=(2\pi)^{-1}\int_{\boldsymbol{R}}\int_{\boldsymbol{R}^n}f(x)\exp\{is\langle x,\xi\rangle\}\,ds\,dx,$$

$$f(x)=\int_{\boldsymbol{R}}\int_{\boldsymbol{H}^n}\hat{f}(\xi)\exp\{-is\langle x,\xi\rangle\}|s|^{n-1}\,ds\,d\xi.$$

Similar formulas hold for the transform $\varphi\to\check{\varphi}$.

Consider now the case of a compact symmetric space. It was shown by Funk in 1916 that a function $f\in\mathscr{E}(\boldsymbol{S}^2)$, symmetric with respect to the center, can be determined from the integrals of f over the great circles. We shall now see how this can be generalized to compact symmetric spaces X_+ of rank one [**20d**, §5]. For $x\in X_+$ let A_x denote the corresponding *antipodal manifold*, that is the set of points in X_+ at maximum distance L from x. A_x is an orbit of the isotropy subgroup at x and is therefore a submanifold of X_+. It is actually a totally geodesic submanifold and, with the Riemannian structure induced by that of X_+, A_x is another symmetric space of rank 1. We have $x\neq y\Rightarrow A_x\neq A_y$ and $x\in A_y\Leftrightarrow Y\in A_x$. Let Ξ_+ be the set of antipodal manifolds with the Riemannian structure determined such that the mapping $\tau:x\to A_x$ is an isometry. For $x\in X_+$ let $\check{x}=\{\xi\in\Xi_+\mid x\in\xi\}$, and let m and μ be the measures in ξ and $\check{x}$ induced by the Riemannian structures of X_+ and Ξ_+, respectively. The *Radon transform* $f\to\hat{f}$ is defined by

$$\hat{f}(\xi)=\int_{\xi}f(x)\,dm(x)$$

and the dual transform $\phi \to \check{\phi}$ by

$$\check{\phi}(x) = \int_{\check{x}} \phi(\xi)\, d\mu(\xi).$$

Here f and ϕ are any continuous functions on X_+ and Ξ_+ respectively. By the above, $\check{x} = \{\tau(y) \mid y \in \tau(x)\}$ which implies

$$\check{\phi}(x) = (\phi \circ \tau)\hat{}\,(\tau(x)).$$

Let Δ and $\hat{\Delta}$ denote the Laplace-Beltrami operators on X_+ and Ξ_+, respectively.

THEOREM 1.1. *The Radon transform $f \to \hat{f}$ is a bijection of $\mathscr{E}(X_+)$ onto $\mathscr{E}(\Xi_+)$ and*

$$(\Delta f)\hat{} = \hat{\Delta}\hat{f}. \tag{1.5}$$

Except for the case when X_+ is an even-dimensional real projective space we have the inversion formula

$$f = P(\Delta)\,((\hat{f})\check{}\,), \qquad f \in \mathscr{E}(X_+), \tag{1.6}$$

where P is a certain polynomial independent of f, explicitly given by the multiplicities of the restricted roots for X_+.

Let us sketch a proof of this ([**20a**, p. 284], [**20d**, §6]). Let G_+ be the group of isometries of X_+, fix a point $o \in X_+$ and let K_+ be the subgroup of G_+ leaving o fixed, dk the normalized Haar measure on K_+. Let d denote the distance in X_+, and fix an antipodal manifold ξ_0 through o. Then if $g \in G_+$, and c a suitable constant

$$(\hat{f})\check{}\,(g \cdot o) = c \int_{K_+} \left(\int_{\xi_0} f(gk \cdot y)\, d\mu(y) \right) dk = c \int_{\xi_0} (M^{d(o,\, y)} f)(g \cdot o)\, d\mu(y),$$

where $(M^r f)(x)$ is the mean value of f over a sphere of radius r centered at x. Now if $d(o, y) < L$ there is a unique geodesic in X_+ of length $d(o, y)$ joining o and y and since ξ_0 is totally geodesic, $d(o, y)$ is also the distance between o and y in ξ_0. Thus, if we use geodesic polar coordinates in the last integral and put $x = g \cdot o$, we obtain

$$(\hat{f})\check{}\,(x) = \int_o^L A_1(r)\, (M^r f)(x)\, dr$$

where $A_1(r)$ is the area of the sphere of radius r in ξ_0. But by the Darboux equation, $\Delta M^r f = \Delta_r(M^r f)$ where Δ_r is the radial part of Δ,

$$\Delta_r = \frac{d^2}{dr^2} + \frac{1}{A(r)}\frac{dA}{dr}\frac{d}{dr},$$

and $A(r)$ is the spherical area in X_+. But then

$$(1.7) \qquad (\Delta(\hat{f})^{\vee})(x) = \int_o^L A_1(r)\left(\frac{d^2}{dr^2} + A(r)^{-1}\frac{dA}{dr}\frac{d}{dr}\right)(M^r f(x))\,dr.$$

The functions $A(r)$ and $A_1(r)$ are explicitly given by the multiplicities of the restricted roots for X_+ and for ξ_0, and the inversion formula (1.6) is proved by iteration of (1.7) followed by a reduction, using the classification.

For the even-dimensional real projective spaces, (1.6) still holds if $P(\Delta)$ is replaced by a certain integral operator [**44**]. Some use of Theorem 1.1 for a geometric problem is shown by R. Michel [**39**]. A satisfactory generalization of Theorem 1.1 to spaces X_+ of higher rank has not been obtained.

2. The space of horocycles in X. We return now to the case of a symmetric space and follow the notational conventions of I, §2 without explicit comments. A *horocycle* in $X = G/K$ is an orbit of a subgroup of G conjugate to N. (This is a special case of the concept of horosphere (cf. [**13b**]) where K is replaced by an arbitrary closed subgroup and N is also allowed to be more general.) If X is a real hyperbolic space with its usual realization as an open ball $\boldsymbol{B}^n$ in $\boldsymbol{R}^n$, the horocycles are the $(n-1)$-spheres tangential to the boundary of $\boldsymbol{B}^n$. Since N in this case is abelian, the horocycles with the Riemannian structure induced by that of X are isometric to a Euclidean space, a fact known to Bolyai and Lobachevsky. The Iwasawa decomposition easily implies that G permutes the horocycles transitively. Let o denote the origin in X (the fixed point for K) and ξ_0 the horocycle $N \cdot o$. Then it is not hard to see that MN is the subgroup of G fixing ξ_0 so the space Ξ of all horocycles in X admits the identification $\Xi = G/MN$ which turns Ξ into a manifold. Using again the Iwasawa decomposition we see that the map $(kM, a) \to ka \cdot \xi_0$ is a diffeomorphism of $K/M \times A$ onto Ξ. If $\xi = ka \cdot \xi_0$, the point $kM \in K/M$ is called the *normal* to ξ and a the *complex distance* from o to ξ.

We also need the algebra $\boldsymbol{D}(G/MN)$ of G-invariant differential operators on Ξ. If $U \in \boldsymbol{D}(A)$ we can define a differential operator D_U on Ξ by

$$(2.1) \qquad (D_U\phi)(g \cdot \xi_0) = \{U_a(\phi(ga \cdot \xi_0))\}_{a=e}, \qquad \phi \in \mathscr{E}(\Xi).$$

Since A normalizes MN, D_U is indeed a well-defined differential operator on Ξ and it is obviously invariant under the action of G.

THEOREM 2.1. *The mapping $U \to D_U$ is an isomorphism of $\mathbf{D}(A)$ onto $\mathbf{D}(G/MN)$.*

This is established in [**20c**] by determining the invariant polynomials for the action of MN on the tangent space $(G/MN)_{\xi_0}$. A simpler proof using the Bruhat decomposition is indicated in [**20g**].

3. The Plancherel formula and Paley-Wiener theorems. The mapping $n \to n \cdot o$ is a diffeomorphism of N onto ξ_0 so the measure dn induces a measure ds on ξ_0. Since ds is MN-invariant, we obtain by translation a measure on each ξ, also denoted ds, which then differs only by a fixed constant factor from the volume element corresponding to the Riemannian structure on ξ induced by that of X. The *Radon transform* of a function f on X is defined by

$$\hat{f}(\xi) = \int_{\xi} f(x)\, ds(x) \tag{3.1}$$

for all $\xi \in \Xi$ for which this integral exists. (Since ξ is a closed subset of X, $f \in \mathcal{D}(X)$ is of course sufficient.) In analogy with (1.3) the *dual transform* $\phi \to \check{\phi}$ is defined for each continuous ϕ on Ξ by

$$\check{\phi}(x) = \int_{\check{x}} \phi(\xi)\, d\mu(\xi), \tag{3.2}$$

where $\check{x}$ is the set of horocycles passing through x, which has a natural measure $d\mu$ induced by the isotropy group at x. These transforms are related by

$$\int_{\Xi} \hat{f}(\xi)\, \phi(\xi)\, d\xi = \int_X f(x)\, \check{\phi}(x)\, dx \tag{3.3}$$

if $f \in \mathcal{D}(X)$, $\phi \in \mathcal{E}(\Xi)$ where dx and $d\xi$ are the invariant measures dg_K, dg_{MN} on X and Ξ, respectively. The following result [**20c**, p. 681] in conjunction with Theorem 2.1 will be quite useful for applications to differential equations.

THEOREM 3.1. *For a certain isomorphism $D \to \hat{D}$ of $\mathbf{D}(G/K)$ into $\mathbf{D}(G/MN)$ we have*

$$(Df)\hat{} = \hat{D}\hat{f}, \qquad f \in \mathcal{D}(X). \tag{3.4}$$

It is not difficult to conclude from (3.3) and (3.4) that

$$(\hat{D}\phi)\check{} = D\check{\phi}, \qquad \phi \in \mathcal{E}(\Xi) \tag{3.5}$$

(a weaker version of (3.5) is given in [**5**]).

The analog of (3.4) for $\boldsymbol{R}^n$ is the formula

$$(\Delta f)\hat{}(\omega, p) = d^2(\hat{f}(\omega, p))/dp^2,$$

and if we specialize it to radial functions f we see that the Radon transform is a "transmutation operator," transforming the operator $d^2/dr^2 + ((n-1)/r)(d/dr)$ into its leading term. Formula (3.4) has a similar feature in that the leading term of $\hat{D}$ is the same as the leading term in the radial part of D.

We now state an analog of the Radon-John inversion formula.

THEOREM 3.2. *For a certain integral operator Λ on Ξ, Λ^* its adjoint, w order of W,*

$$f = w^{-1}(\Lambda\Lambda^*\hat{f})\check{},$$

(3.6)
$$\int_X |f(x)|^2\, dx = w^{-1} \int_\Xi |\Lambda\hat{f}(\xi)|^2\, d\xi, \qquad f \in \mathcal{D}(X).$$

In case all Cartan subalgebras of $\mathfrak{g}$ are conjugate we have, for a certain differential operator $\Box$ on X,

(3.7)
$$f = \Box((\hat{f})\check{}), \qquad f \in \mathcal{D}(X).$$

The proof ([**20c**], [**20f**]) is not difficult to carry out on the basis of Harish-Chandra's inversion formula for the spherical transform on G (II, Theorem 4.1). In fact the function F_f in (4.9) is essentially the Radon transform of f, viewed as a K-invariant function on X. The operators Λ and $\Box$ are constructed from the $\boldsymbol{c}$-function; they reduce to differential operators exactly when $\boldsymbol{c}(\lambda)$ is a polynomial which in turn happens just when all the Cartan subalgebras of $\mathfrak{g}$ are conjugate. For the hyperbolic spaces this happens just when the dimension is odd, so we recover the parity feature of the Euclidean space. A formula similar to (3.6) was given by Gel'fand-Graev [**14**, §5] for complex G and in [**15**] for the hyperbolic spaces.

As for $\boldsymbol{R}^n$ the transforms $f \to \hat{f}$, $\varphi \to \check{\varphi}$ and their inversion formulas can be expressed by means of Fourier integral operators with respect to the kernels

$$\exp\{\pm(i\lambda - \varrho)\langle x, \xi\rangle\} \quad \text{and} \quad \exp\{\mp(i\lambda - \varrho)\langle x, \xi\rangle\}\, |\boldsymbol{c}(\lambda)|^{-2}$$

where $\langle x, \xi\rangle \in \mathfrak{a}$ is log of the complex distance from x to ξ.

If a function $f \in \mathcal{D}(X)$ vanishes outside a closed ball V in X then obviously $\hat{f}(\xi) = 0$ for all $\xi \in \Xi$ disjoint from V. For applications to differential equations it is important to establish the converse. Although inversion formulas (3.6) and (3.7) seem of no use for this problem the property in question does hold.

THEOREM 3.3. *Let $f \in \mathcal{D}(X)$ and let V be a closed ball in X. Assume $\hat{f}(\xi) = 0$ for all horocycles ξ disjoint from V. Then $f(x) = 0$ for $x \notin V$.*

We will indicate a proof later. The proof actually gives a Paley-Wiener theorem for the Radon transform, that is an intrinsic characterization of the image $\mathscr{D}(X)^\wedge$ (cf. [**20h**, Theorem 8.4]), but we will not need that result.

IV. The Fourier transform on X

1. The definition and the inversion formula. We now define a Fourier transform for "arbitrary" functions on X with formula (1.5) in Chapter I as motivation. In that formula the inner product (x, ω) is the signed distance from the origin to the hyperplane through x with normal ω. Now given $x \in X$, $b \in B$, there exists a unique horocycle $\xi(x, b)$ through x with normal b. In fact if $x = gK$, $b = kM$, and $H(g)$ as in §2 of Chapter I,

$$\xi(x, b) = k \exp(-H(g^{-1}k)) \cdot \xi_0. \tag{1.1}$$

Let $a(x, b)$ be the complex distance from o to $\xi(x, b)$ and put $A(x, b) = \log a(x, b)$. If $f \in \mathscr{D}(X)$ (or if f is just measurable and suitably behaved at ∞) we define the *Fourier transform* $\tilde{f}$ by

$$\tilde{f}(\lambda, b) = \int_X f(x) \exp\{(-i\lambda + \varrho)(A(x, b))\}\, dx \tag{1.2}$$

for all $(\lambda, b) \in \mathfrak{a}_c^* \times B$ for which this integral converges absolutely [**20i**].

Let $\pi: G \to G/K$ be the natural mapping and $\times$ the convolution on X induced by the convolution on G,

$$(f_1 \times f_2) \circ \pi = (f_1 \circ \pi) * (f_2 \circ \pi), \tag{1.3}$$

f_1 and f_2 being functions on X for which the right-hand side exists. Of course the spherical functions ϕ_λ on G can be viewed as functions on X which by abuse of notation we also denote by ϕ_λ.

LEMMA 1.1. *If $f \in \mathscr{D}(X)$, $\lambda \in \mathfrak{a}_c^*$, then*

$$(f \times \phi_\lambda)(x) = \int_B \exp\{(i\lambda + \varrho)(A(x, b))\}\, \tilde{f}(\lambda, b)\, db. \tag{1.4}$$

PROOF. From (1.1) we have

$$A(gK, kM) = -H(g^{-1}k). \tag{1.5}$$

The group G acts not only on X but on B because $B=K/M=G/MAN$. We denote the two actions by $(g, x)\to g\cdot x$ and $(g, b)\to g(b)$. Then $g(kM)=k(gk)\,M$. Since A normalizes N, we have

$$H(g_1g_2k)=H(g_1k(g_2k))+H(g_2k),$$

which by (1.5) implies the identity

$$A(g\cdot x, g(b))=A(x, b)+A(g\cdot o, g(b)). \tag{1.6}$$

As used earlier, $\phi_{-\lambda}(g)=\phi_\lambda(g^{-1})$, so

$$\phi_\lambda(x)=\int_B \exp\{(i\lambda+\varrho)(A(x, b))\}\,db, \tag{1.7}$$

whence, by (1.6),

$$\phi_\lambda(g^{-1}\cdot x)=\int_B \exp\{(i\lambda+\varrho)(A(x, g(b))+A(g^{-1}\cdot o, b))\}\,db.$$

Again, by (1.6), $A(g^{-1}\cdot o, b)=-A(g\cdot o, g(b))$, so we get

$$\phi_\lambda(g^{-1}\cdot x)=\int_B \exp\{(i\lambda+\varrho)(A(x, g(b))-A(g\cdot o, g(b)))\}\,db. \tag{1.8}$$

But by [**19d**, p. 294], $d(k(gk))=\exp\{-2\varrho(H(gk))\}\,dk$, that is

$$d(g(b))/db=\exp\{2\varrho(A(g^{-1}\cdot o, b))\}=\exp\{-2\varrho(A(g\cdot o, g(b)))\}, \tag{1.9}$$

which by (1.8) gives

$$\phi_\lambda(g^{-1}\cdot x)=\int_B \exp\{(i\lambda+\varrho)(A(x, b))\}\exp\{(-i\lambda+\varrho)(A(g\cdot o, b))\}\,db.$$

Since

$$(f\times\phi_\lambda)(x)=\int_G f(g\cdot o)\,\phi_\lambda(g^{-1}\cdot x)\,dg,$$

the lemma is proved.

If f is a K-invariant function on X, then $\tilde{f}(\lambda, b)$ is independent of b and because of (1.7) it coincides with the spherical transform of the function $g \to f(g \cdot o)$ which is bi-invariant under K. We shall now see that Lemma 1.1 makes it possible to use the inversion formula for the spherical transform (Theorem 4.1 in Chapter II) to invert the Fourier transform (1.2).

THEOREM 1.2. *For each $f \in \mathcal{D}(X)$,*

$$f(x) = w^{-1} \int_{\mathfrak{a}^*} \int_B \exp\{(i\lambda + \varrho)(A(x, b))\} \tilde{f}(\lambda, b) |c(\lambda)|^{-2} \, d\lambda \, db, \qquad x \in X.$$

PROOF. Fix $g \in G$ and define f_1 on G by

$$f_1(g_1) = \int_K f(gkg_1 \cdot o) \, dk.$$

Then $f_1 \in \mathcal{D}(G)$ and is bi-invariant under K so by Theorem 4.1, Chapter II,

$$f_1(e) = w^{-1} \int_{\mathfrak{a}^*} \tilde{f}_1(\lambda) |c(\lambda)|^{-2} \, d\lambda.$$

But

$$\tilde{f}_1(\lambda) = \int_G \left(\int_K f(gkg_1 \cdot o) \, dk \right) \phi_{-\lambda}(g_1) \, dg_1$$

$$= \int_G f(gg_1 \cdot o) \, \phi_{-\lambda}(g_1) \, dg_1 = \int_G f(gg_1 \cdot o) \, \phi_\lambda(g_1^{-1}) \, dg_1,$$

so, by the lemma,

$$\tilde{f}_1(\lambda) = \int_B \exp\{(i\lambda + \varrho)(A(g \cdot o, b))\} \tilde{f}(\lambda, b) \, db.$$

Since $f(g \cdot o) = f_1(e)$, the theorem follows.

Let T_X denote the natural representation of G on $L^2(X)$. For $\lambda \in \mathfrak{a}^*$ let $\mathcal{H}_\lambda$ denote the Hilbert space

$$\mathcal{H}_\lambda = \left\{ \phi(x) = \int_B \exp\{(i\lambda + \varrho)(A(x, b))\} F(b) \, db \;\middle|\; F \in L^2(B) \right\},$$

the norm of ϕ being defined as the L^2-norm of F (the mapping $F \to \phi$ being one-to-one). The natural representations T_λ of G on $\mathscr{H}_\lambda$ are unitary and can be shown [**20f**, Chapter III, §5] to be equivalent to the so-called spherical principal series, and are therefore, by well-known results (the most general being those of Kostant [**31**]), irreducible. Combining Theorem 1.2 with the following Plancherel formula (2.3) we have, in terms of direct integral theory,

$$L^2(X) = \int_{\mathfrak{a}^*/W} \mathscr{H}_\lambda \, |c(\lambda)|^{-2} \, d\lambda, \qquad T_X = \int_{\mathfrak{a}^*/W} T_\lambda \, |c(\lambda)|^{-2} \, d\lambda$$

in close analogy to the Euclidean case (Chapter I, §1).

2. The Plancherel formula and the Paley-Wiener theorem. Next we consider the problem of determining the range of the Fourier transform. In view of Lemma 1.1 we have the identities

$$\text{(2.1)} \qquad \int_B \exp\{(is\lambda+\varrho)(A(x,b))\} \, \tilde{f}(s\lambda, b) \, db \equiv \int_B \exp\{(i\lambda+\varrho)(A(x,b))\} \tilde{f}(\lambda, b) \, db$$

which every function in the range has to satisfy for each $s \in W$.

If we multiply (1.4) by $(f(x))^-$ we obtain

$$\text{(2.2)} \qquad \int_X f \times \phi_\lambda(x) \, (f(x))^- dx = \int_B |\tilde{f}(\lambda, b)|^2 \, db.$$

Multiplying here by $|c(\lambda)|^{-2}$ and integrating over $\mathfrak{a}^*$ we obtain, by (1.4) and Theorem 1.2,

$$\text{(2.3)} \qquad \int_X |f(x)|^2 \, dx = w^{-1} \int_{\mathfrak{a}^*} \int_B |\tilde{f}(\lambda, b)|^2 \, |c(\lambda)|^{-2} \, d\lambda.$$

For the interchange of the integrations over $\mathfrak{a}^*$ and over X it suffices to verify that, for each x, the function $\lambda \to f \times \phi_\lambda(x) \, |c(\lambda)|^{-2}$ is integrable on $\mathfrak{a}^*$. But this is clear since $|c(\lambda)|^{-2}$ has at most polynomial growth and secondly, Lemma 1.1 and the formula

$$\text{(2.4)} \qquad \tilde{f}(\lambda, kM) = \int_A \hat{f}(ka \cdot \xi_0) \exp\{(-i\lambda+\varrho)(\log a)\} \, da$$

taken together imply that the function $\lambda \to (f \times \phi_\lambda)(x)$ is rapidly decreasing on $\mathfrak{a}^*$. But (2.3) holds in a sharper form. Let $\mathfrak{a}^*_+$ denote the positive Weyl chamber in $\mathfrak{a}^*$, that is the preimage of $\mathfrak{a}^+$ under the mapping $\lambda \to H_\lambda$.

THEOREM 2.1. *The Fourier transform $f(x) \to \tilde{f}(\lambda, b)$ extends to an isometry of $L^2(X)$ onto $L^2(\mathfrak{a}_+^* \times B)$ (with the measure $|\mathbf{c}(\lambda)|^{-2}\, d\lambda\, db$ on $\mathfrak{a}_+^* \times B$).*

For a proof see [**20f**, p. 120]. Actually, for each $\lambda \in \mathfrak{a}^*$, the function space $\{\tilde{f}(\lambda, b) \mid f \in \mathscr{D}(X)\}$ is dense in $L^2(B)$.

REMARK. The "polar coordinate" mapping $(kM, a) \to kaK$ is a bijection of $K/M \times A^+$ onto the set of regular points in X. Thus Theorem 2.1 shows that just as is the Euclidean space $\boldsymbol{R}^n$, the symmetric space X is self-dual under the Fourier transform.

We turn next to the characterization of $\mathscr{D}(X)^{\sim}$. We call a C^∞ function $\psi(\lambda, b)$ on $\mathfrak{a}_c^* \times B$ a *holomorphic function of uniform exponential type* if it is holomorphic in λ and if there exists a constant $R \geqq 0$ such that, for each integer $N \geqq 0$,

$$\sup_{\lambda \in \mathfrak{a}_c^*;\, b \in B} \exp\{-R\, |\mathrm{Im}\, \lambda|\}\, (1 + |\lambda|)^N\, |\psi(\lambda, b)| < \infty \tag{2.5}$$

where $\mathrm{Im}\, \lambda = \eta$, $|\lambda|^2 = |\xi|^2 + |\eta|^2$ if $\lambda = \xi + i\eta$ $(\xi, \eta \in \mathfrak{a}^*)$.

THEOREM 2.2. *The Fourier transform $f(x) \to \tilde{f}(\lambda, b)$ is a bijection of $\mathscr{D}(X)$ onto the set of holomorphic functions $\psi(\lambda, b)$ of uniform exponential type satisfying the identities*

$$\int_B \exp\{(is\lambda + \varrho)(A(x, b))\}\, \psi(s\lambda, b)\, db \equiv \int_B \exp\{(i\lambda + \varrho)(A(x, b))\}\, \psi(\lambda, b)\, db \tag{2.6}$$

for all $s \in W$, $\lambda \in \mathfrak{a}_c^*$.

Because of formula (2.4) relating the Radon and the Fourier transform this theorem has a close relationship to Theorem 3.3. A proof of both will be indicated later.

In the announcement [**6**], Eguchi-Okamoto define the Schwartz space on X (analogous to $I^2(G)$ in Chapter II, §4) and give a characterization of its Fourier transforms as the rapidly decreasing functions on $\mathfrak{a}^* \times B$ satisfying (2.6).

3. Matrix-valued functions. Let δ be a unitary representation of K on a vector space V_δ of dimension $d(\delta)$. For $f \in \mathscr{E}(X)$ put

$$f^\delta(x) = d(\delta) \int_K f(k \cdot x)\, \delta(k^{-1})\, dk.$$

Then f^δ is a C^∞ map from X to $\mathrm{Hom}(V_\delta, V_\delta)$ and

(3.1) $$f^\delta(k\cdot x)=\delta(k)\, f^\delta(x).$$

We define also, for $f\in\mathscr{D}(X)$,

(3.2) $$\tilde{f}^\delta(\lambda, b)=\int_X f^\delta(x)\exp\{(-i\lambda+\varrho)(A(x, b))\}\, dx.$$

Then it is easy to deduce from Theorem 1.2 the following inversion formula.

THEOREM 3.1. *With the notation above, we have, for $a\in A$,*

$$f^\delta(a\cdot o)=w^{-1}\int_{\mathfrak{a}^*}\left[\int_K \exp\{(i\lambda+\varrho)(A(a\cdot o, kM))\}\,\delta(k)\, dk\right]\tilde{f}^\delta(\lambda)\,|\mathbf{c}(\lambda)|^{-2}\, d\lambda$$

where $\tilde{f}^\delta(\lambda)=\tilde{f}^\delta(\lambda, eM)$.

The knowledge of the functions $f^\delta(x)$ determines the function f itself. In fact, if as usual $\hat{K}$ denotes the set of equivalence classes of irreducible representations of K we have, by Harish-Chandra [**19g**, §13],

(3.3) $$f=\sum_{\delta\in\hat{K}}\operatorname{Trace}(f^\delta).$$

Let V_δ^M be the subspace of vectors in V_δ which are fixed under $\delta(M)$. If $a\in A$, then, by (3.1), $f^\delta(a\cdot o)$ maps V_δ into V_δ^M and if $E_\delta=\int_M\delta(m)\, dm$ then

$$\operatorname{Trace}(f^\delta(a\cdot o))=\operatorname{Trace}(E_\delta f^\delta(a\cdot o))=\operatorname{Trace}(f^\delta(a\cdot o)\, E_\delta)$$

so if | denotes restriction,

(3.4) $$f(a\cdot o)=\sum_{\delta\in\hat{K}}\operatorname{Trace}(f^\delta(a\cdot o)\mid V_\delta^M),\qquad f\in\mathscr{E}(X).$$

V. Invariant differential equations on X

1. An existence theorem. As an application of the results in preceding chapters we shall now outline a proof of the fact that each $D\in\mathbf{D}(G/K)$ is surjective on $\mathscr{E}(X)$, that is, each differential equation $Du=f$ $(f\in\mathscr{E}(X))$ has a solution $u\in\mathscr{E}(X)$. For full details see [**20h**].

THEOREM 1.1. *Let $D\neq 0$ in $\mathbf{D}(G/K)$. Then*

$$D\mathscr{E}(X)=\mathscr{E}(X).$$

Since the C^∞ functions on X correspond via π to the C^∞ functions on G, right invariant under K, it is rather easy to prove that there is a homomorphism μ of $\boldsymbol{D}_0(G)$ onto $\boldsymbol{D}(G/K)$ given by $(\mu(D)F)\circ\pi = D(f\circ\pi)$ which induces an isomorphism of $\boldsymbol{D}_0(G)/(\boldsymbol{D}_0(G)\cap\boldsymbol{D}(G)\mathfrak{k})$ onto $\boldsymbol{D}(G/K)$ [**20a**, p. 265]. We extend the convolution $\times$ on X, defined in IV, §1, to distributions T_1, T_2 on X provided at least one of them has compact support. Then $T_1 \times \delta = T_1$ if δ is the delta distribution at o and if $D \in \boldsymbol{D}(G/K)$,

$$D(T_1 \times T_2) = T_1 \times DT_2 \tag{1.1}$$

as a consequence of the associative law. While the symmetry of G/K played no role for (1.1), the analog of (1.1) for the first factor makes strong use of the symmetry.

LEMMA 1.2. *Let S, $T \in \mathscr{D}'(X)$ at least one of compact support. Then if $D \in \boldsymbol{D}(G/K)$,*

$$D(S \times T) = DS \times T = S \times DT.$$

In the proof one first assumes S to be K-invariant and makes use of $\theta(g) \in Kg^{-1}K$. The general case then follows by approximation by linear combination of translates of K-invariant S.

LEMMA 1.3. *Given $D \neq 0$ in $\boldsymbol{D}(G/K)$ there exists a K-invariant distribution $J \in \mathscr{D}'(X)$ such that*

$$DJ = \delta. \tag{1.2}$$

PROOF. Select $E \in D_0(G)$ such that $\mu(E) = D$. Because of (4.10) and Theorem 4.7 in Chapter II, the mapping $f \to F_f$ is a bijection of $\mathscr{I}^2(G)$ onto $\mathscr{I}(A)$, and actually a homeomorphism with the customary topology of $\mathscr{I}(A)$ and the topology of $\mathscr{I}^2(G)$ given by the seminorms (4.2) in Chapter II. Moreover $F_{Ef} = \gamma(E)F_f$ where γ is as in (1.3) and $\gamma(E)$ is now viewed as a differential operator. The mapping $f \to F_f$ has a transpose, mapping the dual $\mathscr{I}'(A)$ of $\mathscr{I}(A)$ onto the dual $\mathscr{I}^2(G)'$ of $\mathscr{I}^2(G)$. Under that isomorphism, the differential equation (1.2) (viewed as a differential equation in $\mathscr{I}^2(G)'$) is transformed into a differential equation for tempered distributions on A, and this last differential equation has constant coefficients since $\gamma(E)$ does. Now according to a theorem of Hörmander [**23a**] and Lojasiewicz [**34**] a constant coefficient differential operator in $\boldsymbol{R}^n$ maps the space of tempered distributions onto itself. This then gives a solution J to (1.2), even a "tempered" one (cf. [**20j**]).

LEMMA 1.4. *Let $D \neq 0$ in $\boldsymbol{D}(G/K)$. Then the mapping $D: \mathscr{E}'(X) \to \mathscr{E}'(X)$ is injective.*

In fact, using (1.2) and Lemma 1.2,

$$T = DT \times J, \qquad T \in \mathscr{E}'(X), \tag{1.3}$$

which proves the lemma.

LEMMA 1.5. *Let $D \in \boldsymbol{D}(G/K)$ and assume that, for each closed ball $V \subset X$,*

$$f \in \mathscr{D}(X),\ \operatorname{supp}(Df) \subset V \Rightarrow \operatorname{supp}(f) \subset V, \tag{1.4}$$

'supp' *denoting support. Then*

$$D\mathscr{E}(X) = \mathscr{E}(X). \tag{1.5}$$

To prove this we first prove by approximation, using Lemma 1.2, that if (1.4) holds, then it holds in a stronger form, namely

$$T \in \mathscr{E}'(X),\ \operatorname{supp}(D^*T) \subset V \Rightarrow \operatorname{supp}(T) \subset V. \tag{1.6}$$

To show that (1.6) implies (1.5) we can invoke the same functional analysis tools as Malgrange does in [**37**] for the constant coefficient case (cf. also [**7**]). Using a theorem of Banach on the Fréchet space $\mathscr{E}(X)$ it suffices in view of Lemma 1.4 to prove that for every bounded set $B' \subset \mathscr{E}'$, $(D^*\mathscr{E}') \cap B'$ is closed in $\mathscr{E}'$. Here B' can be taken as a set $\mathscr{E}'_V$ of distributions with support in a fixed ball $V \subset X$. So let $T_j = D^*S_j$ $(S_j \in \mathscr{E}')$ be a net in $D^*\mathscr{E}' \cap \mathscr{E}'_V$ converging to $T \in \mathscr{E}'$. Then $T \in \mathscr{E}'_V$, and by (1.6), $S_j \in \mathscr{E}'_V$. Let J^* be a K-invariant fundamental solution for D^*; so, by (1.3),

$$S_j = T_j \times J^* \tag{1.7}$$

and taking limits $\operatorname{supp}(T \times J^*) \subset V$. But (1.7) implies, by applying D^*, $T_j = D^*(T_j \times J^*)$ so $T = D^*(T \times J^*) \in D^*\mathscr{E}'$ as desired.

Now our problem is to prove that (1.4) actually holds. For this we proceed to prove Theorem 3.3 of Chapter III. Since the Radon transform commutes with the action of G, we may assume V centered at the origin. Let R be its radius. The horocycle $\xi = ka \cdot \xi_0$ has distance $|\log a|$ from the origin [**20e**, Lemma 5.4] so our assumption in Theorem 3.3 is

$$\hat{f}(ka \cdot \xi_0) = 0 \quad \text{for } k \in K,\ |\log a| > R. \tag{1.8}$$

But by (2.4) in Chapter IV and the classical Paley-Wiener theorem there exists, for each $N \in \boldsymbol{Z}^+$, a constant C_N such that

$$|\tilde{f}(\xi + i\eta, kM)| \leq C_N (1 + |\xi + i\eta|)^{-N} e^{R|\eta|} \qquad (k \in K) \tag{1.9}$$

where $\xi, \eta \in \mathfrak{a}^*$ and $|\xi + i\eta| = (|\xi|^2 + |\eta|^2)^{1/2}$. We would like to conclude from Theorem 3.1 of Chapter IV that, for each $\delta \in \hat{K}$, $f^\delta(a \cdot o)\, v = 0$ for $v \in V_\delta^M$ and $a \in A^+$, $|\log a| > R$, because then by (3.4) of Chapter IV, $f(a \cdot o) = 0$ for $a \in A^+$, $|\log a| > R$ which would prove Theorem 3.3. Now we take $\sigma_1 = \delta$ and $\sigma_2 \equiv 1$ in Theorem 5.1 of Chapter II. Then a simple reformulation gives, for $v \in V_\delta^M$, $a \in A^+$,

$$\exp\{\varrho(\log a)\} \int_K \exp\{(i\lambda + \varrho)(A(a \cdot o, kM))\}\, \delta(k)\, v\, dk = \sum_{s \in W} \Phi(s\lambda : a)\, \mathbf{C}_s(\lambda)\, v$$

so, by Theorem 3.1, if $H \in \mathfrak{a}^+$,

$$e^{\varrho(H)} f^\delta(\exp H \cdot o)\, v$$

$$= w^{-1} \sum_{s \in W} \int_{\mathfrak{a}^*} \sum_{\mu \in L} e^{-\mu(H)} \Gamma_\mu(s\lambda)\, \mathbf{C}_s(\lambda) \exp\{is\lambda(H)\}\, \tilde{f}^\delta(\lambda)\, |\mathbf{c}(\lambda)|^{-2}\, d\lambda\, v. \tag{1.10}$$

Now we need a basic identity from [**20f**, p. 95]. For $s \in W$ fix a representative $m_s \in M'$ and let

$$\bar{N}_s = \bar{N} \cap (m_s^{-1} N m_s).$$

LEMMA 1.6. *With a suitable interpretation by analytic continuation,*

$$\exp\{(is\lambda + \varrho)(A(x, b))\} = \mathbf{c}_s(\lambda)^{-1} \int_B \exp\{(i\lambda + \varrho)(A(x, b))\}\, dS'_{\lambda, s}(b) \tag{1.11}$$

where

$$\mathbf{c}_s(\lambda) = \int_{\bar{N}_s} \exp\{-(i\lambda + \varrho)(H(\bar{n}_s))\}\, d\bar{n}_s,$$

and the distribution $S'_{\lambda, s}$ on B is given by

$$S'_{\lambda, s}(F) = \int_{\bar{N}_s} F(m_s k(\bar{n}_s) M) \exp\{-(i\lambda + \varrho)(H(\bar{n}_s))\}\, d\bar{n}_s, \tag{1.12}$$

$d\bar{n}_s$ being a Haar measure on $\bar{N}_s$.

On the basis of (1.11) we can derive a formula for $\mathbf{C}_s$:

$$C_{s^{-1}}(s\lambda)\, v = c(\lambda)\, c_s(\lambda)^{-1} \int_{\bar{N}_s} \exp\{-(i\lambda+\varrho)(H(\bar{n}_s))\}\, \delta(k(\bar{n}_s)^{-1} m_s^{-1})\, v\, d\bar{n}_s \tag{1.13}$$

valid for $v \in V_\delta^M$ and $\lambda \in {}^*\mathfrak{a}'_c$ by analytic continuation. Next we note that

$$\tilde{f}^\delta(\lambda) = d(\delta) \int_K \tilde{f}(\lambda, kM)\, \delta(k^{-1})\, dk. \tag{1.14}$$

Combining the identities (2.1) of Chapter IV with (1.11), taking (1.13) into account, we obtain the following identities for $\tilde{f}^\delta(\lambda)$:

$$\tilde{f}^\delta(s\lambda)\, v = c(-\lambda)^{-1} C_{s^{-1}}(s\bar{\lambda})^* \tilde{f}^\delta(\lambda)\, v, \tag{1.15}$$

the asterisk denoting adjoint. On the basis of the recursion formulas for $\Gamma_\mu(\lambda)$ ((5.4) in Chapter III) one can estimate them well enough to interchange the integration over $\mathfrak{a}^*$ and the summation over L on the right-hand side of (1.10). This being done, our problem is to prove that, for each $\mu \in L$,

$$\int_{\mathfrak{a}^*} \Gamma_\mu(s\lambda)\, C_s(\lambda) \exp\{is\lambda(H)\}\, \tilde{f}^\delta(\lambda)\, |c(\lambda)|^{-2}\, d\lambda v = 0 \tag{1.16}$$

provided $H \in \mathfrak{a}^+$, $|H| > R$. Here we write $|c(\lambda)|^2 = c(\lambda)\, c(-\lambda) = c(s\lambda)\, c(-s\lambda)$ and use (1.15). Then the integral (1.16) can be rewritten as

$$\int_{\mathfrak{a}^*} e^{-i\lambda(H)} \Gamma_\mu(-\lambda)\, c(\lambda)^{-1} \left\{ \frac{C_s(-s^{-1}\lambda)}{c(-\lambda)} \frac{C_s(-s^{-1}\bar{\lambda})^*}{c(\lambda)} \right\} \tilde{f}^\delta(-\lambda)\, d\lambda\, v. \tag{1.17}$$

Except for the expression inside the braces this is reminiscent of the integral (4.19) in Chapter II. The function $\tilde{f}^\delta(-\lambda)$ is a matrix-valued holomorphic function each of whose entries satisfies an inequality like (1.9), the function $\Gamma_\mu(-\lambda)$ is a matrix-valued rational function which, by the recursion formula, is holomorphic in the tube $\mathfrak{a}^* + i(-\mathfrak{a}^*_+)$. The growth of $\Gamma_\mu(-\lambda)\, c(\lambda)^{-1}$ is cancelled out by the rapid decrease of $\tilde{f}^\delta(-\lambda)$. In order to imitate the proof of Lemma 4.6 of Chapter II, one must first show that the expression inside the braces in (1.17) is holomorphic in the above-mentioned tube. But it turns out that this expression is actually the identity operator. This is proved by reducing (1.13) to rank one integrals and, for the rank one case, one requires the following lemma.

LEMMA 1.7. *Assume X has rank one and suppose s is the nontrivial Weyl* *group element. Given* $\bar{n} \in \bar{N}$ *there exists a unique* $\bar{n}_0 \in \bar{N}$ *such that*

$$m_s k(\bar{n}_0)\, M = k(\bar{n})^{-1}\, m_s M.$$

It satisfies

$$H(\bar{n}_0) = H(\bar{n}).$$

This can be proved by reduction to the case $G = \boldsymbol{SU}(2, 1)$ by the method of [**20f**, p. 54], and for this case can be proved by direct computation.

Now that the proof of Lemma 4.6 of Chapter III can be imitated we do arrive, via (1.10) and (3.4) of Chapter IV, at the desired conclusion $f(a \cdot o) = 0$ for $a \in A^+$, $|\log a| > R$. This then proves Theorem 3.3 in Chapter III, which is the crux of the Paley-Wiener theorem for the Radon transform.

In order to verify that the implication (1.4) holds, we use Theorem 3.1 of Chapter III. Since $f \in \mathcal{D}(X)$ and $\operatorname{supp}(Df) \subset V$, we conclude

$$(Df)\hat{}\,(\xi) = 0 \quad \text{if } \xi \cap V = \emptyset.$$

But $(Df)\hat{} = \hat{D}\hat{f}$ where $\hat{D} \in \boldsymbol{D}(G/MN)$ and under the diffeomorphism $(kM, a) \to ka \cdot \xi_0$ of $K/M \times A$ onto Ξ, the operator $\hat{D}$ corresponds to a constant coefficient operator U on A (Theorem 2.1, Chapter III). Consequently

$$U_a(\hat{f}(ka \cdot \xi_0)) = 0, \qquad k \in K,\ |\log a| > R. \tag{1.18}$$

The function $a \to \hat{f}(ka \cdot \xi_0)$ has compact support so (1.18) implies, by a special case of the Lions-Titchmarsh convexity theorem, that

$$\hat{f}(ka \cdot \xi_0) = 0 \quad \text{for } k \in K,\ |\log a| > R.$$

Now the cited Theorem 3.3 gives $\operatorname{supp}(f) \subset V$ so, via Lemma 1.5, we obtain $D\mathcal{E}(X) = \mathcal{E}(X)$.

2. Harmonic functions on X. A function $u \in \mathcal{E}(X)$ is called harmonic if $Du = 0$ for all $D \in \boldsymbol{D}(G/K)$ which annihilate the constants (i.e., without 0-order term).

THEOREM 2.1. *The function* $u \in \mathcal{E}(X)$ *is harmonic if and only if*

$$\int_K u(gk \cdot x)\, dk \equiv u(g \cdot o), \qquad g \in G,\ x \in X. \tag{2.1}$$

The definition as well as this theorem were given by Godement in 1952 [**18b**].

It generalizes Gauss' mean value theorem for harmonic functions. A generalization of the Poisson integral formula was obtained by Furstenberg [**11a**] and Karpelevič [**26**].

THEOREM 2.2. *Given a bounded harmonic function u on X there exists a unique* $F \in L^\infty(B)$ *such that*

$$u(g\cdot o) = \int_B F(g(b))\, db. \tag{2.2}$$

Conversely, given $F \in L^\infty(B)$ *the function u defined by* (2.2) *is a harmonic function on X.*

The proof is obtained by considering the set of bounded functions ψ satisfying

$$\int_K \psi(gkg_1)\, dk = u(g\cdot o), \qquad \|\psi\|_\infty \leqq \|u\|_\infty.$$

The group G acts on this convex, weakly compact set of ψ on the right and now a fixed point theorem gives a fixed vector under MAN, which then furnishes the boundary function F. Using martingale theory [**11a**], Furstenberg proved

THEOREM 2.3. *A bounded solution u of the Laplace equation* $Lu = 0$ *is harmonic.*

Another proof is given by [**26**]. The theorem was stated earlier in [**1**] but the proof given seems inadequate.

Formula (2.2) can also be written

$$u(g\cdot o) = \int_B F(b) \frac{d(g^{-1}(b))}{db}\, db, \tag{2.3}$$

and by definition $d(g^{-1}(b))/db$ is the Poisson kernel, and as we saw during the proof of Lemma 1.1 of Chapter IV, it is given by $\exp\{-2\varrho(H(g^{-1}k))\}$ if $b = kM$. Thus the Poisson formula can be written

$$u(x) = \int_B \exp\{2\varrho(A(x, b))\}\, F(b)\, db. \tag{2.4}$$

It was implicit in Furstenberg's boundary definition that each $x \in X$ defines a measure on B and that if in the weak topology of measures on B, $x \to \delta_b$ (the delta function at b) then $u(x) \to F(b)$. In [**22**], Helgason and Korányi proved the following analog of the Fatou theorem.

THEOREM 2.4. *Let u be a bounded harmonic function on X. For almost all unit vectors $Z \in \mathfrak{p}$, the limit*

$$\lim_{t \to +\infty} u(\exp tZ \cdot o) \tag{2.5}$$

exists.

Knapp and Williamson [**28**] dropped the assumption that the convergence in (2.5) should be radial and proved

THEOREM 2.5. *For almost all $\bar{n}_0 \in \bar{N}$,*

$$\lim_{a \to +\infty} u(\bar{n}_0 a \bar{n} \cdot o) = F(k(\bar{n}_0)\, M) \tag{2.6}$$

uniformly for $\bar{n}$ varying in any compact subset of $\bar{N}$.

(As usual $a \to +\infty$ means $\alpha(\log a) \to +\infty$ for each $\alpha \in \Sigma^+$.)

As remarked by Korányi, (2.6) gives the following sharpening of (2.5). For $H \in \mathfrak{a}^+$ put $a_t = \exp tH$ and write by the Iwasawa decomposition $\bar{n}_0 = k(\bar{n}_0)\,(a_1 n_1)^{-1}$. Then

$$k(\bar{n}_0)\, a_t \cdot o = \bar{n}_0 a_1 a_t a(t)\, n(t) \cdot o$$

where $a(t) \to e$, $n(t) \to e$ as $t \to +\infty$. Thus, by (2.6),

$$\lim_{t \to \infty} u(k(\bar{n}_0)\, a_t \cdot o) = F(k(\bar{n}_0)\, M) \tag{2.7}$$

for almost all $\bar{n}_0 \in \bar{N}$. Let us say that a vector $Z \in \mathfrak{p}$ lies in the element $b = kM$ if the centralizer of Z in $\mathfrak{p}$ is abelian and the Weyl chamber in which Z lies is $\mathrm{Ad}(k)\, \mathfrak{a}^+$. Thus (2.7) implies

COROLLARY 2.6. *For almost all $b \in B$, the limit* (2.5) *is the same for all Z in b.*

If F is continuous, the qualification "almost" can of course be dropped [**26**, §18]. For further results on the boundary behaviour of (2.4) for not necessarily bounded F, see Knapp [**27**], Lindahl [**33**] and Korányi [**29a**].

3. Harmonic and holomorphic functions on bounded symmetric domains. In this section we assume $X = G/K$ is a Hermitian symmetric space. This amounts to that $\mathfrak{k}$ is not semisimple. Thus there exists a Cartan subalgebra $\mathfrak{h} \subset \mathfrak{k}$ which is also a Cartan subalgebra of $\mathfrak{g}$. This means that in the root space decomposition $\mathfrak{g}_c = \mathfrak{h}_c + \sum_\gamma \mathfrak{g}_c^\gamma$ we have either $\mathfrak{g}_c^\gamma \subset \mathfrak{k}_c$ or $\mathfrak{g}_c^\gamma \subset \mathfrak{p}_c$. In the first case the root γ is called *compact*, in the second case *noncompact*. Then

$$\mathfrak{k}_c = \mathfrak{h}_c + \sum_\alpha \mathfrak{g}_c^\alpha, \qquad \mathfrak{p}_c = \sum_\beta \mathfrak{g}_c^\beta,$$

where α runs over the compact roots, β over the noncompact roots. For a suitable ordering of the roots (cf. [**20b**, p. 313]) we put

$$\mathfrak{p}_- = \sum_{\beta<0} \mathfrak{g}_c^\beta, \qquad \mathfrak{p}_+ = \sum_{\beta>0} \mathfrak{g}_c^\beta.$$

Then $\mathfrak{p}_-$ and $\mathfrak{p}_+$ are abelian subalgebras. Let G_c be the simply connected Lie group with Lie algebra $\mathfrak{g}_c$ and assume, as we can, that $G \subset G_c$. Let K_c, P_+, P_-, U be the analytic subgroups corresponding to $\mathfrak{k}_c$, $\mathfrak{p}_+$, $\mathfrak{p}_-$, $\mathfrak{k}+i\mathfrak{p}$, respectively. The mapping $uK \to uK_cP_+$ is an analytic diffeomorphism of the compact Hermitian symmetric space $X^* = U/K$ onto G_c/K_cP_+ (Borel [**3**]) which contains G/K (the orbit $G\cdot o_c$ of the origin $o_c \in G_c/K_cP_+$ under G) as an open subset. Moreover, the mapping $\xi: Y \to \exp Y\cdot o_c$ is a holomorphic bijection of $\mathfrak{p}_-$ onto a dense open subset of X^*, and $\xi^{-1}(G\cdot o_c)$ is a bounded domain D in $\mathfrak{p}_-$ [**19c**]. In summary,

$$X = G/K \overset{\xi}{\leftarrow} D \subset \mathfrak{p}_- \overset{\xi}{\rightarrow} G_c/K_cP_+ = U/K = X^* \tag{3.1}$$

(for more details see [**20b**], [**30**, in particular p. 286] and [**49**]). As found by R. Hermann the domain D is a nice convex subset of $\mathfrak{p}_-$ ([**32**, p. 110], [**40**, p. 371]). Moreover when the geodesics in X are carried over on D via ξ^{-1} they exhibit interesting behaviour. Consider the oriented geodesics through the origin in X with tangent vectors at o lying in a fixed Weyl chamber. As shown to me by Korányi, it is easily deduced from [**30**, pp. 268–270] that the images of these geodesics under ξ^{-1} all converge to the same point in $\mathrm{Cl}(D)$, and as the Weyl chamber varies, the limit points obtained form the Bergman-Šilov boundary, S_D. Thus while B is the set of Weyl chambers in the tangent space X_0, S_D is obtained by identifying two Weyl chambers if the limit points coincide. In particular, S_D is a coset space K/L where L is a certain subgroup containing M so G acts on S_D. Let $d\zeta$ denote the K-invariant probability measure on S_D. Then we have the following variation of Theorem 2.2 ([**11a**, p. 371], [**40**, p. 368]). For the classical domains, such results are given in Hua [**24**], Lowdenslager [**35**] with explicit kernels.

THEOREM 3.1. *Let f be harmonic in D, continuous in* $\mathrm{Cl}(D)$. *Then*

$$f(z) = \int_{S_D} f(g\cdot\zeta)\, d\zeta, \qquad z\in D. \tag{3.2}$$

All continous functions on S_D arise in this way.

Since G/K is Kählerian, the Laplacian L has the form

$$L=\sum g^{ij}\,\partial^2/\partial z^i\partial\bar{z}^j.$$

Thus by Theorem 2.3 we have

THEOREM 3.2. *A bounded holomorphic function on G/K is harmonic.*

As shown to me by L. Michelson and proved already by Korányi in [**29b**], the boundedness assumption can be dropped here.

Defining now a new kernel $\mathfrak{P}$ on $X\times S_D$ by

$$\mathfrak{P}(g\cdot o,\zeta)=d(g^{-1}(\zeta))/d\zeta,\qquad g\in G,\ b\in B$$

formula (3.2) can be written

$$f(z)=\int_{S_D}\mathfrak{P}(z,\zeta)\,f(\zeta)\,d\zeta. \tag{3.3}$$

This then is a formula which reproduces a bounded holomorphic function on D, continuous on $\mathrm{Cl}(D)$, from its values on the Bergman-Šilov boundary S_D. Boundary behavior of such Poisson integrals where $f(\zeta)$ is not necessarily continuous has been investigated by Korányi, Stein, Urakawa and Weiss. (See Weiss [**50**] for generalizations of the Fatou theorem.)

An L^2 function on the unit circle can be extended to a holomorphic function on the open disk if and only if its negative Fourier coefficients all vanish. In his paper [**42**], Schmid generalizes this to the domain D with the Bergman-Šilov boundary S_D. By definition, H^2 is the closure in $L^2(S_D)$ of the set of functions which can be continued to holomorphic functions in a neighborhood of $\mathrm{Cl}(D)$. Schmid then characterizes the Fourier series of the functions in H^2. Special cases of this had been given by Hua [**24**].

REFERENCES

1. F. A. Berezin, *An analog of Liouville's theorem to symmetric spaces of negative curvature*, Dokl. Akad. Nauk SSSR **125** (1959), 1187–1189. (Russian) MR **21** #3014.

2a. T. S. Bhanu-Murthy, *Plancherel's measure for the factor space* $SL(n;R)/SO(n;R)$, Dokl. Akad. Nauk SSSR **133** (1960), 503–506 = Soviet Math. Dokl. **1** (1960), 860–862. MR **23** #A2481.

2b. ———, *The asymptotic behaviour of zonal spherical functions on the Siegel upper half plane*, Dokl. Akad. Nauk SSSR **135** (1960), 1027–1030 = Soviet Math. Dokl. **1** (1960), 1325–1239. MR **23** #A967.

3. A. Borel, *Les espaces hermitiens symétriques*, Séminaire Bourbaki, 1952.

4. É. Cartan, *Sur la détermination d'un système orthogonal complet dans un espace de Riemann symétrique clos*, Rend. Circ. Mat. Palermo **53** (1929), 217–252.

5. M. Eguchi, *On the Radon transform of the rapidly decreasing functions symmetric spaces*. II, Mem. Fac. Sci. Kyushu Univ. Ser. A (to appear).

6. M. Eguchi and K. Okamoto, *The Fourier transform of the Schwartz space of a symmetric space*, 1972. (preprint).

7. L. Ehrenpreis, *Solutions of some problems of division*. I. *Division by a polynomial of derivation*, Amer. J. Math. **76** (1954), 883–903. MR **16**, 834.

8. L. Ehrenpreis and F. Mautner, *Some properties of the Fourier transform on semisimple Lie groups*. I, II, III, Ann. of Math. (2) **61** (1955), 406–439; Trans. Amer. Math. Soc. **84** (1957), 1–55; **90** (1959), 431–484. MR **16**, 1017; MR **18**, 745.

9. J. Faraut and K. Harzallah, *Fonctions sphériques de type positif sur les espaces hyperboliques*, C. R. Acad. Sci. Paris Sér. A–B **274** (1972), A1396–A1398.

10. M. Flensted-Jensen, *Paley-Wiener theorems for a differential operator connected with symmetric spaces*, Ark. Mat. **10** (1972), 143–162.

11a. H. Furstenberg, *A Poisson formula for semisimple Lie groups*, Ann. of Math. (2) **77** (1963), 335–386. MR **26** #3820; errata, MR **28**, 1246.

11b. ———, *Translation-invariant cones of functions on semisimple Lie groups*, Bull. Amer. Math. Soc. **71** (1965), 271–326. MR **31** #1326.

12a. R. Gangolli, *On the Plancherel formula and the Paley-Wiener theorem for spherical functions on semisimple Lie groups*, Ann. of Math. (2) **93** (1971), 150–165. MR **44** #6912.

12b. ———, *Spherical functions on semisimple Lie groups*, Short Courses Presented at Washington University, Dekker, New York, 1972.

13a. I. M. Gel'fand, *Spherical functions on symmetric Riemann spaces*, Dokl. Akad. Nauk SSSR **70** (1950), 5–8. (Russian) MR **11**, 498.

13b. ———, *Automorphic functions and the theory of representations*, Proc. Internat. Congr. Math. (Stockholm, 1962), Inst. Mittag-Leffler, Djursholm, 1963, pp. 74–85. MR **31** #273.

14. I. M. Gel'fand and M. I. Graev, *Geometry of homogeneous spaces, representations of groups in homogeneous spaces and related questions of integral geometry*. I, Trudy Moskov. Mat. Obšč. **8** (1959), 321–390; English transl., Amer. Math. Soc. Transl. (2) **37** (1964), 351–429. MR **23** #A4013.

15. I. M. Gel'fand, M. I. Graev and N. Vilenkin, *Generalized functions*. Vol. 5: *Integral geometry and representation theory*, "Nauka", Moscow, 1962; English transl., Academic Press, New York, 1966.

16a. I. M. Gel'fand and M. A. Naĭmark, *Unitary representations of the classical groups*, Trudy Mat. Inst. Steklov. **36** (1950); German transl., Akademie-Verlag, Berlin, 1957. MR **13**, 722; MR **19**, 13.

16b. ———, *Unitary representations of a unimodular group containing the identity representation of the unitary subgroup*, Trudy Moskov. Mat. Obšč. **1** (1952), 423–475. (Russian) MR **14**, 352.

17. S. G. Gindikin and F. I. Karpelevič, *Plancherel measure of Riemannian symmetric spaces of nonpositive curvature*, Dokl. Akad. Nauk SSSR **145** (1962), 252–255 = Soviet Math. Dokl. **3** (1962), 962–965. MR **27** #240.

18a. R. Godement, *A theory of spherical functions*. I, Trans. Amer. Math. Soc. **73** (1952), 496–556. MR **14**, 620.

18b. ———, *Une généralization du théorème de la moyenne pour les fonctions harmoniques*, C. R. Acad. Sci. Paris **234** (1952), 2137–2139. MR **13**, 821.

18c. ———, *Introduction aux travaux de Selberg*, Séminaire Bourbaki, 1957.

19a. Harish-Chandra, *On the Plancherel formula for the right invariant functions on a semisimple Lie group*, Proc. Nat. Acad. Sci. U.S.A. **40** (1954), 200–204. MR **16**, 11.

19b. ———, *Representations of semisimple Lie groups*. II, Trans. Amer. Math. Soc. **76** (1954), 26–65. MR **15**, 398.

19c. ———, *Representations of semisimple Lie groups*. VI, Amer. J. Math. **78** (1956), 564–628. MR **18**, 490.

19d. ———, *Spherical functions on a semisimple Lie group*. I, II, Amer. J. Math. **80** (1958), 241–310; 553–613. MR **20** #925; MR **21** #92.

19e. ———, *Some results on differential equations and their applications*, Proc. Nat. Acad. Sci. U.S.A. **45** (1959), 1763–1764. MR **31** #274.

19f. ———, *Differential equations and semisimple Lie groups*, 1960. (unpublished)

19g. ———, *Discrete series for semisimple Lie groups*. II, Acta Math. **116** (1966), 1–111. MR **36** #2745.

20a. S. Helgason, *Differential operators on homogeneous spaces*, Acta Math. **102** (1959), 239–299. MR **22** #8457.

20b. ———, *Differential geometry and symmetric spaces*, Pure and Appl. Math., vol. 12, Academic Press, New York, 1962. MR **26** #2986.

20c. ———, *Duality and Radon transform for symmetric spaces*, Amer. J. Math. **85** (1963), 667–692. MR **28** #1632.

20d. ———, *The Radon transform on Euclidean spaces, compact two-point homogeneous spaces, and Grassmann manifolds*, Acta Math. **113** (1965), 153–180. MR **30** #2530.

20e. ———, *An analog of the Paley-Wiener theorem for the Fourier transform on certain symmetric spaces*, Math. Ann. **165** (1966), 297–308. MR **36** #6545.

20f. ———, *A duality for symmetric spaces with applications to group representations*, Advances in Math. **5** (1970), 1–154. MR **41** #8587.

20g. ———, *Analysis on Lie groups and homogeneous spaces*, Conf. Board of the Math. Sci., Regional Conf. Ser. Math., no. 14, Amer. Math. Soc., Providence, R.I., 1972.

20h. ———, *The surjectivity of invariant differential operators on symmetric spaces*, Ann. of Math. (to appear).

20i. ———, *Radon-Fourier transforms on symmetric spaces and related group representations*, Bull. Amer. Math. Soc. **71** (1965), 757–763. MR **31** #3543.

20j. ———, *Fundamental solutions of invariant differential operators on symmetric spaces*, Amer. J. Math. **86** (1964), 565–601. MR **29** #2323.

21. S. Helgason and K. Johnson, *The bounded spherical functions on symmetric spaces*, Advances in Math. **3** (1969), 586–593. MR **40** #2787.

22. S. Helgason and A. Korányi, *A Fatou type theorem for harmonic functions on symmetric spaces*, Bull. Amer. Math. Soc. **74** (1968), 258–263. MR **37** #4753.

23a. L. Hörmander, *On the division of distributions by polynomials*, Ark. Mat. **3** (1958), 555–568. MR **23** #A2044.

23b. ———, *Fourier integral operators*. I, Acta Math. **127** (1971), 79–183.

24. L. K. Hua, *Harmonic analysis of functions of several complex variables in the classical domains*, Science Press, Peking, 1958; English transl., Transl. Math. Monographs, vol. 6, Amer. Math. Soc., Providence, R.I., 1963. MR **23** #A3277; MR **30** #2162.

25. F. John, *Bestimmung einer Funktion aus ihren Integralen über gewisse Mannigfaltigkeiten*, Math. Ann. **100** (1934), 488–520.

26. F. I. Karpelevič, *The geometry of geodesics and the eigenfunctions of the Beltrami-Laplace operator on symmetric spaces*, Trans. Moskov. Mat. Obšč. **14** (1965), 48–185 = Trans. Moscow Math. Soc. **1965**, 51–199. MR **37** #6876.

27. A. W. Knapp, *Fatou's theorem for symmetric spaces*. I, Ann. of Math. (2) **88** (1968), 106–127. MR **37** #1528.

28. A. W. Knapp and R. E. Williamson, *Poisson integrals and semisimple groups*, J. Analyse Math. **24** (1971), 53–76.

29a. A. Korányi, *Boundary behavior of Poisson integrals on symmetric spaces*, Trans. Amer. Math. Soc. **140** (1969), 393–409. MR **39** #7132.

29b. ———, *A remark on boundary values of functions of several complex variables*, Springer Lecture Notes, No. 185, 1971.

30. A. Korányi and J. A. Wolf, *Realization of hermitian symmetric spaces as generalized half-planes*, Ann. of Math. (2) **81** (1965), 265–288. MR **30** #4980.

31. B. Kostant, *On the existence aned irreducibility of certain series of representations*, Bull. Amer. Math. Soc. **75** (1969), 627–642. MR **39** #7031.

32. R. P. Langlands, *The dimension of spaces of automorphic forms*, Amer. J. Math. **85** (1963), 99–125. MR **27** #6286.

33. L. A. Lindahl, *Fatou's theorem for symmetric spaces*, Ark. Mat. **10** (1972), 34–47.

34. S. Lojasiewicz, *Division d'une distribution par une fonction analytique de variables réelles*, C. R. Acad. Sci. Paris **246** (1958), 683–686. MR **20** #2616.

35. D. Lowdenslager, *Potential theory in bounded symmetric homogeneous complex domains*, Ann. of Math. (2) **67** (1958), 467–484. MR **21** #2836.

36. G. W. Mackey, *Induced representations of locally compact groups*. I, Ann. of Math. (2) **55** (1952), 101–139. MR **13**, 434.

37. B. Malgrange, *Équations aux dérivées partielles à coefficients constants* 2, C. R. Acad. Sci. Paris **238** (1954), 196–198. MR **15**, 626.

38. F. Mautner, *Fourier analysis on symmetric spaces*, Proc. Nat. Acad. Sci. U.S.A. **37** (1951), 529–533. MR **13**, 434.

39. R. Michel, *Sur certains tenseurs symétriques des projectifs réels*, J. Math. Pures Appl. **51** (1972), 273–293.

40. C. C. Moore, *Compactifications of symmetric spaces*. II. *The Cartan domains*, Amer. J. Math. **86** (1964), 358–378. MR **28** #5147.

41. J. Radon, *Über die Bestimmung von Funktionen durch ihre Integralwerte längs gewisser Mannigfaltigkeiten*, Ber. Verh. Sächs. Wiss. Leipzig Math.-Nat. Kl. **69** (1917), 262–277.

42. W. Schmid, *Die Randwerte holomorpher Funktionen auf Hermitesch symmetrischen Räumen*, Invent. Math. **9** (1969/70), 61–80. MR **41** #3806.

43. L. Schwartz, *Théorie des distributions*, Publ. Inst. Math. Univ. Strasbourg, no. IX–X, Nouvelle édition, entièrement corrigée, refondue et augmentée, Hermann, Paris, 1966. MR **35** #730.

44. V. T. Semjanistyĭ, *Homogeneous functions and some problems of integral geometry in spaces of constant curvature*, Dokl. Akad. Nauk SSSR **136** (1961), 288–291 = Soviet Math. Dokl. **2** (1961), 59–62. MR **24** #A2842.

45. A. I. Štern, *Completely irreducible class* I *representations of real semisimple Lie groups*, Dokl. Akad. Nauk SSSR **188** (1969), 1017–1019 = Soviet Math. Dokl. **10** (1969), 1254–1257. MR **41** #1931.

46. R. Takahashi, *Sur les représentations unitaires des groupes de Lorentz généralisés*, Bull. Soc. Math. France **91** (1963), 289–433. MR **31** #3544.

47. P. Trombi and V. S. Varadarajan, *Spherical transforms on semisimple Lie groups*, Ann. of Math. (2) **94** (1971), 246–303.

48. H. Urakawa, *Radial convergence of Poisson integrals on symmetric bounded domains of tube type*, 1972. (preprint).

49. J. A. Wolf, *Fine structure of hermitian symmetric spaces*, Short Courses Presented at Washington University, Dekker, New York, 1972.

50. N. Weiss, *Fatou's theorem for symmetric spaces*, Short Courses Presented at Washington University, Dekker, New York, 1972.

51. G. Warner, *Harmonic analysis on semisimple Lie groups*. II, Springer-Verlag, Berlin and New York, 1972.

MASSACHUSETTS INSTITUTE OF TECHNOLOGY

QUANTIZATION AND REPRESENTATION THEORY

ROBERT J. BLATTNER*

1. Introduction. In [6], B. Kostant has begun to lay the foundations of a general theory of quantization, one purpose of which is to give a unified treatment of the construction of irreducible representations of Lie groups. The basic setup consists of a sympletic manifold (X, ω) together with a line bundle with connection (L, ∇) and invariant Hermitian structure such that $\mathrm{Curv}(L, \nabla)=\omega$. A group $E(L, \nabla)$ acts on L and the resultant action is called the *prequantization* of $E(L, \nabla)$. Quantization is achieved through introducing a *polarization* F on (X, ω), that is, a foliation by maximally isotropic submanifolds of X. The space of leaves X/F will have dimension $=\frac{1}{2}\dim X$, if X/F is a manifold. Let C_F consist of the smooth sections of L covariant constant with respect to ∇ along the leaves of F, and let $E_F(L, \nabla)$ consist of the $\sigma \in E(L, \nabla)$ leaving F invariant. Then $E_F(L, \nabla)$ acts on C_F. There remains the problem of turning C_F into a Hilbert space $\mathfrak{H}^F$. This can be accomplished using the notion of a half-density.

Experience has shown that somehow the result of the above procedure is independent of the choice of F. Thus, if G is a Lie group acting on (X, ω) leaving two polarizations F_1 and F_2 invariant, one can often construct a unitary isomorphism $U_{F_2F_1} : \mathfrak{H}^{F_1} \rightarrow \mathfrak{H}^{F_2}$ intertwining the action of G. A very general construction of these operators and its ramifications form the main subject of these notes. It will turn out that when F_1 and F_2 are transverse one can write down such a $U_{F_2F_1}$ formally. The question of when this formal definition makes analytical sense is a deep problem, which we largely avoid here.

It is natural to investigate whether the restriction of quantization to $E_F(L, \nabla)$ can be gotten around. In many cases this can be done by clever means (see [2]). Here we attack the problem head-on by trying to use the intertwining operators

AMS (MOS) subject classifications (1970). Primary 53C15, 43A32; Secondary 22E45.

* The preparation of this paper was supported in part by NSF grant GP-33696X.

of the previous paragraph to define quantized actions of those $\sigma \in E(L, \nabla)$ which move F into polarizations transverse to F. It turns out that this can be done, but only by replacing half-densities with new kinds of objects called *half-forms*. This necessitates working with double coverings of certain Lie groups, a phenomenon already encountered by Shale [**10**], Weil [**12**], and Duflo [**2**].

The exposition of the above takes up §§2, 3, 5, and 6. In §4 we work two examples to give enlightenment and motivation, while in §7 we sketch applications, problems, and future lines of attack.

These notes are an exposition without proofs of joint work of B. Kostant, S. Sternberg, and myself. They are based in part on three lectures given at the 1972 AMS Summer Institute by Kostant together with one of my own. However, the responsibility for these notes is mine alone.

Notation and conventions. All manifolds are real and C^∞. If $E \to X$ is a vector bundle over X, ΓE denotes the space of smooth sections of E over X, while $\Gamma_0 E$ denotes the subspace of sections with compact support. Moreover, E_* denotes $E - \{0\text{-section}\}$. If V is a real vector space, $V_{\mathbf{C}}$ denotes its compexification. Almost all covering projections in these notes will be denoted by the same letter: p. If X is a manifold, TX and T^*X denote the tangent and cotangent bundles, respectively. Finally, if G is a topological group, G_0 denotes its identity component.

2. Prequantization and polarizations. We begin by recalling some facts from [**6**].

Our basic object will be a symplectic manifold (X, ω), that is, a real manifold X of dimension $2n$ together with a closed nondegenerate 2-form ω on X.

EXAMPLE 2.1. Let Y be a manifold of dimension n and let $X = T^*Y$. Let π be the projection $X \to Y$. Define a 1-form α on X by

$$\langle \alpha_x, \xi_x \rangle = \langle x, \pi_* \xi_x \rangle \quad \text{for } x \in X,\ \xi_x \in T_x X. \tag{2.2}$$

Setting $\omega = d\alpha$, (X, ω) is a symplectic manifold.

EXAMPLE 2.3. Let G be a real Lie group and $\mathfrak{g}$ its Lie algebra. Let O be an orbit in $\mathfrak{g}^*$ under the contragredient κ of the adjoint representation (the *coadjoint* representation). Let $f \in O$. We define

$$\omega_f(d\kappa(\xi)f, d\kappa(\eta)f) = \langle f, [\eta, \xi] \rangle \quad \text{for } \xi, \eta \in \mathfrak{g}. \tag{2.4}$$

Here $T_f O$ is identified with a vector subspace of $\mathfrak{g}^*$. Then (O, ω) is a symplectic manifold.

For prequantization we also need a complex differentiable line bundle L over X with a connection leaving a Hermitian metric invariant.

DEFINITION 2.5. A *connection* on L is a map ∇ which assigns to each $\xi \in T_x X$ and each $s \in \Gamma L$ a vector $\nabla_\xi s \in L_x$ such that

$$\xi \mapsto \nabla_\xi s \text{ is linear on } T_x X, \tag{2.6}$$

(2.7) $$s \mapsto \nabla_\xi s \text{ is linear on } \Gamma L,$$

(2.8) $$\nabla_\xi(fs) = (\xi f)\, s(x) + f(x)\, \nabla_\xi s \quad \text{for } f \in C^\infty(X).$$

The *connection form* of (L, ∇) is the unique 1-form α on L_* such that

(2.9) $$\nabla_\xi s = 2\pi i \langle s^*\alpha, \xi \rangle\, s(x), \qquad \xi \in T_x X.$$

We remark that ∇ and α determine each other.

When one has a connection, one can parallel translate:

Let $l \in L_x$ and let γ be a smooth map of $[0, 1]$ into X such that $\gamma(0) = x$. Then there exists a unique arc $\hat{\gamma}$: $[0, 1] \to L$ projecting onto γ such that $\hat{\gamma}(0) = l$ and (with some abuse of notation) $\nabla_{\gamma'(t)}\, \hat{\gamma} = 0$ for all $t \in [0, 1]$.

LEMMA 2.10. *(L, ∇) has an invariant Hermitian structure if and only if $2\pi i(\alpha - \bar{\alpha})$ is exact on L_*, where α is the connection form of (L, ∇), in which case the invariant Hermitian structure is unique up to a positive multiplicative constant.*

We want (L, ∇) to be related to the symplectic structure of (X, ω). This is accomplished by introducing the curvature of (L, ∇), denoted by $\operatorname{Curv}(L, \nabla)$:

LEMMA 2.11. *There exists a unique 2-form $\operatorname{Curv}(L, \nabla)$ on X such that $\pi^* \operatorname{Curv}(L, \nabla) = d\alpha$, where π is the projection $L_* \to X$.*

REMARK 2.12. Part of the geometrical significance of the curvature can be seen in the fact that if $\operatorname{Curv}(L, \nabla) = 0$ and if X is simply connected, then parallel translation is independent of path. For the general significance, see [**6**, Theorem 1.8.1].

Our prequantization setup will consist of a symplectic manifold (X, ω) and a line bundle with connection (L, ∇) having an invariant Hermitian structure and satisfying $\operatorname{Curv}(L, \nabla) = \omega$.

THEOREM 2.13. *(X, ω) admits a (L, ∇) of the above sort if and only if $[\omega] \in \operatorname{Im} \varepsilon$, where $[\omega]$ is the de Rham class of ω and ε is the canonical map of $H^2(X, \boldsymbol{Z})$ into $H^2(X, \boldsymbol{R})$. Moreover $c^{-1}(\varepsilon^{-1}[\omega])$ is the set of all possible isomorphism classes of line bundles L for which such a ∇ can be found, where $c(L)$ is the Chern class of L.*

For applications of 2.13 to the group case 2.3, the reader is referred to [**6**, Theorem 5.7.1] and [**7**, Theorem 17].

We next recall the group $E(L, \nabla)$.

DEFINITION 2.14. $E(L, \nabla)$ consists of all diffeomorphisms $\sigma: L \to L$ which send fibres into fibres, are linear on fibres, preserve the Hermitian structure, and satisfy $\sigma^*\alpha = \alpha$, where α is the connection form of (L, ∇). The induced map on X is denoted by $\check{\sigma}$.

Obviously T, the multiplicative group of complex numbers z such that $|z|=1$, is a subgroup of $E(L, \nabla)$.

THEOREM 2.15 *The sequence*

$$1 \to T \to E(L, \nabla) \overset{\vee}{\to} A(X, \omega)$$

is exact, where $A(X, \omega)$ is the group of symplectic automorphisms of (X, ω). The right-hand map sends σ to $\check{\sigma}$ and is surjective if X is simply connected.

Finally, we introduce the notion of polarization.

DEFINITION 2.16. Let (X, ω) be a symplectic manifold of dimension $2n$. A *polarization* of (X, ω) is a complex involutive distribution $E \subseteq (TX)_C$ of dimension n such that

(a) F is isotropic with respect to ω, and

(b) $F+\bar{F}$ is involutive, where $\bar{F}$ is complex conjugation.

F is *real* if $F \subseteq TX$. In this case the Frobenius theorem gives us a foliation of X and we will let X/F denote the space of leaves.

For the rest of these notes all polarizations will be real.

3. Quantization using half-densities. Let (X, ω, F) be a symplectic manifold of dimension $2n$ with real polarization F. Let (L, ∇) be a complex line bundle with connection and invariant Hermitian structure whose curvature is ω. We are going to associate with this data a Hilbert space $\mathfrak{H}^F$ by making use of the notion of a half-density normal to F.

Before doing this we define the general notion of density of order α and note certain properties of such densities.

DEFINITION 3.1. Let V be an m-dimensional vector space over $\boldsymbol{R}$. Let $\alpha>0$. Then $|\bigwedge^m|^\alpha V$ is the (one-dimensional) vector space of all $\boldsymbol{C}$-valued functions v defined on the space $\mathfrak{B}(V)$ of ordered bases of V such that

$$v(v_1, \ldots, v_m) = |\mathrm{Det}\, a_{ij}|^\alpha \, v(w_1, \ldots, w_m) \tag{3.2}$$

whenever $(v_1, \ldots, v_m)$, $(w_1, \ldots, w_m) \in \mathfrak{B}(V)$ and satisfy

$$w_i = \sum_{j=1}^{m} a_{ji} v_j, \qquad i=1, \ldots, m. \tag{3.3}$$

Observe that if $v_i \in |\bigwedge^m|^{\alpha_i} V$ for $i=1, 2$, then the product of functions $v_1 v_2 \in |\bigwedge^m|^{\alpha_1+\alpha_2} V$ and the complex conjugated function $\bar{v}_1 \in |\bigwedge^m|^{\alpha_1} V$. Moreover, if v_1 is nonnegative, we may define $v_1^\beta \in |\bigwedge^m|^{\alpha_1 \beta} V$ for all $\beta>0$. Finally, if $\mu \in \bigwedge^m V$, then, regarding μ as a function on $\mathfrak{B}(V)$ by means of the equation $\mu = \mu(v_1, \ldots, v_n)\,[v_1 \wedge \cdots \wedge v_n]$, we may form $|\mu| \in |\bigwedge^m|\, V$.

If $E \to Y$ is a real vector bundle of fibre dimension m over a manifold Y, the covariant functors $\mathfrak{B}$ and $|\bigwedge^m|^\alpha$ provide us with bundles $\mathfrak{B}(E)$ and $|\bigwedge^m|^\alpha E$ over Y. If $\dim Y = n$, $|\bigwedge^n|^\alpha T^*Y$ is called the bundle of *α-densities* on Y. When $\alpha = 1$, we say simply *densities*. Sections $v \in \Gamma_0 |\bigwedge^n| T^*Y$ may be integrated over Y exactly as n-forms are, with the added simplification that orientation does not enter in.

Using product, complex conjugation, and integration of densities, we turn $\Gamma_0 |\bigwedge^n|^{1/2} T^*Y$ into a pre-Hilbert space by setting $(v_1, v_2) = \int v_1 \bar{v}_2$ for $v_1, v_2 \in \Gamma_0 |\bigwedge^n|^{1/2} T^*Y$. More generally, let L be a complex Hermitian vector bundle over Y. Then $\Gamma_0[(|\bigwedge^n|^{1/2} T^*Y) \otimes L]$ becomes a pre-Hilbert space by setting

$$(v_1 \otimes s_1, v_2 \otimes s_2) = \int (s_1, s_2)\, v_1 \bar{v}_2 \tag{3.4}$$

where $v_1, v_2 \in \Gamma |\bigwedge^n| T^*Y$ and $s_1, s_2 \in \Gamma_0 L$.

Now let (X, ω, F) and (L, ∇) be as in the beginning of this section. $|\bigwedge^n|^{1/2}(TX/F)^*$ is the bundle of $\frac{1}{2}$-densities on X *normal to* F. We want to single out the sections of $|\bigwedge^n|^{1/2}(TX/F)^* \otimes L$ which are *covariant constant* along the leaves of F.

DEFINITION 3.5. A polarization F is *tractable* if the space X/F of leaves of X with respect to F is a manifold and if every F-leaf is simply connected.

Let F be a tractable polarization. $\dim X/F = n$. Let $\check{\varrho}$ be the projection of X onto X/F and suppose $\check{\varrho}(x) = y$. Then $\check{\varrho}^*(T_y^*(X/F)) = (T_x X/F_x)^* \subseteq T_x^* X$, and so we obtain the bijection $\check{\varrho}^* : |\bigwedge^n|^{1/2} T_y^*(X/F) \to |\bigwedge^n|^{1/2}(T_x X/F_x)^*$. Using this isomorphism we canonically identify with each other all $|\bigwedge^n|^{1/2}(T_x X/F_x)^*$ where x runs over a leaf of F.

Since for any F-leaf Λ, $\omega \mid \Lambda \equiv 0$, $\omega = \mathrm{Curv}(L, \nabla)$, and Λ is simply connected, $L \mid \Lambda$ has an absolute parallelism according to 2.12. Let π be the projection of L onto X. If $l_1, l_2 \in L$, we shall say $l_1 \sim l_2$ if and only if $\check{\varrho}(\pi(l_1)) = \check{\varrho}(\pi(l_2))$ and ∇-parallel translation along any arc γ running from $\pi(l_1)$ to $\pi(l_2)$ in the F-leaf containing $\pi(l_1)$ and $\pi(l_2)$ carries l_1 into l_2. $\sim$ is a smooth equivalence relation preserving Hermitian inner products, and the space of $\sim$ equivalence classes is a Hermitian line bundle over X/F which we denote by L/F. Letting ϱ denote the projection of L onto L/F and letting π also denote the projection of L/F onto X/F we have the commutative diagram:

$$\begin{array}{ccc} L & \xrightarrow{\ \varrho\ } & L/F \\ {\scriptstyle \pi}\downarrow & & \downarrow{\scriptstyle \pi} \\ X & \xrightarrow{\ \check{\varrho}\ } & X/F \end{array} \tag{3.6}$$

Note that ϱ is a bijection on any fibre.

If $\check{\varrho}(x) = y$, we define $\varrho^* : (L/F)_y \to L_x$ to be $(\varrho \mid L_x)^{-1}$. Then $\varrho^* \otimes \check{\varrho}^*$ pulls back injectively sections of $|\bigwedge^n|^{1/2} T^*(X/F) \otimes L/F$ to sections of $|\bigwedge^n|^{1/2}(TX/F)^* \otimes L$.

DEFINITION 3.7. Sections in $\mathrm{Im}(\varrho^* \otimes \check{\varrho}^*)$ are called *covariant constant along the* *leaves of F*. Those in $(\varrho^* \otimes \check{\varrho}^*)\, \Gamma_0(|\bigwedge^n|^{1/2}\, T^*(X/F) \otimes L/F)$ have *compact support modulo F*.

Using the bijective correspondence between sections of $|\bigwedge^n|^{1/2}\, T^*(X/F) \otimes L/F$ with compact support and sections of $|\bigwedge^n|^{1/2}(TX/F)^* \otimes L$ covariant constant along the leaves of F and with compact support modulo F, we turn the latter into a pre-Hilbert space $\mathfrak{H}_0(H, F, L, \nabla)$ whose completion $\mathfrak{H}(X, F, L, \nabla)$ may be thought of as consisting of covariant constant measurable sections of $|\bigwedge^n|^{1/2}(TX/F)^* \otimes L$ modulo null sections. We shall usually abbreviate these spaces to $\mathfrak{H}_0^F$ and $\mathfrak{H}^F$, respectively.

Let F_1 and F_2 be two tractable polarizations which are transverse. Let $\check{\varrho}_i : X \to X/F_i$ be the natural projections.

DEFINITION 3.8. F_1 and F_2 are *completely transverse* if

$$\check{\varrho}_1 \times \check{\varrho}_2 : X \to (X/F_1) \times (X/F_2)$$

is proper.

Suppose that F_1 and F_2 are completely transverse. We shall now define a sesquilinear pairing, called the *half-density pairing*, between $\mathfrak{H}_0^{F_1}$ and $\mathfrak{H}_0^{F_2}$. First we recall the isomorphism $\lambda : TX \to T^*X$ defined by $\langle \lambda v, w\rangle = \omega(v, w)$ for $v, w \in T_xX$, $x \in X$. We use λ to transfer ω to $T^*X : \hat{\omega}(\lambda v, \lambda w) = \omega(v, w) = \langle \lambda v, w\rangle$. Let $x \in X$. Now $\hat{\omega}$ establishes a complete duality between $(T_xX/F_{1x})^*$ and $(T_xX/F_{2x})^*$. Thus given $(\alpha_1, \ldots, \alpha_n) \in \mathfrak{B}((T_xX/F_{1x})^*)$, there exists a unique $(\beta_1, \ldots, \beta_n) \in \mathfrak{B}((T_xX/F_{2x})^*)$ such that $\hat{\omega}(\beta_i, \alpha_j) = \delta_{ij}$. Let $v_i \in |\bigwedge^n|^{1/2}(T_xX/F_{ix})^*$. It is easy to see that $v_1(\alpha_1, \ldots, \alpha_n) \times (v(\beta_1, \ldots, \beta_n))^-$ is independent of the choice of $(\alpha_1, \ldots, \alpha_n)$. We define this number to be $(v_1 \times \bar{v}_2)(x)$. If, moreover, $z_i \in L_i$, we may form

$$\langle z_1, z_2\rangle\, (v_1 \times \bar{v}_2)(x)\, |\omega^n|_x \in |\bigwedge^{2n}|\, T_x^*X. \tag{3.9}$$

In this expression $|\omega^n| \in |\bigwedge^{2n}|^{1/2}\, T^*X$ is defined in terms of ω^n, the nth exterior power of ω in $\bigwedge T^*X$. One checks that this defines a sesquilinear map

$$\{|\bigwedge^n|^{1/2}(T_xX/F_{1x})^* \otimes L_x\} \times \{|\bigwedge^n|^{1/2}(T_xX/F_{2x})^* \otimes L_x\} \to |\bigwedge^{2n}|\, T_x^*X.$$

Remembering the complete transitivity of F_1 and F_2, we obtain the sesquilinear map $\langle\cdot, \cdot\rangle : \mathfrak{H}_0^{F_1} \times \mathfrak{H}_0^{F_2} \to \Gamma_0 |\bigwedge^{2n}|\, T^*X$, which we may follow by integration to get the sesquilinear functional $(\cdot, \cdot)$ on $\mathfrak{H}_0^{F_1} \times \mathfrak{H}_0^{F_2}$. The discovery of this functional is due to Sternberg and Kostant. Note that $(\lambda_2, \lambda_1) = (\lambda_1, \lambda_2)^-$ for $\lambda_i \in \mathfrak{H}_0^{F_i}$.

DEFINITION 3.10. Let F_1 and F_2 be tractable polarizations which either are completely transverse or are identical. We say F_1 and F_2 are *unitarily related* providing there is a unitary isomorphism $U_{F_2F_1} : \mathfrak{H}^{F_1} \to \mathfrak{H}^{F_2}$ such that $(U_{F_2F_1}\lambda_1, \lambda_2) = (\lambda_1, \lambda_2)$ for $\lambda_i \in \mathfrak{H}_0^{F_i}$. $U_{F_2F_1}$ will be called the *intertwining isomorphism* for F_1 and F_2.

Note that $U_{F_1F_2} = U_{F_2F_1}^{-1}$ and that $U_{FF} = I$. Thus unitary relatedness is symmetric and reflexive.

Now let F be a tractable polarization of (X, ω). Let (L, ∇) be as above and let $\sigma \in E(L, \nabla)$. Then $\check{\sigma}$ is a symplectic automorphism of (X, ω). Plainly $\check{\sigma}^{*-1}$ sends

$(TX/F)^*$ onto $(TX/\check{\sigma}_* F)^*$ and hence sends $\mathfrak{B}((TX/F)^*)$ onto $\mathfrak{B}((TX/\check{\sigma}_* F)^*)$. This induces a map $\sigma: |\bigwedge^n|^{1/2}(TX/F)^* \to |\bigwedge^n|^{1/2}(TX/\check{\sigma}_* F)^*$ by means of the equation

$$(3.11) \qquad (\sigma v)(b) = v(\sigma^{-1} b), \qquad v \in |\bigwedge^n|^{1/2}(T_x X/F_x)^*,\ b \in \mathfrak{B}((TX/F)^*)_{\check{\sigma}x}.$$

Thus we obtain the map $\sigma: |\bigwedge^n|^{1/2}(TX/F)^* \otimes L \to |\bigwedge^n|^{1/2}(TX/\check{\sigma}_* F)^* \otimes L$ which gives rise to a map σ of sections $(\sigma s)(x) = \sigma s(\check{\sigma}^{-1} x)$. Note that $\check{\sigma}_* F$ is tractable. It is easy to check that σ maps $\mathfrak{H}^F$ unitarily onto $\mathfrak{H}^{\check{\sigma}_* F}$.

If F and $\check{\sigma}_* F$ are unitarily related, we have $U = U_{\check{\sigma}_* F, F}$ as in 3.10. From this we manufacture the unitary automorphism $\sigma^F = U^{-1}\sigma$ of $\mathfrak{H}^F$. In this case we have intrinsically associated to σ a unitary operator on $\mathfrak{H}^F$; that is, we have *quantized* σ.

Quantization heretofore has dealt almost exclusively with the situation where $F = \check{\sigma}_* F$. (See [1] and [7] and the references there.)

4. Two examples. (a) Let V be a real n-dimensional vector space with dual V^*. Let $(p_1, \dots, p_n)$ be a basis of V and let $(q^1, \dots, q^n)$ be the basis of V^* dual to this. Then $q^1, \dots, q^n$ (resp. $p_1, \dots, p_n$) is a coordinate system on V (resp. V^*). Let $X = V \times V^*$ and set $\omega = \sum_{j=1}^n dp_j \wedge dq^j$, an invariantly defined 2-form on X making (X, ω) into a symplectic manifold. Let $L = X \times \boldsymbol{C}$ and define ∇ by means of the invariantly defined connection form $\alpha = \sum_{j=1}^n p_j dq^j + (2\pi i)^{-1}(dz/z)$ on L_* where z is the natural coordinate on $\boldsymbol{C}$. We identify sections s of L with $\boldsymbol{C}$-valued functions f on X by means of the equation $s(x) = (x, f(x))$ for $x \in X$.

Let F_1 be the "vertical" polarization given by the vector fields $\partial/\partial p_1, \dots, \partial/\partial p_n$. Then $\bigwedge^n (TX/F_1)^*$ is everywhere spanned by $dq^1 \wedge \cdots \wedge dq^n$, which we abbreviate to dq, and hence $|\bigwedge^n|^{1/2}(TX/F_1)^*$ is spanned by $|dq|^{1/2}$. Now $\nabla_\xi f = \xi f + 2\pi i \langle \sum_{j=1}^n p_j dq^j, \xi \rangle f(x)$ for $\xi \in T_x X$. Therefore f is F_1-covariant constant if and only if it is independent of $p_1, \dots, p_n$: We write $f = g \circ \check{\varrho}_1$. Plainly F_1 is tractable. We see that $\mathfrak{H}_0^{F_1}$ is just $\{(g \circ \check{\varrho}_1)|dq|^{1/2} : g \in C_c^\infty(V)\}$. Since X/F_1 may be identified with V by means of $\check{\varrho}_1 : (v, v^*) \mapsto v$, we see that $\|g \circ \check{\varrho}_1 |dq|^{1/2}\| = (\int_V |g|^2\, |dq|)^{1/2}$; in other words $\mathfrak{H}^F$ "is" just $L_2(V, |dq|)$.

Similarly if F_2 is the "horizontal" polarization given by $\partial/\partial q^1, \dots, \partial/\partial q^n$, then f is F_2-covariant constant if and only if it is of the form $f = \exp\{-2\pi i \sum p_j q^j\} \cdot (h \circ \check{\varrho}_2)$. Thus

$$\mathfrak{H}_0^{F_2} = \{\exp\{-2\pi i \textstyle\sum p_j q^j\}(h \circ \check{\varrho}_2)|dp|^{1/2} : h \in C_c^\infty(V^*)\},$$
$$\|\exp\{-2\pi i \textstyle\sum p_j q^j\}(h \circ \check{\varrho}_2)|dp|^{1/2}\| = (\int_{V^*} |h|^2\, |dp|^{1/2}),$$

and $\mathfrak{H}^{F_2}$ "is" just $L_2(V^*, |dp|)$.

Plainly F_1 and F_2 are completely transverse. Now $|\omega^n| = |dp \wedge dq|$ and $|dp|^{1/2} \times (|dq|)^{1/2} = 1$. Thus, in terms of coordinates,

$$(4.1) \qquad \begin{aligned} &(\exp\{-2\pi i \textstyle\sum p_j q^j\}(h \circ \check{\varrho}_2)|dp|^{1/2}, (g \circ \check{\varrho}_1)|dq|^{1/2}) \\ &= \iint \exp\{-2\pi i \textstyle\sum p_j q^j\}\, h(p)\,(g(q))^-\, |dp \wedge dq|. \end{aligned}$$

It follows immediately from 3.10 that if F_1 and F_2 are unitarily related, then the unitary isomorphism $U=U_{F_1F_2}:\mathfrak{H}^{F_2}\to\mathfrak{H}^{F_1}$ must be given by

$$U[\exp\{-2\pi i\sum p_j q^j\}\,(h\circ\varrho_2)\,|dp|^{1/2}]=\hat{h}\circ\varrho_1\,|dq|^{1/2},$$

where

$$\hat{h}(q)=\int\exp\{-2\pi i\sum p_j q^j\}\,h(p)\,|dp|. \tag{4.2}$$

The Plancherel theorem tells us that we have indeed defined a unitary U, so that F_1 and F_2 are unitarily related.

This example shows that the intertwining isomorphism of the half-density pairing generalizes the Fourier transform.

(b) In the setup of (a), let $V=\boldsymbol{R}$, identify V^* with $\boldsymbol{R}$ in the usual way, and let $F=F_1$. To aid in understanding, we write everything in terms of coordinates q and p. Consider the one-parameter group of symplectic automorphisms $\check{\sigma}_t$ given by $\check{\sigma}_t(q,p)=(q-tp,p)$. The unique $\sigma_t\in E(L,\nabla)$ lying above $\check{\sigma}_t$ such that $\sigma_t|_{L_{(0,0)}}$ is the identity is defined by $\sigma_t(x,z)=(\check{\sigma}_t x,\exp\{i\pi tp^2\}\,z)$. σ_t is a one-parameter subgroup of $E(L,\nabla)$. Using the action of σ_t on sections given by $(\sigma_t s)(x)=\check{\sigma}_t s(\sigma_t^{-1}x)$, we calculate that

$$(\sigma_t f)(q,p)=\exp\{i\pi tp^2\}\,f(q+tp,p), \tag{4.3}$$

where sections have been identified with functions as in (a).

As in (a), $\mathfrak{H}_0^F=\{g|dq|^{1/2}:g\in C_c^\infty(\boldsymbol{R})\}$, where g is identified with a function f on X by $f(q,p)=g(q)$. We see that

$$(\sigma_t[g|dq|^{1/2}])(q,p)=\exp\{i\pi tp^2\}\,g(q+tp)\,|dq+tdp|^{1/2},$$

that $\sigma_t[g|dq|^{1/2}]\in\mathfrak{H}_0^{\check{\sigma}_{t*}F}$, and that $\check{\sigma}_{t*}F$ is the polarization given by $\partial/\partial p-t\partial/\partial q$. Thus F and $\check{\sigma}_{t*}F$ are completely transverse when $t\neq 0$.

We now compute the pairing between $\mathfrak{H}^{\check{\sigma}_{t*}F}$ and $\mathfrak{H}^F$ and show that F and $\check{\sigma}_{t*}F$ are unitarily related, $t\neq 0$. One easily calculates that $|dq+tdp|^{1/2}\times(|dq|^-)^{1/2}=|t|^{1/2}$. Therefore, if $g|dq|^{1/2}$, $h|dq|^{1/2}\in\mathfrak{H}_0^F$, we have

$$\begin{aligned}(\sigma_t[g|dq|^{1/2}],\,h|dq|^{1/2})&\\ &=\iint\exp\{i\pi tp^2\}\,g(q+tp)\,(h(q))^-\,(|dq+tdp|^{1/2}\times(|dq|^-)^{1/2})\,|\omega^n|\\ &=|t|^{1/2}\iint\exp\{i\pi tp^2\}\,g(q+tp)\,(h(q))^-\,|dp\wedge dq|\\ &=|t|^{-1/2}\int\exp\{i\pi t^{-1}p^2\}\,(g*h^*)(p)\,|dp|,\end{aligned} \tag{4.4}$$

where $g*h^*$ is the convolution of g and h^*, $h^*(q)=(h(-q))^-$. We define the Fourier transform $\mathfrak{F}$ by

$$(\mathfrak{F}f)(k)=\int \exp\{-2\pi ipk\}\, f(p)\, |dp|. \tag{4.5}$$

Using a calculation from the method of stationary phase (see Hörmander [**3**, pp. 144–145]), we find that the right side of (4.4) may be evaluated using $\mathfrak{F}$ to give

$$\begin{aligned}(\sigma_t[g|dq|^{1/2}], h|dq|^{1/2}) &= \exp\left\{\frac{i\pi}{4}\operatorname{sgn} t\right\}\int \exp\{-i\pi tk^2\}\,\mathfrak{F}(g*h^*)(k)\,dk \\ &= \exp\left\{\frac{i\pi}{4}\operatorname{sgn} t\right\}\int \exp\{-i\pi tk^2\}\,(\mathfrak{F}g)(k)\,((\mathfrak{F}h)(k))^-\,dk.\end{aligned} \tag{4.6}$$

The Plancherel theorem tells us that we can define the unitary isomorphism $U_t=U_{F,\check{\sigma}_{t*}F}$ of 3.10 as follows:

$$U_t\sigma_t=\exp\left\{\frac{i\pi}{4}\operatorname{sgn} t\right\}\mathfrak{F}^{-1}M_t\mathfrak{F}, \tag{4.7}$$

where $\mathfrak{H}^F$ is identified with $L^2(\boldsymbol{R})$ and M_t is multiplication by $k\to\exp\{-i\pi tk^2\}$.

Except for the factor $c_t=\exp\{i\pi(\operatorname{sgn} t)/4\}$, the quantization $\sigma_t^F=U_t\sigma_t$ of σ_t is a one-parameter group of unitary operators on $\mathfrak{H}^F$.

5. Half-forms. Insofar as it is possible, we should like the quantization of a one-parameter subgroup of $E(L,\nabla)$ to be a one-parameter unitary group. Example (b) of §4 shows that in the case of the particular group σ_t we have very nearly achieved our goal. The only problem is the factor c_t. Let us try to get rid of this problem by modifying the definition of the half-density pairing by replacing (λ_1,λ_2) $(\lambda_i\in\mathfrak{H}_0^{F_i})$ with $c_{F_1F_2}(\lambda_1,\lambda_2)$, where $c_{F_1F_2}$ is some complex constant, $|c_{F_1F_2}|=1$, *depending only on* F_1 and F_2. This has the effect of replacing the $U_{F_1F_2}$ of 3.10 by $c_{F_1F_2}U_{F_1F_2}$. The constants $c_{F_1F_2}$ can obviously be chosen so that the modified σ_t^F, σ_t as in §4, form a one-parameter group. It would *not* be possible to do this for the rotation group

$$\check{\tau}_t:(q,p)\to(q\cos t-p\sin t,\, q\sin t+p\cos t).$$

Indeed, choose $c_{F,\check{\tau}_{t*}F}$ to make the quantization τ_t^F of τ_t a unitary group for $|t|<\pi$. Then τ_t^F is determined for all t. Yet it is *still* not a group for all t. This can be shown by calculations similar to, but more complicated than, those of §4.

The trouble is that half-densities are the wrong objects with which to do quantization if we want representations, and not merely projective representations, of groups that move polarizations. The correct objects are half-forms, which we now proceed to define along the lines of §3. An alternative approach to half-forms due to Kostant will be given in §7(d).

To help make our notions precise, we rephrase Definition 3.1 as follows: Let V be an n-dimensional vector space over $\boldsymbol{R}$, $\dim V = m$. Recall that $GL(m, \boldsymbol{R})$ operates freely and transitively on $\mathfrak{B}(V)$ on the right:

$$(v_1, \ldots, v_m)\, g = (w_1, \ldots, w_m) \quad \text{where } g = (a_{ij}) \in GL(n, \boldsymbol{R}) \quad \text{and } w_i = \sum_{j=1}^{m} a_{ji} v_j. \tag{5.1}$$

In this notation (3.2) becomes (for $\alpha = \frac{1}{2}$)

$$v(bg) = |\operatorname{Det} g|^{-1/2}\, v(b), \qquad b \in \mathfrak{B}(V),\ g \in GL(m, \boldsymbol{R}). \tag{5.2}$$

We want objects v that transform according to $(\operatorname{Det} g)^{-1/2}$ instead of $|\operatorname{Det} g|^{-1/2}$. Since $(\operatorname{Det})^{-1/2}$ is double valued, we must pass to some sort of double covering of $GL(m, \boldsymbol{R})$. So let $ML(m, \boldsymbol{C})$ denote the double covering group of $GL(m, \boldsymbol{C})$ with covering map p. Let $ML(m, \boldsymbol{R}) = p^{-1} GL(m, \boldsymbol{R})$. $ML(m, \boldsymbol{C})$ [resp. $ML(m, \boldsymbol{R})$] is called the *complex* [resp. *real*] $m \times m$ *metalinear group*. Let χ be the unique holomorphic square root of the complex character $\operatorname{Det} \circ p$ of $ML(m, \boldsymbol{C})$ such that $\chi(1) = 1$.

LEMMA 5.3. (a) *$ML(m, \boldsymbol{R})_0$ projects isomorphically under p onto $GL(m, \boldsymbol{R})_0$.*
(b) *$|\chi|^{-1}\chi$ sets up an isomorphism of $ML(m, \boldsymbol{R})/ML(m, \boldsymbol{R})_0$ with the multiplicative group of powers of $i = (-1)^{1/2}$.*

DEFINITION 5.4. Let $E \to Y$ be a real vector bundle over Y, fibre-dim $E = m$. By means of (5.1), $\mathfrak{B}(E)$ becomes a right principal $GL(m, \boldsymbol{R})$ bundle. A *metalinear frame bundle* for E is a right principal $ML(m, \boldsymbol{R})$ bundle $\tilde{\mathfrak{B}}(E)$ over Y together with a map $p: \tilde{\mathfrak{B}}(E) \to \mathfrak{B}(E)$ such that

$$\begin{array}{ccc} \tilde{\mathfrak{B}}(E) \times ML(m, \boldsymbol{R}) & \longrightarrow & \tilde{\mathfrak{B}}(E) \\ \big\downarrow {\scriptstyle p \times p} & & \big\downarrow {\scriptstyle p} \\ \mathfrak{B}(E) \times GL(m, \boldsymbol{R}) & \longrightarrow & \mathfrak{B}(E) \end{array} \tag{5.5}$$

commutes, where the horizontal arrows are the right group actions. A $ML(m, \boldsymbol{R})$-equivariant equivalence class of metalinear frame bundles over Y is a *metalinear structure* on E. A *metalinear structure* on Y is a metalinear structure on T^*Y. Y together with a metalinear structure on Y will be called a *metalinear manifold.*

As usual, the existence of a metalinear structure on E is equivalent to the vanishing of a class in $H^2(X, \boldsymbol{Z}_2)$ characteristic of $\mathfrak{B}(E)$; and, when one metalinear structure is fixed, the set of all metalinear structures on E is parametrized by $H^1(X, \boldsymbol{Z}_2)$.

LEMMA 5.6. *If Y is oriented, then T^*Y supports a metalinear structure.*

DEFINITION 5.7. Let E, Y, and $\tilde{\mathfrak{B}}(E)$ be as in 5.4. Then $(\bigwedge^m)^{1/2} E$ is the bundle over Y where $(\bigwedge^m)^{1/2} E_y$ is the set of all $\nu:\tilde{\mathfrak{B}}(E)_y \to \boldsymbol{C}$ such that

$$\nu(\tilde{b}g)=(\chi(g)^-)^{-1}\,\nu(\tilde{b}), \quad \tilde{b}\in\tilde{\mathfrak{B}}(E)_y,\ g\in ML(m, \boldsymbol{R}). \tag{5.8}$$

The bundle structure on $(\bigwedge^m)^{1/2} E$ is the obvious one. If Y is equipped with a metalinear structure and $\dim Y=n$, then $(\bigwedge^n)^{1/2}\ T^*Y$ is the bundle of *half-forms* on Y.

REMARK 5.9. The reader will note that we have used $\chi(g)^-$ instead of $\chi(g)$ in (5.8). This is done to make constants arising from the method of stationary phase disappear correctly when we quantize in §6. It is also reflective of the fact that in metalinear geometry $(\bigwedge^n)^{1/2}\ T^*Y$ and $(\bigwedge^n)^{1/2}\ TY$ seem to be conjugate duals of each other (cf. §7(d)).

Suppose Y is a metalinear manifold, $\dim Y=n$. Let $\nu_1,\ \nu_2\in\Gamma_0(\bigwedge^n)^{1/2}\ T^*Y$. Then $(\nu_1\bar{\nu}_2)(\tilde{b}g)=|\mathrm{Det}(pg)|\,(\nu_1\bar{\nu}_2)(\tilde{b})$ for $\tilde{b}\in\tilde{\mathfrak{B}}(T^*Y)$, $g\in ML(n, \boldsymbol{R})$. Therefore there exists a unique $\mu\in\Gamma_0\,|\bigwedge^n|\ T^*Y$ such that $\nu_1\bar{\nu}_2=\mu\circ p$. If L is a complex Hermitian vector bundle over Y, we can turn $\Gamma_0[\{(\bigwedge^n)^{1/2}\ T^*Y\}\otimes L]$ into a pre-Hilbert space exactly as in (3.4), replacing $\nu_1\bar{\nu}_2$ by μ.

6. Quantization using half-forms. Let (X, ω) and (L, ∇) be as in §3. We want to define the bundle of half-forms *normal* to F and, using this bundle in place of the bundle of half-densities, define Hilbert spaces $\mathfrak{H}^F$ associated to tractable polarizations F, investigate the existence of pairings between spaces $\mathfrak{H}^{F_1}$ and $\mathfrak{H}^{F_2}$, and try to quantize groups of symplectic automorphisms. To do this we need the notion of a metaplectic manifold.

DEFINITION 6.1. Let (X, ω) be a symplectic manifold of dimension $2n$. The *symplectic frame bundle* is the bundle $\mathfrak{B}(X, \omega)$ over X of all ordered bases $(\alpha_1,\dots, \alpha_n, \beta_1,\dots, \beta_n)$ of T_x^*X, $x\in X$, such that $\hat{\omega}(\alpha_i, \alpha_j)=\hat{\omega}(\beta_i, \beta_j)=0$ and $\hat{\omega}(\beta_i, \alpha_j)=\delta_{ij}$.

We observe that $\mathfrak{B}(X, \omega)$ is a right principal $Sp(n, \boldsymbol{R})$ bundle in the obvious way (cf. (5.1)), where $Sp(n, \boldsymbol{R})$ is realized explicitly as the set of all matrices in $GL(2n, \boldsymbol{R})$ of block form

$$\begin{bmatrix} T_1 & T_2 \\ T_3 & T_4 \end{bmatrix} \tag{6.2}$$

where $T_i\in GL(n, \boldsymbol{R})$, ${}^tT_4T_1-{}^tT_2T_3=I$, and ${}^tT_3T_1$ and ${}^tT_4T_2$ are symmetric.

Let $Mp(n, \boldsymbol{R})$, the $n\times n$ *metaplectic group*, be the double covering group of $Sp(n, \boldsymbol{R})$ with covering map p.

DEFINITION 6.3. A *metaplectic frame bundle* of a symplectic manifold (X, ω), $\dim X = 2n$, is a right principal $Mp(n, \boldsymbol{R})$ bundle $\tilde{\mathfrak{B}}(X, \omega)$ over X together with a map $p: \tilde{\mathfrak{B}}(X, \omega) \to \mathfrak{B}(X, \omega)$ such that

$$(6.4) \qquad \begin{array}{ccc} \tilde{\mathfrak{B}}(X, \omega) \times Mp(n, \boldsymbol{R}) & \longrightarrow & \tilde{\mathfrak{B}}(X, \omega) \\ \Big\downarrow p \times p & & \Big\downarrow p \\ \mathfrak{B}(X, \omega) \times Sp(n, \boldsymbol{R}) & \longrightarrow & \mathfrak{B}(X, \omega) \end{array}$$

commutes, where the horizontal arrows are the right group actions. A $Mp(n, \boldsymbol{R})$-equivariant equivalence class of metaplectic framebundles over X is a *metaplectic structure* on X. X together with a metaplectic structure on X is called a *metaplectic manifold*.

The description of all metaplectic structures carried by (X, ω) is given in the same cohomological manner as for metalinear structures.

Now $GL(n, \boldsymbol{R})$ is embedded in $Sp(n, \boldsymbol{R})$ by means of the map

$$(6.5) \qquad A \mapsto \begin{bmatrix} A & 0 \\ 0 & {}^tA^{-1} \end{bmatrix}, \quad A \in GL(n, \boldsymbol{R});$$

and $U(n)$ is embedded in $Sp(n, \boldsymbol{R})$ *via* the map

$$(6.6) \qquad A + iB \mapsto \begin{bmatrix} A & -B \\ B & A \end{bmatrix}, \quad A + iB \in U(n).$$

We shall identify $GL(n, \boldsymbol{R})$ and $U(n)$ with their images in $Sp(n, \boldsymbol{R})$ for the rest of this article. Now $U(n)$ is a maximal compact subgroup of both $GL(n, \boldsymbol{C})$ and $Sp(n, \boldsymbol{R})$, and hence its double covering $MU(n)$ is a maximal compact subgroup of both $ML(n, \boldsymbol{C})$ and $Mp(n, \boldsymbol{C})$. Let G be the complete inverse image of $GL(n, \boldsymbol{R})$ under $p: Mp(n, \boldsymbol{R}) \to Sp(n, \boldsymbol{R})$.

PROPOSITION 6.7. *There is a unique isomorphism of G with $ML(n, \boldsymbol{R})$ which is the identity on $G \cap MU(n)$.*

We shall henceforth identify G with $ML(n, \boldsymbol{R})$. It is this identification that now allows us to define a canonical metalinear structure on $(TX/F)^*$, where F is a polarization of the metaplectic manifold X. To do this we must first construct an auxiliary bundle.

Let N be the subgroup of $Sp(n, \boldsymbol{R})$ consisting of matrices of the form

$$\begin{bmatrix} I & S \\ 0 & I \end{bmatrix}, \quad S = {}^tS. \tag{6.8}$$

Since N is simply connected, $p: Mp(n, \boldsymbol{R}) \to Sp(n, \boldsymbol{R})$ maps the identity component $\tilde{N}$ of $p^{-1}N$ isomorphically onto N. We henceforth identify $\tilde{N}$ with N. Our auxiliary bundle is $\tilde{\mathfrak{B}}(X, \omega)/N$. The two-fold covering projection $p: \tilde{\mathfrak{B}}(X, \omega) \to \mathfrak{B}(X, \omega)$ induces a two-fold covering projection $p: \tilde{\mathfrak{B}}(X, \omega)/N \to \mathfrak{B}(X, \omega)/N$.

Now the bundle $\mathfrak{B}(X, \omega)/N$ has the following geometrical significance: A linear independent n-tuple $(\alpha_1, \ldots, \alpha_n)$, $\alpha_i \in T_x^*X$, such that $\hat{\omega}(\alpha_i, \alpha_j) = 0$ is an ordered basis of a maximally isotropic (under $\hat{\omega}$) subspace of T_x^*X. Such subspaces are called *Lagrangian*, as are such ordered bases. Now the bundle $\mathfrak{F}(X, \omega)$ of Lagrangian ordered bases in T^*X admits a right free action by $GL(n, \boldsymbol{R})$ in the manner of (5.1), and a right $GL(n, \boldsymbol{R})$-orbit in $\mathfrak{F}(X, \omega)$ is simply the set of all Lagrangian ordered bases of a fixed Lagrangian subspace at some $x \in X$. Moreover, $GL(n, \boldsymbol{R})$ normalizes N in $Sp(n, \boldsymbol{R})$ and hence operates on $\mathfrak{B}(X, \omega)/N$ on the right. We have

PROPOSITION 6.9. *$\zeta: (\alpha_1, \ldots, \alpha_n, \beta_1, \ldots, \beta_n) \mapsto (\alpha_1, \ldots, \alpha_n)$ maps $\mathfrak{B}(X, \omega)$ onto $\mathfrak{F}(X, \omega)$ and induces a $GL(n, \boldsymbol{R})$-isomorphism $\bar{\zeta}$ of $\mathfrak{B}(X, \omega)/N$ with $\mathfrak{F}(X, \omega)$.*

Similarly, $ML(n, \boldsymbol{R})$ normalizes N in $Mp(n, \boldsymbol{R})$ and so operates on $\tilde{\mathfrak{B}}(X, \omega)/N$ on the right. Letting $\tilde{\mathfrak{F}}(X, \omega)$ denote $\tilde{\mathfrak{B}}(X, \omega)/N$, it is obvious that

$$\begin{array}{ccc} \tilde{\mathfrak{F}}(X, \omega) \times ML(n, \boldsymbol{R}) & \longrightarrow & \tilde{\mathfrak{F}}(X, \omega) \\ \Big\downarrow {\scriptstyle p \times p} & & \Big\downarrow {\scriptstyle p} \\ \mathfrak{F}(X, \omega) \times GL(n, \boldsymbol{R}) & \longrightarrow & \mathfrak{F}(X, \omega) \end{array} \tag{6.10}$$

commutes, where the horizontal arrows are the right group actions.

LEMMA 6.11. *The action of $ML(n, \boldsymbol{R})$ on $\tilde{\mathfrak{F}}(X, \omega)$ is free. Moreover, p projects $ML(n, \boldsymbol{R})$-orbits in $\tilde{\mathfrak{F}}(X, \omega)$ onto $GL(n, \boldsymbol{R})$-orbits in $\mathfrak{F}(X, \omega)$.*

All of this justifies regarding $\tilde{\mathfrak{F}}(X, \omega)$ as the bundle of *metalinear* Lagrangian frames on X.

Now let F be a polarization of the metaplectic manifold X. Then $\mathfrak{B}((TX/F)^*) \subseteq \mathfrak{F}(X, \omega)$. We define $\tilde{\mathfrak{B}}((TX/F)^*)$ to be $p^{-1}\,\mathfrak{B}((TX/F)^*)$, where $p: \tilde{\mathfrak{F}}(X, \omega) \to \mathfrak{F}(X, \omega)$.

THEOREM 6.12. *$\tilde{\mathfrak{B}}((TX/F)^*)$ together with $p \mid \tilde{\mathfrak{B}}((TX/F)^*)$ is a metalinear frame bundle for $(TX/F)^*$. The metalinear structure it defines on $(TX/F)^*$ depends only on the metaplectic structure on X and not on the particular choice of metaplectic frame bundle.*

Suppose that F is tractable and form X/F and $\check{\varrho}$ as in §3.

THEOREM 6.13. *There exist an $ML(n, \boldsymbol{R})$-bundle $\tilde{\mathfrak{B}}(T^*(X/F))$, a projection $p: \tilde{\mathfrak{B}}(T^*(X/F)) \to \mathfrak{B}((TX/F)^*)$, and an $ML(n, \boldsymbol{R})$-isomorphism $\check{\varrho}^*: \tilde{\mathfrak{B}}(T^*(X/F))_{\check{\varrho}(x)} \to \tilde{\mathfrak{B}}((TX/F)^*)_x$ depending smoothly on $x \in X$ such that*

$$\begin{array}{ccc} \tilde{\mathfrak{B}}((TX/F)^*)_x & \xleftarrow{\ \check{\check{\varrho}}^*\ } & \tilde{\mathfrak{B}}(T^*(X/F))_{\check{\varrho}(x)} \\ \Big\downarrow p & & \Big\downarrow p \\ \mathfrak{B}((TX/F)^*)_x & \xleftarrow{\ \check{\varrho}^*\ } & \mathfrak{B}(T^*(X/F))_{\check{\varrho}(x)} \end{array} \tag{6.14}$$

commutes. $\tilde{\mathfrak{B}}(T^*(X/F))$, p, and $\check{\varrho}^*$ are essentially unique subject to (6.14). $\mathfrak{B}(T^*(X/F))$ and p define a metalinear structure on X/F which depends only on the metaplectic structure of X. Conversely, each metalinear structure on X/F arises from a unique metaplectic structure on X in this way.

Let (L, ∇) be as in §3. We can now form $(\bigwedge^n)^{1/2}\,(TX/F)^* \otimes L$, define covariant constant sections of this bundle as in 3.7, and form the pre-Hilbert space $\mathfrak{H}_0^F$ from these covariant constant sections exactly following the program of §3.

Now let F_1 and F_2 be completely transverse tractable polarizations. We need to define a sesquilinear pairing, called the *half-form pairing*, between $\mathfrak{H}_0^{F_1}$ and $\mathfrak{H}_0^{F_2}$. The problem, following §3, comes down to defining $v_1 \times \bar{v}_2$ for $v_i \in (\bigwedge^n)^{1/2}(T_xX/F_{ix})^*$. The key to the solution is in noting that

(1) if $(\alpha_1, \dots, \alpha_n) \in \mathfrak{B}((T_xX/F_{1x})^*)$ and $(\beta_1, \dots, \beta_n) \in \mathfrak{B}((T_xX/F_{2x}))^*$ satisfy $\hat{\omega}(\beta_i, \alpha_j) = \delta_{ij}$, then $b = (\alpha_1, \dots, \alpha_n, \beta_1, \dots, \beta_n) \in \mathfrak{B}(X, \omega)_x$; and

(2) if $j = \left[\begin{smallmatrix} 0 & -I \\ I & 0 \end{smallmatrix}\right] \in Sp(n, \boldsymbol{R})$, then $bj = (\beta_1, \dots, \beta_n, -\alpha_1, \dots, -\alpha_n)$.

Therefore $\zeta b = (\alpha_1, \dots, \alpha_n)$ and $\zeta(bj) = (\beta_1, \dots, \beta_n)$.

To lift these observations to our metastructures, set

$$u(t) = \begin{bmatrix} I\cos t & -I\sin t \\ I\sin t & I\cos t \end{bmatrix},$$

let $\tilde{u}$ be the unique one-parameter subgroup of $ML(n, \boldsymbol{R})$ projecting onto u, and set $\tilde{j} = \tilde{u}(\pi/2)$. We then have

LEMMA 6.15. *For each $\tilde{a} \in \tilde{\mathfrak{B}}((TX/F_1)^*)_x$ there exists a unique $\tilde{b} \in \tilde{\mathfrak{B}}(X, \omega)_x$ such that $\tilde{\zeta}\tilde{b} = \tilde{a}$ and $\tilde{\zeta}(\tilde{b}\tilde{j}) \in \tilde{\mathfrak{B}}((TX/F_2)^*)_x$, where $\tilde{\zeta}$ is the canonical projection of $\tilde{\mathfrak{B}}(X, \omega)$ onto $\tilde{\mathfrak{F}}(X, \omega)$.*

Now if $g \in ML(n, \boldsymbol{R})$, then $\tilde{\zeta}(\tilde{b}g) = \tilde{a}g$ while $\tilde{\zeta}(\tilde{b}g\tilde{j}) = \tilde{\zeta}(\tilde{b}\tilde{j})\,(\tilde{j}^{-1}g\tilde{j})$. It is easily seen that $g \mapsto \tilde{j}^{-1}g\tilde{j}$ is the restriction of the Cartan involution of $ML(n, \boldsymbol{C})$ to $ML(n, \boldsymbol{R})$, and one calculates that $\chi(j^{-1}gj) = (\chi(g)^-)^{-1}$. Thus the following definition makes sense.

DEFINITION 6.16. Let F_1 and F_2 be completely transverse tractable polarizations of the metaplectic manifold (X, ω). Let $v_i \in (\bigwedge^n)^{1/2} (T_x X/F_{ix})^*$, $x \in X$. Then $(v_1 \times \bar{v}_2)(x)$ is defined to be $\exp\{i\pi n/4\}\, v_1(\tilde{\zeta}\tilde{b})\, (v_2(\tilde{\zeta}(\widetilde{bj})))^-$ for any $\tilde{b} \in \tilde{\mathfrak{B}}(X, \omega)$ such that $\tilde{\zeta}\tilde{b} \in \tilde{\mathfrak{B}}((TX/F_1)^*)_x$ and $\tilde{\zeta}(\widetilde{bj}) \in \tilde{\mathfrak{B}}((TX/F_2)^*)_x$.

Using 6.16, we may now proceed to define the *half-form pairing* $(\cdot, \cdot)$ on $\tilde{\mathfrak{H}}_0^{F_1} \times \tilde{\mathfrak{H}}_0^{F_2}$ by taking over (3.9) exactly and integrating. We will have $(\lambda_2, \lambda_1) = (\lambda_1, \lambda_2)^-$ again: Here the constant $\exp\{i\pi n/4\}$ of 6.16 is crucial. Definition 3.10 can be taken over *verbatim* with $\tilde{\mathfrak{H}}^{F_i}$ replacing $\mathfrak{H}^{F_i}$ and $\tilde{U}_{F_2 F_1}$ replacing $U_{F_2 F_1}$.

DEFINITION 6.17. A *metaplectic automorphism* of X is a smooth bijective map $\sigma: \tilde{\mathfrak{B}}(X, \omega) \to \tilde{\mathfrak{B}}(X, \omega)$ which commutes with the action of $ML(n, \boldsymbol{R})$ and such that

$$\begin{array}{ccc} \tilde{\mathfrak{B}}(X, \omega) & \xrightarrow{\sigma} & \tilde{\mathfrak{B}}(X, \omega) \\ \downarrow p & & \downarrow p \\ \mathfrak{B}(X, \omega) & \xrightarrow{\check{\sigma}^{*-1}} & \mathfrak{B}(X, \omega) \end{array} \tag{6.18}$$

commutes, where $\check{\sigma}: X \to X$ is the map induced by σ on the base X. The group of metaplectic automorphisms of X is denoted by $\tilde{A}(X, \omega)$.

DEFINITION 6.18. Let X be a metaplectic manifold with (X, ω) and (L, ∇) as in §3. Then $\tilde{E}(L, \nabla)$ is the subgroup of all $\sigma = (\sigma_1, \sigma_2) \in \tilde{A}(X, \omega) \times E(L, \nabla)$ such that $\check{\sigma}_1 = \check{\sigma}_2$. We set $\check{\sigma} = \check{\sigma}_1 = \check{\sigma}_2$ for $\sigma \in \tilde{E}(L, \nabla)$.

Let F be a tractable polarization of (X, ω) and let $\sigma \in \tilde{E}(L, \nabla)$. σ induces a map $\sigma: (\bigwedge^n)^{1/2} (TX/F)^* \to (\bigwedge^n)^{1/2} (TX/\check{\sigma}_* F)^*$ exactly as in (3.11). From this we obtain a unitary isomorphism $\sigma: \tilde{\mathfrak{H}}^F \to \tilde{\mathfrak{H}}^{\check{\sigma}_* F}$. If F_1 and F_2 are unitarily related in the metaplectic sense, we can form $\sigma^F = \tilde{U}^{-1}_{\check{\sigma}_* F, F} \sigma$ on $\tilde{\mathfrak{H}}^F$ thereby quantizing σ.

7. Justifications and prospects. (a) All of the foregoing machinery finds its justification in the production of smooth quantizations in some situations.

Let (X, ω) and $p_1, \ldots, p_n, q^1, \ldots, q^n$ be as in §4(a). Then the trivial right principal $Sp(n, \boldsymbol{R})$-bundle $X \times Sp(n, \boldsymbol{R})$ is identified with $\mathfrak{B}(X, \omega)$ by means of the map $(x, g) \mapsto \varepsilon_x g$, where ε_x is $(dq^1, \ldots, dq^n, dp_1, \ldots, dp_n)$ at x. Setting $\tilde{\mathfrak{B}}(X, \omega) = X \times Mp(n, \boldsymbol{R})$ and defining $p: \tilde{\mathfrak{B}}(X, \omega) \to \mathfrak{B}(X, \omega)$ by $p(x, \sigma) = (x, p\sigma)$, we give (X, ω) a metaplectic structure. $Mp(n, \boldsymbol{R})$ is embedded in $\tilde{A}(X, \omega)$ by setting, for $\sigma \in Mp(n, \boldsymbol{R})$, $\sigma(x, g) = (\check{\sigma}x, \tilde{j}^{-1}\sigma\tilde{j}g)$, where $\check{\sigma}$ is defined as the linear automorphism of $X = V \times V^*$ which sends the ordered basis $b = (p_1, \ldots, p_n, q^1, \ldots, q^n)$ of X into $b(p\sigma)$. (Recall that $p\sigma \in Sp(n, \boldsymbol{R})$.)

Now let (L, ∇) and F_1 be as in §4(a) and set $F = F_1$. Then $\tilde{\mathfrak{H}}^F$ is isomorphic to $L_2(V, |dq|)$. If $\sigma \in Mp(n, \boldsymbol{R})$, then $\check{\sigma}$ lifts uniquely to a member of $E(L, \nabla)$ which leaves the fibre $L_{(0,0)}$ pointwise fixed. Thus $Mp(n, \boldsymbol{R})$ is naturally embedded in $\tilde{E}(L, \nabla)$.

THEOREM 7.1. *Let S be the set of $\sigma \in Mp(n, \boldsymbol{R})$ such that either $F = \check{\sigma}_* F$ or F*

and $\check{\sigma}_* F$ are transverse. If $\sigma \in S$, then $\sigma^F : \mathfrak{H}^F \to \mathfrak{H}^F$ is defined. Moreover, $\sigma \mapsto \sigma^F$, $\sigma \in S$, is the restriction to S of the conjugate of the metaplectic representation of van Hove-Shale-Segal-Weil (*cf.* **[11]**, **[10]**, **[9]**, **[12]**).

As a second example, one may try quantizing the geodesic flow on $X = T^*Y$ where Y is a metalinear Riemannian manifold. By 6.13, X has unique metaplectic structure associated to the given metalinear structure on Y. If Y satisfies certain geometrical assumptions, we have a globally defined one-parameter subgroup $\check{\sigma}_t$ of $A(X, \omega)$ whose generating function (in the sense of symplectic geometry) is the Riemannian metric on TY transferred to X via $\lambda : TY \to T^*Y$ and which lifts uniquely to a one-parameter subgroup σ_t of $\tilde{E}(L, \nabla)$ leaving L_α pointwise fixed for all α in the O-section of X. Let F be the polarization of (X, ω) determined by the fibration $X \to Y$. Then the $\check{\sigma}_{t*}F$ and F are completely transverse for $t \neq 0$ and we can quantize σ_t to get $\sigma_t^F : \mathfrak{H}^F \to \mathfrak{H}^F$. One can show that σ_t^F is C^∞ on a dense subspace $\mathfrak{D}$ of $\mathfrak{H}^F$ and that $D_t \sigma_t^F|_{t=0} = i\Delta$ on $\mathfrak{D}$, where Δ is the operator on $\mathfrak{H}^F$ induced by the Laplace-Beltrami differential operator. In any case σ_t^F does *not* seem to be a one-parameter group, except in special instances.

(b) One can try to construct intertwining operators for group representations using the half-density or half-form pairings. As an example, let us consider the continuous principal series for $SL(2, \boldsymbol{R})$. We begin by following [**6**, Corollary to Theorem 5.7.1] and the terminology therein. (See also [7].) Let $\mathfrak{g}$ be the Lie algebra of $G = SL(2, \boldsymbol{R})$ and let O be a hyperbolic orbit of $\mathfrak{g}^*$. Let $f \in O$ and let G_f be the stabilizer x in G. Since O is integral we can choose $\Lambda \in G_f^\#$ and this determines a class of Hermitian line bundles with connection. Let (L, ∇) be a member of this class. Now O has two G-invariant polarizations F_1 and F_2 satisfying the Pukanszky condition ([**1**, 1.5] and [7]): The integral manifolds of these polarizations are the straight lines lying in O. F_1 and F_2 are both tractable, and so we form $\mathfrak{H}^{F_1}$ and $\mathfrak{H}^{F_2}$ as in §3. (We need not consider metastructures, since we will not be moving these polarizations.) F_1 and F_2 are transverse but *not* completely transverse; in fact, $O/F_i \simeq S^1$ but O is noncompact. Even if one passes to the universal covering spaces of O and G, the lifted polarizations are not completely transverse. However if one calculates *formally* the half-density pairing between $\mathfrak{H}^{F_1}$ and $\mathfrak{H}^{F_2}$, one obtains an expression that can be interpreted as a singular integral which leads exactly to the intertwining operator of Knapp and Stein [**4**]. This shows that development of the transverse but not completely transverse case will lead to very difficult analytic problems. The $SL(2, \boldsymbol{R})$ example is also important because it is the simplest case where F_1 and F_2 fail to be *Heisenberg related*: One cannot choose local coordinates $q^1, \ldots, q^n, p_1, \ldots, p_n$ so that F_1 is spanned by $\{\partial/\partial q^1, \ldots, \partial/\partial q^n\}$ and F_2 is spanned by $\{\partial/\partial p_1, \ldots, \partial/\partial p_n\}$.

A nontransverse theory needs also to be developed in order to handle the independence of polarization question for arbitrary semisimple groups, as well as to be able to quantize all of a Lie subgroup of $\tilde{E}(L, \nabla)$ and not just some subset as in 7.1.

(c) We make brief mention here of the application of the theory to Fourier integral operators. See [3] for a treatment involving half-densities but not the half-density pairing. By using metastructures and the half-form pairing one can account for, and indeed absorb, the Maslov bundle [3, p. 148 and §3.3] in an entirely natural way.

(d) One can reverse the procedure of these notes and derive the spaces $(\bigwedge^n)^{1/2}(TX/F)^*$ of 6.12 and 5.7 from the metaplectic representation. This construction is due to Kostant. Indeed, let (X, ω) be a metaplectic manifold of dimension $2n$. Then TX is canonically isomorphic to the associated bundle $\tilde{\mathfrak{B}}(X, \omega) \times_\gamma \boldsymbol{R}^{2n}$, where

$$(7.2) \qquad \gamma(\sigma)\, u = {}^t(p\sigma)^{-1}\, u \quad \text{for } \sigma \in Mp(n, \boldsymbol{R}),\ u \in \boldsymbol{R}^{2n},$$

and $p: Mp(n, \boldsymbol{R}) \to Sp(n, \boldsymbol{R})$ is the covering map.

DEFINITION 7.3. Let ξ be the representation of $Mp(n, \boldsymbol{R})$ on $L_2(\boldsymbol{R}^n)$ which is the composition of $\sigma \mapsto \tilde{j}^{-1}\sigma\tilde{j}$ with the conjugate of the metaplectic representation of $Mp(n, \boldsymbol{R})$ as realized in 7.1. The associated bundle $\tilde{\mathfrak{B}}(X, \omega) \times_\xi L_2(\boldsymbol{R}^n)$ is called the *metaplectic bundle* of (X, ω) and is denoted by $\tilde{\mathfrak{H}}(X, \omega)$. It is a bundle of Hilbert spaces over X.

Now the additive group $\boldsymbol{R}^{2n} = \boldsymbol{R}^n \times \boldsymbol{R}^n$ has the Weyl projective representation π on $L_2(\boldsymbol{R})$ given by

$$(7.4) \qquad (\pi_{(r,s)} f)(t) = \exp\{-2\pi i \langle s, t - \tfrac{1}{2}r\rangle\}\, f(t-r) \quad \text{for } r, s, t \in \boldsymbol{R}^n.$$

Then

$$(7.5) \qquad \pi(\gamma(\sigma)\, u) = \xi(\sigma)\, \pi(u)\, \xi(\sigma)^{-1} \quad \text{for } \sigma \in Mp(n, \boldsymbol{R}),\ u \in \boldsymbol{R}^{2n}.$$

Now it is obvious that $p(\tilde{j}^{-1}\sigma\tilde{j}) = {}^t(p\sigma)^{-1}$. Therefore π induces a projective representation π_x of T_xX on $\tilde{\mathfrak{H}}(X, \omega)_x$ for each $x \in X$. π_x satisfies

$$(7.6) \qquad \pi_x(v_1)\, \pi_x(v_2) = \exp\{-\pi i \omega(v_1, v_2)\}\, \pi_x(v_1 + v_2) \quad \text{for } v_1, v_2 \in T_xX.$$

Let $\mathscr{S}_x$ be the space of C^∞ vectors for π_x. Then $\mathscr{S}_x = \tilde{\mathfrak{B}}(X, \omega) \times_\xi \mathscr{S}(\boldsymbol{R}^n)$, where $\mathscr{S}(\boldsymbol{R}^n)$ is the complex Schwartz space over $\boldsymbol{R}^n$. Give $\mathscr{S}_x$ the weakest topology stronger than the topology inherited from $\tilde{\mathfrak{H}}(X, \omega)_x$ such that the operators $D_t \pi_x(tv)|_{t=0}$ on $\mathscr{S}_x$ are continuous for all $v \in T_xX$. Using the Hermitian inner product on $\tilde{\mathfrak{H}}(X, \omega)_x$, the complex conjugate $\mathscr{S}_x^*$ of the topological dual of $\mathscr{S}_x$ corresponds to a space of distributions on $\boldsymbol{R}^n$. We have $\mathscr{S}_x \subseteq \tilde{\mathfrak{H}}(X, \omega)_x \subseteq \mathscr{S}_x^*$. π_x extends to a continuous representation on $\mathscr{S}_x^*$.

Let F be a polarization of (X, ω). Then F_x is a Lagrangian subspace of T_xX. Let $(\mathscr{S}_x^*)^F = \{f \in \mathscr{S}_x^* \mid \pi_x(v) f = f \text{ for all } v \in F_x\}$. It is easy to see that this space is one dimensional: If one looks at $\mathscr{S}_x$ with respect to a frame $\tilde{b} \in \tilde{\mathfrak{B}}(X, \omega)$ such that

$\zeta\tilde{b}\in\mathfrak{B}((T_xX/F_x)^*)$, one sees $(\mathscr{S}_x^*)^F$ as constant multiples of the unit mass δ_0 at $O\in\boldsymbol{R}^n$.

Now we assert that $(\mathscr{S}^*)^F$ is isomorphic to the *conjugate dual* of $(\bigwedge^n)^{1/2}(TX/F)$. Indeed, if $v\in(\mathscr{S}_x^*)^F$ and $\tilde{b}\in\mathfrak{B}(X,\omega)_x$ satisfies $\zeta\tilde{b}\in\mathfrak{B}((T_xX/F_x)^*)$, then we may define $v(\zeta\tilde{b})$ by

$$\tilde{v}=\tilde{b}\times_{\xi}v(\zeta\tilde{b})\,\delta_0, \tag{7.7}$$

since $\xi(n)\,\delta_0=\delta_0$ for $n\in N$. Now it is easy to see that $\xi(g)\,\delta_0=\chi(g)^{-1}\,\delta_0$ for $g\in ML(n,\boldsymbol{R})$. It follows that

$$v((\zeta\tilde{b})\,g)=\chi(g)\,v(\zeta\tilde{b})\quad \text{for } g\in ML(n,\boldsymbol{R}). \tag{7.8}$$

This establishes the desired isomorphism.

If F_1 and F_2 are transverse, then the inner product on $\mathfrak{H}(X,\omega)_x$ extends to give a sesquilinear pairing between $(\mathscr{S}^*)^{F_1}$ and $(\mathscr{S}^*)^{F_2}$, which agrees with the pairing 6.16 when transferred to the conjugate dual bundles from $(\bigwedge^n)^{1/2}(TX/F_1)$ and $(\bigwedge^n)^{1/2}(TX/F_2)$. This is basically because if $v\in(\mathscr{S}^*)^{F_1}$ looks like a multiple of δ_0 with respect to $\tilde{b}$ such that $\zeta\tilde{b}\in\mathfrak{B}((TX/F_1)^*)$, it looks like a constant function with respect to $\tilde{b}$ such that $\zeta\tilde{b}\in\mathfrak{B}((TX/F_2)^*)$.

Shale [**10**] has shown ways in which the metaplectic representation of $Mp(n,\boldsymbol{R})$ resembles the spin representation of Spin$(n,\boldsymbol{R})$. This resemblance can be exploited to give the nontransverse pairing mentioned in (b) above, when we use Kostant's approach to half-forms. This approach also makes accessible the treatment of nonreal polarizations, a development needed in the theory. The existence of a pairing between Hilbert spaces coming from real and complex polarizations is suggested by Segal's duality transform [**8**, §4]. Much work needs to be done here, even on the formal level.

REFERENCES

1. L. Auslander and B. Kostant, *Polarization and unitary representations of solvable Lie groups*, Invent. Math. **14** (1971), 255–354. MR **45** # 2092.

2. M. Duflo, *Sur les extensions des représentations irréductible des groupes de Lie nilpotents*, Ann. Sci. École Norm. Sup. **5** (1972), 71–120.

3. L. Hörmander, *Fourier integral operators*. I, Acta Math. **127** (1971), 79–183.

4. A. W. Knapp and E. M. Stein, *Intertwining operators for semisimple groups*, Ann. of Math. (2) **93** (1971), 489–578.

5. B. Kostant, *On certain unitary representations which arise from a quantization theory*, Group Representations in Math. and Phys. (Battelle, Seattle, 1969 Rencontres), Lecture Notes in Phys., vol. 6, Springer, Berlin, 1970, pp. 237–253. MR **43** #2160.

6. ———, *Quantization and unitary representations*, Lecture Notes in Math., vol. 170, Springer-Verlag, Berlin, 1970, pp. 87–208.

7. C. C. Moore, *Representations of solvable and nilpotent groups and harmonic analysis on nil and solvmanifolds*. Proc. Sympos. Pure Math., vol. 26, Amer. Math. Soc., Providence, R.I., 1974, pp. 1–44.

8. I. E. Segal, *Tensor algebras over Hilbert spaces*. I, Trans Amer. Math. Soc. **81** (1956), 106–134. MR **17**, 880.

9. ———, *Transforms for operators and symplectic automorphisms over a locally compact abelian group*, Math. Scand. **13** (1963), 31–43. MR **29** # 486.

10. D. Shale, *Linear symmetrics of free boson fields*, Trans. Amer. Math. Soc. **103** (1962), 149–167. MR **25** #956.

11. L. van Hove, *Sur certaines représentations unitaires d'un groupe infine de transformations*, Acad. Roy. Belg. Cl. Sci. Mém. Coll. in-8° 26, (1951), no. 6, 102 pp. MR **15**, 198.

12. A. Weil, *Sur certains groupes d'opérateurs unitaires*, Acta Math. **111** (1964), 143–211. MR **29** # 2324.

UNIVERSITY OF CALIFORNIA, LOS ANGELES

HARMONIC ANALYSIS ON REDUCTIVE p-ADIC GROUPS

HARISH-CHANDRA

1. Introduction. The object of this article is to present fresh evidence in support of what I call the Lefschetz principle, which says that whatever is true for real reductive groups is also true for p-adic groups. It is a well-established fact in number theory that global arithmetic problems can often be analyzed into local questions at the individual primes. This is so, for example, in class-field theory, the arithmetic theory of algebras and the theory of quadratic forms. However, the local results at infinity are usually much too simple to provide an adequate insight into the complexity of the problem at finite primes. Therefore, we should regard it as a very fortunate fact that the theory of representations of real groups has a sufficiently rich structure so as to serve as a useful guide in our search for corresponding results for p-adic groups.

The work of Chevalley, Borel, Borel-Tits and Bruhat-Tits has given us a fairly good understanding of the structure of reductive p-adic groups. Harmonic analysis on such groups was started by Mautner and then pursued by Bruhat, Satake, Gelfand-Graev, Sally-Shalika and Macdonald. It was clear from the beginning that the theory of induced representations plays an important role here. In his thesis [**2**(a)] Bruhat had applied the theory of distributions on real groups to prove the irreducibility of such representations (see also [**2**(b)]). A few years ago [**3**(d)] I observed that the same ideas, when combined with the theory of cusp forms, give rather sharp results for finite groups. We shall now adapt this method to the p-adic case.

In their work on $GL(2)$, Jacquet and Langlands [**6**] introduced the notion of

AMS (MOS) subject classifications (1970). Primary 22E50; Secondary 22E35.

Key words and phrases. Admissible representations, theory of the constant term, supercusp forms, Maass-Selberg relations, functional equations of the Eisenstein integrals, Steinberg character, Plancherel measure.

an admissible representation. They noticed that one could work with such representations in an entirely algebraic fashion. (The analogue in the real case is the representation of the Lie algebra on the space of K-finite vectors.) Jacquet [**5**] then pushed this theory to $GL(n)$ and obtained the remarkable result that every admissible and irreducible representation of $GL(n)$ is contained in one which is induced from a supercuspidal representation of a parabolic subgroup. Jacquet's ideas enables us to obtain results about the asymptotic behavior of the matrix coefficients of admissible representations and to develop the theory of the constant term.

The whole subject now begins to show a strong resemblance to the theory of automorphic forms [**3**(c)]. In particular, we get the Maass-Selberg relations (§10) which then imply the functional equations of the Eisenstein integrals (§§11, 12). The c-functions appear in the asymptotic formula for the Eisenstein integral and the Plancherel measure turns out to be directly related to their absolute value (§17). The analogy with the real case (see [**3**(e)], [**3**(f)]) is therefore quite exact.

2. Smooth functions. By a totally disconnected (t.d.) space we mean a Hausdorff space X with the following property: Given a point $x\in X$ and a neighborhood U of x in X, we can choose an open and compact subset ω of X such that $x\in\omega\subset U$. Clearly a t.d. space is locally compact.

Let X be a t.d. space and S a set. A mapping $f: X\to S$ is called smooth if it is locally constant. Let V be a complex vector space. By $C^\infty(X:V)$ we mean the space of all smooth functions $f: X\to V$ and by $C_c^\infty(X:V)$ the subspace of those f which have compact support. We omit V in this notation when $V=\mathbf{C}$. One can identify $C_c^\infty(X:V)$ with $C_c^\infty(X)\otimes V$ by means of the mapping $i: C_c^\infty(X)\otimes V\to C_c^\infty(X:V)$ defined as follows: If $f\in C_c^\infty(X)$ and $v\in V$, then $i(f\otimes v)$ is the function $x\mapsto f(x)\,v$ $(x\in X)$ from X to V.

Let μ be a Radon measure on X. Then, for any $f\in C_c^\infty(X:V)$, define

$$\int_X f\,d\mu=\sum_{1\leqq i\leqq r}\mu(f^{-1}(v_i))\,v_i$$

where $v_1,\dots,v_r$ are all the distinct nonzero values of f. (The right side is understood to be zero if $f=0$.) Moreover if μ has compact support, we define

$$\int_X f\,d\mu=\int_X \varphi f\,d\mu \qquad (f\in C^\infty(X:V))$$

where φ is any function in $C_c^\infty(X)$ such that $\varphi=1$ on $\operatorname{Supp}\mu$.

Let End V be the algebra of all endomorphisms of V and π a mapping from X to End V such that, for any $v\in V$, the function $x\mapsto\pi(x)\,v$ from X to V is smooth. If μ is a Radon measure on X with compact support, we define an element $E(\pi,\mu)\in$

End V by

$$E(\pi, \mu)\, v = \int_X \pi(x)\, v\, d\mu \qquad (x \in V).$$

We shall often write $\int_X \pi(x)\, d\mu$ for $E(\pi, \mu)$.

As usual $C(X)$ denotes the space of all (complex-valued) continuous functions f on X and $C_c(X)$ the subspace of those f which have compact support. A distribution on X is a linear mapping of $C_c^\infty(X)$ into $\mathbf{C}$.

3. Admissible representations. Let G be a t.d. group and V a vector space over $\mathbf{C}$. By a representation of G on V, we mean a mapping $\pi: G \to \operatorname{End} V$ such that $\pi(1)=1$ and $\pi(xy)=\pi(x)\pi(y)$ $(x, y \in G)$. A vector $v \in V$ is called π-smooth if the mapping $x \mapsto \pi(x)\, v$ of G into V is smooth. Let V_∞ be the subspace of all smooth vectors. Then V_∞ is π-stable. Let π_∞ denote the restriction of π on V_∞. We say that π is smooth if $V = V_\infty$. Note that π_∞ is always smooth.

Let H be a closed subgroup of G and σ a smooth representation of H on V. Then we define a smooth representation $\pi = \operatorname{Ind}_H^G \sigma$ as follows: Let B denote the space of all smooth functions $\beta: G \to V$ such that

(1) $\beta(hx) = \sigma(h)\, \beta(x)$ $(h \in H, x \in G)$,

(2) $\operatorname{Supp} \beta$ is compact mod H.

Then π is the representation of G on B given by $(\pi(y)\beta)(x) = \beta(xy)$ $(x, y \in G, \beta \in B)$.

A representation π of G on V is said to be admissible if

(1) π is smooth,

(2) for any open subgroup H of G, $\dim V_H < \infty$.

Here V_H is the space of all $v \in V$ such that $\pi(h)v = v$ for all $h \in H$.

LEMMA 1. *Let H be a closed subgroup of G and σ an admissible representation of H. Then if G/H is compact,* $\operatorname{Ind}_H^G \sigma$ *is also admissible.*

Let π be a smooth representation of G on V and V' the (algebraic) dual of V. Then the dual representation π' of G on V' is given by

$$\langle \pi'(x)\, \lambda, v \rangle = \langle \lambda, \pi(x^{-1})\, v \rangle \qquad (x \in G, \lambda \in V', v \in V).$$

Put $\tilde{V} = (V')_\infty$ and $\tilde{\pi} = (\pi')_\infty$. Then $\tilde{\pi}$ is a smooth representation which is said to be contragredient to π.

LEMMA 2. *π is admissible if and only if $\tilde{\pi}$ is admissible. Moreover if π is admissible then* $(\tilde{\pi})^\sim = \pi$.

Let π be an admissible representation of G on V. By a matrix coefficient of π, we mean a function on G of the form

$$x \mapsto \langle \tilde{v}, \pi(x)\, v \rangle \qquad (x \in G),$$

where v and $\tilde{v}$ are fixed elements in V and $\tilde{V}$ respectively. Let $\mathscr{A}(\pi)$ denote the space spanned over $\boldsymbol{C}$ by all matrix coefficients of π.

Let $\mathfrak{T} = \operatorname{End}^{\circ} V$ denote the space of all $T \in \operatorname{End} V$ such that the mappings $x \mapsto \pi(x)\, T$, $x \mapsto T\pi(x)$ $(x \in G)$ are both smooth. Clearly $\mathfrak{T}$ is a subalgebra of $\operatorname{End} V$ and $\pi(x)\, T\pi(y) \in \mathfrak{T}$ for $x, y \in G$ and $T \in \mathfrak{T}$. Moreover since π is admissible, $\dim TV < \infty$ and therefore $\operatorname{tr} T$ is defined for $T \in \mathfrak{T}$. Put $f_T(x) = \operatorname{tr} T\pi(x)$ $(x \in G)$.

LEMMA 3. *Let π be an admissible representation of G. Then π is irreducible*[1] *if and only if the mapping $T \mapsto f_T$ of $\mathfrak{T}$ into $\mathscr{A}(\pi)$ is bijective.*

Let π be a smooth representation of G on V. We say that π is unitary, if there exists a positive-definite hermitian form H on V such that

$$H(\pi(x)\, v_1, \pi(x)\, v_2) = H(v_1, v_2) \qquad (v_1, v_2 \in V;\ x \in G).$$

A double representation π of G on V is a pair (π_1, π_2) where π_1 is a left-representation and π_2 a right-representation of G on V and the action of G on the left commutes with its action on the right. π is called smooth if π_1, π_2 are both smooth. Similarly π is unitary if there exists a pre-hilbertian structure on V with respect to which both π_1 and π_2 are unitary.

Let π_1, π_2 be two representations of G on V_1 and V_2 respectively. We write $\pi_1 \subset \pi_2$ if there exists a linear injection $T: V_1 \to V_2$ such that

$$\pi_2(x)\, T = T\pi_1(x) \qquad (x \in G).$$

Moreover π_1, π_2 are said to be equivalent if T can be taken to be a bijection.

We denote by $\mathscr{E}_c(G)$ the set of all equivalence classes of irreducible admissible representations of G and by $\mathscr{E}(G)$ the subset of those classes which are unitary.

Let $d_l x$ and $d_r x$ respectively denote the left- and right-invariant Haar measures on G. We assume that $d_r x = d_l x^{-1}$. Then

$$d_r x = \delta_G(x)\, d_l x.$$

Here $\delta = \delta_G$ is a homomorphism of G into the multiplicative group of positive real numbers, which is called the module of G. For $f, g \in C_c^\infty(G)$ define their convolution $f * g$ by

$$(f * g)(x) = \int f(y)\, g(y^{-1} x)\, d_l y = \int f(xy^{-1})\, g(y)\, d_r y.$$

Under this convolution product, $C_c^\infty(G)$ becomes an associative algebra. If π is an admissible representation of G on V, put

[1] Here irreducibility is meant in the algebraic sense.

$$\pi(f)=\int f(x)\,\pi(x)\,\delta(x)^{-1/2}\,d_r x=\int f(x)\,\pi(x)\,\delta(x)^{1/2}\,d_l x.$$

Then $\pi(f)\in\mathfrak{T}=\mathrm{End}^\circ V$ and $f\mapsto\pi(f)$ is a homomorphism of $C_c^\infty(G)$ into $\mathfrak{T}$. Put

$$\Theta_\pi(f)=\operatorname{tr}\pi(f)\qquad(f\in C_c^\infty(G)).$$

Then Θ_π is a distribution on G which is called the character of π.

Now assume G is unimodular so that $d_l x=d_r x=dx$ (say). Suppose K is an open compact subgroup and P a closed subgroup of G such that $G=KP$. We can normalize the Haar measures dk and $d_l p$ on K and P respectively in such a way that

$$\int\alpha(x)\,dx=\int\alpha(pk)\,d_l p\,dk\qquad(\alpha\in C_c(G))$$

and the total measure of K is 1. Let σ be an admissible representation of P on U. Then $\pi=\mathrm{Ind}_P^G(\delta_P^{1/2}\sigma)$ is an admissible representation of G.

THEOREM 1. *Let θ_σ and Θ_π denote the characters of σ and π respectively. Then*

$$\Theta_\pi(\alpha)=\theta_\sigma(\bar\alpha)\qquad(\alpha\in C_c^\infty(G))$$

where $\bar\alpha(p)=\int_K\alpha(kpk^{-1})\,dk\quad(p\in P)$.

Let $\mathfrak{H}$ denote the representation space of π. Then $\mathfrak{H}$ consists of all smooth functions $h:K\to U$ such that $h(pk)=\delta_P(p)^{1/2}\sigma(p)\,h(k)$ for $p\in P\cap K$ and $k\in K$. Now suppose σ is unitary with respect to a suitable pre-hilbertian structure on U. Define

$$\|h\|^2=\int_K|h(k)|^2\,dk\qquad(h\in\mathfrak{H}).$$

Then π is unitary with respect to this norm on $\mathfrak{H}$.

Let Z be a t.d. group which is abelian. By a quasi-character χ of Z we mean a continuous homomorphism of Z into $\boldsymbol{C}^\times$. χ is called a character if $|\chi(z)|=1$ $(z\in Z)$. We denote by $\mathfrak{X}(Z)$ the group of all quasi-characters of Z and by Z the subgroup of all characters.

Let G be a unimodular t.d. group and π an admissible and irreducible representation of G on $\mathfrak{H}$. Let Z_G denote the center of G and Z a closed subgroup of Z_G such that Z_G/Z is compact. Since π is irreducible, there exists an element $\chi_\pi\in\mathfrak{X}(Z)$ such that $\pi(z)=\chi_\pi(z)$ for all $z\in Z$.

THEOREM 2. *Suppose $\chi_\pi\in\hat Z$ and there exists an element $f\neq 0$ in $\mathscr{A}(\pi)$ such that*

$$\int_{G/Z} |f(x)|^2 \, dx^* < \infty.$$

Then π is unitary. Fix a pre-hilbertian structure on $\mathfrak{H}$ which is invariant under π. Then there exists a number $d(\pi) > 0$ such that

$$\int_{G/Z} |f_T(x)|^2 \, dx^* = d(\pi)^{-1} \|T\|^2$$

for all $T \in \operatorname{End}^\circ \mathfrak{H}$.

Here $f_T(x) = \operatorname{tr} T\pi(x)$ and dx^* is the Haar measure on G/Z. Moreover $\|T\|^2 = \operatorname{tr}(T^*T)$ where T^* is the adjoint of T with respect to the pre-hilbertian structure on $\mathfrak{H}$. (It is easy to see that T^* exists and lies in $\operatorname{End}^\circ \mathfrak{H}$.)

COROLLARY (SCHUR ORTHOGONALITY RELATIONS). *Let S, $T \in \operatorname{End}^\circ \mathfrak{H}$. Then*[2]

$$\int_{G/Z} \operatorname{conj} f_S(x) \cdot f_T(x) \, dx^* = d(\pi)^{-1} \operatorname{tr} S^*T.$$

Let ω denote the class of π in $\mathscr{E}_c(G)$. We say that π (or ω) is square-integrable, if it satisfies the conditions of Theorem 2. $d(\pi)$ is then called the formal degree of π. Since it depends only on ω, we can also denote it by $d(\omega)$. We shall denote by $\mathscr{E}_2(G)$ the set of all square-integrable elements of $\mathscr{E}(G)$.

Given an open compact subgroup K of G, let $C(G//K)$ denote the space of all functions in $C(G)$ which are constant on double cosets of K. Put $C_c(G//K) = C(G//K) \cap C_c(G)$. Then $H_K = C_c(G//K)$ is an algebra under convolution. An element $f \in C(G)$ is said to be H_K-finite if the space spanned by all functions of the form $h_1 * f * h_2$ ($h_1, h_2 \in H_K$) has finite dimension. Moreover f is said to be Hecke finite if it is H_K-finite for every open compact subgroup K of G.

Let $\mathscr{A}(G)$ denote the space of all functions $f \in C(G)$ such that

(1) $f \in C(G//K)$ for some open compact subgroup K of G,

(2) f is Hecke finite.

THEOREM 3. *$\mathscr{A}(G) = \bigcup_\pi \mathscr{A}(\pi)$ where π runs over all admissible representations of G.*

4. Intertwining forms. Let G be a t.d. group and π_1, π_2 admissible representations of G on V_1 and V_2 respectively. By an intertwining form (between π_1, π_2) we mean a bilinear form B on $V_1 \times V_2$ such that

[2] conj c denotes the conjugate of a complex number c.

$$B(\pi_1(x)\, v_1, \pi_2(x)\, v_2) = B(v_1, v_2) \qquad (x \in G,\ v_i \in V_i,\ i=1, 2).$$

Let $\mathscr{B}(\pi_1, \pi_2)$ denote the space of all intertwining forms and put

$$I(\pi_1, \pi_2) = \dim \mathscr{B}(\pi_1, \pi_2).$$

Similarly an intertwining operator from π_1 to π_2 is a linear mapping $T: V_1 \to V_2$ such that $T\pi_1(x) = \pi_2(x)\, T$. Let $\mathfrak{T}(\pi_2 \mid \pi_1)$ denote the space of all such operators and put

$$J(\pi_2 \mid \pi_1) = \dim \mathfrak{T}(\pi_2 \mid \pi_1).$$

LEMMA 4. *Let π_1, π_2 be two admissible representations of G. Then*

$$I(\tilde{\pi}_2, \pi_1) = J(\pi_2 \mid \pi_1).$$

COROLLARY. *Let π be an admissible unitary representation of G. Then π is irreducible if and only if $I(\tilde{\pi}, \pi) = J(\pi \mid \pi) = 1$.*

Let π be a representation of G on V. A subspace W of V is called π-admissible if W is π-stable and the restriction of π on W is admissible. Moreover π is called quasi-admissible if V is a union of π-admissible subspaces.

5. A theorem of Jacquet. Let Ω be a $\mathfrak{p}$-adic field (i.e., a t.d. field which is not discrete). We shall now use the terminology of [3(g), p. 8]. Let $\boldsymbol{G}$ be a connected reductive Ω-group and G the subgroup of all Ω-rational points in $\boldsymbol{G}$. Then G, being a closed subgroup of $GL(n, \Omega)$, is a t.d. group. Moreover since $\boldsymbol{G}$ is reductive, G is unimodular.

Fix a parabolic pair (p-pair) (P, A) in G and let $P = MN$ be the corresponding Levi decomposition.

LEMMA 5. *There exists a one-one correspondence between p-pairs $({}^*P, {}^*A)$ in M and p-pairs (P', A') in G such that $(P, A) \succ (P', A')$. It is given by*

$$({}^*P, {}^*A) = (M \cap P', A').$$

Moreover if $P' = M'N'$ and ${}^*P = {}^*M\,{}^*N$ are the corresponding Levi decompositions, then

$$M' = {}^*M, \quad A' = {}^*A, \quad N' = {}^*N \cdot N.$$

π being a representation of G on V, we denote by $V(N)$ the subspace of V spanned by all elements of the form $\pi(n)\, v - v\ (n \in N, v \in V)$. Then $V(N)$ is stable under $\pi(P)$. Hence we get a representation of P on $V/V(N)$. Clearly N lies in the kernel of this representation. Since $M \simeq P/N$, we get a representation σ of M on $V/V(N)$. The following theorem of Jacquet plays an important role in harmonic analysis.

THEOREM 4 (JACQUET). *Suppose π is quasi-admissible. Then the same holds for σ.*

6. The constant term and its applications. Fix a p-pair (P, A) in G $(P=MN)$. For any $t>0$, let $A^+(t)$ denote the set of all $a\in A$ such that[3] $|\xi_\alpha(a)|_p \geqq t$ for every simple root α of (P, A). (We denote by ξ_α the rational character of A corresponding to α.)

THEOREM 5.[4] *Fix $f\in\mathcal{A}(G)$. Then there exists exactly one element $f_P\in\mathcal{A}(M)$ with the following property: Given a compact subset ω in M, we can choose $t\geqq 1$ such that $\delta_P(ma)^{1/2} f(ma)=f_P(ma)$ for $m\in\omega$ and $a\in A^+(t)$.*

We call f_P the constant term of f along P.

Let $(*P, *A)$ be a p-pair in M and (P', A') the p-pair in G which corresponds to it under Lemma 5.

LEMMA 6. *Let $f\in\mathcal{A}(G)$. Then $(f_P)_{*P}=f_{P'}$.*

Let ${}^\circ\mathcal{A}(G)$ denote the space of all $f\in\mathcal{A}(G)$ such that $f_P=0$ for every p-pair (P, A) in G with $P\neq G$. Let Z be the split component of G. An element $g\in C(G)$ is called a supercusp form if

(1) Supp g is compact mod Z;

(2) let N be the radical of a parabolic subgroup (psgp) $P\neq G$, then

$$\int_N g(xn)\,dn=0 \qquad (x\in G).$$

(Here dn is the Haar measure of N.)

THEOREM 6. *Let f be an element in $\mathcal{A}(G)$. Then the following three conditions on f are mutually equivalent:*

(1) $f\in{}^\circ\mathcal{A}(G)$.

(2) Supp f *is compact* mod Z.

(3) f *is a supercusp form.*

Let π be an admissible representation of G. We say that π is supercuspidal if $\mathcal{A}(\pi)\subset{}^\circ\mathcal{A}(G)$.

LEMMA 7. *Let π be an admissible representation of G on V. Then the following two conditions are equivalent:*

[3] $|\ |_p$ denotes the usual p-adic absolute value.

[4] Cf. [3(f), Theorem 2].

(1) *π is supercuspidal.*

(2) *N being the radical of any psgp $P \neq G$, define $V(N)$ as in §5. Then $V(N)=V$.*

We note that $V(N)$ is also the space of all $v \in V$ with the following property: There exists an open compact subgroup U of N such that

$$\int_U \pi(n)\, v\, dn = 0.$$

Let π be an admissible and irreducible representation of G. By the central exponent of π, we mean the quasi-character $\chi_\pi \in \mathfrak{X}(Z)$ such that $\pi(z)=\chi_\pi(z)$ for $z \in Z$. χ_π depends only on the class ω of π in $\mathscr{E}_c(G)$. Hence we may denote it by χ_ω.

Let ${}^\circ\mathscr{E}_c(G)$ denote the set of all supercuspidal elements in $\mathscr{E}_c(G)$. Put ${}^\circ\mathscr{E}(G)=\mathscr{E}(G) \cap {}^\circ\mathscr{E}_c(G)$. It is clear from Theorem 2 that an element $\omega \in {}^\circ\mathscr{E}_c(G)$ lies in ${}^\circ\mathscr{E}(G)$ if and only if $\chi_\omega \in \hat{Z}$.

Let π be a quasi-admissible representation (§4) of G on V. For any $\chi \in \mathfrak{X}(Z)$ let $V(\chi)$ denote the space of all $v \in V$ with the following property: There exists an integer $n \geqq 1$ such that $(\pi(z)-\chi(z))^n v=0$ for all $z \in Z$. Since π is quasi-admissible, it follows without difficulty that

$$V = \sum_{\chi \in \mathfrak{X}(Z)} V(\chi)$$

where the sum is direct.

Let r denote the right-regular representation of G on $\mathscr{A}(G)$. It follows from the definition of $\mathscr{A}(G)$ that r is quasi-admissible. Hence

$$\mathscr{A}(G) = \sum_{\chi \in \mathfrak{X}(Z)} \mathscr{A}(G, \chi).$$

If $f \in \mathscr{A}(G)$, we denote by f_χ the component of f in $\mathscr{A}(G, \chi)$. Fix a p-pair (P, A) $(P=MN)$ and an element $\eta \in \mathfrak{X}(A)$. Then if $f \in \mathscr{A}(G)$, $f_{P,\eta}$ is the component of f_P in $\mathscr{A}(M, \eta)$.

Let π be an admissible and irreducible representation of G. By an exponent of π (with respect to (P, A)) we mean an element $\chi \in \mathfrak{X}(A)$ such that $f_{P,\chi} \neq 0$ for some $f \in \mathscr{A}(\pi)$. Let $\mathfrak{X}_\pi(P, A)$ denote the set of all such exponents. Then this set depends only on the class ω of π and so we may denote it by $\mathfrak{X}_\omega(P, A)$.

THEOREM 7. *Fix $\omega \in \mathscr{E}_c(G)$ and a p-pair (P, A) in G. Then $\mathfrak{X}_\omega(P, A)$ is a finite set.*

Fix $\chi \in \mathfrak{X}_\omega(P, A)$ and let r denote the right-regular representation of M on $\mathscr{A}(M)$. Then there exists an integer $d \geqq 1$ such that $(r(a)-\chi(a))^d f_{P,\chi}=0$ for all $f \in \mathscr{A}(\pi)$ and $a \in A$. By the multiplicity $d(\chi)$ of χ, we mean the least such integer d.

Moreover we say that χ is simple if $d(\chi)=1$.

Notice that $\mathfrak{X}_\omega(G, Z)=\{\chi_\omega\}$ where χ_ω is the central exponent of ω. Moreover $\mathfrak{X}_\omega(P, A)=\emptyset$ if $\omega\in{}^\circ\mathscr{E}_c(G)$ and $P\neq G$.

7. The complex structure on $^\circ\mathscr{E}_c(M)$. By a special torus in G, we mean a split component of some psgp of G. Given a special torus A, we denote by $\mathscr{P}(A)$ the set of all psgps P of G such that (P, A) is a p-pair. Let M be the centralizer of A in G. Then if $P\in\mathscr{P}(A)$, $P=MN$ where N is the radical of P. Moreover $\mathscr{P}(A)$ is a finite set.

Let A_1, A_2 be two special tori. By $\mathfrak{w}(A_2 \mid A_1)$ we mean the set of all homomorphisms $s: A_1\to A_2$ with the following property: There exists an element $y\in G$ such that $a^s=yay^{-1}$ for all $a\in A_1$. y is then called a representative of s in G.

Let A be a special torus. Then $\mathfrak{w}(A \mid A)$ is a group which we denote by $\mathfrak{w}(A)$ or sometimes by $\mathfrak{w}(G/A)$. If M is the centralizer and $\tilde{M}$ the normalizer of A in G, then $\mathfrak{w}(G/A)\simeq\tilde{M}/M$. Hence it is obvious that $\mathfrak{w}=\mathfrak{w}(A)$ operates on $\mathscr{E}_c(M)$. We say that an element $\omega\in\mathscr{E}_c(M)$ is unramified (in G) if $\omega^s\neq\omega$ for $s\neq 1$ in $\mathfrak{w}$.

Let $X(M)$ and $X(A)$ respectively denote the groups of all rational characters of M and A which are defined over Ω. Then the inclusion $i: A\to M$ defines an injective homomorphism $i^*: X(M)\to X(A)$ and $X(A)/i^*(X(M))$ is a finite group (see [**1**]). Moreover if $l=\dim A$, $X(A)$ is a free abelian group of rank l. We identify $X(M)$ with its image under i^* and put

$$\mathfrak{a}^*=X(A)\otimes \boldsymbol{R}=X(M)\otimes\boldsymbol{R},$$
$$\mathfrak{a}=\operatorname{Hom}(X(A), \boldsymbol{R})=\operatorname{Hom}(X(M), \boldsymbol{R}).$$

These are real vector spaces of dimension l which are dual to each other. We denote the corresponding complexifications by $\mathfrak{a}_c^*$ and $\mathfrak{a}_c$.

Define a homomorphism $H_M=H$ of M into $\operatorname{Hom}(X(M), \boldsymbol{Z})$ as follows. If $\chi\in X(M)$ and $m\in M$, then $\langle\chi, H(m)\rangle\in\boldsymbol{Z}$ is given by[5]

$$|\chi(m)|_{\mathfrak{p}}=q^{\langle\chi, H(m)\rangle}.$$

Let $^\circ M=\ker H$. Then $^\circ M$ contains every compact subgroup of M and $^\circ MA$ has finite index in M.

If $\omega\in{}^\circ\mathscr{E}_c(M)$ and $\nu\in\mathfrak{a}_c^*$, we define $\omega_\nu\in{}^\circ\mathscr{E}_c(M)$ as follows. Fix a representation σ of M in the class ω and put

$$\sigma_\nu(m)=\sigma(m)\, q^{i\langle\nu, H(m)\rangle} \qquad (m\in M).$$

Then ω_ν is the class of σ_ν. The mapping $(\nu, \omega)\mapsto\omega_\nu$ defines an action of the additive group $\mathfrak{a}_c^*$ on $^\circ\mathscr{E}_c(M)$. Now $L=H(M)$ is a lattice in $\mathfrak{a}$. Let L^* denote the dual lattice consisting of all $\lambda\in\mathfrak{a}^*$ such that

[5] q is the number of elements in the residue field of Ω.

$$\langle \lambda, H(m)\rangle \in 2\pi(\log q)^{-1}\mathbf{Z} \qquad (m\in M).$$

It is easy to see that L^* is exactly the stabilizer in $\mathfrak{a}_c^*$ of every $\omega\in {}^\circ\mathscr{E}_c(M)$. Hence the orbit $\mathfrak{o}_c$ of ω under $\mathfrak{a}_c^*$ may be identified with the complex Lie group $\mathfrak{a}_c^*/L^*$. We now introduce a complex structure on ${}^\circ\mathscr{E}_c(M)$ by demanding that each orbit be an open submanifold. It is easy to see that every orbit intersects ${}^\circ\mathscr{E}(M)$.

We shall sometimes call $\mathfrak{a}$ the real Lie algebra of A.

8. Some consequences of Bruhat's theory. Let (P_i, A_i) $(i=1, 2)$ be two p-pairs in G and $P_i=M_iN_i$ their Levi decompositions. Let σ_i be an admissible representation of M_i on V_i. Since $M_i\simeq P_i/N_i$, we may regard σ_i as a representation of P_i. Put

$$\pi_i=\mathrm{Ind}_{P_i}^G(\delta_{P_i}^{1/2}\sigma_i).$$

THEOREM 8. *Assume σ_1, σ_2 are both supercuspidal. Then $I(\pi_1, \pi_2)=0$ unless A_1 and A_2 are conjugate in G.*

Now suppose $A_1=A_2$. Then $M_1=M_2$ and so we can drop the subscripts and write M and A.

THEOREM 9. *Suppose σ_1, σ_2 are irreducible and supercuspidal representations of M and ω_1, ω_2 the corresponding elements in ${}^\circ\mathscr{E}_c(M)$. Let $\mathfrak{w}_0$ denote the set of all $s\in\mathfrak{w}=\mathfrak{w}(G/A)$ such that $\omega_1^s=\omega_2$. Then*[6]

$$I(\tilde{\pi}_1, \pi_2)\leqq[\mathfrak{w}_0].$$

COROLLARY. *Fix $\omega\in{}^\circ\mathscr{E}(M)$ and assume that ω is unramified in G. Then if $\sigma\in\omega$ and $P\in\mathscr{P}(A)$, $\pi=\mathrm{Ind}_P^G\delta_P^{1/2}\sigma$ is a unitary, admissible and irreducible representation of G.*

This is proved by combining Lemma 7 with Bruhat's theory (cf. [**3**(d), Theorem 1]).

9. Induced representations. Let π be an admissible and irreducible representation of G. A p-pair (P, A) in G $(P=MN)$ is said to be π-minimal if $\mathfrak{X}_\pi(P, A)\neq\emptyset$ but $\mathfrak{X}_\pi(P', A')=\emptyset$ for every p-pair $(P', A')\prec(P, A)$ $(P'\neq P)$. Let $\bar{P}$ denote the unique element in $\mathscr{P}(A)$ such that α is a root of $(\bar{P}, A)$ if and only if $-\alpha$ is a root of (P, A).

THEOREM 10.[7] *Fix a π-minimal p-pair (P, A) in G and $\chi\in\mathfrak{X}_\pi(P, A)$. Then there exists an irreducible, admissible, supercuspidal representation σ of M such that*

---

[6] $[S]$ stands for the number of elements in a set S.

[7] This is a generalization of a result of Jacquet [**5**].

(1) *χ is the central exponent of σ,*
(2) $\pi \subset \mathrm{Ind}_P^G(\delta_P^{1/2}\sigma)$.

COROLLARY. *Suppose $\chi \in \hat{A}$. Then π is unitary.*

We keep to the notation of the above theorem.

THEOREM 11. *If (P', A') is any p-pair in G, then every element of $\mathfrak{X}_\pi(P', A')$ is of the form $\chi \circ s$ where $s \in \mathfrak{w}(A \mid A')$. Moreover if (P', A') is also π-minimal, then A' is conjugate to A in G.*

For any $\chi \in \mathfrak{X}(A)$, put

$$\chi^*(a) = \mathrm{conj}\, \chi(a^{-1}) \qquad (a \in A).$$

Then $\chi \mapsto \chi^*$ is an automorphism of the group $\mathfrak{X}(A)$ and $\hat{A}$ is exactly the set of all fixed points of this automorphism.

LEMMA 8. *Suppose π is unitary and (P, A) is any p-pair in G. Then $\mathfrak{X}_\pi(\bar{P}, A) = \mathfrak{X}_\pi(P, A)^*$.*

Let A be a special torus in G and M its centralizer. Fix $\omega \in {}^\circ\mathscr{E}(M)$ and, for any $P \in \mathscr{P}(A)$, put

$$\pi_P = \mathrm{Ind}_P^G(\delta_P^{1/2}\sigma)$$

where $\sigma \in \omega$. Then π_P is unitary. Let $C(P, \omega)$ denote its class. If ω is unramified in G, $C(P, \omega) \in \mathscr{E}(G)$ (§8).

THEOREM 12. *Suppose $P_1, P_2 \in \mathscr{P}(A)$ and $s \in \mathfrak{w} = \mathfrak{w}(G/A)$. Then*

$$C(P_1, \omega) = C(P_2, \omega^s).$$

Hence we may write $C(\omega)$ (or $C_M^G(\omega)$) instead of $C(P, \omega)$.

THEOREM 13. *Suppose $\omega \in {}^\circ\mathscr{E}(M)$ is unramified in G. Then if $\pi \in C(\omega)$ and $P \in \mathscr{P}(A)$,*

$$\mathfrak{X}_\pi(P, A) = \{\chi \circ s\}_{s \in \mathfrak{w}}$$

where χ is the central exponent of ω. Moreover if (P', A') is any p-pair in G, every exponent $\eta \in \mathfrak{X}_\pi(P', A')$ is simple.

10. Maass-Selberg relations. Fix a maximal split torus A_0 in G and a maximal compact subgroup K of G satisfying the conditions of Bruhat and Tits [**3**(g), p. 16] with respect to A_0. By a standard torus we mean a special torus contained in A_0. A p-pair (P, A) in G is called semistandard if $A \subset A_0$.

Let τ be a smooth and unitary double representation of K on V. For any $\omega \in \mathscr{E}_c(G)$ put $\mathscr{A}(\omega) = \mathscr{A}(\pi)$ where $\pi \in \omega$. Let $C(G, \tau)$ denote the space of all functions $f: G \to V$ such that $f(k_1 x k_2) = \tau(k_1) f(x) \tau(k_2)$ $(k_1, k_2 \in K, x \in G)$, and $C_c(G, \tau)$ the subspace of all f with compact support. Moreover we denote by ${}^\circ C_c(G, \tau)$ the space of all $f \in C_c(G, \tau)$ such that

$$\int_N f(xn)\, dn = 0 \qquad (x \in G)$$

where N is the radical of a psgp $P \neq G$. Put

$$\mathscr{A}(G, \tau) = (\mathscr{A}(G) \otimes V) \cap C(G, \tau).$$

If (P, A) is a semistandard p-pair in G $(P = MN)$, let τ_M denote the restriction of τ on $K_M = K \cap M$. Then by Theorem 5 we have the mapping $f \mapsto f_P$ of $\mathscr{A}(G, \tau)$ into $\mathscr{A}(M, \tau_M)$. Fix $f \in \mathscr{A}(G, \tau)$. Then we write $f_P \sim 0$ if

$$\int_M (\varphi(m), f_P(m))\, dm = 0$$

for every $\varphi \in {}^\circ C_c(M, \tau_M)$. (Here dm is the Haar measure of M and the scalar product is in V.)

LEMMA 9. *Let f be an element in $\mathscr{A}(G, \tau)$ such that $f_P \sim 0$ for every semistandard p-pair (P, A) in G (including $(P, A) = (G, Z)$). Then $f = 0$.*

If π is an admissible representation of G and C its equivalence class, we put

$$\mathscr{A}(C, \tau) = \mathscr{A}(\pi, \tau) = C(G, \tau) \cap (\mathscr{A}(\pi) \otimes V).$$

Let M be the centralizer in G of a standard torus A. Fix $\omega \in {}^\circ\mathscr{E}(M)$. Then $\mathscr{A}(\omega, \tau_M)$ has a natural pre-hilbertian structure given by

$$\|g\|^2 = \int_{M/A} |g(m)|^2\, dm^* \qquad (g \in \mathscr{A}(\omega, \tau_M)),$$

where dm^* is the Haar measure on M/A. Put $\mathfrak{w} = \mathfrak{w}(G/A)$.

THEOREM 14. *Let ω be an element in ${}^\circ\mathscr{E}(M)$ which is unramified in G and put $C(\omega) = C_M^G(\omega)$. Fix $f \in \mathscr{A}(C(\omega), \tau)$. Then the following statements are true:*

(1) *If (P', A') is a semistandard p-pair in G, then $f_{P'} \sim 0$ unless A' is conjugate to A in G.*

(2) *If $P \in \mathscr{P}(A)$, then*

$$f_P \in \sum_{s \in \mathfrak{w}} \mathscr{A}(\omega^s, \tau_M)$$

where the sum is direct.

Let $f_{P,s}$ *denote the component of* f_P *in* $\mathscr{A}(\omega^s, \tau_M)$.

(3) $\|f_{P_1, s_1}\| = \|f_{P_2, s_2}\|$ *for* $P_1, P_2 \in \mathscr{P}(A)$ *and* $s_1, s_2 \in \mathfrak{w}$.

By the Maass-Selberg relations we mean the equalities in (3) (cf. [**3**(f), §7]).

COROLLARY. *Fix* $P \in \mathscr{P}(A)$ *and* $s \in \mathfrak{w}$. *Then the mapping* $f \mapsto f_{P,s}$ *of* $\mathscr{A}(C(\omega), \tau)$ *into* $\mathscr{A}(\omega^s, \tau_M)$ *is injective.*

11. The functional equations. We keep to the notation of Theorem 14. If $P_i = MN_i$ $(i=1, 2)$ are two elements of $\mathscr{P}(A)$, let $V(P_1 \mid P_2)$ denote the subspace of all $v \in V$ such that $\tau(n_1) v \tau(n_2) = v$ $(n_i \in N_i \cap K,\ i=1, 2)$. Then $V(P_1 \mid P_2)$ is stable under τ_M. Let $\tau_{P_1 \mid P_2}$ denote the restriction of τ_M on $V(P_1 \mid P_2)$. Put

$$L(\omega, P) = \mathscr{A}(\omega, \tau_{P \mid P}), \qquad \mathscr{L}(\omega, P) = \mathscr{A}(\omega, \tau_{P \mid \bar{P}})$$

for $\omega \in {}^\circ\mathscr{E}(M)$ and $P \in \mathscr{P}(A)$. For $\psi \in L(\omega, P)$, define an element $E(P:\psi) \in \mathscr{A}(G, \tau)$ as follows. Extend ψ and δ_P to functions on G by setting

$$\psi(kmn) = \tau(k)\,\psi(m), \quad \delta_P(kp) = \delta_P(p) \qquad (k \in K,\ m \in M,\ n \in N,\ p \in P)$$

and put

$$E(P:\psi:x) = \int \psi(xk)\,\tau(k^{-1})\,\delta_P(xk)^{-1/2}\,dk \qquad (x \in G)$$

where dk is the Haar measure on K so normalized that the total measure of K is 1.

We shall now assume that ω is unramified in G.

LEMMA 10. *Fix* $P \in \mathscr{P}(A)$. *Then* $\psi \mapsto E(P:\psi)$ *is a bijective mapping of* $L(\omega, P)$ *onto* $\mathscr{A}(C(\omega), \tau)$.

THEOREM 15. *Fix* $P_1, P_2 \in \mathscr{P}(A)$. *Then there exist unique linear mappings* $c_{P_2 \mid P_1}(s:\omega)$ $(s \in \mathfrak{w})$ *of* $L(\omega, P_1)$ *into* $\mathscr{L}(\omega^s, P_2)$ *such that*

$$E_{P_2}(P_1:\psi) = \sum_{s \in \mathfrak{w}} c_{P_2 \mid P_1}(s:\omega)\,\psi$$

for all $\psi \in L(\omega, P_1)$. *Moreover* $c_{P_2 \mid P_1}(s:\omega)$ *is bijective and there exists a number* $\mu(\omega) > 0$ *such that*

$$\mu(\omega)\,\|c_{P_2|P_1}(s:\omega)\,\psi\|^2=\|\psi\|^2$$

for all $P_1, P_2 \in \mathscr{P}(A)$, $s\in\mathfrak{w}$ *and* $\psi\in L(\omega, P_1)$. *Finally* $\mu(\omega)=\mu(\omega^s)$ $(s\in\mathfrak{w})$.

Put

$$^{\circ}c_{P_2|P_1}(s:\omega)=c_{P_2|P_2}(1:\omega^s)^{-1}\,c_{P_2|P_1}(s:\omega),$$
$$c^{\circ}_{P_2|P_1}(s:\omega)=c_{P_2|P_1}(s:\omega)\,c_{P_1|P_1}(1:\omega)^{-1}.$$

Then $^{\circ}c_{P_2|P_1}(s:\omega)$ is a unitary bijection of $L(\omega, P_1)$ onto $L(\omega^s, P_2)$. Similarly $c^{\circ}_{P_2|P_1}(s:\omega)$ is a unitary bijection of $\mathscr{L}(\omega, P_1)$ onto $\mathscr{L}(\omega^s, P_2)$.

THEOREM 16. *We have the functional equations*

$$E(P_2:{}^{\circ}c_{P_2|P_1}(s:\omega)\,\psi)=E(P_1:\psi)\qquad(\psi\in L(\omega, P_1))$$

and

$$^{\circ}c_{P_3|P_1}(st:\omega)={}^{\circ}c_{P_3|P_2}(s:\omega^t)\,{}^{\circ}c_{P_2|P_1}(t:\omega)$$

for $P_i\in\mathscr{P}(A)$ $(i=1, 2, 3)$ *and* $s, t\in\mathfrak{w}$.

Now put

$$E^{\circ}(P:\psi)=E(P:c_{P|P}(1:\omega)^{-1}\,\psi)$$

for $P\in\mathscr{P}(A)$ and $\psi\in\mathscr{L}(\omega, P)$.

THEOREM 17. *The following functional equations hold:*

$$E^{\circ}(P_2:c^{\circ}_{P_2|P_1}(s:\omega)\,\psi)=E^{\circ}(P_1:\psi)\qquad(\psi\in\mathscr{L}(\omega, P_1))$$

and

$$c^{\circ}_{P_3|P_1}(st:\omega)=c^{\circ}_{P_3|P_2}(s:\omega^t)\,c^{\circ}_{P_2|P_1}(t:\omega)$$

for $P_i\in\mathscr{P}(A)$ $(i=1, 2, 3)$ *and* $s, t\in\mathfrak{w}$.

Let P_1, P_2 be two elements in $\mathscr{P}(A)$ and (P', A') a p-pair in G $(P'=M'N')$ such that

$$(P', A')\succ(P_i, A)\qquad(i=1, 2).$$

Put $^*P_i=M'\cap P_i$. Then $(^*P_i, A)$ is a p-pair in M' and $\mathscr{L}(\omega, P_1)\subset\mathscr{L}(\omega, {}^*P_1)$. Let $^*\mathfrak{w}$ be the subgroup of those elements of $\mathfrak{w}$ which leave A' pointwise fixed. Then $^*\mathfrak{w}=\mathfrak{w}(M'/A)$.

THEOREM 18. *Let* $s\in{}^*\mathfrak{w}$. *Then* $c^{\circ}_{P_2|P_1}(s:\omega)=c^{\circ}_{^*P_2|^*P_1}(s:\omega)$ *on* $\mathscr{L}(\omega, P_1)$.

We now make the group $\mathfrak{w}$ operate on $\mathscr{A}(M, \tau_M)$ as follows. Fix $s \in \mathfrak{w}$ and choose a representative y for s in K. Then if $\varphi \in \mathscr{A}(M, \tau_M)$ we define $s\varphi$ to be the function $m \mapsto \tau(y)\, \varphi(y^{-1}my)\, \tau(y^{-1})\, (m \in M)$. Also if $P \in \mathscr{P}(A)$, we define $P^s = yPy^{-1}$.

LEMMA 11. *Let $P_1, P_2 \in \mathscr{P}(A)$ and $s, t \in \mathfrak{w}$. Then*

$$s c_{P_2|P_1}(t:\omega) = c_{P_2{}^s|P_1}(st:\omega),$$
$$c_{P_2|P_1}(t:\omega)\, s^{-1} = c_{P_2|P_1{}^s}(ts^{-1}:\omega^s).$$

Similar statements hold for $c^\circ_{P_2|P_1}(t:\omega)$ and ${}^\circ c_{P_2|P_1}(t:\omega)$ instead of $c_{P_2|P_1}(t:\omega)$.

12. Extension to the complex domain. Now assume that $\dim V < \infty$. Then $\dim \mathscr{A}(\omega, \tau_M) < \infty$ for any $\omega \in {}^\circ\mathscr{E}(M)$. For $P \in \mathscr{P}(A)$, $\psi \in L(\omega, P)$ and $\nu \in \mathfrak{a}_c^*$, define

$$E(P:\psi:\nu:x) = \int_K \psi(xk)\, \tau(k^{-1})\, q^{\langle i\nu - \varrho_P,\, H_P(xk)\rangle}\, dk \qquad (x \in G).$$

Here ψ is extended to a function on G as in § 11, the mapping $H_P: G \to \mathfrak{a}$ is defined by (see § 7) $H_P(kmn) = H(m)$ $(k \in K, m \in M, n \in N)$ and $\varrho_P \in \mathfrak{a}^*$ is given by $q^{\langle \varrho_P, H(m)\rangle} = \delta_P(m)^{1/2}$ $(m \in M)$. It is easy to see that $E(P:\psi:\nu) \in \mathscr{A}(G, \tau)$.

Let $\mathfrak{F}_c'(\omega)$ be the set of all $\nu \in \mathfrak{a}_c^*$ such that $\chi_{\omega,\nu} \circ s \neq \chi_{\omega,\nu}$ for any $s \neq 1$ in $\mathfrak{w} = \mathfrak{w}(G/A)$. (Here $\chi_{\omega,\nu}$ is the central exponent of ω_ν.) Then $\mathfrak{F}_c'(\omega)$ is an open, connected and everywhere dense subset of $\mathfrak{a}_c^*$. We observe that $\mathfrak{w}$ acts on $\mathfrak{a}_c$ and therefore by duality also on $\mathfrak{a}_c^*$.

THEOREM 19. *Fix $P_1, P_2 \in \mathscr{P}(A)$. Then, for any $\nu \in \mathfrak{F}_c'(\omega)$, there exist unique linear mappings*

$$c_{P_2|P_1}(s:\omega:\nu): L(\omega, P_1) \to \mathscr{L}(\omega^s, P_2) \qquad (s \in \mathfrak{w})$$

such that

$$E_{P_2}(P_1:\psi:\nu:m) = \sum_{s \in \mathfrak{w}} (c_{P_2|P_1}(s:\omega:\nu)\, \psi)\,(m)\, q^{i\langle s\nu,\, H(m)\rangle}$$

for $\psi \in L(\omega, P_1)$ and $m \in M$. Moreover the functions $\nu \mapsto c_{P_2|P_1}(s:\omega:\nu)$ are meromorphic on $\mathfrak{a}_c^*$.

We have seen in § 10 that the space $\mathscr{A}(\omega, \tau_M)$ has a natural hilbertian structure. Let

$$(c_{P_2|P_1}(s:\omega:\nu))^*: \mathscr{L}(\omega^s, P_2) \to L(\omega, P_1)$$

denote the adjoint of $c_{P_2|P_1}(s:\omega:\nu)$.

THEOREM 20. *There exists a complex-valued meromorphic function $\nu \to \mu(\omega:\nu)$ on $\mathfrak{a}_c^*$ such that*

$$\mu(\omega:\nu)\,(c_{P_2|P_1}(s:\omega:\bar{\nu}))^* c_{P_2|P_1}(s:\omega:\nu)$$

is the identity on $L(\omega, P_1)$ for all $P_1, P_2 \in \mathscr{P}(A)$ and $s \in \mathfrak{w}$. Moreover $\mu(\omega^s : s\nu) = \mu(\omega:\nu)$ and $\mu(\omega:\nu)$ is holomorphic and nonnegative on $\mathfrak{a}^*$.

As usual $\bar{\nu}$ stands for the complex conjugate of ν with respect to $\mathfrak{a}^*$.

Put $\mathfrak{F}'(\omega) = \mathfrak{a}^* \cap \mathfrak{F}'_c(\omega)$. Then $\mathfrak{F}'(\omega)$ is an open and dense subset of $\mathfrak{a}^*$. Fix $\nu \in \mathfrak{F}'(\omega)$. Then ω_ν is unramified in G, $\mu(\omega:\nu) = \mu(\omega_\nu)$ in the notation of §11 and $c_{P_2|P_1}(s:\omega:\nu)$ is bijective. Hence

$$^\circ c_{P_2|P_1}(s:\omega:\nu) = c_{P_2|P_2}(1:\omega^s:s\nu)^{-1}\, c_{P_2|P_1}(s:\omega:\nu),$$
$$c^\circ_{P_2|P_1}(s:\omega:\nu) = c_{P_2|P_1}(s:\omega:\nu)\, c_{P_1|P_1}(1:\omega:\nu)^{-1}$$

are defined as meromorphic functions of ν on $\mathfrak{a}_c^*$. They are holomorphic and unitary on $\mathfrak{a}^*$. Put

$$E^\circ(P:\psi:\nu) = E(P: c_{P|P}(1:\omega:\nu)^{-1}\,\psi:\nu) \qquad (\psi \in \mathscr{L}(\omega, P)).$$

Then $E^\circ(P:\psi:\nu)$ is also holomorphic in ν on $\mathfrak{a}^*$.

THEOREM 21. *The following functional equations hold:*

(1) $E(P_2: {}^\circ c_{P_2|P_1}(s:\omega:\nu)\,\psi: s\nu) = E(P_1:\psi:\nu)$ $(\psi \in L(\omega, P_1))$,

(2) $E^\circ(P_2: c^\circ_{P_2|P_1}(s:\omega:\nu)\,\psi: s\nu) = E^\circ(P_1:\psi:\nu)$ $(\psi \in \mathscr{L}(\omega, P_1))$,

(3) ${}^\circ c_{P_3|P_1}(st:\omega:\nu) = {}^\circ c_{P_3|P_2}(s:\omega^t:t\nu)\, {}^\circ c_{P_2|P_1}(t:\omega:\nu)$,

(4) $c^\circ_{P_3|P_1}(st:\omega:\nu) = c^\circ_{P_3|P_2}(s:\omega^t:t\nu)\, c^\circ_{P_2|P_1}(t:\omega:\nu)$,

for $P_i \in \mathscr{P}(A)$ $(i = 1, 2, 3)$, $s, t \in \mathfrak{w}$ and $\nu \in \mathfrak{a}_c^*$.

Let $\mathfrak{a}_0$ be the real Lie algebra of A_0 (see §10). Fix a positive-definite quadratic form on $\mathfrak{a}_0$ which is invariant under $\mathfrak{w}_0 = \mathfrak{w}(G/A_0)$. This defines a scalar product in $\mathfrak{a}_0$ and therefore also in $\mathfrak{a}$. Hence we can speak of the reflexion in $\mathfrak{a}$ corresponding to a given hyperplane. Fix any root α of A, let s_α denote the reflexion corresponding to the hyperplane defined by the equation $\alpha = 0$. Fix $P \in \mathscr{P}(A)$. A root α of (P, A) is called reduced if $r\alpha$ is not a root for $0 < r < 1$ $(r \in \mathbf{R})$. For any $\omega \in {}^\circ\mathscr{E}(M)$, let $\mathfrak{w}(\omega)$ denote the subgroup of all $s \in \mathfrak{w}$ such that $\omega^s = \omega$.

Fix a connected component $\mathfrak{o}_c$ of ${}^\circ\mathscr{E}_c(M)$ (see §7) and put $\mathfrak{o} = \mathfrak{o}_c \cap {}^\circ\mathscr{E}(M)$. Let $\Sigma_r(\mathfrak{o})$ denote the set of all reduced roots α of (P, A) such that $s_\alpha \in \mathfrak{w}(\omega)$ for some $\omega \in \mathfrak{o}$.

THEOREM 22. *Fix $P_1, P_2 \in \mathscr{P}(A)$, $s \in \mathfrak{w}$, $\omega \in \mathfrak{o}$ and for every $\alpha \in \Sigma_r(\mathfrak{o})$ choose an element $a_\alpha \in A$. Then the function*

$$\nu \mapsto \prod_{\alpha \in \Sigma_r(\mathfrak{o})} (\chi_{\omega,\nu}(a_\alpha^{s_\alpha}) - \chi_{\omega,\nu}(a_\alpha))\, c_{P_2|P_1}(s:\omega:\nu)$$

is holomorphic on $\mathfrak{a}_c^*$.

We recall that $\chi_{\omega, \nu}$ is the central exponent of ω_ν.

13. The product formula for $\mu(\omega:\nu)$. Let dh be a Haar measure on a closed and unimodular subgroup H of G. We shall say that dh is normalized if

$$\int_{H\cap K} dh = 1.$$

Fix $P\in\mathscr{P}(A)$ and choose a complete set of representatives $\{p_i\}_{i\in I}$ for $K\cap P\backslash P$ in P. Since $G=KP$, every element $x\in G$ can be written uniquely in the form $x=kp_i$ where $k\in K$ and $i\in I$. Let $p_i=mn$ $(m\in M, n\in N)$. We define $\kappa(x)=k$, $\mu(x)=m$. Then κ and μ are continuous mappings of G into K and M respectively.

Recall that we have a positive-definite scalar product in $\mathfrak{a}$ and hence also in $\mathfrak{a}^*$ (see § 12). Let ${}^+\mathfrak{a}^*$ denote the set of all $\nu\in\mathfrak{a}^*$ such that $\langle\nu, \alpha\rangle>0$ for every root α of (P, A). If $\nu\in\mathfrak{a}_c^*$, we write $\nu=\nu_R+(-1)^{1/2}\nu_I$ with ν_R and ν_I in $\mathfrak{a}^*$. Let $\mathfrak{F}_c(P)$ be the set of all $\nu\in\mathfrak{a}_c^*$ such that $\nu_I\in{}^+\mathfrak{a}^*$.

THEOREM 23. *Fix $\omega\in{}^\circ\mathscr{E}(M)$, $\nu\in\mathfrak{F}_c(P)$, $\psi\in\mathscr{A}(\omega, \tau_M)$ and put*

$$J(m)=\int_{\bar{N}} \tau(\kappa(\bar{n}))\,\psi(\mu(\bar{n})\,m)\,q^{\langle(-1)^{1/2}\nu-\varrho, H(\bar{n})\rangle}\,d\bar{n} \qquad (m\in M),$$

where $\varrho=\varrho_P$, $H=H_P$ and $d\bar{n}$ is the normalized Haar measure on $\bar{N}$. Then the above integral is convergent and

$$J=\gamma c_{\bar{P}\mid P}(1:\omega:\nu)\,\psi \qquad \text{if } \psi\in L(\omega, P),$$
$$=\gamma(c_{P\mid\bar{P}}(1:\omega:\bar{\nu}))^*\psi \quad \text{if } \psi\in\mathscr{L}(\omega, P).$$

Here

$$\gamma=\int_{\bar{N}} q^{-2\langle\varrho, H(\bar{n})\rangle}\,d\bar{n}.$$

It turns out that γ is actually independent of $P\in\mathscr{P}(A)$. Hence we may denote it by $\gamma(G/M)$.

Let Φ be the set of all reduced roots of (P, A). For $\alpha\in\Phi$, let A_α be the maximal torus lying in the kernel of ξ_α and M_α the centralizer of A_α in G. Then

$$({}^*P_\alpha, A)=(M_\alpha\cap P, A)$$

is a maximal p-pair in M_α and ${}^*P_\alpha=MN_\alpha$ where $N_\alpha=M_\alpha\cap N$. Let $\mu_\alpha(\omega:\nu)$ have the same meaning for (M_α, A) as $\mu(\omega:\nu)$ has for (G, A).

THEOREM 24. *Put* $\gamma=\gamma(G/M)$ *and* $\gamma_\alpha=\gamma(M_\alpha/M)$ $(\alpha\in\Phi)$. *Then*

$$\gamma^{-2}\mu(\omega:\nu)=\prod_{\alpha\in\Phi}\gamma_\alpha^{-2}\mu_\alpha(\omega:\nu)$$

for all $\omega\in{}^\circ\mathscr{E}(M)$ *and* $\nu\in\mathfrak{a}_c^*$.

14. The Schwartz space $\mathscr{C}(G)$. Fix $P_0\in\mathscr{P}(A_0)$, extend δ_{P_0} on G as in §11 and put

$$\Xi_G(x)=\Xi(x)=\int_K \delta_{P_0}(xk)^{-1/2}\,dk.$$

Then Ξ is independent of the choice of P_0, it lies in $\mathscr{A}(G)$ and it is constant on double cosets of KZ.

We recall that $G\subset GL(n,\Omega)$. Put $|x|=\max_{i,j}|x_{ij}|_{\mathfrak{p}}$ for any $n\times n$ matrix with coefficients in Ω. Then

$$\|x\|=\max(|x|,|x^{-1}|)\geqq 1 \qquad (x\in G).$$

Put $\sigma(x)=\log_q\|x\|$ and $\sigma_*(x)=\inf_{z\in Z}\sigma(xz)$ $(x\in G)$.

THEOREM 25. *Let* A_0^+ *be the set of all* $a\in A_0$ *such that* $|\xi_\alpha(a)|_{\mathfrak{p}}\geqq 1$ *for every root* α *of* (P_0,A_0). *Then we can choose an integer* $d\geqq 0$ *and numbers* $c_2\geqq c_1>0$ *such that*

$$c_1\leqq\delta_{P_0}(a)^{1/2}\,\Xi(a)\leqq c_2(1+\sigma_*(a))^d$$

for all $a\in A_0^+$.

COROLLARY 1. *We can choose* $r\geqq 0$ *such that*

$$\int_G \Xi(x)^2(1+\sigma(x))^{-r}\,dx<\infty,\qquad \int_{G/Z}\Xi(x)^2(1+\sigma_*(x))^{-r}\,dx^*<\infty.$$

Let (P,A) be a semistandard p-pair $(P=MN)$. Then A_0 is a maximal split torus contained in M. Hence we obtain the function Ξ_M on M by replacing (G,A_0,K) by (M,A_0,K_M) in the above definition.

COROLLARY 2. *Fix* $r\geqq 0$ *and* $\varepsilon>0$. *Then we can choose* $c=c(r,\varepsilon)>0$ *such that*

$$\delta_P(m)^{1/2}\int_{\bar N}\Xi(\bar n m)\,(1+\sigma(\bar n m))^{-(2d+r+\varepsilon)}\,d\bar n\leqq c\,\Xi_M(m)\,(1+\sigma(m))^{-r}\qquad (m\in M).$$

For any open compact subgroup K_0 of G, let $\mathscr{C}_{K_0}(G)$ denote the space of all functions $f\in C(G/\!/K_0)$ such that

$$\nu_r(f)=\sup_G |f(x)|\,\Xi(x)^{-1}(1+\sigma(x))^r<\infty$$

for every $r\geqq 0$. We topologize $\mathscr{C}_{K_0}(G)$ by means of the seminorms $\{\nu_r\}_{r\geqq 0}$. Put

$$\mathscr{C}(G)=\bigcup_{K_0}\mathscr{C}_{K_0}(G)$$

where K_0 runs over all open compact subgroups of G. Let $\mathfrak{S}$ be the collection of all seminorms ν on $\mathscr{C}(G)$ with the following property. For any K_0, the restriction of ν on $\mathscr{C}_{K_0}(G)$ is continuous. We topologize $\mathscr{C}(G)$ by means of $\mathfrak{S}$. Then $\mathscr{C}(G)$ becomes a complete locally convex Hausdorff space and it is easy to verify that $C_c^\infty(G)$ is a dense subspace of $\mathscr{C}(G)$.

LEMMA 12. *For $f\in\mathscr{C}(G)$, put*

$$f^{(P)}(m)=\delta_P(m)^{1/2}\int_N f(mn)\,dn \qquad (m\in M).$$

Then this integral converges and $f^{(P)}\in\mathscr{C}(M)$. Moreover $f\mapsto f^{(P)}$ is a continuous mapping of $\mathscr{C}(G)$ into $\mathscr{C}(M)$.

Let $l_0=\dim Z$ and $\chi_1,\dots,\chi_{l_0}$ be a set of generators of $X(G)$ (see §7). Define $\nu(x)$ by

$$q^{\nu(x)}=\max_{1\leqq i\leqq l_0}(|\chi_i(x)|,|\chi_i(x)|^{-1}) \qquad (x\in G).$$

For any $T\geqq 0$, let G_T denote the set of all $x\in G$ such that $\nu(x)\leqq T$.

THEOREM 26. *Fix $f\in\mathscr{A}(G)$. Then the following two conditions on f are equivalent.*

(1) *For any $T\geqq 0$,*

$$\int_{G_T}|f(x)|^2\,dx<\infty.$$

(2) *Given $T\geqq 0$ and $r\geqq 0$, we can choose $c\geqq 0$ such that*

$$|f(x)|\leqq c\,\Xi(x)\,(1+\sigma(x))^{-r}$$

for all $x\in G_T$.

A distribution on G is said to be tempered if it extends (uniquely) to a continuous linear function on $\mathscr{C}(G)$.

COROLLARY. *Let Θ be the character of an element in $\mathscr{E}_2(G)$. Then Θ is tempered.*

Let $\mathscr{E}'(G)$ denote the set of all elements $\omega \in \mathscr{E}(G)$ with the following property. Let (P, A) be any p-pair in G. Then (see §6) $\mathfrak{X}_\omega(P, A) \cap \hat{A} = \emptyset$ unless $P = G$. The following result is a simple consequence of Theorem 26.

THEOREM 27. $\mathscr{E}_2(G) \subset \mathscr{E}'(G)$.

It is now possible to generalize the results of §8 and replace ${}^\circ\mathscr{E}(M)$ by $\mathscr{E}'(M)$ in Theorem 9. In particular we have the following theorem.

THEOREM 28. *Fix $\omega \in \mathscr{E}'(M)$ and assume that ω is unramified in G. Then if $\sigma \in \omega$ and $P \in \mathscr{P}(A)$, $\pi = \mathrm{Ind}_P^G \delta_P^{1/2} \sigma$ is a unitary, admissible and irreducible representation of G.*

Fix $C \in \mathscr{E}_2(G)$ and $f \in \mathscr{A}(C)$. Then if (P, A) is a p-pair in G $(P = MN)$, it follows from Theorem 26 that the integral $\int_N f(xn)\, dn$ $(x \in G)$ converges absolutely.

THEOREM 29. *Suppose $P \neq G$ and $C \in \mathscr{E}_2(G)$. Then*

$$\int_N f(xn)\, dn = 0 \qquad (x \in G)$$

for $f \in \mathscr{A}(C)$.

15. The Steinberg character. We fix $P_0 \in \mathscr{P}(A_0)$ and call a p-pair (P, A) standard if $(P, A) \succ (P_0, A_0)$. Let $\mathfrak{S}$ be the set of all standard p-pairs. Then $\mathfrak{S}$ is a finite set. For $(P, A) \in \mathfrak{S}$, put

$$\pi_P = \mathrm{Ind}_P^G\, 1_P$$

where 1_P denotes the trivial representation of P. It follows from Theorem 1 that its character θ_P is a locally summable function on G.

Let l be the (absolute) rank of G and t an indeterminate. We denote by $D_G(x)$ the coefficient of t^l in $\det(t - \mathrm{Ad}(x) + 1)$ $(x \in G)$. As usual let G' be the set of all points $x \in G$ where $D_G(x) \neq 0$. It is easy to see that θ_P is locally constant on G'.

Put

$$\Theta = \sum_{(P, A) \in \mathfrak{S}} (-1)^{\dim A} \theta_P.$$

Then $(-1)^{\dim A_0}\Theta$ is called the Steinberg character of G. Borel and Serre have proved that it is, in fact, the character of an element in $\mathscr{E}_2(G)$.

Let Γ be a Cartan subgroup of G and A_Γ the maximal split torus contained in Γ. Let M_Γ be the centralizer of A_Γ in G and $\mathfrak{g}$ and $\mathfrak{m}_\Gamma$ the Lie algebras (over Ω) of G and M_Γ respectively. Let

$$D_{G/M_\Gamma}(m)=\det(1-\mathrm{Ad}(m))_{\mathfrak{g}/\mathfrak{m}_\Gamma} \qquad (m\in M_\Gamma).$$

THEOREM 30. *Put $\Gamma'=G'\cap\Gamma$ and $\Phi(\gamma)=|D_{G/M_\Gamma}(\gamma)|^{1/2}\Theta(\gamma)$ $(\gamma\in\Gamma')$. Let $\mathfrak{Q}(\Gamma)$ be the set of all p-pairs (P, A) in G with $A\subset A_\Gamma$. Then $\mathfrak{Q}(\Gamma)$ is a finite set,*

$$\Phi(\gamma)=\sum_{(P,A)\in\mathfrak{Q}(\Gamma)}(-1)^{\dim A}\delta_P(\gamma)^{1/2}|D_{M/M_\Gamma}(\gamma)|_{\mathfrak{p}}^{1/2}$$

and[6]

$$|\Phi(\gamma)|\leqq[\mathfrak{Q}(\Gamma)] \qquad (\gamma\in\Gamma').$$

A Cartan subgroup Γ of G is called elliptic if Γ/Z is compact. Moreover a point $x\in G$ is called elliptic if x lies in some elliptic Cartan subgroup. Let G_e denote the set of all elliptic points in G'. Then G_e is open in G. Let φ_e denote the characteristic function of G_e.

It is easy to see that $\Theta=1$ on G_e and $\int_K \Theta(xk)\,dk=0$. These two facts, together with Theorem 30, give us the following result (cf. [**3**(a), Theorem 5]).

THEOREM 31. *There exists a number $c>0$ such that $\int_K \varphi_e(xk)\,dk\leqq c\Xi(x)$ for all $x\in G$.*

COROLLARY 1. *Let $f\in\mathscr{C}(G)$. Then $\int_{G_e}|f(x)|\,dx<\infty$.*

COROLLARY 2. *Let Θ_π be the character of a square-integrable representation π of G. Then Θ_π coincides with a locally summable function on G_e.*

16. A theorem of Howe. Let K_0 be an open compact subgroup of G. For $\mathfrak{d}\in\mathscr{E}(K_0)$, let $\xi_\mathfrak{d}$ denote the character of $\mathfrak{d}$. We extend it to a function on G by defining it to be zero outside K_0.

Let K_1, K_2 be two open compact subgroups of G. Fix $\mathfrak{d}_i\in\mathscr{E}(K_i)$ $(i=1, 2)$ and a subset ω of G. We say that ω intertwines $\mathfrak{d}_1$ with $\mathfrak{d}_2$ if there exists a function $f\in C(G)$ such that

$$\omega\cap\mathrm{Supp}(\xi_{\mathfrak{d}_1}*f*\xi_{\mathfrak{d}_2})\neq\emptyset.$$

The following theorem of Howe ([**4**(a)], [**4**(b)]) plays an important role in harmonic analysis on G. Its proof has so far been worked out only in the case of characteristic zero. So we assume in this section that char $\Omega=0$.

THEOREM 32 (HOWE). *Let Γ be a Cartan subgroup of G and ω a compact subset of $\Gamma'=\Gamma\cap G'$. Then there exists a compact open subgroup K_1 of G with the following property. Fix an open compact subgroup K_2 of G and an element $\mathfrak{d}_2\in\mathscr{E}(K_2)$ and let F denote the set of all $\mathfrak{d}_1\in\mathscr{E}(K_1)$ such that*

(1) *G intertwines $\mathfrak{d}_1$ with $\mathfrak{d}_2$,*
(2) *ω intertwines $\mathfrak{d}_1$ with itself.*
Then F is a finite set.

COROLLARY. *Let π be an admissible representation of G on V and suppose that V is a finite G-module under π. Then the character of π coincides with a locally constant function on G'.*

Combining this with the results of §15 we get the following lemma.

LEMMA 13. *Let Γ be a Cartan subgroup of G and dx^* the invariant measure on G/Γ. Then for any $f\in\mathscr{C}(G)$ and $\gamma\in\Gamma'=\Gamma\cap G'$, the integral $\int_{G/\Gamma}|f(x\gamma x^{-1})|\,dx^*$ converges. Put*

$$F_f(\gamma)=|D_G(\gamma)|^{1/2}\int_{G/\Gamma} f(x\gamma x^{-1})\,dx^* \qquad (\gamma\in\Gamma')$$

and let Γ'' be the set of all points $\gamma\in\Gamma$ where $D_{M_\Gamma}(\gamma)\neq 0$. Then F_f extends to a locally constant function on Γ''.

One would actually like to prove that

$$\sup_{\gamma\in\Gamma'}|F_f(\gamma)|<\infty.$$

This has been done so far only for $f\in C_c^\infty(G)$ (see [**3**(g), p. 82]). Of course the crucial case is when Γ is elliptic.

17. Wave packets and the Plancherel measure. We return to the notation of §12. Fix $\omega\in{}^\circ\mathscr{E}(M)$, $P\in\mathscr{P}(A)$ and $\psi\in L(\omega, P)$. Then $\mu(\omega:\nu)$ and $E(P:\psi:\nu)$ depend only on $\nu \bmod L^*$ (see §7). We consider the space $C_c^\infty(\mathfrak{a}^*)$ with its usual topology.

LEMMA 14. *Fix $\omega\in{}^\circ\mathscr{E}(M)$, $P\in\mathscr{P}(A)$, $\alpha\in C_c^\infty(\mathfrak{a}^*)\otimes L(\omega, P)$ and put*

$$\varphi_\alpha=\int_{\mathfrak{a}^*}\mu(\omega:\nu)\,E(P:\alpha(\nu):\nu)\,d\nu$$

where $d\nu$ is the Euclidean measure on $\mathfrak{a}^*$. Then $\varphi_\alpha\in\mathscr{C}(G,\tau)$ and $\alpha\mapsto\varphi_\alpha$ is a continuous mapping of $C_c^\infty(\mathfrak{a}^*)\otimes L(\omega, P)$ into $\mathscr{C}(G,\tau)$.

Here $\mathscr{C}(G,\tau)=(\mathscr{C}(G)\otimes V)\cap C(G,\tau)$.

LEMMA 15. *Fix $P_1, P_2\in\mathscr{P}(A)$ and put*

$$\varphi_\alpha = \int_{\mathfrak{a}^*} \mu(\omega:\nu)\, E(P_1:\alpha(\nu):\nu)\, d\nu$$

for $\alpha \in C_c^\infty(\mathfrak{a}^*) \otimes L(\omega, P_1)$. *Then*

$$\varphi_\alpha^{(P_2)} = \gamma \sum_{s \in \mathfrak{w}} \int_{\mathfrak{a}^*} {}^\circ c_{P_2|P_1}(s:\omega:s^{-1}\nu)\, \alpha(s^{-1}\nu) \cdot \chi_\nu\, d\nu$$

where $\chi_\nu(m) = q^{(-1)^{1/2}\langle \nu, H(m)\rangle}$ $(m \in M)$ *and* $\gamma = \gamma(G/M)$.

Let $\mathfrak{o}$ be a connected component of ${}^\circ\mathscr{E}(M)$. Fix $P \in \mathscr{P}(A)$ and let $L(\mathfrak{o}, P)$ denote the set of all functions ψ which assign to every $\omega \in \mathfrak{o}$ an element $\psi(\omega) \in L(\omega, P)$ in such a way that

$$\psi(\omega_\nu) = \psi(\omega) \cdot \chi_\nu \qquad (\nu \in \mathfrak{a}^*).$$

Then $L(\mathfrak{o}, P)$ is a complex vector space and in fact the mapping $\psi \mapsto \psi(\omega)$ defines a bijection of $L(\mathfrak{o}, P)$ onto $L(\omega, P)$ for every $\omega \in \mathfrak{o}$.

Put ${}^\circ c_{P_2|P_1}(s:\omega) = {}^\circ c_{P_2|P_1}(s:\omega:0)$ in the notation of §12 (see also §11). Put $\mu(\omega) = \mu(\omega:0)$ and let $d\omega$ denote the unique measure on $\mathfrak{o}$ which is invariant under the action of $\mathfrak{a}^*$ and such that the total measure of $\mathfrak{o}$ is 1. Then Lemma 15 may be restated in the following form.

THEOREM 33. *Fix* $P_1, P_2 \in \mathscr{P}(A)$ *and put*

$$\varphi_\alpha = \int \mu(\omega)\, E(P_1:\alpha:\omega)\, d\omega$$

for $\alpha \in C^\infty(\mathfrak{o}) \otimes L(\mathfrak{o}, P_1)$. *Then* $\varphi_\alpha \in \mathscr{C}(G, \tau)$ *and*

$$\varphi_\alpha^{(P_2)} = \gamma \sum_{s \in \mathfrak{w}} \int_{\mathfrak{o}} {}^\circ c_{P_2|P_1}(s:\omega)\, \alpha(\omega:\omega)\, d\omega.$$

Here $E(P_1:\alpha:\omega) = E(P_1:\alpha(\omega:\omega))$.

For any $\omega \in {}^\circ\mathscr{E}(M)$, let Θ_ω denote the character of the class $C(\omega) = C_M^G(\omega)$ (§9). Then Θ_ω is a tempered distribution.

Let T be a distribution on G. Then we put[2]

$$(T, f) = (T, f)_G = \operatorname{conj} T(g) \qquad (f \in C_c^\infty(G)),$$

where $g = \operatorname{conj} f$. Moreover if U is a finite-dimensional complex vector space, we define $(T, f \otimes u) = (T, f)\, u$ for $f \in C_c^\infty(G)$ and $u \in U$. So if T is tempered, we obtain in this way a continuous mapping $f \mapsto (T, f)$ of $\mathscr{C}(G) \otimes U$ into U.

Let F denote the projection in V given by

$$Fv = \int_K \tau(k)\, v\tau(k^{-1})\, dk \qquad (v \in V).$$

Then if $\omega_0 \in {}^\circ\mathcal{E}(M)$, we have

$$\begin{aligned}(\Theta_{\omega_0}, \varphi_\alpha) &= F(\theta_{\omega_0}, \varphi_\alpha^{(P_2)})_M \\ &= \gamma' \sum_{s \in \mathfrak{w}(\mathfrak{o} \mid \omega_0)} F(\theta_{\omega_0}, {}^\circ c_{P_2 \mid P_1}(s^{-1} : s\omega_0)\, \alpha(s\omega_0 : s\omega_0))_{M/A}\end{aligned}$$

in the notation of Theorem 33 where $\gamma' = \gamma[M : {}^\circ MA]^{-1}$ (see §7). Here θ_{ω_0} is the character of ω_0 and $\mathfrak{w}(\mathfrak{o} \mid \omega_0)$ is the set of all $s \in \mathfrak{w}$ such that $s\omega_0 = \omega_0^s \in \mathfrak{o}$. Now observe that

$$\begin{aligned}(\theta_{\omega_0}, \psi_0)_{M/A} &= d(\omega_0)^{-1}\, \psi_0(1), \\ E(P : \psi_0 : 1) &= F\psi_0(1) \qquad (P \in \mathcal{P}(A),\ \psi_0 \in L(\omega_0, P))\end{aligned}$$

and

$$E(P_2 : {}^\circ c_{P_2 \mid P_1}(s^{-1} : s\omega_0)\, \psi) = E(P_1 : \psi) \qquad (\psi \in L(s\omega_0, P_1)).$$

From this it follows that

$$d(\omega_0)\, (\Theta_{\omega_0}, \varphi_\alpha) = \gamma' \sum_{s \in \mathfrak{w}(\mathfrak{o} \mid \omega_0)} F\alpha(s\omega_0 : s\omega_0 : 1).$$

On the other hand

$$\varphi_\alpha(1) = \int \mu(\omega)\, F\alpha(\omega : \omega : 1)\, d\omega.$$

Therefore we obtain the following result (cf. [**3**(e), §§12, 13]).

THEOREM 34. *Put* $f = \varphi_\alpha$ *and* $\hat{f}(\omega) = (\theta_\omega, f)$ $(\omega \in {}^\circ\mathcal{E}(M))$. *Then*

$$f(1) = [\mathfrak{w}]^{-1}\, \gamma(G/M)^{-1}\, [M : {}^\circ MA] \int_{{}^\circ\mathcal{E}(M)} d(\omega)\, \mu(\omega)\, \hat{f}(\omega)\, d\omega.$$

Here $d\omega$ is the measure on ${}^\circ\mathcal{E}(M)$ which coincides on each connected component of ${}^\circ\mathcal{E}(M)$ with the measure defined above.

REFERENCES

1. A. Borel and J. Tits, *Groupes réductifs*, Inst. Hautes Etudes Sci. Publ. Math. No. 27 (1965), 55–150. MR **34** #7527.

2. F. Bruhat, (a) *Sur les représentations induites des groupes de Lie*, Bull. Soc. Math. France **84** (1956), 97–205. MR **18**, 907.

(b) *Distributions sur un groupe localement compact et applications à l'étude des représentations des groupes p-adiques*, Bull. Soc. Math. France **89** (1961), 43–75. MR **25** #4354.

3. Harish-Chandra, (a) *Two theorems on semisimple Lie groups*, Ann. of Math. (2) **83** (1966), 74–128. MR **33** #2766.

(b) *Discrete series for semisimple Lie groups*. II. *Explicit determination of the characters*, Acta Math. **116** (1966), 1–111. MR **36**# 2745.

(c) *Automorphic forms on semisimple Lie groups*, Lecture Notes in Math., vol. 62, Springer-Verlag, Berlin and New York, 1968. MR **38**# 1216.

(d) *Eisenstein series over finite fields*, Functional Analysis and Related Fields, Springer-Verlag, Berlin and New York, 1970, pp. 76–88.

(e) *Harmonic analysis on semisimple Lie groups*, Bull. Amer. Math. Soc. **78** (1970), 529–551. MR **41** #1933.

(f) *On the theory of the Eisenstein integral*, Proc. Intern. Conf. on Harmonic Analysis, University of Maryland, College Park, Md., 1971; Lecture Notes in Math., vol. 266, Springer-Verlag, Berlin and New York, 1971.

(g) *Harmonic analysis on reductive p-adic groups*, Lecture Notes in Math., vol. 162, Springer-Verlag, Berlin and New York, 1970.

4. R. E. Howe, (a) *Kirillov theory for compact p-adic groups* (preprint).

(b) *Some qualitative results on the representation theory of GL (n) over a p-adic field*, 1972. (preprint).

5. H. Jacquet, *Représentations des groupes linéaires p-adiques*, Theory of Group Representations and Fourier Analysis (C.I.M.E., II Ciclo, Montecatini Terme, 1970), Edizioni Cremonese, Rome, 1971, pp. 119–220. MR **45**# 453.

6. H. Jacquet and R. P. Langlands, *Automorphic forms on GL* (2), Lecture Notes in Math., vol. 114, Springer-Verlag, Berlin and New York, 1970.

INSTITUTE FOR ADVANCED STUDY

BOUNDARY THEORY AND STOCHASTIC PROCESSES ON HOMOGENEOUS SPACES

HARRY FURSTENBERG

Introduction. Associated with a connected Lie group G is a compact homogeneous space of G which we denote $B(G)$ and which plays a number of interesting roles. $B(G)$ is called the "universal boundary" of G. When G is semisimple the space $B(G)$ is important for determining the so-called "principal series" of irreducible unitary representations of G. It also plays a crucial role in the theory of harmonic functions on the Riemannian symmetric space associated with G. In this article we will find various characterizations of $B(G)$ as a G-space and tie up these properties with boundary theory of harmonic functions. A basic tool for analyzing the manner in which a group G acts on a G-space will be the introduction of certain G-related stochastic processes on the G-space.

A brief sketch of what we do is the following. In the first chapter we study a kind of "contractive" action of a group G on a G-space. This means that there exist elements in G which "compress" the space more and more in a manner to be described. This gives rise to the notion of boundary. The first task is to determine all the G-spaces with this type of behavior. We then notice that this type of space occurs in the theory of harmonic functions on a symmetric space G/K. Knowledge of all the spaces of this type enables one to obtain a representation of a harmonic function in terms of its boundary values.

After a summary of the relevant probabilistic prerequisites we develop the notion of a μ-boundary. Here one obtains a similar type of group action on the space, but not for arbitrary group elements, rather for almost all sequences arising in a certain way dictated by the distribution μ on G. Having developed an analogous theory we can turn to the corresponding mean value relation, which now yields the more general notion of a μ-harmonic function.

AMS (MOS) subject classifications (1970). Primary 22E40, 60J50; Secondary 43A85, 22D12.

In a final chapter we show that the space $B(G)$ plays not only a significant role as a G-space but also as a Γ-space, where Γ is a lattice subgroup of G.

I. Groups and their boundaries

1. G-spaces. G will denote a σ-compact locally compact group. A *G-space* will be a locally compact space M together with a continuous map $G\times M\to M$ denoted by $(g, x)\to gx$ satisfying

$$\text{(i)}\qquad ex=x \quad \text{for } e=\text{identity},$$

$$\text{(ii)}\qquad (g_1g_2)\,x=g_1(g_2x).$$

If M and M' are G-spaces and $\phi:M\to M'$, we say that ϕ is *equivariant* if $\phi(gx)=g\phi(x)$ for $x\in M$, $g\in G$.

A G-space is homogeneous if, for x, $y\in M$, there exist $g\in G$ with $gx=y$. If we fix $x_0\in M$ and set $H=\{g\in G \mid gx_0=x_0\}$ then the coset space G/H can be identified with M if the latter is homogeneous. H will be a closed subgroup and G/H carries a natural Hausdorff topology. Under our assumptions the identification of G/H with M will be topological as well as set theoretic. H is called the stability group of G at x_0.

A G-space M is called *minimal* if no proper closed subset of M is invariant under G. This is clearly equivalent to requiring that, for each $x\in M$, the orbit Gx is dense in M. Clearly a homogeneous G-space is also minimal.

A G-space is called *proximal* if it is endowed with a uniform structure (e.g., a metric) and if for all pairs of points x, y in the space there exists $g\in G$ with (gx, gy) arbitrarily close. If M is metric then it is proximal if for each x, $y\in M$ there is a sequence $\{g_n\}\subset G$ with $d(g_nx, g_ny)\to 0$.

EXAMPLES. $G=GL(m, \boldsymbol{R})$, $M=\boldsymbol{R}^m$, $M'=\boldsymbol{R}^m-\{0\}$, $M''=P^{m-1}$. M, M', and M'' are G-spaces, the latter being the space of lines through the origin of $\boldsymbol{R}^m$. If $\phi:M'\to M''$ denotes the map that assigns to a nonzero element in $\boldsymbol{R}^m$ the unique line joining it to the origin, then ϕ is an equivariant map. $M'=G/H'$, $M''=G/H''$ where

$$H'=\left\{\begin{pmatrix} 1 & a_{12} & \cdots & a_{1m} \\ 0 & a_{22} & \cdots & a_{2m} \\ 0 & a_{32} & \cdots & a_{3m} \\ \vdots & & \cdots & \end{pmatrix}\right\}, \quad H''=\left\{\begin{pmatrix} t & a_{12} & \cdots & a_{1m} \\ 0 & a_{22} & \cdots & a_{2m} \\ 0 & a_{32} & \cdots & a_{3m} \\ \vdots & & \cdots & \end{pmatrix}\right\}.$$

M'' being compact has a unique uniform structure, and one checks that it is a proximal G-space. With the usual Euclidean metric on $\boldsymbol{R}^m$, both M and M' are also proximal.

2. Probability measures on locally compact spaces. Let M be a compact topological space. By $\mathscr{C}(M)$ we denote the Banach space of continuous complex valued functions on M. The term *measure* denotes a regular, σ-additive measure defined on Borel sets of M. Such a measure determines a continuous linear functional on $\mathscr{C}(M)$ by

$$\mu(f)=\int_M f(x)\,d\mu(x).$$

Since conversely every continuous linear functional on $\mathscr{C}(M)$ arises in this way, we may identify the measure with the functional. A measure μ is positive if $\mu(f)\geqq 0$ whenever $f\geqq 0$. μ is a *probability measure* if it is positive and $\mu(M)=1$. We denote the set of these by $\mathscr{P}(M)$. We endow $\mathscr{P}(M)$ with the weak* topology it inherits as a subset of $\mathscr{C}(M)^*$. As a weakly closed bounded subset of the letter, $\mathscr{P}(M)$ is compact. It is, of course, also convex.

To each $x\in M$ we can assign a probability measure on M by setting $\delta_x(f)=f(x)$. These "point measures" constitute the set of extremals of the compact convex set $\mathscr{P}(M)$.

If M is locally compact then let $M'=M\cup\{\infty\}$ be its one-point compactification. $\mathscr{P}(M)$ can then be defined as the subset of probability measures on M' which assigns measure 0 to the point at ∞.

If $\phi: M\to M'$ is a continuous map then $f\to f\circ\phi$ takes $\mathscr{C}(M')\to\mathscr{C}(M)$ and the adjoint map takes $\mathscr{P}(M)\to\mathscr{P}(M')$. We write this map again as $\mu\to\phi\mu$. Note that $\phi\delta_x=\delta_{\phi(x)}$. Note that by definition one has

$$\int_M f(\phi x)\,d\mu(x)=\int_{M'} f(y)\,d\phi\mu(y). \tag{2.1}$$

If M_1 and M_2 are two locally compact spaces and $\mu_i\in\mathscr{P}(M)_i$, then the product measure $\mu_1\times\mu_2$ is defined on $M_1\times M_2$ by requiring

$$\mu_1\times\mu_2(f_1\otimes f_2)=\mu_1(f_1)\,\mu_2(f_2)$$

where $f_1\otimes f_2(x_1,x_2)=f_1(x_1)\,f_2(x_2)$, $x_i\in M_i$, $f_i\in\mathscr{C}(M_i)$.

Now let M be a G-space, $\mu\in\mathscr{P}(G)$ and $\nu\in\mathscr{P}(M)$. Denote by α the map of $G\times M\to M$. The product $\mu\times\nu$ is then defined on $G\times M$ and $\alpha(\mu\times\nu)$ is defined on M. This measure is called the *convolution* of μ and ν and denoted $\mu*\nu$. By (2.1) we have

$$\int_M f(x)\, d\mu * \nu(x) = \int_G \int_M f(gy)\, d\mu(g)\, d\mu(y). \tag{2.2}$$

In particular, we may take $M = G$, since G is itself a G-space. So for $\mu_1, \mu_2 \in \mathscr{P}(G)$ the convolution $\mu_1 * \mu_2$ is defined. Using (2.2) one shows that if $\mu_1, \mu_2 \in \mathscr{P}(G)$ and $\nu \in \mathscr{P}(M)$ where M is once again an arbitrary G-space,

$$(\mu_1 * \mu_2) * \nu = \mu_1 * (\mu_2 * \nu). \tag{2.3}$$

3. Amenable groups. There are a number of equivalent definitions of *amenability* (see [**3**]). The following will be most convenient for our purposes.

DEFINITION 3.1. A topological group G is *amenable* if whenever M is a compact G-space there exists a G-invariant measure in $\mathscr{P}(M)$.

We remark that, from the previous section, the action of G on M induces one on $\mathscr{P}(M)$ so that, in the foregoing, the notion of a G-invariant measure is well defined.

It is not hard to show that if G is a group, H a normal closed subgroup such that H and G/H are amenable then G is amenable. Also one sees readily that $\boldsymbol{Z}$ is amenable. (Let $\boldsymbol{Z}$ act on M by $(n, x) \to \tau^n x$. If $x_0 \in M$ let $\nu \in \mathscr{P}(M)$ belong to the closure of $\{n^{-1}(\delta_{x_0} + \delta_{\tau x_0} + \cdots + \delta_{\tau^{n-1} x_0})\}$. Then $\tau\nu = \nu$.) Putting all this together one finds that every solvable group is amenable. One can also show that compact groups are amenable. In fact, let G be compact and m_G its Haar measure. Then, for any $\nu \in \mathscr{P}(M)$, where M is a G-space, we can form $m_G * \nu$ which is G-invariant (by (2.3)). It follows that G is amenable if G contains a normal closed subgroup H which is solvable and such that G/H is a compact group. In [**1**] it is shown that the converse is true for connected Lie groups. It is not hard to see that not every Lie group is amenable. For example, take $G = SL(m, \boldsymbol{R})$. The sphere S^{m-1} is a G-space by identifying S^{m-1} with rays in $\boldsymbol{R}^m$. If G were amenable it would have a fixed measure on S^{m-1}. But $SO(m)$ operates transitively on S^{m-1} and is a subgroup of $SL(m, \boldsymbol{R})$. So the fixed measure would have to be the usual measure on the sphere. But this measure is not invariant under all of $SL(m, \boldsymbol{R})$.

4. Boundaries. In this section we describe a phenomenon which might be considered the opposite of the behavior encountered for amenable groups.

DEFINITION 4.1. A compact G-space M is said to be a *boundary* of G if

(i) M is minimal,

(ii) for any $\nu \in \mathscr{P}(M)$ the closure of the orbit $G\nu \in \mathscr{P}(M)$ contains point measures.

PROPOSITION 4.1. *A boundary of G is a proximal G-space.*

PROOF. Take $\nu=\frac{1}{2}(\delta_x+\delta_y)$. Then $g\nu=\frac{1}{2}(\delta_{gx}+\delta_{gy})$. If $g\nu$ is close to δ_z we must have both gx and gy close to z. Hence if the space is a boundary, the orbit of any pair of points comes arbitrarily close to the diagonal.

EXAMPLES. (A) P^{m-1} is a boundary of $GL(m, \boldsymbol{R})$. For let $\nu\in\mathscr{P}(P^{m-1})$. It is not hard to show that, for some hyperplane $\Pi\subset P^{m-1}$, $\nu(\Pi)=0$. Now let $\tilde{\Pi}$ be the hyperplane through the origin of $\boldsymbol{R}^m$ corresponding to Π and let $g_0\in GL(m, \boldsymbol{R})$ be defined by $g_0u=u$ for $u\in\tilde{\Pi}$, $g_0v_0=2v_0$ for some $v_0\notin\tilde{\Pi}$. Then, for any $v\in\boldsymbol{R}^m$, $v=av_0+u$, $u\in\tilde{\Pi}$, and $g_0^n v=2^n av_0+u$. It follows that the direction of the vector $g_0^n v$ converges to that of v_0 provided $a\neq 0$. Hence, for each $x\in P^{m-1}-\Pi$, $g_0^n x\to x_0$ where x_0 corresponds to v_0. Now let $f\in\mathscr{P}(P^{m-1})$.

$$g_0^n\nu(f)=\int f(g_0^n x)\,d\nu(x)\to f(x_0)=\delta_{x_0}(f),$$

since $g_0^n x\to x_0$ almost everywhere with respect to ν. This verifies (ii) of Definition 4.1. The first condition is satisfied inasmuch as P^{m-1} is homogeneous for $GL(m, \boldsymbol{R})$.

The same holds for $G=SL(m, \boldsymbol{R})$. Results to be obtained later will show that P^{m-1} is a boundary also for a discrete subgroup Γ of $SL(m, \boldsymbol{R})$ provided $SL(m, \boldsymbol{R})/\Gamma$ has finite measure. In this case the space is minimal without being homogeneous.

(B) Take G to be the free group on r generators, $r>1$. We obtain a compact G-space as follows. Let the generators of G be $\gamma_1,\ldots,\gamma_r$. Let Σ consist of all sequences $\omega=(\omega_1,\omega_2,\omega_3,\ldots)$ where each ω_i is either a generator γ_j or its inverse γ_j^{-1}, and subject to the condition that a generator and its inverse never follow in succession in either order in ω. Σ is a closed subset of the compact space consisting of all sequences with entries from $\{\gamma_1^{\pm 1},\ldots,\gamma_r^{\pm 1}\}$. Hence Σ is itself compact. The action of G on Σ will be determined if we describe how the generators γ_i act on Σ. We set

$$\begin{aligned}\gamma_i(\omega_1,\omega_2,\ldots,\omega_{n1},\ldots)&=(\gamma_i,\omega_1,\omega_2,\ldots) &&\text{if } \omega_1\neq\gamma_i^{-1},\\ &=(\omega_2,\omega_3,\omega_4,\ldots) &&\text{if } \omega_1=\gamma_i^{-1}.\end{aligned}$$

We now claim that Σ is a boundary for the free group G. Namely consider the sequence of group elements γ_1^n. Unless $\omega=(\gamma_1^{-1},\gamma_1^{-1},\ldots)\equiv\gamma_1^{-\infty}$ we will have $\gamma_1^n\omega\to(\gamma_1,\gamma_1,\gamma_1,\ldots)\equiv\gamma_1^{\infty}$. On the other hand $\gamma_1^n(\gamma_1^{-\infty})=\gamma_1^{-\infty}$. It follows that by applying γ_1^n to any $\nu\in\mathscr{P}(\Sigma)$ and letting $n\to\infty$ we obtain $\lim\gamma_1^n\nu=a\delta_{\gamma_1^{-\infty}}+(1-a)\,\delta_{\gamma_1^{\infty}}$. Applying γ_2^n to this and letting $n\to\infty$ we obtain $\delta_{\gamma_2^{\infty}}$. Hence the orbit $G\nu$ of any $\nu\in\mathscr{P}(\Sigma)$ has a point measure in its closure. It is not hard to see that Σ is a minimal G-space so that Σ turns out to be a boundary.

Returning to the general case, let M be a G-space; then $\mathscr{P}(M)$ is also a G-space. We shall denote by $\delta_M\subset\mathscr{P}(M)$ the subset of point measures of $\mathscr{P}(M)$. δ_M may be identified with M. We now have

PROPOSITION 4.2. *If M' is a minimal G-space and M is a boundary, then any equivariant map from M' to $\mathscr{P}(M)$ has δ_M as its range. Moreover there is at most one equivariant map from M' to M $(=\delta_M)$.*

PROOF. Let $\alpha: M' \to \mathscr{P}(M)$ be an equivariant map. $\alpha(M')$ is a closed G-invariant set in $\mathscr{P}(M)$ and since M is a boundary, $\alpha(M')$ contains point measures. So $M' \cap \alpha^{-1}(\delta_M) \neq \emptyset$. But M' is minimal and $\alpha^{-1}(\delta_M)$ is a closed G-invariant subset; hence $M' \subset \alpha^{-1}(\delta_M)$ or $\alpha(M') \subset \delta_M$.

Now suppose β_1 and β_2 are two equivariant maps of M' to M. Then $\alpha(x) = \frac{1}{2}(\delta_{\beta_1(x)} + \delta_{\beta_2(x)})$ is an equivariant map of M' to $\mathscr{P}(M)$. Since $\alpha(M) \subset \delta_M$ we must have $\beta_1(x) = \beta_2(x)$. This proves the proposition.

The following is obvious from the definitions:

PROPOSITION 4.3. *If G is amenable then a boundary of G must consist of a single point.*

More generally we have

PROPOSITION 4.4. *Let G contain a closed subgroup H which is amenable and such that G/H is compact. Then every boundary of G is an equivariant image of the homogeneous G-space G/H.*

PROOF. Let M be a boundary of G. Since G acts on M, so does H. By Definition 3.1, H has a fixed measure λ on M. Let $A \subset G$ be a compact set with $AH = G$. Then $G\lambda = A\lambda$. But $A\lambda$ is closed and $G\lambda$ contains in its closure a point measure. Hence some $g\lambda$, $g \in A$, is a point measure. But then λ is a point measure. We conclude that H leaves fixed a point $x_0 \in M$. Then the map $gH \to gx_0$ is well defined and takes G/H into a closed G-invariant subset of M. Since M is minimal, this map is onto and this proves the proposition.

This proposition is very useful in the Lie group case, since it turns out that every connected Lie group satisfies the hypotheses of this proposition.

LEMMA 4.5. *If G is a connected Lie group it contains a solvable subgroup S such that G/S is compact.*

PROOF. This is true for G semisimple and connected with finite center, for then $G = K \cdot A \cdot N$ where K is compact and $A \cdot N$ is solvable. For general connected Lie groups G, we first divide by the radical of G to obtain a semisimple group; we then divide by the center to arrive at the case previously handled. Inasmuch as a solvable extension of a solvable group is solvable, the lemma follows.

DEFINITION 4.2. A G-space M is called a "universal boundary" of G if it is a boundary of G and every boundary of G is an equivariant image of M.

Note that an equivariant image of a boundary is always a boundary. An initial element in the lattice of boundaries of G is a universal element.

By Proposition 4.2 an equivariant map from a minimal G-space to a boundary is unique. This implies that the universal boundaries for G are isomorphic as G-spaces. We next observe that some universal boundary always exists. Namely we can index all the boundaries $\{M_\alpha\}$ of G so that up to isomorphism all boundaries occur. (This is because the minimality of a G-space places a bound on its cardinality.) Now form $\prod_\alpha M_\alpha$ which will be a compact G-space, and let $B(G)$ denote a minimal G-invariant subset. It is not hard to see that $B(G)$ is a boundary and clearly it is a minimal boundary. We then have

PROPOSITION 4.6. *Every group G has a universal boundary $B(G)$ which is unique up to isomorphism.*

When G is a connected Lie group we can write $B(G)=G/H(G)$ where the subgroup $H(G)$ is uniquely determined up to conjugacy. The argument of Proposition 4.4 shows that any amenable cocompact subgroup of G has a fixed point on $B(G)$. In particular if we suppose G semisimple with finite center and with Iwasawa decomposition $G=K\cdot A\cdot N$, then $S=A\cdot N$ is a solvable cocompact subgroup and we may suppose $S\subset H(G)$. It may be shown that $H(G)$ is exactly the normalizer of S in G. Hence $H(G)/S$ is a compact group so that $H(G)$ itself is amenable. (See [1].) This yields

THEOREM 4.7. *If G is a connected Lie group, it contains an amenable subgroup $H(G)$ for which $G/H(G)$ is the universal boundary of G.*

REMARK. It is easy to see that the two properties (i) '$H(G)$ is amenable' and (ii) '$G/H(G)$ is a boundary' already characterize $H(G)$ up to conjugacy. In fact $H(G)$ is a maximal subgroup with respect to the property (i) and a minimal subgroup with respect to (ii).

The following proposition will be used later. Its proof is left to the reader.

PROPOSITION 4.8. *Let M be a metrizable compact G-space. A necessary and sufficient condition for M to be a boundary is that, for every $\nu\in\mathscr{P}(M)$ and for a dense set of $x\in M$, there exists a sequence $\mu_n\in\mathscr{P}(G)$ with $\mu_n*\nu\to\delta_x$.*

5. Harmonic functions on groups.

DEFINITION 5.1. A function $f(g)$ on a locally compact group G is left-uniformly-continuous (LUC) if, for every $\varepsilon>0$, there exists a neighborhood U of e with $|f(g'g)-f(g)|<\varepsilon$ for all $g\in G$ provided $g'\in U$.

Let G be any locally compact group and let K denote a compact subgroup.

DEFINITION 5.2. A locally bounded function $f(g)$ on G is *harmonic* if, for every $g, g' \in G$,

$$f(g) = \int_K f(gkg')\, dk. \tag{5.1}$$

Note that (5.1) implies many more mean value relations. For example, if ψ is a Borel function of bounded support and $\psi(kg) = \psi(g)$ for $k \in K$, and f is harmonic, then

$$\left(\int_G \psi(g')\, dg'\right) f(g) = \int_G f(gg')\, \psi(g')\, dg'. \tag{5.2}$$

Moreover, (5.1) implies that $f(gk) = f(g)$ for $k \in K$, so that a harmonic function on G may be considered a function on G/K. (When G/K is a Riemannian symmetric space then the foregoing harmonic functions coincide with those defined by the Laplace operator provided one assumes they are bounded.)

To obtain a theory of arbitrary harmonic functions on G it suffices to have one for LUC harmonic functions. The reason is that if $f(g)$ is harmonic and ψ has compact support then

$$f_\psi(g) = \int f(g'g)\, \psi(g')\, dg'$$

is also harmonic. Now if ψ is continuous then f_ψ is LUC. Moreover by an appropriate choice of ψ_n we can obtain $f_{\psi_n} \to f$ weakly (in the topological linear space of locally bounded functions).

We remark that our nomenclature is not quite precise, and we should refer to these functions as (G, K)-harmonic functions. Let us suppose however that G and K are fixed in the discussion.

THEOREM 5.1. *Assume G is an arbitrary locally compact, σ-compact group and K a compact subgroup. If $f(g)$ is an LUC bounded harmonic function then there exist a boundary M of G and a continuous function $\hat{f}(\xi)$ on M with*

$$f(g) = \int_M \hat{f}(g\xi)\, dm(\xi) \tag{5.3}$$

where $m \in \mathcal{P}(M)$ is any K-invariant probability measure on M.

PROOF. Since $f(g)$ is LUC, the family of functions $f(gg')$ of g is equicontinuous. Since moreover f is bounded, this family has compact closure in the topology of uniform convergence on compact sets. Let $Q(f)$ denote the closed convex hull of this family. $Q(f)$ is a compact convex set. Let M denote the closure of the extremals of $Q(f)$. Now G acts on $Q(f)$ by $F(g)\to F(gg')$ so that M is also invariant under the action of G. M is a compact G-space. We claim that M is a boundary. In fact M is metrizable, since G is σ-compact. So we may use the criterion of Proposition 4.8. Suppose $\nu\in\mathscr{P}(M)$ and let $\eta\in M$ be an extremal of $Q(f)$. Let m_K denote Haar measure in K, and form $\bar\nu=m_K*\nu$.

Now from (5.1) we see that if $\phi\in Q(f)$ we will have

$$\int_K \phi(gk)\,dk=f(g).$$

Hence

$$\int \phi_\xi(g)\,d\bar\nu(\xi)=\iint \phi_{k\xi}(g)\,d\nu(\xi)\,dk=\int \phi_\xi(gk)\,dk\,d\nu(\xi)=f(g) \tag{5.4}$$

For any $\phi\in Q(f)$ we can find a sequence $\mu_n\in\mathscr{P}(G)$ with $\int f(gg')\,d\mu_n(g')\to\phi(g)$. Take $\phi=\phi_\eta$ where η is an extremal of $Q(f)$ and form $\mu_n*\bar\nu$. Our assertion that M is a boundary will be proved if we show that $\mu_n*\bar\nu\to\delta_\eta$. Suppose that some subsequence $\mu_{n'}*\bar\nu\to\theta$. Consider

$$\begin{aligned}\int \phi_\xi(g)\,d\theta(\xi)&=\lim_{n'}\int \phi_\xi(g)\,d\mu_{n'}*\bar\nu(\xi)=\lim_{n'}\iint \phi_{h\xi}(g)\,d\mu_{n'}(h)\,d\bar\nu(\xi)\\&=\lim_{n'}\iint \phi_\xi(gh)\,d\mu_{n'}(h)\,d\bar\nu(\xi)=\lim_{n'}\int f(gh)\,d\mu_{n'}(h)=\phi_\eta(g).\end{aligned}$$

But ϕ_η is extremal. Hence $\theta=\delta_\eta$. So M is a boundary.

Returning to (5.4) and recalling that ν was arbitrary, we find

$$f(g)=\int \phi_\xi(gk)\,dk. \tag{5.5}$$

Define $\hat f(\xi)=\phi_\xi(e)$. Then (5.5) becomes

$$f(g)=\int \phi_{gk\xi}(e)\,dk=\int \hat f(gk\xi)\,dk.$$

Letting $\xi \in M$ and integrating we see that this yields (5.3) and completes the proof of the theorem.

We conclude this section by noting that if G is a semisimple Lie group with finite center then a maximal compact subgroup K of G is transitive on its universal boundary, and therefore on any boundary. Hence the K-invariant probability measure on M is unique. In this case (5.3) becomes the Poisson representation of a harmonic function on the symmetric space G/K.

II. μ-processes

6. Probabilistic background. We shall sketch the aspects of probability theory needed for our subsequent discussion. Part of our purpose is to establish notation that we will find useful.

(A) *Probability spaces.* We consider first the category of pairs $(\Omega, \mathcal{F})$ where Ω is a set and $\mathcal{F}$ some σ-algebra of subsets of Ω. A map $\phi:(\Omega, \mathcal{F}) \to (\Omega', \mathcal{F}')$ is *measurable* if ϕ takes Ω to Ω' and $\phi^{-1}(A) \in \mathcal{F}$ for $A \in \mathcal{F}'$. If Ω' is a topological space, then a map $\phi: \Omega \to \Omega'$ is measurable with respect to $\mathcal{F}$ if it is measurable when $\mathcal{F}'$ is taken to be the σ-algebra of Borel sets in Ω'. We use this notion in particular when Ω' is $\boldsymbol{R}$ or $\boldsymbol{C}$ or a topological group G or a G-space M. In this case we call ϕ a *random variable*.

Given a family of pairs $(\Omega_\alpha, \mathcal{F}_\alpha)$ and maps ϕ_α of space Ω to each Ω_α, there is a least σ-algebra $\mathcal{F}$ with respect to which all the ϕ_α are measurable. We write $\mathcal{F} = \mathcal{F}(\{\phi_\alpha\})$. In particular, if $x_1, x_2, x_3, \dots$ is a sequence of random variables in $(\Omega, \mathcal{F})$, the subalgebra $\mathcal{F}(x_1, x_2, x_3, \dots)$ is well defined.

A *probability space* is determined by a σ-additive probability measure P defined on the elements of $\mathcal{F}$ for a pair $(\Omega, \mathcal{F})$. We speak of a probability triple $(\Omega, \mathcal{F}, P)$. Let $(\Omega, \mathcal{F}, \mu)$ be a probability triple and ϕ a measurable map of $(\Omega, \mathcal{F})$ to $(\Omega', \mathcal{F}')$ then $\phi\mu$ is defined by $\phi\mu(A) = \mu(\phi^{-1}A)$ for $A \in \mathcal{F}'$. $(\Omega', \mathcal{F}', \phi\mu)$ is again a probability triple. In particular if x is an M-valued random variable defined on $(\Omega, \mathcal{F}, P)$ then xP is a Borel measure on M. It is customary to call xP the *distribution* of the random variable x.

DEFINITION 6.1. If x is a random variable defined on $(\Omega, \mathcal{F}, P)$ we denote the distribution of x by $E^*(x)$.

For suitably bounded numerically- ($\boldsymbol{R}$- or $\boldsymbol{C}$-) valued random variables one has the notion of integration

$$E(x) = \int_\Omega x(\omega)\, dP(\omega)$$

with which we assume the reader is familiar. The space of functions, or random variables, for which this integral exists is $L^1(\Omega, \mathscr{F}, P)$. As usual we denote by $L^p(\Omega, \mathscr{F}, P)$ the space $\mathbf{C}$-valued random variables with $|x|^p \in L^1$, and $L^\infty(\Omega, \mathscr{F}, P)$ denotes the bounded random variables. The following lemma is often useful.

LEMMA 6.1. *Let $\mathscr{F}_0$ be an algebra of sets in Ω and $\mathscr{F}$ the σ-algebra it generates. The space of functions*

$$\phi(\omega) = \sum_i a_i x_{A_i}(\omega), \quad A_i \in \mathscr{F}_0,$$

is then a dense subset of each $L^p(\Omega, \mathscr{F}, P)$, $1 \leqq p < \infty$.

A typical example would be to let $\{x_n\}$ be an infinite sequence of random variables, $\mathscr{F}_0 = \bigcup_n \mathscr{F}(x_1, \ldots, x_n)$ and $\mathscr{F} = \mathscr{F}(x_1, x_2, x_3, \ldots)$.

Using Lemma 6.1 one obtains the important formula

$$(6.1) \qquad \int_\Omega f(\phi(\omega))\, dP(\omega) = \int_\Omega f(\omega')\, d\phi P(\omega')$$

for $\phi: (\Omega, \mathscr{F}) \to (\Omega', \mathscr{F}')$ measurable. In particular we find, for random variables x,

$$(6.2) \qquad E(f(x)) = \int_M f(\xi)\, dE^*(x)(\xi).$$

If μ is a probability measure on $(\Omega, \mathscr{F})$ and ν is any σ-additive measure on $(\Omega, \mathscr{F})$ we say ν is *absolutely continuous* with respect to μ, $\nu \prec \mu$, if $\mu(A) = 0 \Rightarrow \nu(A) = 0$. The Radon-Nikodym theorem asserts that when this happens there exists a unique (up to measure 0) function in $L^1(\Omega, \mathscr{F}, \mu)$ with

$$\nu(A) = \int_A f(\omega)\, d\mu(\omega).$$

If ν is itself a probability measure we shall have

$$\int \phi(\omega)\, d\nu(\omega) = \int \phi(\omega) f(\omega)\, d\mu(\omega)$$

whenever either of these integrals makes sense. We denote $f(\omega) = d\nu/d\mu(\omega)$.

PROPOSITION 6.2. *Let ϕ be a measurable map of $(\Omega, \mathscr{F})$ to $(\Omega', \mathscr{F}')$, and let μ, ν be probability measures on $(\Omega, \mathscr{F})$ with $\nu \prec \mu$. Then $\phi\nu \prec \phi\mu$ and if ϕ is 1-1 we will have*

(6.3)
$$\frac{d\phi\nu}{d\phi\mu}(\omega) = \frac{d\nu}{d\mu}(\phi^{-1}\omega).$$

PROPOSITION 6.3. *If $\nu_1 \prec \nu_2 \prec \nu_3$, then*

(6.4)
$$\frac{d\nu_1}{d\nu_3} = \frac{d\nu_1}{d\nu_2}\frac{d\nu_2}{d\nu_3}.$$

Both equalities above are to be understood as valid almost everywhere with respect to the "stronger" measure in question.

EXAMPLE. Let G be a group and M a locally compact G-space. If $\nu \in \mathscr{P}(M)$ then ν is defined on all Borel sets of M. Letting $\mathscr{B}$ denote the σ-algebra of Borel sets of M, $(M, \mathscr{B}, \nu)$ will be a probability triple. Any homeomorphism of M is a measurable map; in particular for $g \in G$ the measure $g\nu$ will be defined. This can be seen to agree with the transform of a measure defined in Chapter I. Assume now that M is a C^1-manifold and that G operates by diffeomorphisms. Moreover assume that ν is given in local coordinates by

$$d\nu(\xi) = \psi(\xi_1, \ldots, \xi_n)\, d\xi_1 \cdots d\xi_n$$

with $\psi > 0$ and continuous. Then each $g\nu$ will have the same property, and $g\nu \prec \nu$ for all $g \in G$. Moreover one sees that $dg\nu/d\nu$ is a continuous function on M. Set

$$\sigma(g, x) = dg^{-1}\nu(x)/d\nu, \quad g \in G, \quad x \in M.$$

LEMMA 6.4. *σ is a "multiplier" function in the sense that*

$$\sigma(g_1 g_2, x) = \sigma(g_1, g_2 x)\, \sigma(g_2, x).$$

PROOF. For

$$\frac{dg_2^{-1}g_1^{-1}\nu}{d\nu}(x) = \frac{dg_2^{-1}g_1^{-1}\nu}{dg_2^{-1}\nu}(x)\,\frac{dg_2^{-1}\nu}{d\nu}(x)$$
$$= \frac{dg_1^{-1}\nu}{d\nu}(g_2 x)\,\frac{dg_2^{-1}\nu}{d\nu}(x) = \sigma(g_1, g_2 x)\,\sigma(g_2, x)$$

by (6.3) and (6.4). Note that in this case identity holds everywhere, since the Radon-Nikodym derivatives have continuous versions.

(B) *Conditional expectation.* Now let $(\Omega, \mathscr{F}, P)$ be a probability triple and $\mathscr{G} \subset \mathscr{F}$ a sub-σ-algebra. $L^2(\Omega, \mathscr{G}, P)$ will be a closed subspace of $L^2(\Omega, \mathscr{F}, P)$, and we denote by $E(\cdot \mid \mathcal{G})$ the orthogonal projection from the latter to the former. If $x \in L^2(\Omega, \mathcal{F}, P)$, we call $E(x \mid \mathcal{G})$ the conditional expectation of x with respect to $\mathcal{G}$. The following follows without much difficulty from the definition:

PROPOSITION 6.5. *Let $z \in L^2(\Omega, \mathcal{F}, P)$.*

(i) *$E(z \mid \mathcal{G})$ is $\mathcal{G}$-measurable.*

(ii) *$E(E(z \mid \mathcal{G})\, w) = E(zw)$ for $w \in L^2(\Omega, \mathcal{G}, P)$.*

(iii) *$E(\cdot \mid \mathcal{G})$ is a linear operator.*

(iv) *If $z \in L^2(\Omega, \mathcal{F}, P)$ is measurable with respect to $\mathcal{G}$, then $E(z \mid \mathcal{G}) = z$.*

(v) *If $z \geqq 0$ then $E(z \mid \mathcal{G}) \geqq 0$.*

(vi) *If ϕ is convex, $\phi(z) \in L^2(\Omega, \mathcal{F}, P)$, then $\phi(E(z \mid \mathcal{G})) \leqq E(\phi(z) \mid \mathcal{G})$.*

(vii) *$E(\cdot \mid \mathcal{G})$ is a contraction on $L^2 \cap L^p$ in the L^p-norm for each p.*

(viii) *If $\mathcal{G}_1 \subset \mathcal{G}_2 \subset \mathcal{F}$ then $E(E(x \mid \mathcal{G}_2) \mid \mathcal{G}_1) = E(z \mid \mathcal{G}_1)$.*

We remark that (iv) and (v) follow from the fact that in an orthogonal projection in Hilbert space, the distance between an element and its image is minimized. Also (v) may be used to prove (vi) using the fact that a convex function may be defined as the supremum of a family of affine functions. (vi) implies (vii) and the latter implies that the conditional expectation is defined on each L^p.

When $\mathcal{G} = \mathcal{F}(\{x_\alpha\})$ we shall denote $E(\cdot \mid \mathcal{G})$ by $E(\cdot \mid \{x_\alpha\})$.

(C) *Independence and product measures.*

DEFINITION 6.2. Let $(\Omega, \mathcal{F}, P)$ be a probability triple and $\{\mathcal{F}_\alpha\}$ a family of sub-σ-algebras. We say the $\mathcal{F}_\alpha$ are *independent* if $A_i \in \mathcal{F}_{\alpha_i}$ for distinct $\alpha_1, \ldots, \alpha_k$ gives $P(A_1 \cap A_2 \cap \cdots \cap A_k) = P(A_1)\, P(A_2) \cdots P(A_k)$. If $\{x_\alpha\}$ are random variables they are independent if the σ-algebras $\mathcal{F}(x_\alpha)$ are independent.

Using Lemma 6.1 one shows

PROPOSITION 6.6. *If $x, y \in L^1(\Omega, \mathcal{F}, P)$ are independent, then $xy \in L^1$ and $E(xy) = E(x)\, E(y)$.*

EXAMPLE OF INDEPENDENT σ-ALGEBRAS. Let $(\Omega_\alpha, \mathcal{F}_\alpha, P_\alpha)$ be a probability triple. Form $\Omega = \prod \Omega_\alpha$, $x_\alpha : \Omega \to \Omega_\alpha$ the natural projection. Let $\mathcal{F} = \mathcal{F}(\{x_\alpha\})$ and on Ω define $P = \prod P_\alpha$ by

$$P(x_{\alpha_1}^{-1}(A_1) \cap x_{\alpha_2}^{-1}(A_2) \cap \cdots \cap x_{\alpha_k}^{-1}(A_k)) = P_{\alpha_1}(A_1) \cdots P_{\alpha_k}(A_k).$$

A fundamental extension theorem in measure theory shows that P extends to the full σ-algebra $\mathcal{F}$. It is evident that the variables x_α are independent over $(\Omega, \mathcal{F}, P)$.

This proves

PROPOSITION 6.7. *If $\{M_\alpha\}$ is a family of topological spaces and $\{\mu_\alpha\}$ a corre-*

sponding family of probability measures, $\mu_\alpha \in \mathscr{P}(M_\alpha)$, there exists a probability space on which are defined independent random variables $\{x_\alpha\}$ with $E^*(x_\alpha)=\mu_\alpha$.

PROPOSITION 6.8. *Assume x, y are independent random variables, then $E^*(x, y) = E^*(x) \times E^*(y)$. (The distribution of the pair is the product of the distributions.)*

PROOF. It suffices to show that, for $\phi(x, y) \in L^1$,

$$E(\phi(x, y)) = \iint \phi(\xi, \eta)\, dE^*(x)(\xi)\, dE^*(y)(\eta).$$

This is true for $\phi(x, y) = \psi_1(x)\psi_2(y)$ and using Lemma 6.1 one sees that these functions span a dense set in L^1.

The following is a version of Fubini's theorem.

PROPOSITION 6.9. *If x, y are independent random variables and $\phi(x, y) \in L^1$, then*

$$E(\phi(x, y) \mid x) = \int \phi(x, \eta)\, dE^*(y)(\eta).$$

The proof is similar to that of the foregoing proposition. It also follows from Proposition 6.8 using the classical Fubini theorem for product measures, and the fact that properties (i) and (ii) in Proposition 6.5 characterize the conditional expectation.

Another application of Proposition 6.8 is

PROPOSITION 6.10. *Let x be G-valued and y M-valued where M is a G-space. Assume x and y are independent. Then xy is M-valued and $E^*(xy) = E^*(x) * E^*(y)$.*

This is true in particular if $M = G$. This generalizes the well-known fact that the distribution of the sum of independent random variables is the convolution of the two distributions. The proof follows immediately from the definition of convolution in §2.

(D) *Conditional distribution.* Assume M is a compact metric topological space and that x is an M-valued random variable on $(\Omega, \mathscr{F}, P)$. Let $\mathscr{G}$ be a sub-σ-algebra of $\mathscr{F}$. For each $f \in \mathscr{C}(M)$ we have defined $E(f(x) \mid \mathscr{G}) \in L^\infty(\Omega, \mathscr{G}, P)$. Using the fact that a measure on M is determined by integrating a countable dense set of continuous functions with respect to it, one proves

PROPOSITION 6.11. *There exists a measurable function μ_ω from $(\Omega, \mathscr{G}, P)$ to $\mathscr{P}(M)$ satisfying*

$$E(f(x) \mid \mathscr{G})(\omega) = \int_M f(\xi)\, d\mu_\omega(\xi) \quad a.e.$$

for all $f \in \mathscr{C}(M)$. *We denote* μ_ω *by* $E^*(x \mid \mathscr{G})$.

Thus $E^*(x \mid \mathscr{G})(f) = E(f(x) \mid \mathscr{G})$. When $\mathscr{G}$ is the trivial algebra then $E^*(x \mid \mathscr{G}) = E^*(x)$.

(E) *Martingales.* A sequence $w_1, w_2, w_3, \ldots$ of numerically valued random variables forms a *martingale* if $E(w_n \mid w_1, \ldots, w_{n-1}) = w_{n-1}$. One checks using Proposition 6.5(viii) that if $\mathscr{F}_1 \subset \mathscr{F}_2 \subset \mathscr{F}_3 \subset \cdots \subset \mathscr{F}$ and $x \in L^1(\Omega, \mathscr{F}, P)$ then $w_n = E(x \mid \mathscr{F}_n)$ forms a martingale. We need the following version of the martingale convergence theorem:

THEOREM 6.12. *If* $\{w_n\}$ *is a martingale with each* $w_n \in L^1$ *then with probability* 1, $\lim w_n$ *exists. If each* $w_n \in L^2$ *and* $E(|w_n|^2)$ *is bounded, then* $w = \lim w_n \in L^2$ *and* $w_n = E(w \mid w_1, \ldots, w_n)$.

One application is the following:

PROPOSITION 6.13. *If* $\{x_n\}$ *is a sequence of random variables and* z *is any bounded random variable, then*

$$E(z \mid x_1, x_2, x_3, \ldots) = \lim_{n \to \infty} E(z \mid x_1, \ldots, x_n)$$

almost everywhere.

PROOF. Assume first that z is measurable with respect to $\{x_1, x_2, \ldots\}$. The left-hand side is then z. That $E(z \mid x_1, \ldots, x_n) \to z$ in $L^2(\Omega, \mathscr{F}, P)$ follows from the fact that this is true for $z \in L^2(\Omega, \mathscr{F}\{x_1, \ldots, x_n\}, P)$ and Lemma 5.1. Now by Theorem 6.12 the right-hand side converges a.e. Since z is bounded, the two limits must coincide. When z is arbitrary we make use of the result for $z' = E(z \mid x_1, x_2, x_3, \ldots)$ and the fact that $E(z \mid x_1, \ldots, x_n) = E(z' \mid x_1, \ldots, x_n)$.

(F) *Stationary processes.* A *stochastic process* is a collection $\{x_\alpha\}$ of random variables on a probability space $(\Omega, \mathscr{F}, P)$. Generally one is not concerned with the underlying space, but rather the joint distributions of the variables $\{x_\alpha\}$. We make this precise by calling two processes $\{x_\alpha\}$, $\{x'_\alpha\}$ defined on spaces $(\Omega, \mathscr{F}, P)$ and $(\Omega', \mathscr{F}', P')$ *isomorphic* if $E^*(x_{\alpha_1}, \ldots, x_{\alpha_k}) = E^*(x'_{\alpha_1}, \ldots, x'_{\alpha_k})$ for each finite collection $\alpha_1, \ldots, \alpha_k$ of indices.

DEFINITION 6.3. Let $\{x_n\}$ be a sequence of M-valued random variables either for all n or all $n \geqq a$, $a \in \boldsymbol{Z}$. We say $\{x_n\}$ is *stationary* if the processes $\{x_n\}$ and $\{x_{n+1}\}$ are isomorphic.

The following proposition is an easy consequence of a basic existence theorem of Kolmogoroff.

PROPOSITION 6.14. *Let* $\{x'_n\}$ *be a stationary process defined for* $n \geqq 0$. *There exists a stationary process* $\{x_n\}$ *defined for all* n *with* $\{x_n\}_{n>0} \cong \{x'_n\}_{n \geqq 0}$.

The ergodic theorem can be formulated in the context of stationary processes. It states

THEOREM 6.15. *If x_n is a* **C**-*valued stationary process with $x_n \in L^1$ then, with probability* 1,

$$\lim \frac{x_1 + x_2 + \cdots + x_n}{n}$$

exists.

7. μ-processes. In this section G is a locally compact topological group, M a locally compact G-space and $\mu \in \mathscr{P}(G)$. Using Proposition 6.7 we may construct a sequence of G-valued independent random variables x_n with $E^*(x_n) = \mu$. We will always reserve the notation $\{x_n\}$ for a sequence of this type. Clearly $\{x_n\}$ is a stationary process.

We let $(\Omega', \mathscr{F}', P')$ be a probability space on which the $\{x_n\}_{n \geqq 1}$ are defined and we form $\Omega = M \times \Omega'$. Let ν be any probability measure on M and set $P = \nu \times P'$. Let $w_0 : \Omega \to M$ be the projection on M. w_0 is a random variable with distribution ν. Now set

$$w_n = x_n x_{n-1} \cdots x_1 w_0 .$$

w_n is M-valued. Since all the variables $x_n, x_{n-1}, \ldots, x_1, w_0$ are independent, we find

$$E^*(w_n) = \underbrace{\mu * \mu * \cdots * \mu}_{n} * \nu .$$

DEFINITION 7.1. A μ-process on M consists of a sequence of random variables $\{(x_{n+1}, w_n)\}$ defined either for $n \geqq 0$ or for all n, and satisfying

(i) $E^*(x_n) = \mu$,

(ii) w_n is M-valued,

(iii) $w_{n+1} = x_{n+1} w_n$,

(iv) x_{n+1} is independent of $\{x_n, x_{n-1}, \ldots, w_n, w_{n-1}, \ldots\}$.

The foregoing construction proves

PROPOSITION 7.1. *Up to isomorphism there exists a unique μ-process $\{(x_{n+1}, w_n)\}_{n \geqq 0}$ with $E^*(w_0) = \nu$.*

It is not clear how such a process may be defined for negative indices n. However in one case this can be done.

DEFINITION 7.2. A measure $\nu \in \mathscr{P}(M)$ is called *stationary* (with respect to μ) if $\mu * \nu = \nu$.

When ν is stationary then $E^*(w_1) = \nu = E^*(w_0)$. Now the family $\{(x_{n+1}, w_n)\}$ is uniquely determined by the conditions (i)–(iv) of Definition 7.1 and the condition $E^*(w_0) = \nu$. Inasmuch as the family $\{(x_{n+2}, w_{n+1})\}$ satisfies the same conditions provided $E^*(w_1) = \nu$ we have

PROPOSITION 7.2. *The μ-process $\{(x_{n+1}, w_n)\}_{n \geq 0}$ is stationary iff $E^*(w_0) = \nu$ is a stationary measure for μ.*

We may now invoke Proposition 6.14.

PROPOSITION 7.3. *If $\mu * \nu = \nu$ there exists a stationary μ-process $\{(x_{n+1}, w_n)\}$ defined for all n and satisfying $E^*(w_n) = \nu$.*

An example of a μ-process is obtained by taking $M = G$ and $\nu = \delta_e$. Then $w_0 = e$ and $w_n = x_n x_{n-1} \cdots x_1$.

This G-valued process is known as the *random walk on G generated by μ*. Clearly if $\mu \neq \delta_e$ this process will not be stationary. We shall see later that unless μ has its support in a compact subgroup of G there will not exist a stationary measure for μ on G. On the other hand we have

PROPOSITION 7.4. *If M is compact there always exists a stationary measure for μ on M.*

PROOF. By standard arguments there exist fixed points in $\mathscr{P}(M)$ for the operation $\nu \to \mu * \nu$.

It is also possible that M possesses a stationary measure without being compact. For example if G is a semisimple group and Γ a lattice in G so that G/Γ has finite measure, then this measure is stationary for any $\mu \in \mathscr{P}(G)$.

The following theorem is basic to our considerations.

THEOREM 7.5. *Assume M is metrizable, locally compact, and σ-compact and suppose ν is a stationary measure for μ on M. Let $\{(x_{n+1}, w_n)\}$ denote the corresponding stationary μ-process. Then, with probability 1,*

$$E^*(w_k \mid x_k, x_{k-1}, \ldots) = \lim_{n \to \infty} x_k x_{k-1} \cdots x_{k-n} \nu. \tag{7.1}$$

In particular, the limit to the right exists.

PROOF. The hypotheses on M ensure that its one-point compactification is a compact metric space so that $E^*(w \mid \mathscr{G})$ will be defined for M-valued random variables w. To prove (7.1) it suffices to show that if f is a bounded continuous function on M then

$$E(f(w_k) \mid x_k, x_{k-1}, \ldots) = \lim_{n\to\infty} x_k x_{k-1} \cdots x_{k-n} \nu(f). \tag{7.2}$$

By Proposition 6.13 the left-hand side of (7.2) is

$$\lim_{n\to\infty} E(f(w_k) \mid x_k, x_{k-1}, \ldots, x_{k-n})$$
$$= \lim_{n\to\infty} E(f(x_k x_{k-1} \cdots x_{k-n} w_{k-n-1}) \mid x_k, x_{k-1}, \ldots, x_{k-n}).$$

Since $x_k, \ldots, x_{k-n}$ are independent of w_{k-n-1}, we may use Proposition 6.9 to obtain

$$E(f(x_k \cdots x_{k-n} w_{k-n-1}) \mid x_k, \ldots, x_{k-n}) = \int_M f(x_k \cdots x_{k-n} u)\, d\nu(u) = x_k \cdots x_{k-n} \nu(f).$$

This proves the theorem.

8. Proper μ-processes and μ-boundaries. Let $\{(x_{n+1}, w_n)\}$ be a stationary μ-process with stationary measure ν. It may happen that w_k is measurable over the σ-algebra $(x_k, x_{k-1}, \ldots)$ (if this is true for one k it is true for all by stationarity). Then $E^*(w_k \mid x_k, x_{k-1}, \ldots) = \delta_{w_k}$; for, $f(w_k)$ is measurable over $(x_k, x_{k-1}, \ldots)$, whence by Theorem 7.5, $x_k x_{k-1} \cdots x_{k-n} \nu$ converges to a point measure. The converse is readily seen to be true: If $E^*(w_k \mid \mathscr{G})$ is a point measure, w_k must be measurable over $\mathscr{G}$.

PROPOSITION 8.1. *If $\{(x_{n+1}, w_n)\}$ is a stationary μ-process with stationary measure ν then w_k is measurable with respect to $x_k, x_{k-1}, \ldots$ iff, with probability* 1, $\lim x_{-1} x_{-2} \cdots x_{-n} \nu$ *is a point measure. In this case we say $\{(x_{n+1}, w_n)\}$ is a* **proper** *μ-process.*

We remark that the equivalence of these two properties depends on the metrizability of the space M. In general we take, as a definition of a proper process, the condition that w_n is a function of $x_n, x_{n-1}, x_{n-2}, \ldots$.

Note that we could also write $\lim x_1 x_2 \cdots x_n \nu =$ point measure, since this property depends on the distribution of $(x_1, x_2, \ldots) \cong (x_{-1}, x_{-2}, \ldots)$.

The following is one way of obtaining proper μ-processes. In a later chapter we shall see another way.

Let $\{(x_{n+1}, w_n)\}$ be a stationary μ-process. Let $w_n^* = E^*(w_n \mid x_n, x_{n-1}, \ldots)$. By

Theorem 7.5, $w_n^* = \lim x_n x_{n-1} \cdots x_{n-l} \nu$. From this it follows that $x_{n+1} w_n^* = w_{n+1}^*$. Note that w_n^* takes values in $\mathscr{P}(M)$ which is also a G-space. We see that w_n^* is a function of $\{x_n, x_{n-1}, \ldots\}$ which implies that x_{n+1} is independent of $\{x_n, x_{n-1}, \ldots; w_n^*, w_{n-1}^*, \ldots\}$. We thus have fulfilled all the conditions for a μ-process on $\mathscr{P}(M)$. It is clearly stationary and moreover this time w_k^* is measurable with respect to $x_k, x_{k-1}, \ldots$. Hence $\{(x_{n+1}, w_n^*)\}$ is a proper μ-process.

The notion of a proper μ-process is vaguely related to that of a boundary because here too we have the situation where a sequence of group elements from G contracts the measure ν to a point. This suggests the following definitions. Here G and $\mu \in \mathscr{P}(G)$ are fixed.

DEFINITION 8.1. A (G, μ)-space is a G-space M together with a measure $\nu \in \mathscr{P}(M)$ with $\nu = \mu * \nu$.

DEFINITION 8.2. A (G, μ)-space (M, ν) is called a *μ-boundary* if the associated μ-process is proper; i.e., if, with probability 1, $x_1 x_2 \cdots x_n \nu \to$ a point measure, where $\{x_n\}$ is a sequence of independent random variables with distribution μ.

EXAMPLE. Let G consist of projection transformations of $P^1 = R \cup \{\infty\}$ and let $M = P^1$. Let μ assign measure p to the transformation $t \to t+1$ and measure q to the transformation $t \to 1/(t+1)$, $p+q=1$. The stationarity equation $\nu = \mu * \nu$ for ν on P' gives

$$\nu(A) = p\nu(A-1) + q\nu(A^{-1}-1). \tag{8.1}$$

The existence of some stationary ν follows by Proposition 7.4. (8.1) can be used to show that $\nu(-\infty, 0) = 0$ and that ν is nonatomic. Hence $\nu \in \mathscr{P}([0, \infty])$. Now consider the associated μ-process $\{(x_{n+1}, w_n)\}$ and consider w_0. In the sequence $x_0, x_{-1}, x_{-2}, \ldots$ let the operation $t \to t+1$ occur u_1 times before the first occurrence of $t \to 1/(t+1)$. After the first occurrence of the latter let $t \to t+1$ occur u_2 times. Define $\{u_n\}$ in this way, $u_n \geqq 0$. It is easy to see that

$$w_0 = u_1 + \cfrac{1}{1 + u_2 + \cfrac{1}{1 + u_3 + \ddots + \cfrac{1}{1 + u_k + w_{-l}}}}$$

for $l = u_1 + u_2 + \cdots + u_k + k - 1$. Since $w_l \geqq 0$, we have

$$w_0 = u_1 + \cfrac{1}{1 + u_2 + \cfrac{1}{1 + u_3 + \ddots}} \tag{8.2}$$

which is some function of $x_1, x_{-1}, x_{-2}, \ldots$. Hence $\{(x_{n+1}, w_n)\}$ is a proper μ-process. Moreover its stationary measure is unique being the distribution of the continued fraction (8.2). (It may be verified that the u_n are independent exponentially distributed random variables.)

9. Application to transience. Associated with μ we have defined the "left-handed" random walk $x_n x_{n-1} \cdots x_1$. We can also define a "right-handed" random walk $x_1 \cdots x_n$. Let us write $y_n = x_1 \cdots x_n$. The random walk y_n on the locally compact group is called *transient* if, with probability 1, $y_n \to \infty$. Using the results of the previous section we can show

THEOREM 9.1. *If (G, μ) has a μ-boundary (M, ν) for which ν is not concentrated at one point, then the random walk y_n is transient.*

PROOF. For by Definition 8.2, $y_n \nu \to$ a point measure. Now if some subsequence of y_n were contained in a compact set we would have $\lim y_n \nu = g' \nu$ for some $g' \in G$. But $g' \nu$ is a point measure only if ν is a point measure. This proves the theorem.

THEOREM 9.2. *If (G, μ) has a (G, μ)-space (M, ν) for which not all $g \in \operatorname{support} \mu$ preserve ν, then $\{y_n\}$ is transient.*

PROOF. Let $\{(x_{n+1}, w_n)\}$ be the μ-process associated with (M, ν). Form the $\mathscr{P}(M)$ valued μ-process $\{(x_{n+1}, w_n^*)\}$ as in §8. This is a proper μ-process. So we may apply Theorem 9.1 to deduce that $\{y_n\}$ is transient unless $E^*(w_k^*)$ is a point, i.e., w_k^* is constant. Assume w_k^* is constant. Then $E^*(w_k \mid x_k, x_{k-1}, \ldots)$ is constant and from this it follows that this same constant is the absolute distribution $\nu = E^*(w_k)$. So $w_k^* = \nu$ and $g \in \operatorname{support} \mu$ preserve ν.

COROLLARY. *Let G be a nonamenable group and assume that $\mu \in \mathscr{P}(G)$ is such that no proper subgroup of G supports μ. Then the random walk $\{y_n\}$ is transient.*

PROOF. For by definition there exists a compact G-space M on which G preserves no measure. But there exists some stationary $\nu \in \mathscr{P}(M)$ so the conditions of Theorem 9.2 are fulfilled.

Another corollary is the following:

COROLLARY. *If $\mu \in \mathscr{P}(G)$, then the equation $\mu * \nu = \nu$ for $\nu \in \mathscr{P}(G)$ has no solution unless μ is supported on a compact subgroup of G. In this case the Haar measure of that group is a solution.*

PROOF. Assume (G, ν) is a (G, μ)-space. Either all $g \in \operatorname{support} \mu$ preserve the measure ν or $\{y_n\}$ is transient. But $y_n \nu$ converges by Theorem 7.5. It is easy to see that if $g_n \to \infty$ we cannot have $g_n \nu$ convergent in $\mathscr{P}(G)$. So $g\nu = \nu$ for $g \in \text{support}\, \mu$. Let $K = \{g \in G \mid g\nu = \nu\}$ then $\mu(K) = 1$. Now let Q be a compact set in G with $\nu(Q) > \frac{1}{2}$. Then $\nu(g^{-1}Q) > \frac{1}{2}$ for $g \in K$ and so $g^{-1}Q \cap Q \neq \emptyset$ for $g \in K$. Hence $K \subset QQ^{-1}$ whence K is compact.

We notice in connection with the first corollary that what has been shown is that if G is nonamenable and the support of μ generates G then there exists a nontrivial μ-boundary. This is analogous to the fact that every nonamenable group G possesses a nontrivial boundary in the sense of Chapter I. We omit the proof of this latter characterization here.

It should not be thought, however, that amenable groups cannot have μ-boundaries for appropriate μ. For example, if G consists of the solvable group of affine transformations of $\boldsymbol{R}: t \to at + b$, and μ concentrates its mass at the two transformations $t \to t+1$, $t \to \frac{1}{2}t$, then it is not hard to show that $R \cup \{\infty\}$ is a μ-boundary. Nonetheless it seems reasonable to

CONJECTURE. *G possesses a measure μ whose support is all of G and for which no nontrivial μ-boundary exists iff G is amenable.*

10. Universal μ-boundaries. By analogy with the theory of Chapter I we might expect that for each pair (G, μ) there exists a universal μ-boundary. To make this meaningful we first give the following definition:

DEFINITION 10.1. A (G, μ)-space (M', ν') is called an *equivariant* image of a (G, μ)-space (M, ν) if there is an equivariant map $\phi: M \to M'$ with $\phi\nu = \nu'$.

We note that since ϕ is equivariant, $\phi(\mu * \nu) = \mu * \phi\nu = \mu * \nu'$; so since $\mu * \nu = \nu$, we will also have $\mu * \nu' = \nu'$.

LEMMA 10.1. *If (M', ν') is an equivariant image of (M, ν) and if the latter is a μ-boundary, so is the former.*

PROOF. For if $\{(x_{n+1}, w_n)\}$ is the μ-process associated with ν then $\{(x_{n+1}, \phi(w_n))\}$ will be a version of the μ-process associated with ν'. Now if (M, ν) is a μ-boundary, w_n will be a function of $(x_n, x_{n-1}, x_{n-2}, \ldots)$. But then the same is true of $\phi(w_n)$. This proves the lemma.

DEFINITION 10.2. A (G, μ)-space (M_0, ν_0) is a *universal μ-boundary* if it is a μ-boundary and if every μ-boundary is an equivariant image of (M_0, ν_0).

REMARK. Lemma 10.1 is valid if (M', ν') is only a measurable but not continuous equivariant image of (M, ν). We could therefore also introduce the notion of a *measurable* universal μ-boundary. In certain situations the measurable universal μ-boundary is a familiar space whereas the μ-boundary of Definition 10.2 is too large to be identifiable.

The following definition is made in analogy with the concept of minimality of Chapter I.

DEFINITION 10.3. A (G, μ)-space (M, ν) is called a *minimal* (G, μ)*-space* if no proper closed G-invariant subset of M has ν-measure $=1$.

THEOREM 10.1. *For each pair* (G, μ) *there exists a universal* μ*-boundary* (M_0, ν_0). *We may require* (M_0, ν_0) *to be minimal and then it is unique up to isomorphism.*

PROOF. We sketch the proof. It is easy to see that one can confine the discussion entirely to minimal (G, μ)-spaces. One may also show that the cardinality of the family of equivalence classes of minimal (G, μ)-spaces that are μ-boundaries is bounded by 2^{2^a}, a being the cardinality of G. Hence one may form a set $\{(M_\alpha, \nu_\alpha)\}$ of representatives of all these equivalence classes. For each member we form the process $\{w_n^\alpha\}$ with values in M_α and associated to ν_α. Then $\{w_n^\alpha\}$ is a "function" of $(x_n, x_{n-1}, \ldots, x_{n-k}, \ldots)$. Now let $M = \prod M_\alpha$ and set $w_n = \{\ldots, w_n^\alpha, \ldots\}$. It is easy to see that $\{(x_{n+1}, w_n)\}$ is a proper μ-process, and letting M_0 be the minimal G-invariant subset supporting $\{w_n\}$ and $\nu_0 = E^*(w_n)$ it is evident that (M_0, ν_0) will be universal.

11. Triviality of μ-boundaries in special cases.

THEOREM 11.1. *If G is abelian then a* μ*-boundary is trivial* (*the stationary measure is a point measure*).

PROOF. Suppose (M, ν) is a μ-boundary for (G, μ). Let $Q = \{\nu' \in \mathscr{P}(M) \mid \mu * \nu' = \nu'\}$. Q is a compact convex set and so is spanned by its extremals. Note that because of the commutativity of G, Q is G-invariant. It follows that if $\nu' \in Q$ and $\mu' \in \mathscr{P}(G)$ then $\mu' * \nu' \in Q$. So if we can write $\mu = \alpha\mu' + (1-\alpha)\,\mu''$ with $\mu', \mu'' \in \mathscr{P}(G)$ then $\nu' = \mu * \nu' = \alpha(\mu' * \nu') + (1-\alpha)\,(\mu'' * \nu')$ is a convex combination of elements of Q. Thus ν' cannot be extremal unless $\mu' * \nu' = \nu'$ for all $\mu' \prec \mu$. This may be seen to imply that $g\nu' = \nu'$ for all g in the group generated by the support of μ. Since this is true for all extremals in Q, it is true for every stationary measure on M.

We now assume M is metrizable. Then we shall have

$$x_0 x_{-1} \cdots x_{-n} \nu \to \text{point measure}$$

if (M, ν) is a μ-boundary. By the foregoing this implies that ν is itself a point measure, so that a μ-boundary is trivial. For the general case we observe that if (M, ν) is a (G, μ)-space then the points of M may be separated by equivariant maps $\phi\colon (M, \nu) \to (M', \nu')$ where M' is metrizable. This proves the theorem.

DEFINITION 11.1. Let $\mu \in \mathscr{P}(G)$. We say μ satisfies condition A if for each

$g \in G$ there exists a power $\mu_n = \mu * \mu * \cdots * \mu$ such that the measures μ_n and $g\mu_n$ are *not* mutually singular.

In particular if μ is absolutely continuous on G and is not supported by an open subgroup, this condition is verified.

THEOREM 11.2. *If G is nilpotent and μ satisfies condition* A *then every μ-boundary (G, μ) is trivial.*

We omit this proof which is similar to that of Theorem 11.1.

We do not know if condition A is necessary in this theorem. The nilpotence of the group is indispensable for as we saw in the end of §9 there are nontrivial μ-boundaries for solvable G.

OPEN PROBLEM. Determine the universal μ-boundary for an absolutely continuous measure μ on a solvable group G.

This is open even for

$$G = \left\{ \begin{pmatrix} a & b \\ 0 & 1 \end{pmatrix} \right\}.$$

In the next chapter we will resolve a similar problem for semisimple groups G.

III. μ-boundaries and μ-harmonic functions

12. μ-harmonic functions.

DEFINITION 12.1. A complex valued function $f(g)$ defined on the group G is called μ-harmonic ($\mu \in \mathscr{P}(G)$) if each translate $f(g_1 g)$ belongs to $L^1(G, \mu)$ and

$$f(g) = \int f(gg')\, d\mu(g').$$

Recalling Definition 5.2 we see that a (G, K)-harmonic function is μ-harmonic whenever μ is a K-invariant measure on G.

We shall be mainly concerned with bounded μ-harmonic functions and just as in §5 it suffices to develop a theory of LUC μ-harmonic functions. (See Definition 5.1.)

PROPOSITION 12.1. *Let (M, ν) be a (G, μ)-space and let $\phi \in \mathscr{C}(M)$. Then the function*

$$f(g) = \int_M \phi(g\xi)\, d\nu(\xi) \tag{12.1}$$

is a μ-harmonic function which is LUC.

For

$$\int f(gg')\,d\mu(g') = \iint \phi(gg'\xi)\,d\mu(g')\,d\nu(\xi)$$
$$= \int \phi(g\eta)\,d\mu*\nu(\eta) = \int \phi(g\eta)\ d\nu(\eta') = f(g).$$

Thus each (G, μ)-space gives rise to a family of μ-harmonic functions. We sometimes refer to a representation (12.1) of a μ-harmonic function as a *Poisson representation.*

THEOREM 12.2. *Let G be a locally compact, σ-compact metrizable group, $\mu \in \mathscr{P}(G)$. Every LUC bounded μ-harmonic function admits a representation* (12.1). *In fact all LUC bounded μ-harmonic functions admit a representation on the* (G, μ)-*space which is the universal μ-boundary of* (G, μ).

PROOF. We note first that if (M', ν') is an equivariant image of (M, ν) and if

$$f(g) = \int_{M'} \phi(g\xi)\,d\nu'(\xi),$$

then setting $\psi = \phi \circ \alpha$ where $\alpha: (M, \nu) \to (M', \nu')$, we shall have

$$f(g) = \int_{M} \psi(g\eta)\,d\nu(\eta).$$

Hence to prove that every LUC bounded μ-harmonic function admits a Poisson representation on the universal μ-boundary of (G, μ) it suffices to show that such a function admits a Poisson representation on *some* μ-boundary.

Write ${}^{\gamma}f(g) = f(g\gamma)$. Assume f is a bounded LUC μ-harmonic function. The set $\{{}^{\gamma}f \mid \gamma \in G\}$ is equicontinuous and has compact closure in the space of continuous functions on G in the topology of uniform convergence on compact sets. Let $M(f)$ denote the compact set. Then G operates on $M(f)$ by $(\gamma, F) \to {}^{\gamma}F$ and $M(f)$ is a compact G-space.

Let $\{x_n\}$ be the usual sequence of independent G-valued random variables with distribution μ. Define $z_n(g) = f(gx_0x_{-1}\cdots x_{-n})$. By Proposition 6.9 we have

$$E(z_n(g) \mid x_0, \dots, x_{-n+1}) = \int f(gx_0x_{-1} \cdots x_{-n+1}g')\, d\mu(g')$$
$$= f(gx_0x_1 \cdots x_{-n+1}) = z_{n-1}(g),$$

using the μ-harmonicity of f. Hence

$$E(z_n(g)\, z_0(g), \dots, z_{n-1}(g)) = z_{n-1}(g)$$

and $z_n(g)$ forms a bounded martingale for each g. By Theorem 6.12, $w_0(g) = \lim_{n\to\infty} z_n(g)$ exists with probability 1, and $E(w_0(g)) = E(z_0(g)) = f(g)$.

In a similar way we set

$$w_n(g) = \lim_{l\to\infty} f(gx_nx_{n-1} \cdots x_{n-l}).$$

We now consider w_n as an $M(f)$-valued random variable. (The simultaneous definition of the almost everywhere defined $w_n(g)$ for all g is taken care of by the fact that f is LUC and G has a countable dense set.)

Clearly w_n is a function of $x_n, x_{n-1}, x_{n-2}, \dots$. Also

$$x_{n+1}w_n(g) = w_n(gx_{n+1}) = w_{n+1}(g)$$

whence $x_{n+1}w_n = w_{n+1}$. It follows that w_n is a proper μ-process. Letting $v = E^*(w_n)$, $(M(f), v)$ will be a μ-boundary.

Finally

$$f(g) = E(w_0(g)) = \int_{M(f)} F(g)\, dv(F) = \int_{M(f)} \phi(g\xi)\, dv(\xi)$$

where $\phi(F) = F(e)$. This completes the proof of the theorem.

13. μ-harmonic functions and μ-boundaries for G semisimple. In this section we assume G is a noncompact semisimple group with finite center and that μ is an absolutely continuous measure on G. We shall describe the universal μ-boundary.

Let K be a maximal compact subgroup of G and let $X = G/K$ be the associated symmetric space. The main tool in our discussion is the following lemma:

LEMMA 13.1. *Let $f(x)$ be a bounded continuous function on X satisfying*

$$f(x) = \int_G f(gx)\, d\mu(g)$$

for all $x \in X$*; then* $f(x)$ *is a constant.*

The proof is given in [**1**], and we refer the reader to that paper for the details. The proof depends on the fact that, for any two points $x, y \in X$, the measures $\mu_n * \delta_x, \mu_n * \delta_y$ $(\mu_n = \mu * \mu * \cdots * \mu)$ eventually "overlap" as $n \to \infty$. Using martingale theory one deduces from this that $f(x)$ is a constant.

This lemma now implies

LEMMA 13.2. *If* $f(g)$ *is a bounded* μ*-harmonic function satisfying* $f(g) = f(kg)$ *for all* $k \in K$*, then* $f(g)$ *is a constant.*

PROOF. $\tilde{f}(g) = f(g^{-1})$ is a function of $gK \in X$ and satisfies

$$\tilde{f}(g) = \int \tilde{f}(g'g)\, d\tilde{\mu}(g')$$

where $\tilde{\mu}(A) = \mu(A^{-1})$. By Lemma 13.1, $\tilde{f}(g)$ is a constant.

PROPOSITION 13.3. *If* K *is a maximal compact subgroup of* G *and* (M, ν) *is a minimal* μ*-boundary then* K *is transitive on* M.

PROOF. If K is not transitive on M there will exist two distinct K-orbits on M. It is then possible to construct a function $\phi \in \mathscr{C}(M)$ satisfying $\phi(k\xi) = \phi(\xi)$ for $k \in K$, $\xi \in M$, and ϕ not constant. But by the foregoing proposition together with Proposition 12.1 we must have

$$F(g) = \int \phi(g\xi)\, d\nu(\xi) = \text{constant}.$$

Assume M is metrizable. We then have $x_0 x_{-1} \cdots x_{-n} \nu \to$ point measure $= \delta_{z_0}$. This gives $\phi(gz_0) =$ constant, which contradicts the choice of ϕ. Here z_0 is a random variable with distribution ν. Finally we reduce the case of a general M to the metrizable case. This proves the proposition.

DEFINITION 13.1. We denote by Π_μ the underlying space of the minimal universal μ-boundary.

Recall the notation $B(G) = G/H(G)$ for the universal boundary.

THEOREM 13.4. Π_μ *is a finite-sheeted covering space of* $B(G)$*. If* μ *or some power of* μ *has the identity in the interior of its support, then* $\Pi_\mu = B(G)$.

PROOF. Let $\Pi_\mu = G/H_\mu$. To represent Π_μ in this form we use the fact that K is transitive on Π_μ and so, *a fortiori*, G is transitive on Π_μ.

Now in §8 we saw that whenever a G-space M supports a stationary μ-process, the space $\mathscr{P}(M)$ will support a proper μ-process; i.e., $\mathscr{P}(M)$ contains a μ-boundary. In particular $\mathscr{P}(B(G))$ contains a μ-boundary and so there exists an equivariant map $\phi: \Pi_\mu \to \mathscr{P}(B(G))$. But, by Proposition 4.2, inasmuch as Π_μ is homogeneous, ϕ corresponds to an equivariant map of Π_μ to $B(G)$. Thus we may assume $H_\mu \subset H(G)$. We shall show that $H(G)/H_\mu$ is finite.

Since $H(G)$ is amenable, it leaves invariant a measure $\lambda \in \mathscr{P}(\Pi_\mu)$. Under the map of $\Pi_\mu \to B(G)$, λ maps into the unique $H(G)$-invariant measure on $B(G)$, namely $\delta_{H(G)}$. It follows that λ is concentrated on this fibre $H(G)/H_\mu \subset \Pi_\mu$, and is the unique $H(G)$-invariant measure.

Let ν_0 be the stationary measure for μ on Π_μ. Since (Π_μ, ν_0) is a μ-boundary, there is some sequence $\{g_n\}$ with $g_n \nu_0 \to$ point measure. From the fact that μ is absolutely continuous one deduces that ν_0 is absolutely continuous. It is not hard to deduce from this that if $\omega \in \mathscr{P}(\Pi_\mu)$ has small enough support then there exists a sequence $\{g_n\}$ with $g_n \omega \to$ point measure. In particular some limit of translates of λ dominates a point measure. But this implies that λ itself is discrete which implies that $H(G)/H_\mu$ is finite.

Finally we see that if μ or some power of μ includes the identity in the interior of its support, then ν_0 is equivalent to the K-invariant measure on Π_μ. In this case the same property as before holds for *all* $\omega \in \mathscr{P}(\Pi_\mu)$ which implies that Π_μ is a boundary. Hence $\Pi_\mu \cong B(G)$.

COROLLARY. *Let $\psi \in L^1(G)$, $\psi \geqq 0$, $\psi(g) > 0$, for g close to the identity and $\int \psi(g)\, dg = 1$. Then if*

$$f(g) = \int f(gg')\, \psi(g')\, dg'$$

where f is bounded and LUC,

$$f(g) = \int_{B(G)} \tilde{f}(g\xi)\, d\nu(\xi)$$

for some continuous $\tilde{f} \in \mathscr{C}(B(G))$. Here ν is the unique solution to $\mu * \nu' = \nu'$ in $\mathscr{P}(B(G))$.

What has not been shown here is that the stationary measure on $B(G)$ is unique. But what appeared in the foregoing proof that, for any stationary measure ν, $(B(G), \nu)$ is an equivariant image of (Π_π, ν_0). But, by Proposition 4.2, there is a unique equivariant map from Π_μ to $B(G)$. Hence there is a unique stationary measure.

IV. μ-proximal spaces

14. μ-proximality. We would like to attach to a countable group Γ a Γ-space that has universal properties similar to those of $B(G)$ for a Lie group G. When Γ is countable the space $B(\Gamma)$ is generally too large to be useful; e.g., it is generally not metrizable. More specifically we will show that if Γ is a lattice in a semisimple group G, then $B(G)$ has a certain universal property as a Γ-space.

In the following, $\mu\in\mathcal{P}(G)$ as usual and M is a compact metric G-space. $\mathcal{P}(M)$ is metrizable and we use $d(\cdot,\cdot)$ to denote a metric in M, or in $\mathcal{P}(M)$, or the distance between a point and a set in $\mathcal{P}(M)$. We let $\mu^{(n)}=\mu*\mu*\cdots*\mu$, the n-fold product, and we set

$$\mu_n=n^{-1}(\mu+\mu^{(2)}+\cdots+\mu^{(n)}).$$

DEFINITION 14.1. M is a *μ-proximal* G-space if, for each $x,y\in M$, $\mu_n\{g\mid d(gx,gy)>\varepsilon\}\to 0$ as $n\to\infty$ for any $\varepsilon>0$.

The set function $\lim_{n\to\infty}\mu_n(A)$, $A\subset G$, may be thought of as a kind of density. The foregoing definition requires that the set of $g\in G$ for which gx and gy are not within ε of each other has 0 density. This notion is, of course, relative to μ.

THEOREM 14.1. *The following are equivalent for a compact metric G-space M*:

(a) *M is μ-proximal.*

(b) *Any solution $\nu\in\mathcal{P}(M\times M)$ to $\mu*\nu=\nu$ is concentrated on the diagonal $\Delta(M)\subset M\times M$.*

(c) *For any stationary measure $\nu\in\mathcal{P}(M)$, i.e., solution to $\mu*\nu=\nu$, the (G,μ)-space (M,ν) is a μ-boundary.*

(d) *For any $\theta\in\mathcal{P}(M)$ and $\varepsilon>0$,*

$$\lim_{n\to\infty}\mu_n\{g\in G\mid d(g\theta,\delta_M)>\varepsilon\}=0.$$

PROOF. (d)$\Rightarrow$(a). Take $\theta=\frac{1}{2}(\delta_x+\delta_y)$.

(a)$\Rightarrow$(b). $\mu*\nu=\nu$ implies that $\mu_n*\nu=\nu$. Consider

$$\int_{M\times M} d(x,y)\,d\nu(x,y)=\int_G\int_{M\times M} d(gx,gy)\,d\mu_n(g)\,d\nu(x,y).$$

Now the right-hand side converges to 0 as $n\to\infty$, since

$$\int d(gx,gy)\,d\mu_n(g)\to 0$$

for all x, y. It follows that ν is concentrated on $\Delta(M)$.

To show that (b)$\Rightarrow$(c) we need the notion of the expectation of a $\mathscr{P}(M)$-valued random variable. This makes sense because $\mathscr{P}(M)$ is a convex compact set. Specifically we may define $E(Z)$ for Z a $\mathscr{P}(M)$-valued random variable by

$$E(Z)(\phi)=E(Z(\phi)), \qquad \phi\in\mathscr{C}(M). \tag{14.1}$$

LEMMA 14.2. *If X is a G-valued random variable and Z a $\mathscr{P}(M)$-valued random variable and X and Z are independent, then*

$$E(XZ)=E^*(X)*E(Z).$$

The proof is straightforward and is left to the reader. We proceed with the proof of Theorem 14.1.

(b)$\Rightarrow$(c). Let $\{(x_{n+1}, z_n)\}$ be a stationary μ-process on M and form $z_n^* = E^*(z_n \mid x_n, x_{n-1}, \ldots)$. Then $z_{n+1}^* = x_{n+1} z_n^*$. Hence

$$x_{n+1}(z_n^* \times z_n^*) = z_{n+1}^* \times z_{n+1}^* \tag{14.2}$$

where $z_n^* \times z_n^*$ is the product random measure on $M \times M$. Take expectations of both sides of (14.2) and by Lemma 14.2 we have

$$\mu * E(z_n^* \times z_n^*) = E(z_n^* \times z_n^*).$$

So by (b), $E(z_n^* \times z_n^*)$ is concentrated on $\Delta(M)$. It follows that with probability 1, $z_n^* \times z_n^*$ is concentrated on $\Delta(M)$ which implies that z_n^* is a point measure. This implies that $\{(x_{n+1}, z_n)\}$ is proper and that $(M, E^*(z_n))$ is a μ-boundary.

LEMMA 14.3. *Let M be a compact metric G-space, M' a closed G-invariant subspace. Assume every M-valued stationary μ-process on M is already M'-valued. Then, for each $x\in M$ and $\varepsilon>0$,*

$$\lim_{n\to\infty} \mu_n\{g \mid d(gx, M')>\varepsilon\}=0.$$

PROOF. Suppose $\mu_{n_k}\{g \mid d(gx, M')>\varepsilon\}>\delta>0$. Let $\nu_k=\mu_{n_k}*\delta_x\in\mathscr{P}(M)$. We can assume $\nu_k\to\nu\in\mathscr{P}(M)$. Clearly $\mu*\nu=\nu$ so ν is concentrated on M'. So, for large k, $\nu_k\{y \mid d(y, M')>\varepsilon\}<\delta$. But this is a contradiction.

With this lemma we show that (b)$\Rightarrow$(d). Let $\{(x_{n+1}, z_n)\}$ be a stationary μ-process with z_n taking values in $\mathscr{P}(M)$. Form $z_n \times z_n$. The above argument shows that by (b) this is concentrated on the diagonal so that z_n is a point measure. Thus every $\mathscr{P}(M)$-valued stationary μ-process is already δ_M-valued. So by the above lemma we have (d).

Finally we need

(c)⇒(b). Let $w_n = (w'_n, w''_n)$ be $M \times M$-valued so that (x_{n+1}, w_n) is a stationary μ-process. We wish to show that (c) implies that $w'_n = w''_n$. If Ω is the underlying probability space, form $\tilde{\Omega} = \Omega \times \{1, 2\}$. Define $\tilde{w}_n(\omega, 1) = w'_n(\omega)$, $\tilde{w}_n(\omega, 2) = w''_n(\omega)$. It is easy to check that $\{(x_{n+1}, \tilde{w}_n)\}$ is a stationary μ-process which cannot be a function of $x_n, x_{n-1}, \ldots$ alone unless it does not depend on ε in $\tilde{\omega} = (\omega, \varepsilon)$. This means that if every stationary μ-process on M is to be proper, we must have $w'_n = w''_n$. This completes the proof of the theorem.

COROLLARY. *If M is a μ-proximal G-space then every $\mathscr{P}(M)$-valued stationary μ-process is δ_M-valued.*

EXAMPLE. If G is a semisimple Lie group and μ is an absolutely continuous measure on G then every boundary is μ-proximal. Conversely any μ-proximal space will contain a boundary. We cannot do any better than this since, for example, the space $M \times M$ is always μ-proximal if M is, and if M is a boundary, $M \times M$ contains $\Delta(M)$ which is a boundary.

To see this let M be a boundary of G. We know $\mathscr{P}(M)$ contains a μ-boundary so there exists an equivariant map $\Pi_\mu \to \mathscr{P}(M)$. But Π_μ is minimal and so by Proposition 4.2 this induces a map $\Pi_\mu \to M$. Now if (x_{n+1}, w_n) is M-valued $(x_{n+1}, E^*(w_n \mid x_n, x_{n-1}, \ldots))$ is $\mathscr{P}(M)$-valued and by the foregoing it is δ_M-valued. This implies that (x_{n+1}, w_n) is a proper μ-process and that each (M, ν) is a μ-boundary. This is property (b) of Theorem 14.1.

15. Lattices of G and $B(G)$. If G is a connected Lie group and Γ a discrete subgroup, we call Γ a lattice if G/Γ possesses a finite G-invariant measure. This includes the case that G/Γ is compact.

THEOREM 15.1. *Let Γ be a lattice in the semisimple connected group G, and let M be any compact Γ-space. Then there exists a measurable Γ-equivariant map from $B(G)$ to $\mathscr{P}(M)$.*

("Measurability" is with respect to the compact topology of $\mathscr{P}(M)$ and the manifold measure class on $B(G)$. Precisely, the assertion is that there exists a Γ-equivariant map $\beta: B(G) \to \mathscr{P}(M)$ defined almost everywhere with respect to "Lebesgue measure" on the manifold $B(G)$ and such that the inverse of an open set in $\mathscr{P}(M)$ is Lebesgue measurable.)

PROOF. Form $G \times M$ and let Γ operate on this space by $\gamma(g, x) = (g\gamma^{-1}, \gamma x)$. Also G operates on $G \times M$ by $g'(g, x) = (g'g, x)$. Let $G \times_\Gamma M = \Gamma \backslash G \times M$. $G \times_\Gamma M$ is a locally compact G-space and there exists an equivariant map $\psi: G \times_\Gamma M \to G/\Gamma$ with $\psi(g, x) = g\Gamma$. Let $m_{G/\Gamma}$ denote the G-invariant probability measure on G/Γ.

It is not hard to show that the set of probability measures on $G \times_{\Gamma} M$ which map into $m_{G/\Gamma}$ under ψ is compact and convex. Inasmuch as G takes this set to itself, and $H(G) \subset G$ is amenable, there exists an $H(G)$-invariant measure $\pi \in \mathscr{P}(G \times_{\Gamma} M)$ with $\psi(\pi) = m_{G/\Gamma}$. Lifting π from $\Gamma \backslash G \times M$ to $G \times M$ we obtain a σ-finite measure $\tilde{\pi}$ invariant under Γ and the induced action of $H(G) \subset G$, and mapping onto Haar measure m_G under the projection from $G \times M$ to G. We may write

$$\tilde{\pi} = \int_G \delta_g \times \lambda_g \, dg$$

where $\lambda_g \in \mathscr{P}(M)$. (This is by a general decomposition theorem valid for measures on product spaces.) The invariance under $H(G)$ implies

$$\int \delta_g \times \lambda_g \, dg = \int \delta_{hg} \times \lambda_g \, dg = \int \delta_g \times \lambda_{h^{-1}g} \, dg$$

for $h \in H(G)$, so that $\lambda_g = \lambda_{H(G)\,g}$. The invariance under the action of Γ implies

$$\int \delta_g \times \lambda_g \, dg = \int \delta_{g\gamma^{-1}} \times \gamma\lambda_g \, dg = \int \delta_g \times \gamma\lambda_{g\gamma} \, dg$$

whence $\lambda_{g\gamma} = \gamma^{-1}\lambda_g$ for $\gamma \in \Gamma$. Set $\beta(gH(G)) = \lambda_{g^{-1}}$, then β is a measurable function and

$$\beta(\gamma g H(G)) = \lambda_{g^{-1}\gamma^{-1}} = \gamma\lambda_{g^{-1}} = \gamma\beta(gH(G))$$

so that β is Γ-equivariant.

THEOREM 15.2. *Let Γ be a lattice in the semisimple connected group G. Let μ_0 be an absolutely continuous measure on G given by $d\mu_0(g) = \psi_0(g)\, dg$ where $\psi_0(e) > 0$ and ψ_0 is continuous and has compact support. Let $\nu_0 \in \mathscr{P}(B(G))$ with $\mu_0 * \nu_0 = \nu_0$. Then there exists $\mu \in \mathscr{P}(\Gamma)$ with $\mu(\gamma) > 0$ for all $\gamma \in \Gamma$ and $\mu * \nu_0 = \nu_0$.*

We begin with

LEMMA 15.3. *If $g_1, g_2 \in G$ there exists a positive integer n and $\varepsilon > 0$ with $g_1\mu_0 * \mu_0 * \cdots * \mu_0 > \varepsilon\, g_2\mu_0$ (μ_0 n times on the left-hand side).*

PROOF. The support of $g_2\mu_0$ is compact and that of $\mu_0 * \mu_0 * \cdots * \mu_0$ includes arbitrarily large powers of neighborhoods of the identity. Since both measures are given by continuous densities, the lemma follows.

LEMMA 15.4. *If $g \in G$ and $\gamma \in \Gamma$ there exist $\varepsilon(g, \gamma) > 0$ and $\omega_g \in \mathscr{P}(G)$ with*

$$g\nu_0 = \varepsilon\gamma\nu_0 + (1-\varepsilon)\int g'\nu_0\, d\omega_g(g').$$

PROOF. Write $g\mu_0 * \mu_0 * \cdots * \mu_0 > \varepsilon\gamma\mu_0$. Then

$$\begin{aligned} g\nu_0 &= g\mu_0 * \cdots * \mu_0 * \nu_0 = \varepsilon\gamma\mu_0 * \nu_0 + (g\mu_0 * \mu_0 * \cdots * \mu_0 - \varepsilon\gamma\mu_0) * \nu_0 \\ &= \varepsilon\gamma\nu_0 + (1-\varepsilon)\int g'\nu_0\, d\omega_g(g'). \end{aligned}$$

Now define

$$L(g) = \sup\{l : 0 \leqq l \leqq 1 \text{ and } g\nu_0 = l\mu' * \nu_0 + (1-l)\,\mu'' * \nu_0\}$$

where $\mu' \in \mathscr{P}(\Gamma)$ and is everywhere positive and $\mu'' \in \mathscr{P}(G)$. From Lemma 15.4 it is easily seen that $L(g) > 0$ for each $g \in G$. Note that $L(\gamma g) = L(g)$ for $\gamma \in \Gamma$. We also have

$$g\nu_0 = g\mu_0 * \nu_0 = \int gg'\nu_0\, d\mu_0(g')$$

for which it is easy to deduce that

$$L(g) \geqq \int L(gg')\, d\mu_0(g').$$

Setting $L'(\Gamma g) = 1 - L(g)$ we find L' defined on $\Gamma \backslash G$ with

$$L'(x) \leqq \int L'(xg)\, d\mu_0(g). \tag{15.1}$$

So

$$\int_{\Gamma\backslash G} L'(x)^2\, dx \leqq \int_{\Gamma\backslash G}\int_G L'(xg)^2\, d\mu_0(g)\, dx = \int_{\Gamma\backslash G} L'(x)^2\, dx.$$

This implies that $L'(x)$ is constant almost everywhere, and by (15.1) we see that uniformly $L'(x) \leqq 1-\varepsilon$ or $L(g) \geqq \varepsilon$. Now choose $0 < l < \varepsilon$. We can write

$$\nu_0 = l\mu_1 * \nu_0 + (1-l)\,[l\mu_2 * \nu_0 + (1-l)\,[l\mu_3 * \nu_0 + \cdots]].$$

Since $l+(1-l)\,l+(1-l)^2 l+\cdots=1$, we have obtained a proof of the theorem.

Now let us choose μ_0 in Theorem 15.2 with $\psi_0(g)$ everywhere >0. Then ν_0 is equivalent to the K-invariant measure on $B(G)$. Let $\{y_n\}$ be a sequence of Γ-valued independent random variables with the distribution μ of the theorem. Let $\{(y_{n+1}, w_n)\}$ be the stationary μ-process corresponding to the measure $\nu_0 \in \mathscr{P}(B(G))$. Recall the setup of Theorem 15.1 and let β be the Γ-equivariant map from $B(G)$ to $\mathscr{P}(M)$. Since β is defined and measurable with respect to $(B(G), \nu_0)$, the variables $\beta(w_n)$ are defined and together with u_{n+1} form a μ-process on $\mathscr{P}(M)$. Recalling the corollary to Theorem 14.1 we find that $\beta(w_n)$ must take values in δ_M if M happens to be μ-proximal. That means that β comes from an equivariant map of $B(G)$ to M itself. This proves

THEOREM 15.5. *Let Γ be a lattice subgroup of the connected semisimple group G. There exists a measure $\mu \in \mathscr{P}(\Gamma)$, $\mu(\gamma)>0$ for all $\gamma \in \Gamma$, such that if M is any μ-proximal compact Γ-space then there exists a measurable Γ-equivariant map from $B(G)$ to M. This map is unique modulo sets of measure 0 in $B(G)$.*

The uniqueness follows from the fact that an equivariant map to $\mathscr{P}(M)$ has an image in δ_M. If there were two maps to M, i.e., to δ_M, their average would be in $\mathscr{P}(M)$ and would not be in δ_M unless they coincide.

16. Mean-proximal spaces.

DEFINITION 16.1. A G-space M is called *mean-proximal* if it is μ-proximal for every $\mu \in \mathscr{P}(G)$ with support $\mu = G$.

The existence of such spaces is ensured by the following result.

THEOREM 16.1. *Let $G \subset SL(m+1, \mathbf{R})$ satisfy one of the following conditions:*

(a) *G acts irreducibly on $\mathbf{R}^{m+1}$ and proximally on P^m.*

(b) *Every subgroup of finite index in G acts irreducibly on $\mathbf{R}^{m+1}$ and G contains a matrix whose eigenvalue of maximal absolute value is real and occurs without multiplicity.*

Then P^m is a mean-proximal G-space.

(See §1 for the definition of proximality.)

We prove this in a series of steps.

LEMMA 16.2. *Under the hypotheses of the theorem, if V is a proper subspace of $\mathbf{R}^{m+1}$, then $\{gV \mid g \in G\}$ is an infinite set.*

PROOF. This is obvious if every subgroup of finite index of G is irreducible.

To prove it under hypothesis (a), let V be a proper subspace of minimal dimension with $\{gV \mid g\in G\}$ finite. We must have $g_1V\cap g_2V=g_1V$ or 0, for otherwise it would be a subspace of smaller dimension having only finitely many transforms under G. Since G acts irreducibly, there must be distinct g_1V, g_2V. The images of g_1V and g_2V in P^m are disjoint linear subvarieties having only finitely many transforms under G. It is easy to see that G cannot act proximally on P^m. This proves the lemma.

LEMMA 16.3. *Let $\mu\in\mathscr{P}(G)$ with support $\mu=G$, $\mu*\nu=\nu$. Let W be a linear subvariety $\subsetneq P^m$. Then $\nu(W)=0$.*

PROOF. Let W be a linear subvariety of least dimension with $\nu(W)>0$. Only finitely many such W can have $\nu(W)>\delta$, since $\nu(W'\cap W'')=0$ for $W'\neq W''$. Hence we can find W maximizing $\nu(W)$ for that dimension. Now

$$\nu(W)=\int \nu(g^{-1}W)\,d\mu(g)$$

which implies that $\nu(g^{-1}W)=\nu(W)$ for almost all $g\in G$. But then $\{gW \mid g\in G\}$ is finite contrary to Lemma 16.1.

Now let $\{g_n\}$ be any sequence in G. Let $\|g\|$ denote some operator norm in the vector space of linear transformations of $\boldsymbol{R}^{n+1}$. We can choose a subsequence $\{n_k\}$ so that $g_{n_k}/\|g_{n_k}\|$ converges to a matrix σ in $M_{n+1}(\boldsymbol{R})$. If $g_{n_k}\to\infty$ in G then the limit matrix σ is singular. In any case $\|\sigma\|=1$, so that the nullspace of σ is not all of $\boldsymbol{R}^{n+1}$. Let V be this nullspace and let W (which may be $\emptyset$) be the linear subvariety of P^m corresponding to V. For $x\notin V$, $\sigma x\neq 0$, and so for $\bar{x}\in P^m-W$ we can define a transformation $\bar{\sigma}$:

$$\bar{\sigma}\bar{x}=\overline{\sigma x}.$$

We then have $g_{n_k}\bar{x}\to\bar{\sigma}\bar{x}$. Call $\bar{\sigma}$ a *quasi-projective* transformation. It is defined on all of P^m outside of a linear subvariety. We have

LEMMA 16.4. *If $\{g_n\}\subset G$ there exist a quasi-projective transformation τ defined on P^m-W and a subsequence $\{n_k\}$ such that for $y\in P^m-W$, $g_{n_k}y\to\tau y$.*

LEMMA 16.5. *Under the hypotheses of the theorem, if $\mu\in\mathscr{P}(G)$ with support all of G and $\nu\in\mathscr{P}(P^m)$ with $\mu*\nu=\nu$, there exists a sequence $\{g_n\}\subset G$ with $g_n\nu\to$ point measure.*

PROOF. We first prove this under hypothesis (b). We have $g\in G$ where g has a unique eigendirection for the eigenvalue of maximal absolute value. It follows that for $y \in P^m$, $g^n y \to$ this eigendirection unless y corresponds to a vector in the subspace $V \subset \boldsymbol{R}^{m+1}$ spanned by remaining eigenspaces. Since, by Lemma 16.3, this subspace corresponds to a linear subvariety of measure 0, we have $g^n \nu \to \delta_{y_0}$.

We now turn to hypothesis (a). It is easily shown that under the assumption of proximality on P^m, if $y_1, \ldots, y_{2m+1}$ are any $2m+1$ points in P^m, there exists a sequence $\{g_n\} \subset G$ with $g_n y_i \to y_0$ for some fixed y_0 and for all i. Now assume the y_i are in general position so that no $m+1$ of them lie in an $(m-1)$-dimensional linear subvariety. By Lemma 16.4 we can find $\{g_{n_k}\}$ with $g_{n_k} y \to \tau y$ outside of some $W \subsetneqq P^m$. At most m of $\{y_1, \ldots, y_{2m+1}\}$ are in W. The remaining $m+1$ correspond to a basis of $\boldsymbol{R}^{m+1}$. We have $g_{n_k} y_i \to \tau y$ for these, so $\tau y = y_0$. Hence $g_{n_k} y \to y_0$ for all $y \notin W$. Since $\nu(W) = 0$, we have $g_{n_k} \nu \to \delta_{y_0}$. This proves the lemma.

Proof of Theorem 16.1. We let $\mu \in \mathscr{P}(G)$ with support all of G and we would like to show that P^m is μ-proximal. We shall use the characteristic property (c) in Theorem 14.1; namely we shall show that a stationary ν-process on P^m is proper. This amounts to showing that a limit measure of the form

$$\lim_{n \to \infty} x_0 x_{-1} \cdots x_{-n} \nu$$

(we know the limit exists by Theorem 7.5) is a point measure. Now suppose that the $\omega \varepsilon$ underlying probability space is such that $\lim_{n \to \infty} x_0(\omega)\, x_{-1}(\omega) \cdots x_{-n}(\omega)\, \nu$ exists. We also assume as we may that the set $\{x_{-n}(\omega)\}$ is dense in the support of μ, i.e., dense in G. Fix $\gamma \in G$ and suppose $x_{-n_k}(\omega) \to \gamma$. Passing to a further subsequence and invoking Lemma 16.4 we assume

$$x_0(\omega) \cdots x_{-n_k+1}(\omega) \to \tau$$

where τ is a quasi-projective transformation. We then have (using the fact that ν assigns measure 0 to linear submanifolds of P^m)

$$\lim x_0(\omega) \cdots x_{-n}(\omega)\, \nu = \theta = \tau \nu = \tau \gamma \nu .$$

Now it is easy to show that any quasi-projective transformation τ can be expressed $\tau = g P_r k$, where $g \in GL(m+1, \boldsymbol{R})$, $k \in O(m+1, \boldsymbol{R})$ and P_r is the quasi-projective map corresponding to the linear transform $(u_1, \ldots, u_{m+1}) \to (u_1, \ldots, u_r, 0, \ldots, 0)$. So we can write $\theta = g_\gamma P_{r_\gamma} k_\gamma \nu = g_\gamma P_{r_\gamma} k_\gamma \gamma \nu$ or $P_{r_\gamma} k_\gamma \nu = P_{r_\gamma} k_\gamma \gamma \nu = P_{r_\gamma} k_\gamma \gamma \nu$. Now choose γ_n with $\gamma_n \nu \to$ point measure. We can assume k_{γ_n} converges and that r_{γ_n} is fixed. $P_{r_\gamma} k_\gamma \gamma \nu$ is converging to a point measure. But this implies that P_{r_γ} must be a projection onto a point and so θ is a point measure. This proves that with probability one, $\lim x_0 x_{-1} \cdots x_{-n} \nu$ is a point measure and that proves the theorem.

Using Theorem 16.1 one may prove

THEOREM 16.6. *If G is a semisimple Lie group with finite center then $B(G)$ is a mean-proximal G-space.*

Namely one may show that there is a representation on $G \to SL(m+1, \boldsymbol{R})$ satisfying hypothesis (b) of Theorem 16.1 and such that some orbit for some $y \in P^m$ the stability group $\{g \mid gy = y\} = H(G)$. Hence $B(G) \to P^m$. However if P^m is mean-proximal so is any G-invariant subset. We omit the details of the construction of this representation of G.

LEMMA 16.7. *If M is a compact mean-proximal G-space and Γ is a lattice in G then M is also a proximal Γ-space.*

PROOF. Assume Γ not proximal on M and let $A \subset M \times M$ be a Γ-invariant closed set, $A \cap \text{diagonal} = \emptyset$. Consider $G/\Gamma \times (M \times M)$ which is a locally compact space.

$$\tilde{A} = \bigcup_{g\Gamma \in G/\Gamma} (g\Gamma \times gA) \subset G/\Gamma \times (M \times M)$$

is a closed subset, invariant under G. The map $\tilde{A} \to G/\Gamma$ is proper and it is easy to show that there exists an $H(G)$-invariant probability measure on $\tilde{A}$; for the convex set of measure on $\tilde{A}$ mapping to the G-invariant $m_{G/\Gamma}$ is compact. Consider the projection λ of this measure on $M \times M$. λ is still $H(G)$-invariant. Now by the definition of μ-proximality

$$\int d(gx, gy)\, d\mu_n(g) \to 0$$

and also

$$\iint d(gx, gy)\, d\mu_n(g)\, d\lambda(x, y) \to 0.$$

So

$$\inf_{g \in G} \int d(gx, gy)\, d\lambda(x, y) = 0.$$

But since $\{g\lambda\}$ is compact this infimum is attained and we must have λ concentrated on the diagonal of $M \times M$.

But this contradicts the fact that each $gA \cap \text{diagonal} = \emptyset$. This proves the lemma.

Combining this with Theorem 16.6 we obtain

THEOREM 16.8. *If Γ is a lattice in the semisimple group G and M is a boundary of G, then M is a mean-proximal space for Γ.*

Finally combining the above with Theorem 15.5, we get

THEOREM 16.9. *If Γ is a lattice in the semisimple group G then $B(G)$ is the universal mean-proximal space for Γ. Namely, $B(G)$ is a mean-proximal Γ-space, and if M is any other mean-proximal Γ-space there exists a unique Γ-equivariant measurable map from $B(G)$ to M.*

In particular, the space $B(G)$ is determined measure-theoretically by any lattice in G. In [**6**], this fact is exploited to provide a proof that $SL(2, \mathbf{R})$ and $SL(n, \mathbf{R})$, $n>2$, cannot contain isomorphic lattices.

REFERENCES

1. H. Furstenberg, *A Poisson formula for semi-simple Lie groups*, Ann. of Math. (2) **77** (1963), 335–386. MR **26** #3820; Errata, **28**, 1246.

2. ———, *Non-commuting random products*, Trans. Amer. Math. Soc. **108** (1963), 377–428. MR **29** #648.

3. F. P. Greenleaf, *Invariant means on groups*, Van Nostrand, Princeton, N.J., 1969.

4. R. Azencott, *Espaces de Poisson des groupes localement compacts*, Lecture Notes in Math., no. 148, Springer-Verlag, Berlin and New York, 1970.

5. C. C. Moore, *Compactifications of symmetric spaces*, Amer. J. Math. **86** (1964), 201–218. MR **28** #5146.

6. H. Furstenberg, *Random walks and discrete subgroups of Lie groups*, Advances in Probability and Related Topics, Vol. I, Dekker, New York, 1971, pp. 1–63. MR **44** #1794.

7. ———, *Translation-invariant cones of functions on semi-simple Lie groups*, Bull. Amer. Math. Soc. **71** (1965), 271–326. MR **31** #1326.

8. ———, *Boundaries of Riemannian symmetric spaces*, Symmetric Spaces, Short courses presented at Washington University, New York, 1972.

HEBREW UNIVERSITY
JERUSALEM, ISRAEL

PART II: SEMINARS

THE PRIMITIVE IDEAL SPACE OF SOLVABLE LIE GROUPS

L. PUKANSZKY

Let $\mathscr{G}$ be a connected and simply connected solvable Lie group with the Lie algebra y. In a recent publication (Ann. Ecole Norm. Sup. **4** (1971), 457–608) we associated with $\mathscr{G}$ a family M of certain geometrical objects, and with each element m of M a unitary equivalence class $\mathscr{F}(m)$ of factor representations. If $\mathscr{G}$ is of type one, our theory reduces essentially to a construction of L. Auslander and B. Kostant (Invent. Math. **14** (1971), 255–354). Here elements of M are orbits of the coadjoint representation taken with a certain multiplicity, and $\mathscr{F}(m)$ is composed of multiples of irreducible representations assigned to m by these authors. Hence, by virtue of one of their results, the map sending the point m of M into the corresponding element of the dual $\hat{\mathscr{G}}$ of $\mathscr{G}$ is a bijection between M and $\hat{\mathscr{G}}$.

In the general case, intuition shows that the representations $\{\mathscr{F}(m): m \in M\}$, although just a small subfamily of all factor representations, are the interesting objects of the representation theory of $\mathscr{G}$, and hence their further characterization is of substantial interest. One suggestion, due to C. C. Moore, in this regard is as follows: We recall that, given a factor representation, the kernel of the associated representation of the group C^* algebra is a primitive ideal, that is, a kernel of an irreducible representation. Let $\operatorname{Prim}(\mathscr{G})$ be the set of all such ideals in the hull-kernel topology. The map F assigning to m in M the kernel of $\mathscr{F}(m)$ is a map from M into $\operatorname{Prim}(\mathscr{G})$. The question is if F is a bijection.

THEOREM 1. *The map F just defined is a bijection between M and* $\operatorname{Prim}(\mathscr{G})$.

In the type one case, there is a canonical identification of $\hat{\mathscr{G}}$ to $\operatorname{Prim}(\mathscr{G})$ and thus our result reduces to that of Auslander and Kostant quoted above.

AMS (MOS) subject classifications (1970). Primary 22E25, 22E45.

THEOREM 2. *Prim$(\mathcal{G})$ is a T_1 topological space if and only if all the roots of y are purely imaginary.*

In the type one case this result was found by C. C. Moore (Mem. Amer. Math. Soc., no. 62, 1966, Theorems 1–2, p. 171).

UNIVERSITY OF PENNSYLVANIA

GROUP EXTENSIONS AND THE PLANCHEREL FORMULA

ADAM KLEPPNER AND RONALD L. LIPSMAN*

1. Introduction. In his fundamental paper [3], Mackey developed a method for describing the representation theory of a separable locally compact group G in terms of the representation theories of a closed normal subgroup N and certain subgroups (little groups) of G/N. In this paper, we show how the same approach can be adopted in order to describe the Plancherel measure of G in terms of corresponding measures for N and the little groups.

We begin with a quick review of the ingredients that go into the "Mackey machine." Let G be a separable locally compact group and $N \subseteq G$ a closed normal subgroup. The dual space $\hat{N}$ is then a Borel G-space. For $\gamma \in \hat{N}$, we denote the corresponding isotropy subgroup by G_γ. We assume henceforth that N is of type I so that every G_γ is closed. Let $\check{G}_\gamma = \{\nu \in \hat{G}_\gamma : \nu|_N$ is a multiple of $\gamma\}$. For $\nu \in \check{G}_\gamma$, the induced representation $\pi_\nu = \mathrm{Ind}_{G_\gamma}^G \nu$ is irreducible. Assume next that the orbit space $\hat{N}/G$ is countably separated. Then we have a disjoint union $\hat{G} = \bigcup_{\hat{N}/G} \{\pi_\nu : \nu \in \check{G}_\gamma\}$.

The machine actually provides a finer description of $\check{G}_\gamma$. Indeed, there exists a unique normalized multiplier ω_γ on G_γ/N (and so on G_γ) such that γ extends to an ω_γ-representation γ' of G_γ, and any $\nu \in \check{G}_\gamma$ is of the form $\nu = \gamma' \otimes \sigma''$, where σ is an irreducible $\bar{\omega}_\gamma$-representation of G_γ/N and σ'' is its lift to G_γ. The class of ν is uniquely determined by that of σ, and if (G_γ/N) has only type I $\bar{\omega}_\gamma$-representations for all γ, then G is of type I and every $\pi \in \hat{G}$ is of the form $\pi = \pi_{\gamma,\sigma} = \mathrm{Ind}_{G_\gamma}^G \gamma' \otimes \sigma''$. Thus $\hat{G}$ is a "fiber space" where the base is $\hat{N}/G$ and the fibers are $(G_\gamma/N)^{\wedge\,\bar{\omega}_\gamma}$. Our main result will be a description of the Plancherel measure of G in terms of this fiber picture.

AMS (MOS) subject classifications (1970). Primary 22D10, 22D30; Secondary 43A30, 43A49, 43A80.

* The latter author was partially supported by NSF grant GP-33039.

2. The Plancherel Theorem. The Plancherel Theorem, as customarily stated for unimodular type I groups, provides a direct integral decomposition of the ordinary regular representation into irreducible constituents. In the next theorem, we allow for nonunimodular groups as well as projective representations.

THEOREM 1. *Let ω be a normalized multiplier on G and suppose G has only type* I *ω-representations. Then there is a positive Borel measure $\mu=\mu_{G,\omega}$ on $\hat{G}^{\omega}$, a measurable field $\xi \to \mathscr{H}_\xi \otimes \bar{\mathscr{H}}_\xi$ on $\hat{G}^{\omega}$, a measurable field of ω-representations $\xi \to \pi_\xi$ acting on $\mathscr{H}_\xi$ such that $\pi_\xi \in \xi$ for μ-almost all ξ, and an isomorphism Ψ of $L_2(G)$ onto $\int_{\hat{G}^\omega}^{\oplus} \mathscr{H}_\xi \otimes \bar{\mathscr{H}}_\xi \, d\mu(\xi)$ which carries the left-regular ω-representation $\lambda_{G,\omega}$ to $\int_{\hat{G}^\omega}^{\oplus} \pi_\xi \otimes 1_\xi \, d\mu(\xi)$ and the right-regular $\bar{\omega}$-representation $\varrho_{G,\omega}$ to $\int_{\hat{G}^\omega}^{\oplus} 1_\xi \otimes \bar{\pi}_\xi \, d\mu(\xi)$. There exists a positive invertible selfadjoint operator M on $L_2(G)$ such that, whenever $f, h \in D_M^\circ$,*

$$\Psi M f(\xi) = \pi_\xi(\Delta^{-1/2} f),$$

$$\int_G Mf(x)\,(Mh(x))^- \, dx = \int_{\hat{G}^\omega} \mathrm{Tr}(\pi_\xi(\Delta^{-1/2} f)\,\pi_\xi(\Delta^{-1/2} h)^*)\, d\mu(\xi),$$

where Δ is the modular function of G and D_M° is a certain subspace (unspecified here) of the domain of M. The measure μ is uniquely determined to within equivalence; M is uniquely determined to within multiplication by a positive invertible selfadjoint operator affiliated with the center of the left ring generated by $\lambda_{G,\omega}$; and μ and M determine each other uniquely. If $\Delta=1$, we can take $M=1$ and then μ is uniquely determined (up to scalars). If in addition ω is trivial, we get the usual Plancherel Theorem.

The proof of Theorem 1 relies crucially on the theory of quasi-Hilbert algebras developed by Dixmier and Pukanszky. We also mention that we can prove orthogonality relations for "discrete series" representations. Basically, they state that the matrix coefficients for irreducible subrepresentations of the regular ω-representation satisfy certain orthogonality relations. However, when $\Delta \neq 1$, one must be careful to account for the unbounded operator M – its effect is that not all, but only a dense collection of such coefficients, are square-integrable. We also can determine a necessary and sufficient condition for $\pi \in \hat{G}$ to be a subrepresentation of λ_G. We must refer the reader to [**2**] for the precise formulation of these results.

3. The Plancherel formula for group extensions. We return to the situation and notation described in § 1. In order to state the main result, we first need

LEMMA 2. *The Plancherel measure μ_N is quasi-invariant under the action of G. Let $\bar{\mu}_N$ be a pseudo-image of μ_N on $\hat{N}/G$. Then for almost all γ there exists a unique quasi-invariant measure ν_γ on G/G_γ such that*

$$\int_{\hat{N}} f(\gamma)\, d\mu_N(\gamma) = \int_{\hat{N}/G} \int_{G/G_\gamma} f(\gamma \cdot g)\, d\nu_\gamma(\bar{g})\, d\bar{\mu}_N(\bar{\gamma}), \quad f \in L_1(\hat{N}, \mu_N).$$

We can now state the Plancherel formula for group extensions.

THEOREM 3. (i) *Under the assumptions of* §1, *we have the following direct integral decomposition of the left-regular representation* λ_G *of* G:

$$\lambda_G = \int_{\hat{N}/G}^{\oplus} \int_{((G_\gamma/N)^\frown)\bar{\omega}_\gamma}^{\oplus} \pi_{\gamma,\sigma} \otimes 1_{\pi_{\gamma,\sigma}}\, d\mu_{G_\gamma/N,\bar{\omega}_\gamma}(\sigma)\, d\bar{\mu}_N(\bar{\gamma}).$$

(ii) *If in addition* G *is unimodular, then almost all the little groups* G_γ/N *are unimodular. Moreover,* μ_G *is uniquely specified as follows: Take any pseudo-image* $\bar{\mu}_N$; *then for almost all* γ *there is a unique choice of the measures* $\mu_{G_\gamma/N,\bar{\omega}_\gamma}$ *so that* $\mu_G = \int \mu_{G_\gamma/N,\omega_\gamma}\, d\bar{\mu}_N(\bar{\gamma})$. (*That is, if we change* $\bar{\mu}_N$ *to* $c \cdot \bar{\mu}_N$, *then almost all of the fiber measures must be switched to* $c(\bar{\gamma})^{-1}\mu_{G_\gamma/N,\bar{\omega}_\gamma}$).

Once one is fluent with the mechanics of the Mackey extension procedure, and once one has the tools provided by Theorem 1, then the method of proof of (i) becomes quite clear. Unfortunately, the actual argument involves rather technical measure-theoretic considerations for which we refer the reader to [1]. The proof of (ii) requires that we add to the arguments in (i) a method for keeping track of the functions and characters that arise in the extension procedure. We mention only that the following general theorem on the character of an induced representation provides the means to satisfy that need.

THEOREM 4. *Let* G *be a separable locally compact group, and* $H \subseteq G$ *a closed subgroup. Let* ϕ *be a continuous function of compact support on* G, $\phi^*(g) = (\phi(g^{-1}))^- \cdot \Delta_G(g^{-1})$, *and set* $\psi = \phi * \phi^*$. *Then, for* $\gamma \in \hat{H}$,

$$\mathrm{Tr}(\mathrm{Ind}_H^G \gamma)(\psi) = \int_{G/H} \Delta_G(g)^{-1} q(g)^{-1} \mathrm{Tr}\left[\int_H \psi(g^{-1}hg)\, \gamma(h)\, \Delta_G(h)^{1/2} \Delta_H(h)^{-1/2}\, dh\right] d\bar{g}$$

in the sense that both sides are finite and equal, or both are $+\infty$. *Here* q *is a strictly positive continuous function on* G *satisfying* $q(e) = 1$, $q(hx) = \Delta_H(h)\, \Delta_G(h)^{-1} q(x)$, $h \in H$, $x \in G$, *and* $d\bar{g}$ *is the quasi-invariant measure on* G/H *defined by*

$$\int_{G/H} \int_H f(hg)\, dh\, d\bar{g} = \int_G f(x)\, q(x)\, dx.$$

4. Examples. We list here very briefly some of the types of groups which can be handled by Theorem 3. The explicit computations can be found in [**1**] and [**2**]. First we give some unimodular examples.

(a) *Compact extensions.* In case G/N is compact and one "knows" μ_N, then our theory gives explicit formulas for μ_G. In particular, we can compute the Plancherel measure for various kinds of motion groups, for central groups, and for groups all of whose irreducible representations are finite dimensional.

(b) *Extensions of the center.* Applying our results to $N = \mathrm{Cent}(G)$, we can compute μ_G for some nilpotent groups, in particular for general Heisenberg-type groups.

(c) *Solvable groups.* For solvable Lie groups G which are almost algebraic, type R, and in which the nilradical is regularly embedded and splits, we can compute μ_G explicitly.

(d) *Reductive groups.* We can compute the Plancherel measure for reductive groups as extensions of semisimple groups. For example, if F is a local field of characteristic 0, then we can write down μ_G for $G = GL(n, F)$ whenever μ_N is known for $N = SL(n, F)$.

(e) *Semidirect products.* We can handle various kinds of groups $G = S \cdot U$ where S is semisimple, U is nilpotent and normal. Two such examples are the inhomogeneous Lorentz groups, and the case where S stabilizes the representations of U corresponding to orbits (in the coadjoint representation) of maximal dimension (the situation where the Weil representation arises).

Finally, we give two nonunimodular examples.

(f) *Parabolic groups.* Groups like the "$ax+b$" group are treated easily with Theorem 3. Using some unpublished results of Moore, it seems likely that one could write down the Plancherel measure of a minimal parabolic subgroup of a connected semisimple Lie group.

(g) *Non-type* I *groups.* Mackey has given an example of a non-type I group G for which λ_G is type I. The Plancherel measure of G is also easily computed using (a slight generalization of) Theorem 3.

References

1. A. Kleppner and R. L. Lipsman, *The Plancherel formula for group extensions*, Ann. Sci. Ecole Norm. Sup. **5** (1972), 459–516.

2. ———, *The Plancherel formula for group extensions.* II, Ann. Sci. Ecole Norm. Sup **6** (1973), 103–132.

3. G. Mackey, *Unitary representations of group extensions.* I, Acta Math. **99** (1958), 265–311. MR **20** #4789.

University of Maryland

SQUARE-INTEGRABLE REPRESENTATIONS OF NILPOTENT GROUPS

JOSEPH A. WOLF AND CALVIN C. MOORE

1. Notion of square-integrable representation. Let G be a unimodular locally compact group and Z the center of G. As usual, $\hat{G}$ denotes the set of all equivalence classes $[\pi]$ of irreducible unitary representations π of G, and H_π denotes the representation space of π. So $\hat{Z}$ is the group of unitary characters on Z. If $[\pi]\in\hat{G}$ then $\pi|_Z$ is a multiple of the *central character* $\zeta_\pi\in\hat{Z}$ of $[\pi]$. If $\zeta\in\hat{Z}$ then $L_2(G/Z, \zeta)$ denotes the space of functions on G that is the representation space for $\mathrm{Ind}_{Z\uparrow G}(\zeta)$.

If $[\pi]\in\hat{G}$ and $\zeta_\pi\in\hat{Z}$ one knows that the following conditions are equivalent: (1) There exist nonzero $\phi, \psi\in H_\pi$ such that $\langle\pi(\cdot)\,\phi, \psi\rangle\in L_2(G/Z, \zeta)$. (2) If $\phi, \psi\in H_\pi$ then $\langle\pi(\cdot)\,\phi, \psi\rangle\in L_2(G/Z, \zeta)$. (3) $[\pi]$ is a discrete summand of $\mathrm{Ind}_{Z\uparrow G}(\zeta)$. Then we say that $[\pi]$ is *square-integrable*. The usual case is the case where Z is compact, but the more general setting is useful for reductive and for nilpotent groups.

If $[\pi]$ is square-integrable, there is a number $d_\pi>0$ such that

$$\int_{G/Z} \langle\pi(g)\,\phi_1, \psi_1\rangle\,\overline{\langle\pi(g)\,\phi_2, \psi_2\rangle}\,d(gZ)=d_\pi^{-1}\langle\phi_1, \phi_2\rangle\,\overline{\langle\psi_1, \psi_2\rangle}$$

for all $\phi_i, \psi_i\in H_\pi$. The number d_π is the *formal degree* of $[\pi]$.

If π_1 and π_2 are inequivalent square-integrable representations with the same central character, one has the *orthogonality relations*

$$\int_{G/Z} \langle\pi(g)\,\phi_1, \psi_1\rangle\,\overline{\langle\pi(g)\,\phi_2, \psi_2\rangle}\,d(gZ)=0 \quad \text{for } \phi_i, \psi_i\in H_{\pi_i}.$$

2. The case of a connected, simply connected nilpotent Lie group. Let N be a

AMS (MOS) subject classifications (1970). Primary 43A80; Secondary 22E25, 43A85.

nilpotent group as just described, $\mathfrak{n}$ its Lie algebra and $\mathfrak{z}$ the center of $\mathfrak{n}$. Then $Z=\exp\mathfrak{z}$ is the center of N and $\log: Z\to\mathfrak{z}$ denotes the inverse of $\exp:\mathfrak{z}\to Z$. Let $\mathfrak{n}^*$ denote the (real) dual space of $\mathfrak{n}$; we denote the representation of N on $\mathfrak{n}^*$ by ad^*. If $f\in\mathfrak{n}^*$ then its orbit $O_f=\mathrm{ad}^*(N)(f)$ determines a class $[\pi_f]=[\pi_{O_f}]$ by the Kirillov theory. The central character of $[\pi_f]$ is $\zeta_{\pi_f}(z)=\exp\{2\pi\, if(\log z)\}$. Here note $f|_{\mathfrak{z}}=f'|_{\mathfrak{z}}$ whenever $f'\in O_f$.

Let $\mathfrak{z}^\perp=\{f\in\mathfrak{n}^*:f(\mathfrak{z})=0\}$. Now O_f is contained in the affine hyperplane $H(f|_{\mathfrak{z}})=f+\mathfrak{z}^\perp$ of $\mathfrak{n}^*$.

Finally recall the antisymmetric form b_f on $\mathfrak{n}$ defined by $b_f(x,y)=f[x,y]$. Evidently b_f can be viewed as an antisymmetric bilinear form on $\mathfrak{n}/\mathfrak{z}$.

THEOREM 1. *Let $f\in\mathfrak{n}^*$ and ζ the central character of π_f. Then the following conditions are equivalent:*

(1) *π_f is square-integrable.*

(2) $\mathrm{Ind}_{Z\uparrow N}(\zeta)$ *is a primary representation of N.*

(3) *The orbit O_f is the hyperplane $H(f|_{\mathfrak{z}})=f+\mathfrak{z}^\perp$.*

(4) *The form b_f is nondegenerate on $\mathfrak{n}/\mathfrak{z}$.*

The proof is an induction on $\dim N$. If N is abelian the equivalence is routine; now suppose N nonabelian. If $\dim Z>1$, choose a subalgebra $\mathfrak{z}^0\subset\mathfrak{z}$ of codimension 1 contained in kernel(f). Then $\pi=\pi_f$ is the lift of a representation $\pi^0=\pi^0_{f^0}$ of $N^0=N/Z^0$, $Z^0=\exp(\mathfrak{z}^0)$. Theorem 1 applies to N^0, π^0 and f^0 by induction on dimension, and it follows for N, π and f.

Now $\dim Z=1$ and $f(\mathfrak{z})\neq 0$. Choose $x\in\mathfrak{n}-\mathfrak{z}$ with $[x,\mathfrak{n}]\subset\mathfrak{z}$ and $f(x)=0$. Let $\mathfrak{n}_0=\{u\in\mathfrak{n}:[u,x]=0\}$, ideal of codimension 1 in $\mathfrak{n}$, and $N_0=\exp(\mathfrak{n}_0)$. Then $f_0=f|_{\mathfrak{n}_0}$ has $\mathrm{ad}^*(N_0)$-orbit $O_0\subset\mathfrak{n}_0^*$ which in turn gives $[\pi_0]\in\hat{N}_0$. The Kirillov machine gives $\pi=\pi_f$ as $\mathrm{Ind}_{N_0\uparrow N}(\pi_0)$. Choose $y\notin\mathfrak{n}_0$ such that $f(z=[x,y])=1$. Then $f_s=\mathrm{ad}^*(\exp(sy))\,f|_{\mathfrak{n}_0}$ has $f_s(x)=s$. Let O_s be the orbit of f_s in $\mathfrak{n}_0^*$. From Kirillov, the projection $p:\mathfrak{n}^*\to\mathfrak{n}_0^*$, kernel $\mathfrak{n}_0^\perp$, satisfies $p^{-1}(O_s)=\{f'\in O_f:f'(x)=s\}$ and $p\{f'\in O_f:f'(x)=s\}=O_s$. Let $\mathfrak{z}_0=(x)+\mathfrak{z}$. With some linear algebra and the fact that unipotent orbits are closed, one sees: $O_f=f+\mathfrak{z}^\perp\Leftrightarrow O_s=f_s+\mathfrak{z}_0^\perp$ *for some real* $s\Leftrightarrow O_s=f_s+\mathfrak{z}_0^\perp$ *for all real s; in that case $\mathfrak{z}_0$ is the center of* $\mathfrak{n}_0$. Induction by stages and some direct integral theory then give: *π is a discrete summand of* $\mathrm{Ind}_{Z\uparrow N}(\zeta)$ $\Leftrightarrow O=f+\mathfrak{z}^\perp$; *in that case* $\mathrm{Ind}_{Z\uparrow N}(\zeta)$ *is primary.* Using the results quoted above in §1, now assertions (1), (2) and (3) of Theorem 1 are equivalent. Some linear algebra proves (3) and (4) equivalent.

3. Square-integrability and the Pfaffian. Retain N, Z, etc., as in §2. Make a definite choice of Haar measure $\mu_{N/Z}$ on N/Z. That gives a volume element (alternating form of degree $\dim(\mathfrak{n}/\mathfrak{z})$), say α, on $\mathfrak{n}/\mathfrak{z}$. Let $f\in\mathfrak{n}^*$ and recall that the alternating bilinear form b_f on $\mathfrak{n}/\mathfrak{z}$ has Pfaffian (relative to α) defined as follows: If

$\dim \mathfrak{n}/\mathfrak{z}$ is odd, then $Pf(b_f)=0$. If $\dim \mathfrak{n}/\mathfrak{z}=2m$ even, then the mth exterior power $b_f^m = Pf(b_f)\,\alpha$. Evidently $P(f)=Pf(b_f)$ is a real polynomial function on $\mathfrak{n}^*$, and $P(f)\neq 0$ precisely when b_f is nondegenerate on $\mathfrak{n}/\mathfrak{z}$. Using Theorem 1, now $P(\cdot)$ is constant on $\mathrm{ad}^*(N)$-orbits and $P(f)$ depends only on $f|_{\mathfrak{z}}$. Thus we view P as a polynomial on $\mathfrak{z}^*$.

If $h \in \mathfrak{z}^*$ with $P(h)\neq 0$, denote $\phi(h)=[\pi_f]$ for any $f\in h+\mathfrak{z}^{\perp}$. Note that $\phi(h)$ is the only class in $\hat{N}$ with central character $\zeta(\exp z)=\exp\{2\pi i h(z)\}$. We often write π_h for $\phi(h)$.

THEOREM 2. *ϕ is a bijection of $\{h\in\mathfrak{z}^* : P(h)\neq 0\}$ onto $\{[\pi]\in\hat{N} : [\pi]$ is square-integrable$\}$. Further, it is a homeomorphism from the natural topology to the Fell (hull-kernel) topology of $\hat{N}$.*

As polynomial function on $\mathfrak{z}^*$, P is in the symmetric algebra $S(\mathfrak{z})$. Let $\mathfrak{Z}$ be the center of the universal enveloping algebra $\mathscr{U}$ of $\mathfrak{n}$. Then $\mathfrak{z}\subset\mathfrak{Z}$ gives $S(\mathfrak{z})\subset\mathfrak{Z}$, so also $P\in\mathfrak{Z}$.

THEOREM 3. *N has square-integrable representations if, and only if, $S(\mathfrak{z})=\mathfrak{Z}$*

4. Formal degree and Plancherel measure. Formal degree and Pfaffian depend in the same way on the normalization of Haar measure on N/Z, so the following is intrinsic.

THEOREM 4. *In the notation of Theorem 2, the square-integrable class $\phi(h)$ has formal degree $|P(h)|$.*

The *infinitesimal character* of a class $[\pi]\in\hat{N}$ is the homomorphism $\chi_\pi:\mathfrak{Z}\to C$ given by $\chi_\pi(z)\,v=d\pi(v)$ on C^∞ vectors. It is related to the central character ζ_π by $\zeta_\pi(\exp z)=\exp\{\chi_\pi(z)\}$ for $z\in\mathfrak{z}\subset\mathfrak{Z}$.

If $[\pi]\in\hat{N}$ is not square-integrable, we understand its formal degree $d_\pi=0$. This is consistent with Theorem 4.

THEOREM 5. *If $[\pi]\in\hat{N}$, then $d_\pi=|\chi_\pi(P)|$.*

The fact $d_\pi^2=|\chi_\pi(P)|^2=\chi_\pi(P\bar{P})$ is in accord with the Weyl formula for the degree of a representation in terms of its highest weight. The analogy may go quite far.

Fix Haar measures μ_N on N and μ_Z on Z consistent with our choice $\mu_{N/Z}$ for N/Z: $d_{\mu_N}(n)=d_{\mu_{N/Z}}(nZ)\,d_{\mu_Z}(z)$. These choices fix Lebesgue measures dn on $\mathfrak{n}$, dz on $\mathfrak{z}$ and $d\dot{n}$ on $\mathfrak{n}/\mathfrak{z}$ with $dn=d\dot{n}\,dz$. In turn Lebesgue measures dn^* on $\mathfrak{n}^*$, dz^* on $\mathfrak{z}^*$ and dv on $(\mathfrak{n}/\mathfrak{z})^*$ are fixed by the condition that Fourier transforms be isometries.

THEOREM 6. *If N has square-integrable representations, then Plancherel measure is concentrated in $\{[\pi]\in\hat{N}:[\pi]$ is square-integrable$\}$. Then the map ϕ of Theorem 2 pulls Plancherel measure back to $c|P(h)|\,dz^*(h)$ where $c=k!\,2^k$ and $k=\frac{1}{2}\dim\mathfrak{n}/\mathfrak{z}$.*

Theorem 4 is proved using the same inductive procedure as that sketched for Theorem 1. Theorem 5 is a corollary. In Theorem 6, the concentration of Plancherel measure comes from induction by stages and the fact that the zeroes of P have Lebesgue measure zero in $\mathfrak{z}^*$. The formula for Plancherel measure then comes by expressing the distribution character Θ_π of a square-integrable class $\pi=\phi(h)$ on a C_c^∞ function f by

$$\Theta_\pi(f)=c^{-1}\int_{O_h}(f\cdot\exp)\hat{}(y)\,d_{\mu_h}(y)$$

where $\hat{}$ is Fourier transform on $\mathfrak{n}^*$ and μ_h is the symplectic measure on $O_h=h+\mathfrak{z}^\perp$.

5. Two examples. First consider the Heisenberg group N_k, simply connected group for the $(2k+1)$-dimensional Lie algebra $\mathfrak{n}_k$:

$$\{x_1,\dots,x_k;\quad y_1,\dots,y_k;\quad z:[x_i,y_i]=z,\quad \text{all others zero}\}.$$

Choose the measure on N_k/Z so that $\mathfrak{n}_k/\mathfrak{z}$ has volume α with $\alpha(x_1,\dots,y_k)=1$. Then $P(tz^*)=t^k$. So the infinite-dimensional classes in $\hat{N}_k$ all are square-integrable, π_{tz^*} having formal degree $|t|^k$, and the Plancherel formula on N_k is

$$f(1)=c\int_{-\infty}^{\infty}\Theta_{\pi_{tz^*}}(f)\,|t|^k\,dt.$$

This is classical.

For the second example, let (a_{ij}) be a $k\times k$ matrix that gives a $(2k+2)$-dimensional Lie algebra $\mathfrak{n}$ with basis $\{x_1,\dots,x_k;y_1,\dots,y_k;z;w\}$ and $[x_i,y_j]=\delta_{ij}z+a_{ij}w$, all others zero. Then $\mathfrak{z}=(z)+(w)$ and we describe $f\in\mathfrak{z}^*$ by $(u,v)=(f(z),f(w))$. Now $P(u,v)=\det(u-va_{ij})$, which is v^k times the value at u/v of the characteristic polynomial of (a_{ij}). In particular P can be any homogeneous polynomial of degree k in two variables.

Suppose $k=1$ and $(a_{ij})=-\lambda$ irrational; then $P(az^*+bw^*)=a+\lambda b$. Let $\Delta=\{\exp(nz+mw):m,n \text{ integers}\}$ and $\bar{N}=N/\Delta$. Then $\bar{N}$ has compact center and $\bar{N}\hat{}$ consists of $\{[\pi]\in\hat{N}:\pi \text{ kills } \Delta\}$. Thus the square integrable classes in $\bar{N}\hat{}$ are the $\pi_{nz^*+mw^*}$, m and n integers. The formal degree of $\pi_{nz^*+mw^*}$ is $n+\lambda m$, which has 0 as limit point when (n,m) varies over $\mathbf{Z}^2-\{0\}$.

6. Square-integrable subrepresentations for $L_2(N/\Gamma)$. Let Γ be a discrete uniform subgroup of N. N acts on $L_2(N/\Gamma)$ by $U=\mathrm{Ind}_{\Gamma\uparrow N}(1_\Gamma)$, and $U=\sum_{\hat{N}} m_\pi\cdot\pi$ discrete sum.

Normalize $\mu_{N/Z}$ by: $N/Z\Gamma$ has volume 1. Then the Pfaffian polynomial P is determined up to sign. $\Gamma\cap Z$ is a lattice in Z and $\log(\Gamma\cap Z)$ a lattice L in $\mathfrak{z}$. Let $L^*\subset\mathfrak{z}^*$ be the lattice dual to $L\subset\mathfrak{z}$.

THEOREM 7. *Let $h\in\mathfrak{z}^*$ with $P(h)\neq 0$. Then the square-integrable representation π_h occurs in U precisely when $h\in L^*$. In that case the multiplicity $m_{\pi_h}=|P(h)|$, formal degree of π_h.*

Γ specifies a rational form $\mathfrak{n}_Q$ of $\mathfrak{n}$. A subalgebra $\mathfrak{h}\subset\mathfrak{n}$ and the group $H=\exp(\mathfrak{h})$ are rational just when $H/H\cap\Gamma$ is compact. One proceeds inductively as in Theorem 1, keeping everything rational. The final induction from N_0 to N is managed by a multiplicity formula of C. C. Moore (Ann. of Math. (2) **82** (1965), 153).

7. Extension to solvable groups. Do not assume G unimodular. The classes in $\hat{G}$ that are subrepresentations of $L_2(G)$ correspond (using results of Auslander, Kostant and Pukanszky) to the simply connected open orbits in $\mathfrak{n}^*$. If $f\in\mathfrak{n}^*$, we expect the associated representations to be square-integrable (in the appropriate sense) just when G_f/Z is compact.

UNIVERSITY OF CALIFORNIA, BERKELEY

MACKEY'S LITTLE GROUP METHOD AND L^2 OF COMPACT HOMOGENEOUS SPACES

J. BREZIN

We are going to be concerned with some problems that arise when one tries to use Mackey's "little group method" to study translation-invariant subspaces in L^2 of a compact homogeneous space. (Such positive results will appear soon in a joint paper with Louis Auslander.)

Let us begin by extracting from Mackey's work the precise result we shall need. We shall use G to denote a separable, unimodular, locally compact group. Let H be a closed normal subgroup of G with G/H abelian. The elements of G act on H via conjugation in G. The action of G on H defines an action of G on the dual $H^\wedge$ of H (= space of equivalence classes of irreducible unitary representations). Given $p \in H^\wedge$, we set $G(p) = \{g \in G : gp = p\}$; then $G(p)$ will be a subgroup of G containing H.

THEOREM 1 (MACKEY). *Let $p \in H^\wedge$; then from among the closed subgroups of $G(p)$ that contain H, we can choose M so that p extends to M but no further. Also, if q is a representation of M that extends p, the induced representation $I_M^G(q)$ of G is irreducible.*

We intend to apply this theorem in the following context.

Let K be a closed, unimodular subgroup of G such that both G/K and $H/H \cap K$ are compact, and set $L = H \cap K$. Each of the spaces G/K and H/L supports a unique translation-invariant probability measure, which shall remain nameless, as no other measures will be considered on either space. Represent G on $L^2(G/K)$ by setting $(\varrho_g F)(Kh) = F(Khg)$ for all $F \in L^2(G/K)$. The corresponding representation of H on $L^2(H/L)$ will be denoted by σ. Set $(G/K)^\wedge$ equal to $\{p \in G^\wedge : p$ is a subrepresentation of $\varrho\}$, and define $(H/L)^\wedge$ similarly. With a little work, Theorem 1 yields

---

AMS (MOS) subject classifications (1970). Primary 22D30.

THEOREM 2. *Let $r \in G/K^\frown$; then we can find a closed subgroup M of G and an element $q \in M^\frown$ so that*

(i) $$H \subseteq M \subseteq G,$$

(ii) $$M/M \cap K \text{ is compact},$$

(iii) $$q \in (M/M \cap K)^\frown \text{ and } q \mid H \in (H/L)^\frown,$$

(iv) $$I_M^G(q) = r.$$

From general principles, we know that there exist closed, ϱ-invariant subspaces $\Theta(r) \subseteq L^2(G/K)$ for each $r \in (G/K)^\frown$ such that $\varrho \mid \Theta(r)$ is a multiple of r and $L^2(G/K) = \sum \oplus_r \Theta(r)$. Let $\sum \oplus_p \Phi(p)$ be the analogous decomposition for $L^2(H/L)$. Theorem 2 shows that to describe the subspaces $\Theta(r)$ in terms of the subspaces $\Phi(p)$ reduces to being able to solve two problems: (1) what happens when $r \mid H = p$, and (2) what happens when $r = I_M^G(p)$.

Let us consider problem (1). We are given $r \in (G/K)^\frown$ with $r \mid H = p \in (H/L)^\frown$, and we want to get our hands on $\Theta(r)$. Using the canonical homeomorphism $H/L \approx KH/K$, we may identify $\Phi(p)$ with a subspace $\Phi^*(p) \subseteq L^2(KH/K)$. Because $p = r \mid H$, it is easy to see that $\Phi^*(p)$ is invariant under translation by elements of KH. Let $\sum \oplus \Omega(q)$ be the primary decomposition of $\Phi^*(p)$ under the action of KH/K, where q runs over some subset of $(KH/K)^\frown$. Let $q_0 = r \mid KH$, and observe that $KH/K \subseteq G/K$, whence

THEOREM 3. *The restriction map defines an isometric isomorphism from $\Theta(r)$ onto $\Omega(q_0)$.*

The proof of Theorem 3 consists of constructing the inverse map $\Omega(q_0) \to \Theta(r)$, thereby solving problem (1). Note that our solution to problem (1) is incomplete precisely insofar as we are unable to compute the decomposition $\sum \oplus \Omega(q)$ of $\Phi^*(p)$. Roger Howe will show later an example of what can happen.

Having solved (as best we can) problem (1), let us turn to problem (2). Here we assume $p \in (H/L)^\frown$ and $r = I(p) \in (G/K)^\frown$. Again, we are after $\Theta(r)$. To begin, we observe that $\sigma \mid \Phi(p)$ is a multiple of p, and hence $I_H^G(\sigma \mid \Phi(p))$ is a multiple of r. We will prove that $I_H^G(\sigma \mid \Phi(p))$ is a subrepresentation of ϱ, and get part of $\Theta(r)$ in the process.

The natural Hilbert space on which to realize $I_H^G(\sigma \mid \Phi(p))$ is $L^2(G/H : \Phi(p))$, square-integrable $\Phi(p)$-valued functions on G/H. Now $\Phi(p) \subseteq L^2(H/L)$, and hence $L^2(G/H : \Phi(p)) \subseteq L^2(G/H : L^2(H/L))$. From "Fubini's theorem" we get an isometric isomorphism T: $L^2(G/H : L^2(H/L)) \to L^2(G/L)$ – we will describe T momentarily. Using T, we can embed $L^2(G/H : \Phi(p))$ in $L^2(G/L)$. But what we want is to embed $L^2(G/H : \Phi(p))$ in $L^2(G/K)$. Thus, we are led to introduce some sort of averaging map A: $L^2(G/L) \to L^2(G/K)$. Here are the precise descriptions of A and T.

Let Δ be a Borel subset of G that meets each H coset precisely once and meets H at the identity element. Define $s: G \to \Delta$ by setting $Hs(g) = Hg$. Then T is given on $F \in L^2(G/H: L^2(H/L))$ by $(TF)(Lg) = [F(Hg)](Lgs(g)^{-1})$. As for A, it turns out (somewhat surprisingly) that what works is to set $(AF)(Kg) = \sum_{x \in K/L} F(xg)$, where F is a bounded function on G/L with compact support – implicit in this is that one can prove that K/L *is countable*.

THEOREM 4. *The composition AT defines an isometry from $L^2(G/H: \Phi(p))$ onto a (possibly proper) subspace $\Phi^{\#}(p)$ of $\Theta(r)$. Further, $\Phi^{\#}(p)$ is ϱ-invariant.*

There are apt to be many elements of $(H/L)\hat{}$ that induce r, and that is why $\Phi^{\#}(p)$ need not be all of $\Theta(r)$. Set $S(r) = \{p \in (H/L)\hat{} : I_H^G(p) = r\}$. It is easy to see that if $k \in K$ and $p \in S(r)$, then $kp \in S(r)$. Hence K acts on $S(r)$.

THEOREM 5. *Let p and q be two elements of $S(r)$. If $p \in Kq$, then $\Phi^{\#}(p) = \Phi^{\#}(q)$. However, if $p \notin Kq$, then $\Phi^{\#}(p) \perp \Phi^{\#}(q)$.*

In view of Theorem 5, we may speak of $\Phi^{\#}(w)$, $w \in S(r)/K$.

THEOREM 6. $\Theta(r) = \sum \oplus_{w \in S(r)/K} \Phi^{\#}(w)$.

Thus, the only obstacle to computing $\Theta(r)$ is computing the orbit space $S(r)/K$. Good examples of what can happen have been available from the beginning of the theory in my paper on L^2 of a nilmanifold (1968). Thus, the situation here is better understood than the situation in problem (1), where the $\Omega(q)$'s remain somewhat of a mystery.

Now that problems (1) and (2) are solved, we know (in principle) how to get from $\sum \oplus \Phi(p)$ to $\sum \oplus \Theta(r)$. This passage can be viewed as the step $n \Rightarrow n+1$ in an induction argument. It is reasonable to ask whether one can actually carry out the induction when, for example, G is solvable. Roger Howe will comment on what is known along these lines in his article.

UNIVERSITY OF CALIFORNIA, BERKELEY

ON THE SPECTRUM OF A COMPACT SOLVMANIFOLD

ROGER HOWE

1. Let G be a Lie group. By the nil-radical of G we will mean the largest connected nilpotent normal subgroup of G. Let S be a Lie group with nil-radical N; we will call S solvable if S/N is abelian. Now let Γ be a discrete subgroup of the solvable Lie group N such that S/Γ is compact. In this case, it is known that S is unimodular and so S/Γ carries a unique probability measure invariant under the action of S operating by left translation on S/Γ. Let $L^2(S/\Gamma)$ denote the L^2 space of this measure space. For $f \in L^2(S/\Gamma)$ we define $U(s) f(g\Gamma) = f(sg\Gamma)$, $s, g \in S$, and note that this defines a unitary representation of S on $L^2(S/\Gamma)$. It is well known that this representation is discretely decomposable; that is, we may write

$$L^2(S/\Gamma) \simeq \sum_{i=1}^{\infty} a_i \pi_i$$

where the π_i are irreducible unitary representations of S, the a_i are positive integers, and π_i is inequivalent to π_j for $i \neq j$.

In this paper we will outline a fairly complete method for determining the π_i and the a_i. The main drawback of this procedure is that it is cumbersome. When $S = N$ there is a direct solution to the problem presented in [**2**] and [**6**]. Our solution to the general problem proceeds in three stages. The major stage, the middle one, is a generalization of the results that hold when $S = N$. The advantage of our results is that they show that there is no essential mystery about how multiplicity arises in $L^2(S/\Gamma)$ and they can be used effectively to compute interesting special cases.

I would like to thank L. Auslander for some stimulating conversations relevant to this paper.

AMS (*MOS*) *subject classifications* (1970). Primary 22E45, 22E50.

2. The major steps follow the pattern familiar to students of representation theory of solvable Lie groups. They are the following:

(1) reduction to the case where N is Heisenberg;

(2) solution in the case where N is Heisenberg.

Step (1) breaks naturally into two main pieces which, together with step (2), constitute the parts mentioned in §1. Not all the details of step (2) have been worked out, but the essential point is to make a link between this problem and the celebrated Weil representation. We will now describe the steps in greater detail.

3. Step 1a. Let $\Delta = N \cap \Gamma$. Mostow [**5**] has shown that N/Δ is compact. Since we know the decomposition of $L^2(N/\Delta)$, we are in a good position to apply methods such as those described by Brezin [**1**]. When we take this approach, we find that it is enough to answer the following question: Suppose μ is an irreducible representation of N occurring in $L^2(N/\Delta)$. Suppose, further, that μ extends to S, in the sense that there is an irreducible representation ν_0 of S such that $\nu_{0|N} \simeq \mu$. Then what is the multiplicity of ν_0 in $L^2(S/\Gamma)$?

Let us be more graphic about the nature of this problem. It is an easy consequence of the general theory of Mackey that if ν is any extension of μ to S, then ν is uniquely of the form $\nu \simeq \nu_0 \otimes \psi$, where $\psi \in (S/N)\hat{}$. Let ν_0' denote the restriction of ν_0 to $N\Gamma$. Then ν_0' is obviously also an extension of μ to $N\Gamma$.

There is a natural identification of the coset spaces N/Δ and $N\Gamma/\Gamma$, and this identification commutes with the action of N on both spaces. Suppose μ occurs a times in $L^2(N/\Delta)$. Then under the action of $N\Gamma$, the μ isotypic subspace of $L^2(N/\Delta)$ will break up into a direct sum

$$\sum_{i=1}^{l} b_i(\nu_0' \otimes \phi_i')$$

where $\phi_i' \in (N\Gamma/N)\hat{}$ and $\sum_{i=1}^{l} b_i = a$. Let $\phi_i \in (S/N)\hat{}$ agree with ϕ_i' on $N\Gamma$. It is not hard to show that the subspace of $L^2(S/\Gamma)$ consisting of functions which are transformed by N according to μ has the decomposition

$$\sum_{i=1}^{l} \sum_{\Psi} b_i(\nu_0 \otimes \phi_i \psi)$$

where ψ runs through $(S/N\Gamma)\hat{}$. Thus, to solve our problem, it is enough to determine the b_i and the ϕ_i'. In particular, we could, if we wished, assume that $S = N\Gamma$.

4. Step 1b. This is the major reduction and proceeds by analogy with the double coset results of Mackey for finite groups. Now in [**2**] and [**6**] the multiplicity formula for $L^2(N/\Delta)$ is stated in terms of the Kirillov orbit picture. Let us now see how to restate our results in a way that brings out the analogy with Mackey's result.

Suppose $R \subseteq S$ is a closed subgroup such that (i) $R/R \cap \Gamma$ is compact, (ii) $M = R \cap N$ is connected, and (iii) $S = R \cdot N$. Let μ and ν_0 be as in §3. Suppose τ is a representation of R such that (a) τ occurs in $L^2(R/R \cap \Gamma)$ and (b) $\nu_0 = \mathrm{ind}_R^S \tau$.

THEOREM 1. *Let all notation be as above. There are finitely many double cosets $Rg_i\Gamma$, $g_i \in S$, $i = 1, \ldots, k$, such that* (i) *$R/R \cap g_i \Gamma g_i^{-1}$ is compact (equivalently, $Rg_i\Gamma$ is closed) and* (ii) *τ occurs in $L^2(R/R \cap g_i \Gamma g_i^{-1})$ with positive multiplicity a_i. The multiplicity of ν_0 in $L^2(S/\Gamma)$ is given by $\sum_{i=1}^k a_i$.*

Thus, except for the necessity of looking only at double cosets $Rg_i\Gamma$ which are closed, the result is entirely parallel to the finite group case.

The proof is in two parts. One constructs subspaces of $L^2(S/\Gamma)$ accounting for the stated multiplicity, then shows that these spaces exhaust the ν_0 isotypic component of $L^2(S/\Gamma)$. The construction uses the standard technique of intertwining operators. We notice that if τ_0 occurs in $L^2(R/R \cap \Gamma)$, then ν_0 occurs in $L^2(S/R \cap \Gamma)$. Averaging over $\Gamma/R \cap \Gamma$ should put ν_0 inside $L^2(S/\Gamma)$. The main observation in proving this is true is the very down-to-earth fact that two irreducible subspaces of $L^2(S/R \cap \Gamma)$ that define inequivalent representations of S are orthogonal.

That the spaces constructed above exhaust $L^2(S/\Gamma)$ is more delicate. The basic element of the proof is expressed in the following result.

LEMMA. *Let all notations be as in Theorem* 1. *Let $\sigma = \tau_{|M}$. Then if $Mn\Delta$ is a closed (M, Δ) double coset in N such that σ occurs in $L^2(M/M \cap n\Delta n^{-1})$, then $Rn\Gamma$ is a closed (R, Γ) double coset in S.*

The proof of this lemma depends on algebraic geometric properties of N, where we view N as a unipotent algebraic group over $\boldsymbol{Q}$. It is my inability to establish such a lemma more generally that forces me to break up step (1) into two parts.

5. Step 2. Once Theorem 1 is established, standard structure-theoretic arguments (see, for example, [3]) allow one to reduce the general problem to the case where N is abelian, in which case it may be considered as completely solved, or to the case where N is Heisenberg. We will now discuss this latter case.

Recall that a Heisenberg Lie group H is a Lie group isomorphic to a group of real-valued matrices of the form

$$\begin{pmatrix} 1 & x_1 & \dots & x_n & z \\ & 1 & & 0 & y_1 \\ & & \ddots & & \vdots \\ & \mathbf{0} & & 1 & y_n \\ & & & & 1 \end{pmatrix}$$

for some n. The center and commutator group Z of H is the one-dimensional subgroup defined by $x_i = y_i = 0$, $1 \leqq i \leqq n$. Further, $V = H/Z$ is a $2n$-dimensional vector space. The commutator operation $(g, h) \to ghg^{-1}h^{-1}$ induces on V a symplectic form. As is well known, we may choose a group Σ of automorphisms of H such that Σ acts trivially on Z, and Σ induces on V the group of this symplectic form.

Recall that to each nontrivial unitary character $\chi \in \hat{Z}$ there corresponds a unique irreducible unitary representation $\mu(\chi)$ of H with the property that, on Z, $\mu(\chi)$ is a multiple of χ. This correspondence establishes a parametrization of all infinite-dimensional irreducible unitary representations of H. The group Σ acts on the unitary dual of H, and since Σ acts trivially on Z, each $\mu(\chi)$ is fixed by Σ. This gives rise, in the usual way, to a projective representation $\varrho(\chi)$ of Σ on the space of $\mu(\chi)$. Weil **[8]** has explicitly computed a cocycle attached to $\varrho(\chi)$ and has shown $\varrho(\chi)$ becomes an actual representation $\varrho'(\chi)$ of a certain two-fold covering M of Σ. The group M is usually referred to as the metaplectic group and $\varrho'(\chi)$ is called the Weil representation.

Let us now return to our group S. As we said, we may assume that the nilradical of S is the Heisenberg group H. With a little fuss we can arrange it so that $S = T \times_S H$, where (i) T is abelian, (ii) $T/T \cap \Gamma$ is compact, (iii) T acts trivially on Z and (iv) T acts effectively and semisimply on V. Following §3, we may also assume that $S = \Gamma H$. The fact that S is a semidirect product guarantees that $\mu(\chi)$ extends to a representation $\nu_0(\chi)$ and, as discussed in §3, an arbitrary extension ν of $\mu(\chi)$ to S has the form $\nu = \nu_0(\chi) \otimes \phi$ with $\phi \in \hat{T}$. Since the difference between ν and $\nu_0(\chi)$ is slight, it is a somewhat subtle matter, given that $\mu(\chi)$ occurs in $L^2(N/\Delta)$, to determine precisely which extensions of $\mu(\chi)$ to S occur in $L^2(S/\Gamma)$. The problem is almost completely solved, however, by using the richness of the automorphism group of H. Specifically, we may imbed T in Σ as a discrete subgroup. We can then find a very large arithmetic subgroup of Σ containing T, and the situation becomes so rigid that the computations become much simpler.

Everything is immediately determined up to characters of order 6, which are in doubt because of the nonperfectness of certain small symplectic groups. I believe a closer examination of the Weil representation will permit the elimination of this small indeterminacy.

REFERENCES

1. J. Brezin, *Mackey's little group method and L^2 of compact homogeneous spaces*, Proc. Sympos. Pure Math., vol. 26, Amer. Math. Soc., Providence, R. I., 1974, pp. 245–247.

2. R. Howe, *Frobenius reciprocity for unipotent algebraic groups over Q*, Amer. J. Math. **93** (1971), 163–172. MR **43** #7556.

3. ———, *On the character of Weil's representations*, Trans. Amer. Math. Soc. **178** (1973), 287–298.

4. C. C. Moore, *Decomposition of unitary representations defined by discrete subgroups of nilpotent groups*, Ann. of Math. (2) **82** (1965), 146–182. MR **31** #5928.

5. G. D. Mostow, *Factor spaces of solvable groups*, Ann. of Math. (2) **60** (1954), 1–27. MR **15**, 853.

6. L. Richardson, *Decomposition of the L^2-space of a general compact nilmanifold*, Amer. J. Math. **93** (1971), 173–190.

7. A. Weil, *Basic number theory*, Die Grundlehren der math. Wissenschaften, Band 144, Springer-Verlag, New York, 1967. MR **38** #3244.

8. ———, *Sur certaines groupes d'opérateurs unitaires*, Acta Math. **111** (1964), 143–211. MR **29** #2324.

SUNY AT STONY BROOK

STUDY OF SOME HOLOMORPHICALLY INDUCED REPRESENTATIONS OF SOLVABLE GROUPS AND RESTRICTION OF THE HOLOMORPHIC DISCRETE SERIES OF A SEMISIMPLE GROUP G TO THE MINIMAL PARABOLIC OF G

H. ROSSI AND M. VERGNE*

Let G be a connected Lie group and O an orbit of G in $\mathfrak{G}^*$ under the coadjoint representation. Roughly speaking, the "orbit theory" associates to a point $f \in O$ and a "polarization" $\mathfrak{h}$ at f and some topological data a unitary representation $\varrho(f;\mathfrak{h})$ of G on a Hilbert space $\mathscr{H}(f;\mathfrak{h})$ (see [1], [10], [13]). In particular, for solvable Type I Lie groups this method using special kinds of polarizations produces all the dual of G [1]; for general polarizations the problem of the nontriviality of $\mathscr{H}(f;\mathfrak{h})$ is unanswered, and can naturally be restated as the problem of finding nonzero square holomorphic sections of a Hermitian line bundle on a Hermitian manifold. In the case studied by Auslander and Kostant, these manifolds are $\boldsymbol{C}^n$.

Another central problem of the theory is the study of the equivalence of the representations $\varrho(f;\mathfrak{h})$ and their irreducibility (see [3]).

The case which we consider is a case of an open orbit $O(f)$ of an element f in the dual of a completely solvable Lie algebra $\mathfrak{b}$, and of a totally complex positive polarization $\mathfrak{b}^-$ at f. It leads to a determination of $\mathscr{H}(f;\mathfrak{b}^-)$ as a Hilbert space of holomorphic functions $\mathscr{H}(\psi)$ on a Siegel domain $D(\Omega; Q)$ of Type II, where ψ is a positive homogeneous function on the cone Ω. We can describe the space $\mathscr{H}(\psi)$, answer the question "$\mathscr{H}(\psi) \neq \{0\}$?" in terms of the function ψ and calculate back to give conditions in terms of f for the space $\mathscr{H}(f;\mathfrak{b}^-)$ to be nontrivial, and identify the representation $\varrho(f;\mathfrak{b}^-)$.

The method is given by a Fourier-Laplace transform and we obtain a "Paley-Wiener theorem."

AMS (MOS) subject classifications (1970). Primary 22E45, 22E25.

* During this work the first-named author was at the University of Washington, and received partial support under NSF contract GP-9606 (0217). The second-named author received partial support from the CNRS (Paris).

We can also realize the holomorphic discrete series of a semisimple Lie group G as a Hilbert space of vector-valued holomorphic functions on a Siegel domain of Type II. Similar analysis of this Hilbert space allows us to give the formula of the restriction of the holomorphic discrete series to the minimal parabolic group MAN of G and to derive the well-known Harish-Chandra condition [9].

In this abstract we shall state only results; details of the proofs will appear elsewhere. We are indebted to R. Godement, whose work on $sp(n; \boldsymbol{R})$ [**8**] is the inspiration of the present work, and to J. Wolf for many valuable conversations while this work was in progress. We use for the study of the geometry of bounded homogeneous domains the results of Gindikin, Piatetskiĭ-Shapiro, Vinberg [7]; the article of Gindikin [**6**] contains much analysis similar to ours.

Let B be a completely solvable Lie group and $\mathfrak{b}$ its Lie algebra. Let $f_0 \in \mathfrak{b}^*$, and assume the following:

(a) The orbit $O(f_0) = B \cdot f_0$ under the coadjoint representation of B on $\mathfrak{b}^*$ is open.

(b) There exists a totally complex polarization $\mathfrak{b}^-$ at the point f_0, i.e., $\mathfrak{b}^-$ is a subalgebra such that $f_0([\mathfrak{b}^-, \mathfrak{b}^-]) = 0$ and that $\mathfrak{b}^- + \bar{\mathfrak{b}}^- = \mathfrak{b}_{\boldsymbol{C}}$. We shall write $\mathfrak{b}^+ = \bar{\mathfrak{b}}^-$. This is a direct sum, since we have $\mathfrak{b}^- \cap \mathfrak{b}^+ \subset \mathfrak{b}(f_0) = \{0\}$.

(c) Assume finally that this polarization is positive, i.e., let $j: \mathfrak{b}^{\boldsymbol{C}} \to \mathfrak{b}^{\boldsymbol{C}}$ the (real) operator defined by

$$j(x) = -ix \quad \text{if } x \in \mathfrak{b}^-,$$
$$j(x) = ix \quad \text{if } x \in \mathfrak{b}^+.$$

Then we ask that $f_0[x, jx] > 0$ if $x \neq 0$.

These conditions (a), (b), (c) on the pair $(\mathfrak{b}, j)$ are just the axioms for $(\mathfrak{b}, j)$ to be a normal j-algebra in the sense of Piateckiĭ-Shapiro. As a result of his theorem [**14**], $B = \exp \mathfrak{b}$ can be realized as a simply transitive group of affine transformations on a Siegel domain $D = D(\Omega; Q)$ of Type II. And, conversely, all homogeneous Siegel domains can be described by that method.

Now a Siegel domain of Type II is a complex domain of the following description. Ω is a convex cone in $\boldsymbol{R}^n$ whose dual cone

$$\Omega^* = \{\xi \in \boldsymbol{R}^n : \langle \xi, y \rangle > 0 \text{ for all } y \in \bar{\Omega} - \{0\}\}$$

is nonempty. $Q: \boldsymbol{C}^m \times \boldsymbol{C}^m \to \boldsymbol{C}^n$ is a real bilinear form with the properties

(1) $Q(u, u') = \overline{Q(u', u)}$, and Q is complex linear in u,

(2) $Q(u, u) \in \bar{\Omega}$ for $u \in \boldsymbol{C}^m$,

(3) $Q(u, u) = 0$ only if $u = 0$.

We define $D(\Omega; Q)$ as

$$D(\Omega; Q) = \{(z, u) \in \boldsymbol{C}^n \times \boldsymbol{C}^m,\ z = x + iy, \text{ and } y - Q(u, u) \in \Omega\}.$$

Furthermore, there exists a point $(it_0, 0) \in \boldsymbol{C}^n \times \boldsymbol{C}^m$ such that the map $\alpha: B \to D(\Omega; Q)$,

$\alpha(b)=b.(it_0, 0)$ is a biholomorphic diffeomorphism of the complex manifold $(b, \mathfrak{b}^-)$ with $D(\Omega; Q)$.

Now let $f \in \mathfrak{b}^*$ such that $f[\mathfrak{b}^-, \mathfrak{b}^-]=0$; associated to f we consider the representation $\varrho(f; \mathfrak{b}^-)$ of B defined as follows: $\mathscr{H}(f; \mathfrak{b}^-)$ is the Hilbert space of C^∞ functions ϕ on B satisfying

(i) $$\phi * x = -i\langle f, x\rangle \phi, \quad x \in \mathfrak{b}^- \subset \mathfrak{b}_C,$$

(ii) $$\int_B |\phi|^2 \, db < +\infty,$$

where

$$(\phi * x)(b) = \frac{d}{dt} \phi(b \exp(tx))\bigg|_{t=0}$$

if $x \in \mathfrak{b}$ and the action of B is by left translations.

This space can be identified with the space of square-integrable sections of a holomorphic line bundle L_f over $D(\Omega; Q)$ varying with f, and so we know that this holomorphic line bundle can be trivialized as $D(\Omega; Q)$ is a contractible Stein manifold [**15**, p. 40]. We construct a canonical homomorphism β of $\mathfrak{b}$ into $\mathfrak{b}^-$ such that a particular solution of (1) is given by the character λ_f of B:

$$\lambda_f(\exp x) = \exp(-i\langle f, \beta x\rangle).$$

In the identification $\alpha: B \to D(\Omega; Q)$, λ_f is written as $\psi_f(y - Q(u, u))$ where ψ_f is a function on the cone Ω.

It results that the space $\mathscr{H}(f; \mathfrak{b}^-)$ can be identified with the Hilbert space $\mathscr{H}(\psi)$ of functions F holomorphic on D such that

(*) $$\|F\|^2 = \int_D |F(x+iy, u)|^2 \, \psi(y - Q(u, u)) \, dx \cdot dy \cdot du < +\infty,$$

$\psi = |\psi_f|^2 \psi_0$ and $(\psi_0 \, dx \cdot dy \cdot du)$ is a Haar measure on $D(\Omega; Q)$ (identified with B).

That this representation is irreducible follows from the theorem of Blattner [**3**]. We must describe $\mathscr{H}(f; \mathfrak{b}^-)$ to determine when $\mathscr{H}(f; \mathfrak{b}^-)$ is nonzero, and we must give the answer in terms of the form f. The first step is provided by our "Paley-Wiener theorem" which we now state.

THEOREM 1. *Let $D(\Omega; Q)$ be a Siegel domain of Type* II, *and ψ a positive continuous function on Ω which is homogeneous in the following sense: There exists a μ real such that $\psi(rt) = r^\mu \psi(t)$ for all $r > 0$ and all $t \in \Omega$.*

Let

$$I_\psi(\xi) = \int_\Omega \exp\{-2\langle \xi, t\rangle\} \, \psi(t) \, dt, \qquad \xi \in \boldsymbol{R}^n.$$

(a) $I_\psi(\xi) = +\infty$ *if* $\xi \notin \bar{\Omega}^*$ *where* Ω^* *is the dual cone of* Ω;

(b) $I_\psi(\xi) < +\infty$ *for all* $\xi \in \Omega^*$ *if and only if* $I_\psi(\xi) < +\infty$ *for some* $\xi_0 \in \Omega^*$. *In this case* I_ψ *is continuous on* Ω^*.

Let $\mathscr{H}(\psi)$ *be the Hilbert space of functions holomorphic on D with the norm* (*).

(c) $\mathscr{H}(\psi) \neq \{0\}$ *if and only if* $I_\psi(\xi_0) < \infty$ *for some* $\xi_0 \in \Omega^*$.

For $\xi \in \Omega^*$, *let* $\mathscr{H}_\xi$ *be the Hilbert space of entire functions F on* $\boldsymbol{C}^q$ *such that*

$$\|F\|^2 = \int_{\boldsymbol{C}^q} |F(u)|^2 \exp\{-2\langle \xi, Q(u,u)\rangle\}\, du < +\infty.$$

Let $\mathscr{H}'_\psi$ *be the space of sections S of the Hilbert fibration* $\mathscr{H}_\xi \to \xi$ *over* Ω^* *which are square-integrable with respect to*

$$I_\psi\colon \|S\|^2 = \int_{\Omega^*} \|S(\xi)\|_\xi^2\, I_\psi(\xi)\, d\xi < \infty.$$

$\mathscr{H}_\psi$ *can also be defined as the closure of the space of functions* $L(\xi, u)$ *of the type*

$$L(\xi, u) = \sum \lambda_i(\xi)\, P_i(u), \quad \lambda_i \in C_0^\infty(\Omega^*), \quad P_i \in \boldsymbol{C}[u],$$

in the norm

$$\|L\|^2 = \int_{\Omega^* \times \boldsymbol{C}^q} |L(\xi, u)|^2 \exp\{-2\langle \xi, Q(u,u)\rangle\}\, I_\psi(\xi)\, d\xi\, du.$$

Then the Fourier transform $F \to \hat{F}$,

$$\hat{F}(\xi, u) = \int_{\boldsymbol{R}^m + i\{y\} \times \{u\}} F(z, u) \exp\{-i\langle \xi, z\rangle\}\, dz,$$

is an isometry of $\mathscr{H}(\psi)$ *with* $\mathscr{H}'(\psi)$.

REMARK. By another Fourier-Laplace transformation in the variable u, one could also identify the space $\mathscr{H}(\psi)$ with a *space* $L^2(\boldsymbol{R}^n \times \boldsymbol{R}^m; d\mu)$. Now in the case of a homogeneous Siegel domain of Type II, and for the function $\psi = |\psi_f|^2 \psi_0$, the integrals $I_\psi(\xi_0)$ for suitable $\xi_0 \in \Omega^*$ are the same as the $\Gamma_v(\varrho)$ of Gindikin (after correctly relating ψ with ϱ), and they can be explicitly calculated as a product of Γ functions. The condition for $I_\psi(\xi_0) < +\infty$ or equivalently $\mathscr{H}(f; \mathfrak{b}^-) \neq \{0\}$ can be reinterpreted by saying that there exists a character δ_0 of $\mathfrak{b}$, a character whose formula, even if completely explicitly given, is still mysterious for us, such that $\mathscr{H}(f; \mathfrak{b}^-) \neq \{0\}$ if and only if $\mathfrak{b}^-$ is a positive polarization for the form $f + \delta_0 \circ j$ of $\mathfrak{b}^*$. This implies that $f + \delta_0 \circ j$ is in the orbit of f_0 in $\mathfrak{b}^*$.

Let us turn now to the representation $\varrho(f; \mathfrak{b}^-)$ of this $\mathscr{H}(f; \mathfrak{b}^-)$. Let f_0 be the given point of $\mathfrak{b}^*$. Then to the orbit $O(f_0)$ on the dual of a completely solvable Lie algebra is associated a canonical irreducible unitary representation $\varrho(O(f_0))$ of B, by the method of Bernat [**2**].

This representation, as shown by Auslander-Kostant, can also be realized using a positive polarization $\mathfrak{h}^-$ at the point f_0 admissible for the maximal nilpotent ideal η of $\mathfrak{b}$ ($\mathfrak{b}^-$ itself is never admissible for the nil-radical) in the Hilbert space $\mathcal{H}(f_0; \mathfrak{h}^-)$.

It is then easy to see that for a natural choice of a positive admissible polarization $\mathfrak{h}^-$ at the point f_0 there exists a natural intertwining operator between $\mathcal{H}(f; \mathfrak{b}^-)$ and $\mathcal{H}(f_0; \mathfrak{h}^-)$ closely related to the Fourier-Laplace transformation $F \to \hat{F}$, and we obtain

THEOREM 2. *Let $f \in \mathfrak{b}^*$ such that $\mathcal{H}(f; \mathfrak{b}^-) \neq \{0\}$. Let $\varrho(f; \mathfrak{b}^-)$ be the representation of B by left translations on $\mathcal{H}(f; \mathfrak{b}^-)$. Then all representations $\varrho(f; \mathfrak{b}^-)$ are equivalent to the representation $\varrho(O(f_0))$.*

The above results can be applied to the study of the holomorphic discrete series of a semisimple group G with finite center. Let K be the maximal compact subgroup of G, and we suppose that G/K is Hermitian symmetric.

Let $\mathfrak{G}$ be the Lie algebra of G, k the Lie algebra of K:

(**) $\qquad \mathfrak{G} = k \oplus \mathfrak{M}$ the Cartan decomposition of $\mathfrak{G}$.

Let $j: \mathfrak{M} \to \mathfrak{M}$ a complex structure on $\mathfrak{M}$ invariant by K.

Let Z_0 be the element of the center of k such that $\mathrm{Ad}_{\mathfrak{M}} Z_0 = j$.

Let $\mathfrak{M}^- \subset \mathfrak{M}_C$ be the subspace of elements of $\mathfrak{M}_C$ such that $[Z_0, X] = -iX$ and let S be the Killing form of $\mathfrak{G}$.

Let $\mathfrak{G} = k \oplus \mathfrak{a} \oplus \eta$ be an Iwasawa decomposition of $\mathfrak{G}$, and let $\mathfrak{b} = \mathfrak{a} \oplus \eta$, and $B = \exp \mathfrak{b}$; so $G = B \cdot K$.

Let us introduce the projection $k: \mathfrak{b}_C \to k_C$ relative to the splitting (**), and define $\mathfrak{b}^- = \{X \in \mathfrak{b}_C : X - k(X) \in \mathfrak{M}^-\}$. k restricted to $\mathfrak{b}^-$ is a homomorphism of $\mathfrak{b}^-$ in k_C.

Let f_0 be the form on $\mathfrak{b}^*$ defined by $f_0(X) = S(Z_0, X)$. Then f_0 has an open orbit on $\mathfrak{b}^*$. Then, as is well known, $(G/K, \mathfrak{M}^-, S)$ inherits the structure of a Siegel domain $D(\Omega; Q)$ of Type II [**12**]. It can easily be checked that the natural isomorphism

$$\begin{array}{ccc} & \overset{\sim}{\nearrow} & (G/K, \mathfrak{M}^-) \\ B & & \wr\Big| \\ & \underset{\sim}{\searrow} & (O(f_0), \mathfrak{b}^-) \end{array}$$

is an isomorphism of Kahler manifolds.

For U an irreducible unitary representation of K on a vector space V, let $\mathcal{H}_U$ be the Hilbert space of V-valued functions on G satisfying

(1) $\phi(g, k)=U(k)^{-1}\,\phi(g)$,
(2) $\phi^* X=0$ for $X \in \mathfrak{M}$,
(3) $\|\phi\|^2=\int_G |\phi|^2\, dg<+\infty$.

G acts unitarily on this space by left translations. We denote this representation by T_U.

$\mathscr{H}(U)$ is describable as a Hilbert space of square-integrable sections of a vector bundle V_U over G/K ($V_U=G\otimes_K V$). A trivialization of this vector bundle will allow us to realize $\mathscr{H}_U$ as a Hilbert space of holomorphic V-valued functions on D. We must find such a trivialization which allows us to compute effectively.

Let β be the homomorphism $\mathfrak{b}\to\mathfrak{b}^-$ introduced above. Finally, let $\mathfrak{u}: k_c\to$ End V be the differential of U and define the finite-dimensional representation Φ of B in $GL(V)$ by $\Phi(\exp X)=\exp(\mathfrak{u}\circ k\circ\beta(X))$, $X\in\mathfrak{b}$. This Φ gives the desired trivialization of $G\otimes_K V$ by mapping (b, v) to $b\otimes\Phi^{-1}(b)\,v$. In the coordinates of D, Φ determines a $GL(V)$-valued function ϱ on D and $\mathscr{H}_U$ and becomes the Hilbert space $\mathscr{H}(\varrho)$ of holomorphic V-valued functions F on D such that

$$\int_D \|\varrho(z; u)^{-1}F(z, u)\|^2\psi_0\, dx\, dy\, du<\infty.$$

It is easily verified that

$$\varrho(z; u)=\varrho_1(u)\,\varrho_2(y-Q(u, u))$$

where ϱ_1 has polynomial entries. Our reduction to the scalar case is this:

THEOREM 3. *$\mathscr{H}(\varrho)\neq\{0\}$ if and only if for every vector $v\in V$ there is a nonzero scalar-valued function F holomorphic on D such that $Fv\in\mathscr{H}(\varrho)$.*

Take Λ to be the highest weight of the representation U on K (with respect to a suitable compact Cartan subgroup τ and a suitable ordering on τ^*). Let V_Λ be an eigenvector for Λ. Then $\mathscr{H}(\Lambda)=\{F\in O(D): FV_\Lambda\in\mathscr{H}(\varrho)\}$ is a nonzero subspace of $\mathscr{H}(\varrho)$ if and only if $\mathscr{H}(\varrho)$ is nonzero. Furthermore, $\mathscr{H}(\Lambda)$ is invariant under left translations by B and is of the form $\mathscr{H}(f; \mathfrak{b}^-)$ studied above.

It is an easy matter to reduce the conditions of Theorem 3 to those of Harish-Chandra, and obtain

THEOREM 4. *$\mathscr{H}_U\neq\{0\}$ if and only if $\langle\Lambda+\varrho, H_\alpha\rangle<0$ for every positive noncompact root α.*

Let us consider M the centralizer of A in K; then the map $(m, b)\to U(m)\,\Phi(b)$ ($\mathfrak{M}\in M$, $b\in B$) is a finite-dimensional representation of the group MAN in the vector space V.

In the isomorphism $\mathscr{H}(U)\leftrightarrow\mathscr{H}(\varrho)$ the restriction of the representation T_U to MAN has a simple description, since M acts on the domain $D(\Omega; Q)$ by linear transformations.

For the coadjoint action of M on $\mathfrak{b}^*$, M leaves the form f_0 stable, and the representation $\varrho(f_0)$ can be extended canonically to MAN as a representation $\nu(f_0)$.

Let us write down the decomposition of the restriction of the irreducible representation U of K to M:

$$U \mid \cdot M = \bigoplus_{i \in I} \tau_i$$

and denote by $\tilde{\tau}_i$ the finite irreducible representation of MAN defined by $\tilde{\tau}_i(man) = \tau_i(m)$. Then we have

THEOREM 5. $T_U \mid MAN = \oplus_{i \in I} (\tilde{\tau}_i \otimes \nu(f_0))$.

It is clear that each of the representations $\tilde{\tau}_i \otimes \nu(f_0)$ is an irreducible representation of MAN and that as a corollary we have

$$T_U \mid AN = (\dim V_U)\, \rho(f_0).$$

REFERENCES

1. L. Auslander and B. Kostant, *Polarization and unitary representations of solvable Lie groups*, Invent. Math. **14** (1971), 255–354.

2. P. Bernat, *Sur les représentations unitaires des groupes de Lie résolubles*, Ann. Sci. Ecole Norm. Sup. (3) **82** (1965), 37–99. MR **33** #2763.

3. R. J. Blattner, *On induced representations*, Amer. J. Math **83** (1961), 79–98, 499–512. MR**23** # A 2757; **26** # 2885.

4. R. J. Blattner, B. Kostant and S. Sternberg, Proc. Conf. on Group Representations, Williamstown, 1972.

5. S. Bochner and W. T. Martin, *Several complex variables*, Princeton Math. Series, vol. 10, Princeton Univ. Press, Princeton, N.J., 1948. MR **10**, 366.

6. S. G. Gindikin, *Analysis in homogeneous domains*, Uspehi Mat. Nauk **19** (1964), no. 4, (118), 3–92 = Russian Math. Surveys **19** (1964), no. 4, 1–89. MR**30** #2167.

7. S. G. Gindikin, I. I. Pjateckiĭ-Šapiro and E. E. Vinberg, *Geometry of homogeneous bounded domains* (C.I.M.E., 3° Ciclo, Urbino, 1967), 3–87; Edizione Cremonese, Rome, 1968. MR **38** #6513.

8. R. Godement, *Séminaire H. Cartan*, Exposé no. 6, 1958.

9. Harish-Chandra, *Representations of semi-simple Lie groups*. V, Amer. J. Math. **78** (1956), 1–41. MR **18**, 490.

10. A. A. Kirillov, *Unitary representations of nilpotent Lie group*, Uspehi Mat. Nauk **17** (1962), no. 4, (106), 57–110 = Russian Math. Surveys **17** (1962), no. 4, 53–104. MR **25** #5396.

11a. ———, *Plancherel's measure for nilpotent Lie groups*, Funkcional. Anal. i Priložen. **1** (1967), no. 4, 84–85. (Russian) MR **37** #347.

b. ———, *Characters of unitary representations of Lie groups*, Funkcional. Anal. i Priložen. **2** (1968), no. 2, 40–55. (Russian) MR **38** #4615.

12. A. Koranyi and J. Wolf, *Generalized Cayley transformations of bounded symmetric domains*, Amer. J. Math. **87** (1965), 899–939. MR **33** #229.

13. B. Kostant, *Quantization and unitary representations*, Lecture Notes in Math., vol. 170, Springer-Verlag, Berlin and New York, 1970.

14. I. I. Pjateckiĭ-Šapiro, *Geometry of classical domains and theory of automorphic functions*, Fizmatgiz, Moscow, 1961; English transl., *Automorphic functions and the geometry of classical domains*, Math. and its Applications, vol. 8, Gordon and Breach, New York, 1969. MR **25** #231; **40** #5908.

15. H. Behnke and H. Grauert, *Analysis in non-compact spaces*, Analytic functions, Princeton Univ. Press, Princeton, N.J., 1960, pp. 11–44. MR **22** #5988.

BRANDEIS UNIVERSITY

UNIVERSITY OF CALIFORNIA, BERKELEY

DETERMINATION OF INTERTWINING OPERATORS

A. W. KNAPP *

The subject is representations of the principal series and complementary series. Let $G=KAN$ be a connected semisimple Lie group of matrices, and let MAN be a minimal parabolic subgroup, where M is the centralizer of A in K. If σ is an irreducible unitary representation of M and λ is a unitary character of A, then

$$U(\sigma, \lambda) = \operatorname*{ind}_{MAN \uparrow G} \; (man \to \lambda(a)\,\sigma(m))$$

is a representation of the *principal series*. The principal series is one of the series contributing to the Plancherel formula and corresponds to a Cartan subgroup as noncompact as possible. The principal series with $\sigma=1$ was investigated by Kostant [**5**], who proved that $U(1, \lambda)$ is irreducible. However, $U(\sigma, \lambda)$ need not be irreducible in general.

It is still possible to define $U(\sigma, \lambda)$ as a nonunitary representation on a Hilbert space when λ is nonunitary. We say $U(\sigma, \lambda)$ is in the *complementary series* if there is an invariant inner product on the C^∞ vectors that is continuous in the C^∞ topology.

We consider two problems: (1) Find the dimension and algebra structure of the commuting ring $C(\sigma, \lambda)$ of $U(\sigma, \lambda)$ when λ is unitary; (2) produce complementary series. At this point we could state our solutions to these problems, but we prefer first to motivate the results by introducing the intertwining operators. The development of these operators was begun by Kunze and Stein [**6**], continued by Schiffmann [**8**] and to an extent by Helgason [**2**], and completed by Knapp and Stein [**3**].

AMS (MOS) subject classifications (1970). Primary 22E30, 22E45; Secondary 17B20, 20G20, 22D30, 22E15.

* Partly in collaboration with E. M. Stein; supported by NSF grant GP-28251.

Let M' be the normalizer of A in K and let $W = M'/M$. If w is in M', we write $[w]$ for the coset wM in W. The group M' operates on λ and σ by conjugation of the A or M variable by w^{-1}. The class of $w\sigma$ depends only on $[w]$. Let

$$W_{\sigma,\lambda} = \{w \in W \mid w\sigma \sim \sigma \text{ and } w\lambda = \lambda\}.$$

Bruhat [1] obtained the results for the principal series that

(1) $\dim C(\sigma, \lambda) \leqq |W_{\sigma,\lambda}|$ (and so $=1$ for almost all λ),

(2) $U(\sigma, \lambda)$ is unitarily equivalent with $U(w\sigma, w\lambda)$.

The intertwining operator that implements (2) is, by (1), unique up to a scalar for almost all λ. By [3] such operators $\mathscr{A}(w, \sigma, \lambda)$ can be chosen so that the following hold:

(i) $U(w\sigma, w\lambda)\, \mathscr{A}(w, \sigma, \lambda) = \mathscr{A}(w, \sigma, \lambda)\, U(\sigma, \lambda)$.

(ii) $\mathscr{A}(w, \sigma, \lambda)$ is unitary and in its action on smooth functions varies real-analytically in λ.

(iii) $\mathscr{A}(w_1 w_2, \sigma, \lambda) = \mathscr{A}(w_1, w_2\sigma, w_2\lambda)\, \mathscr{A}(w_2, \sigma, \lambda)$.

(iv) $\mathscr{A}(w, E\sigma E^{-1}, \lambda) = E\mathscr{A}(w, \sigma, \lambda)\, E^{-1}$.

(v) If $[w]$ is the reflection relative to a simple restricted root α and if $\mathfrak{g}_\alpha$ is the real-rank-one algebra generated by $\mathfrak{n}_\alpha$ and $\theta\mathfrak{n}_\alpha$, then $\mathscr{A}(w, \sigma, \lambda)$ is essentially $\mathscr{A}_\alpha(w, \sigma|_{M_\alpha}, \lambda|_{A_\alpha})$. [Here M_α and A_α denote the M and A subgroups for the group corresponding to $\mathfrak{g}_\alpha$. The representation $\sigma|_{M_\alpha}$ is a multiple of a single irreducible representation of M_α, and consequently there is no difficulty in defining the operator $\mathscr{A}_\alpha$.]

If $w\sigma = \sigma$ and $w\lambda = \lambda$, (i) says $\mathscr{A}(w, \sigma, \lambda)$ is in $C(\sigma, \lambda)$. More generally suppose $w\sigma \sim \sigma$ and $w\lambda = \lambda$. Then it is possible to extend σ to a representation of the group generated by M and w. So $\sigma(w)$ is defined; it is unique up to a root of unity. In this case, (i) and (iv) show that $\sigma(w)\, \mathscr{A}(w, \sigma, \lambda)$ is in $C(\sigma, \lambda)$. This operator depends only on $[w]$ and we may write $\sigma([w])\, \mathscr{A}([w], \sigma, \lambda)$ instead. Then

$$\operatorname{span}\{\sigma(p)\, \mathscr{A}(p, \sigma, \lambda) \mid p \in W_{\sigma,\lambda}\} \subseteq C(\sigma, \lambda).$$

The following unpublished theorem was proved in other notation by Harish-Chandra and translated into this notation by Wallach; it is given here with Harish-Chandra's permission.

THEOREM. $\operatorname{span}\{\sigma(p)\, \mathscr{A}(p, \sigma, \lambda) \mid p \in W_{\sigma,\lambda}\} = C(\sigma, \lambda)$.

In view of the theorem it is of interest to determine a linear basis of the left side of the equality, in particular to determine which operators are scalar. Another reason for wanting this information is given by the next theorem [3].

THEOREM. *Let p be an element of order 2 in $W_{\sigma,1}$. If $\sigma(p)\, \mathscr{A}(p, \sigma, 1)$ is scalar,* *then $U(\sigma, \lambda)$ is in the complementary series for all λ sufficiently close to 1 such that $p\lambda = \bar{\lambda}^{-1}$. "Sufficiently close" depends on G but not σ or p.*

We shall now describe $C(\sigma, \lambda)$. Let Δ be the set of restricted roots and let

$$\Delta' = \{\alpha \in \Delta \mid p_\alpha \in W_{\sigma,\lambda} \text{ and } \sigma(p_\alpha)\, \mathscr{A}(p_\alpha, \sigma, \lambda) = cI\}.$$

From [3] one knows that $\sigma(p_\alpha)\, \mathscr{A}(p_\alpha, \sigma, \lambda)$ is scalar if and only if the real-rank-one Plancherel density satisfies $p_{\sigma|M_\alpha}(\lambda|_{A_\alpha}) = 0$, and so it is an easy matter to determine the members of Δ'. Now Δ' is a root system, and we let $W'_{\sigma,\lambda} \subseteq W_{\sigma,\lambda}$ be its Weyl group. Let

$$R_{\sigma,\lambda} = \{p \in W_{\sigma,\lambda} \mid p\alpha > 0 \text{ for all } \alpha > 0 \text{ in } \Delta'\}.$$

THEOREM. (i) *$W_{\sigma,\lambda}$ is the semidirect product $W_{\sigma,\lambda} = W'_{\sigma,\lambda} R_{\sigma,\lambda}$ with $W'_{\sigma,\lambda}$ normal. The operators $\sigma(w)\, \mathscr{A}(w, \sigma, \lambda)$ are scalar exactly for w in $W'_{\sigma,\lambda}$ and they are linearly independent for w in $R_{\sigma,\lambda}$. Consequently, the operators for $R_{\sigma,\lambda}$ are a basis for $C(\sigma, \lambda)$ and*

$$\dim C(\sigma, \lambda) = |R_{\sigma,\lambda}|.$$

(ii) *For w in W let $p_\sigma^w(\lambda)$ be the product of $p_{\sigma|M_\alpha}(\lambda|_{A_\alpha})$ over all $\alpha > 0$ in Δ such that $\alpha/2$ is not in Δ and $w\alpha$ is <0. Then*

$$\dim C(\sigma, \lambda) = |\{w \in W_{\sigma,\lambda} \mid p_\sigma^w(\lambda) \neq 0\}|.$$

(iii) *$R_{\sigma,\lambda} = \sum \mathbf{Z}_2$ with the number of summands $\leqq \dim A$.*

In the theorem, part (i) is elementary and self-proving, and (ii) comes out of the proof of (i). Part (i) shows that the subgroup of $W_{\sigma,\lambda}$ corresponding to trivial operators is a Weyl group; consequently the elements p in the theorem about complementary series are all given by commuting products of reflections relative to Δ' and are easy to determine. Part (iii) is the part that is hard to prove, and it is the one that gives insight into the nature of $R_{\sigma,\lambda}$. Despite property (iii) of $\mathscr{A}(w, \sigma, \lambda)$, this result falls short of saying that $C(\sigma, \lambda)$ is commutative, saying only that is commutative modulo $\pm$ signs. It seems possible to analyze this matter further and use the methods of proof of (iii) to prove commutativity of $C(\sigma, \lambda)$, but such a proof has yet to be carried out.[1]

We shall discuss one aspect of the proof of (iii). First, to prove (iii) for $\lambda = \lambda_0$, it suffices to prove (iii) for $\lambda = 1$. Then the basic idea is that $R_{\sigma,1}$ can be under-

[1] (Footnote added January, 1973.) The proof of the commutativity of $C(\sigma, \lambda)$ has now been carried out. The algebra structure of $C(\sigma, \lambda)$ is as follows: The ambiguous signs of $\sigma(w)$ for w in $R_{\sigma,\lambda}$ can be chosen so that the operators $\sigma(w)\, \mathscr{A}(w, \pi, \lambda)$ for w in $R_{\sigma,\lambda}$ form both a group isomorphic to $\sum \mathbf{Z}_2$ and a linear basis of $C(\sigma, \lambda)$.

stood provided σ is moved to some "standard position." By such a device the proof for general G is reduced to the case that G is split over $\boldsymbol{R}$, and then this case is considered separately. To simplify the exposition we shall not deal with general G here but will content ourselves with two cases.

Case 1. We assume that, for each simple α in Δ, $\mathfrak{g}_\alpha$ is not isomorphic with $sl(2, \boldsymbol{R})$. This condition implies that M is connected. It is satisfied, for example, if G is complex semisimple or if G is simple and twice some restricted root is again a restricted root. For complex G, the whole principal series is irreducible, by [7] and [11], and there is a corresponding simple computation that one can do to show, without the irreducibility theorem, that all the $\sigma(w)\,\mathscr{A}(w, \sigma, \lambda)$ are scalar. The idea in Case 1 will be to imbed into G as much of the complex case as possible to show that most of the operators are scalar.

Let $\mathfrak{h} \subseteq \mathfrak{m}$ be a maximal abelian subspace, so that $\mathfrak{a}+\mathfrak{h}$ is a Cartan subalgebra of $\mathfrak{g}$ and $\mathfrak{h}$ is a Cartan subalgebra of $\mathfrak{m}$. The roots of $(\mathfrak{g}^{\boldsymbol{C}}, (\mathfrak{a}+\mathfrak{h})^{\boldsymbol{C}})$ are real on $\mathfrak{a}$ and imaginary on $\mathfrak{h}$. The restricted roots are the restrictions to $\mathfrak{a}$ of the roots, and the roots of $(\mathfrak{m}^{\boldsymbol{C}}, \mathfrak{h}^{\boldsymbol{C}})$ are the restrictions to $\mathfrak{h}^{\boldsymbol{C}}$ of the roots that vanish on $\mathfrak{a}$. We may assume that the ordering on the roots is chosen so that the $\mathfrak{a}$ part is more significant than the $\mathfrak{h}$ part.

We say that α in Δ is *essential* if neither α nor 2α is a root when extended to be 0 on $\mathfrak{h}$. Otherwise α is *inessential.* The name refers to the possibilities for how the Weyl group reflection p_α can be extended to $\mathfrak{h}'$; essential restricted roots cannot act trivially on $\mathfrak{h}'$. Specifically, if α is inessential, there is w in M' so that $[w]=p_\alpha$ and $\mathrm{Ad}(w)=1$ on $i\mathfrak{h}'$. If α is essential and $\alpha \pm \beta$ are roots (with β in $i\mathfrak{h}'$), there is w in M' so that $[w]=p_\alpha$ and $\mathrm{Ad}(w)=p_\beta$ on $i\mathfrak{h}'$.

In the complex case every restricted root is essential. Quite generally the idea is that essential restricted roots lead to trivial intertwining operators, and we intend to discard these by imbedding into $i\mathfrak{h}'$ the part of $\mathfrak{a}'$ corresponding to the essential restricted roots. Let

$$\pi_e = \{\text{essential simple restricted roots}\},$$

$$\mathfrak{a}_e' = \text{span of } \pi_e \text{ in } \mathfrak{a}',$$

$$W_e = \text{subgroup of } W \text{ generated by the } p_\alpha \text{ for } \alpha \text{ in } \pi_e.$$

IMBEDDING LEMMA. *It is possible to choose β in $i\mathfrak{h}'$ corresponding to each α in π_e so that $\alpha+\beta$ is a root, so that p_β preserves the set of positive roots of $\mathfrak{m}$, and so that the linear extension of the mapping given by $\alpha \to J(\alpha)=\beta$ is an isometry of $\mathfrak{a}_e'$ into $i\mathfrak{h}'$.*

Fix J as in the lemma and let W_π be the set of simple reflections in W. J defines a map of W_π into the orthogonal group $O(i\mathfrak{h}')$ as follows: If α is in π_e, map p_α into $p_{J\alpha}$. If α is in $\pi - \pi_e$, map p_α into the identity.

THEOREM. *The mapping of W_π into $O(i\mathfrak{h}')$ defined by J extends to a group homomorphism of W into $O(i\mathfrak{h}')$. The resulting action of W on $i\mathfrak{h}'$ has the properties that*

(a) *for w in W_e, $Jw=wJ$ on $\mathfrak{a}_e'$,*

(b) *for w in W, if γ is a positive root of $\mathfrak{m}$, so is $w\gamma$,*

(c) *for w in W, if σ has highest weight Λ, then $w\sigma$ has highest weight $w\Lambda$ (and so $\sigma \sim w\sigma$ if and only if $\Lambda = w\Lambda$).*

The map J is not unique, but (c) in the theorem shows that the action of W on $i\mathfrak{h}'$ is canonical.

We say Λ in $i\mathfrak{h}'$ is *dominant* if $\langle \Lambda, J\alpha\rangle \geqq 0$ for all α in π_e. It is a simple exercise with the theorem to show that σ is conjugate under W to a representation of M whose highest weight is dominant. Now if σ is replaced by $p\sigma$ for some p in M', the whole situation for σ, intertwining operators and all, is conjugated to the situation for $p\sigma$. Thus it is enough to prove that $R_{\sigma,1} = \sum \boldsymbol{Z}_2$ under the assumption that the highest weight Λ of σ is dominant. With this observation, we proceed as follows: Let

$$S = \{w \in W \mid w = 1 \text{ on } i\mathfrak{h}'\}.$$

Then S is normal in W and $W = W_e S$, as a semidirect product. Moreover, $W_{\sigma,1} = (W_{\sigma,1} \cap W_e)\, S$. If Λ is dominant, $W_{\sigma,1} \cap W_e$ is generated by the simple reflections that it contains. [In fact, an easy induction reduces this statement to showing that if w is in $W_{\sigma,1} \cap W_e$ and $w\alpha < 0$ for some α in π_e, then $p_\alpha \Lambda = \Lambda$. This last equality follows from the chain $0 \leqq \langle \Lambda, \alpha\rangle = \langle w^{-1}\Lambda, \alpha\rangle = \langle \Lambda, w\alpha\rangle \leqq 0$, which uses the dominance of Λ twice.] By means of properties (iii) and (v) of the $\mathscr{A}$ operators, we can therefore reduce the operators for $W_{\sigma,1} \cap W_e$ to operators for a real-rank-one group whose simple restricted root is essential. Such a group is a cover of the Lorentz group SO(odd, 1), and the whole principal series for such a group is irreducible, by [**3**]. Consequently all the operators corresponding to $W_{\sigma,1} \cap W_e$ are scalar. Thus all the operators for $R_{\sigma,1}$ are already represented by members of S, except for scalar factors. Under our assumption on G, S is $\sum \boldsymbol{Z}_2$, and it follows easily that $R_{\sigma,1} = \sum \boldsymbol{Z}_2$.

An interesting special case occurs when every α in Δ is essential. Wallach has pointed out that this is exactly the case that G has only one conjugacy class of Cartan subgroups. For such a group, $W = W_e$ and so $S = \{1\}$ and $R_{\sigma,1} = \{1\}$. The principal series is therefore irreducible. This result was obtained earlier by Wallach in an unpublished work (cf. [**10**]).

Case 2. We assume that G is simple and is split over $\boldsymbol{R}$. Then $\mathfrak{a}$ is a Cartan subalgebra and M is a finite abelian group. Each $\mathfrak{g}_\alpha$ is isomorphic with $sl(2, \boldsymbol{R})$, and we let γ_α be the image of $\left(\begin{smallmatrix}-1 & 0\\ 0 & -1\end{smallmatrix}\right)$ under the corresponding map of $SL(2, \boldsymbol{R})$ into G. Then γ_α is in M and $\gamma_\alpha^2 = 1$. The γ_ε's for ε simple generate M.

Thus σ is determined by its values on the γ_ε's, and σ assumes only the values ± 1. It is easy to check that

$$\Delta' = \{\beta \in \Delta \mid \sigma(\gamma_\beta) = +1\}.$$

We assume that σ is not identically 1. In this case we say σ is *dominant* if $\sigma(\gamma_\varepsilon) = -1$ for exactly one simple ε, say $\varepsilon = \varepsilon_k$. Every $\sigma \not\equiv 1$ can be conjugated by W so as to be dominant, and we shall assume that σ is dominant from now on.

It is an easy matter to see that the simple roots of Δ' consist of the ε_i (for $i \neq k$) and at most one other root, say α. The root α exists if and only if there is a root $\beta > 0$ such that $\sigma(\gamma_\beta) = +1$ and $\langle \beta, \varepsilon_i \rangle \leqq 0$ for $i \neq k$. In this case α is the least such β. This fact makes it easy to determine Δ' explicitly in examples.

If α does not exist, a short argument shows that $R_{\sigma,1}$ is $\{1\}$ or $\boldsymbol{Z}_2$. If α does exist, $W'_{\sigma,1}$ is a Weyl group of the same rank as W, and it follows that $R_{\sigma,1}$ must be small. In fact, it need not have 1 or 2 elements, but it is always $\sum \boldsymbol{Z}_2$. Of the two proofs of this statement at present, one is by classification and one is not. The one by classification is shorter.

References

1. F. Bruhat, *Sur les représentations induites des groupes de Lie*, Bull. Soc. Math. France **84** (1956), 97–205. MR **18**, 907.

2. S. Helgason, *A duality for symmetric spaces with applications to group representations*, Advances in Math. **5** (1970), 1–154. MR **41** #8587.

3. A. W. Knapp and E. M. Stein, *Intertwining operators for semisimple groups*, Ann. of Math. (2) **93** (1971), 489–578.

4. ———, *Irreducibility theorems for the principal series*, Conference on Harmonic Analysis, Lecture Notes in Math., vol. 266, Springer-Verlag, Berlin and New York, 1972, pp. 197–214.

5. B. Kostant, *On the existence and irreducibility of certain series of representations*, Bull. Amer. Math. Soc. **75** (1969), 627–642. MR **39** #7031.

6. R. Kunze and E. M. Stein, *Uniformly bounded representations*. III. *Intertwining operators for the principal series on semisimple groups*, Amer. J. Math. **89** (1967), 385–442. MR **38** #269.

7. K. R. Parthasarathy, R. Ranga Rao and V. S. Varadarajan, *Representations of complex semisimple Lie groups and Lie algebras*, Ann. of Math. (2) **85** (1967), 383–429. MR **37** #1526.

8. G. Schiffmann, *Intégrales d'entrelacement et fonctions de Whittaker*, Bull. Soc. Math. France **99** (1971), 3–72.

9. R. Takahashi, *Sur les représentations unitaires des groupes de Lorentz généralisés*, Bull. Soc. Math. France **91** (1963), 289–433. MR **31** #3544.

10. N. Wallach, *Cyclic vectors and irreducibility for principal series representations*, Trans. Amer. Math. Soc. **158** (1971), 107–113. MR **43** #7558.

11. D. P. Želobenko, *The analysis of irreducibility in the class of elementary representations of a complex semisimple Lie group*, Izv. Akad. Nauk SSSR **32** (1968), 105–128 = Math. USSR Izv. **2** (1968), 108–133. MR **37** #2906.

CORNELL UNIVERSITY

KOSTANT'S P^{γ} AND R^{γ} MATRICES AND INTERTWINING INTEGRALS

NOLAN R. WALLACH

1. Introduction. Let G be a connected semisimple Lie group with finite center. Let K be a maximal compact subgroup of G and let $G=KAN$ be an Iwasawa decomposition of G corresponding to K (cf. Helgason [2]). In particular, A is a maximal vector subgroup of G such that the elements $\mathrm{Ad}(a)$, $a\in A$, are simultaneously diagonalizable, N is a maximal unipotent subgroup of G normalized by A. Let M be the centralizer of A in K.

Let $X^{\infty}=\{f\in C^{\infty}(K)\mid f(km)=f(k)\}$. If $x, k\in K$, let $\pi(x)f(k)=f(x^{-1}k)$. Let X be the space of all K-finite elements of X^{∞}. That is, if $f\in X$ then $\pi(K)f$ is contained in a finite-dimensional subspace of X^{∞}.

If (τ, V) is a representation of K let $V^{M}=\{v\in V\mid \tau(m)v=v \text{ for all } m\in M\}$. Let $\hat{K}$ be the set of all equivalence classes of irreducible finite-dimensional representations of K. If $\gamma\in\hat{K}$ fix $(\pi_{\gamma}, V_{\gamma})\in\gamma$. If $\gamma\in\hat{K}$ let $(\pi_{\gamma}^{*}, V_{\gamma}^{*})$ be the contragradient representation.

We identify $V_{\gamma}\otimes(V_{\gamma}^{*M})$ with a subspace X_{γ} of X as follows:

$$(v\otimes\lambda)(k)=\lambda(\pi_{\gamma}(k)^{-1}v)$$

for $v\in V_{\gamma}$, $\lambda\in V_{\gamma}^{*M}$. Frobenius reciprocity says that

$$X=\sum_{\gamma\in\hat{K}} X_{\gamma}=\sum_{\gamma\in\hat{K}} V_{\gamma}\otimes(V_{\gamma}^{*M}).$$

Let $\mathfrak{a}$ be the Lie algebra of A and let $\log: A\to\mathfrak{a}$ be the inverse map to $\exp:\mathfrak{a}\to A$. Let $\mathfrak{a}^{*}$ and $\mathfrak{a}_{c}^{*}$ be respectively the real-valued and complex-valued linear forms in $\mathfrak{a}$. Let $\mathfrak{N}$ be the Lie algebra of N. Define $\varrho(H)=\frac{1}{2}\mathrm{tr}(\mathrm{ad}\,H\mid\mathfrak{N})$ for $H\in\mathfrak{a}$. Let θ be

---

AMS (MOS) subject classifications (1970). Primary 22E45, 22D30.

the Cartan involution of G corresponding to K. Let $\bar{\mathfrak{N}}=\theta(\mathfrak{N})$. Let $\bar{N}$ be the connected subgroup of G corresponding to $\bar{\mathfrak{N}}$.

If $\nu\in\mathfrak{a}_C^*$ and $f\in X^\infty$ define $f_\nu(kan)=\exp\{-(\varrho+i\nu)(\log a)\}f(k)$. If $g\in G, f\in X^\infty$, $\nu\in\mathfrak{a}_C^*$ define

$$(\pi_\nu(g)f)(k)=f_\nu(g^{-1}k).$$

Let $\mathfrak{G}$ denote the Lie algebra of G. If $x\in\mathfrak{G}$ define

$$\pi_\nu(X)f=\frac{d}{dt}\pi(\exp tx)f\bigg|_{t=0}.$$

Then π_ν defines a representation of $\mathfrak{G}$ on X^∞. Let $U(\mathfrak{G})=U$ in the complexified universal enveloping algebra of $\mathfrak{G}$. Then π_ν defines a representation of U on X^∞. Clearly $\pi_\nu(U)X\subset X$. Hence we have a representation (π_ν, X) of U.

Let M^* be the normalizer of A in K. Let $W=M^*/M$. Then W acts naturally as a group of linear transformations of $\mathfrak{a}$, hence $\mathfrak{a}^*$ and $\mathfrak{a}_C^*$. For each $s\in W$ fix $s^*\in M^*$ so that $s^*M=s$. If $f\in X^\infty$, $\nu\in\mathfrak{a}_C^*$, $s\in W$, define (cf. Kunze-Stein [9], Schiffmann [10])

$$(A_s(\nu)f)(k)=\int_{\bar{N}\cap s^{*-1}Ns^*} f(ks^*\bar{n}_s)\,d\bar{n}_s \tag{1}$$

where $d\bar{n}_s$ is some normalization of Haar measure on $\bar{N}\cap s^{*-1}Ns^*$.

Let, for $\lambda\in\mathfrak{a}^*$, $\mathfrak{N}_\lambda=\{x\in\mathfrak{N}\mid[H,x]=\lambda(H)x$ for all $H\in\mathfrak{a}\}$. Let $\Lambda^+=\{\lambda\in\mathfrak{a}^*\mid\mathfrak{N}_\lambda\neq(0)\}$. If $\lambda\in\mathfrak{a}^*$ let $H_\lambda\in\mathfrak{a}$ be defined by $B(H,H_\lambda)=\lambda(H)$ for $H\in\mathfrak{a}$ ($B(X,Y)=\operatorname{tr}\operatorname{ad}X\operatorname{ad}Y$).

THEOREM 1.1 (KUNZE-STEIN [9], HARISH-CHANDRA [1], SCHIFFMANN [10], HELGASON [3], KNAPP-STEIN [5]). *If* $\operatorname{Im}\nu(H_\lambda)<0$ *for all* $\lambda\in\Lambda^+$ *then* (1) *converges for all* $f\in X^\infty$ *and* $s\in W$. *Furthermore under the above conditions the following hold:*

(i) $A_s(\nu)X_\gamma\subset X_\gamma$ *for each* $\gamma\in\hat{K}$.

(ii) $A_s(\nu)\pi_\nu(X)=\pi_{s\nu}(X)A_s(\nu)$ *for* $x\in G$, $s\in W$.

(iii) *The map* $\nu\mapsto A_s(\nu)|_{X_\gamma}$ *extends to a meromorphic function on* $\mathfrak{a}_C^*$ *for each* $\gamma\in\hat{K}$.

In Johnson-Wallach [4], a formula for $A_s(\nu)|_{X_\gamma}$ was announced which in the case of split rank 1 ($\dim A=1$) is a scalar multiplication by ratio of two polynomials defined by Kostant [6]. The purpose of this note is to give a direct relationship between Kostant's polynomial valued matrices and the $A_s(\nu)$.

The results of this note will appear in expanded form in the forthcoming paper of Johnson-Wallach which will also contain the details of [4].

2. The P^γ and R^γ matrices. In Kostant-Rallis [8], Kostant [7] certain canonical subspaces H^*, J^* of U are defined. We will only need some formal properties of these spaces.

THEOREM 2.1 (KOSTANT-RALLIS [8], KOSTANT [7]). *$U = U\mathfrak{k} \oplus H^*J^*$ ($\mathfrak{k}$ the Lie algebra of K) and*

(1) *$U^{\mathfrak{k}} \supset J^*$ ($U^{\mathfrak{k}} = \{u \in U \mid \operatorname{ad}(X)\, u = 0$ for all $X \in \mathfrak{k}\}$),*

(2) *$H^*J^* \cong H^* \otimes J^*$, and*

(3) *H^* is K-invariant and equivalent as a representation of K with X.*

Let, for $\gamma \in \hat{K}$, $E_\gamma = \{A: V_\gamma \to H^* \mid A\pi_\gamma(k) = \operatorname{Ad}(k)\, A\}$.

The Poincaré-Birchoff-Witt theorem implies then $U = (U\mathfrak{k} + \mathfrak{N}U) \oplus U(\mathfrak{a})$. If $u \in U$ let $P_u \in U(\mathfrak{a})$ be the component of u in $U(\mathfrak{a})$ relative to the direct sum decomposition. If $a \in U(\mathfrak{a})$ we look upon a as a polynomial on $\mathfrak{a}_{\mathbf{C}}^*$ ($U(\mathfrak{a}) = S(\mathfrak{a}_{\mathbf{C}}) = S(\mathfrak{a}_{\mathbf{C}}^*)^*$).

LEMMA. *Let $\gamma \in R$ be fixed. There is a basis $\sigma_1, \ldots, \sigma_{l(\gamma)}$ and a basis $v_1, \ldots, v_{l(\gamma)}$ of V_γ^M such that $P_{\sigma_i(v_j)}(-2\varrho) = \delta_{ij}$.*

In general let $P^u(v) = P_u(-(\varrho + iv))$. Thus $P^{\sigma_i(v_j)}(-i\varrho) = \delta_{ij}$. Let $P_{ij}^\gamma(v) = P^{\sigma_i(v_j)}(v)$. $P^\gamma = (P_{ij}^\gamma)$ is Kostant's P^γ matrix.

Let $V_\gamma^* = V_{\gamma^*}$. Let $\sigma_1, \ldots, \sigma_{l(\gamma)}$, $v_1, \ldots, v_{l(\gamma)}$, $\sigma_1^*, \ldots, \sigma_{l(\gamma^*)}^*$, $v_1^*, \ldots, v_{l(\gamma)}^*$ be as in Lemma 2.1 for γ, γ^*. We may assume that $\langle v_i, v_j^* \rangle = \delta_{ij}$. Extend $v_1^*, \ldots, v_{l(\gamma)}^*$ to a basis of V_γ and extend $v_1, \ldots, v_{l(\gamma)}$ to a dual basis of V_γ^*. Let $z(i,j) = \sum_k \sigma_i(v_k)\, \sigma_i^*(v_k^*)$. Set $R_{ij}^\gamma(v) = P^{z(i,j)}(v)$. $R^\gamma = (R_{ij}^\gamma)$ is Kostant's R^γ matrix.

We collect several results on these matrices whose proofs can be found in Kostant [7]. (We use K rather than Kostant's, K_θ, but his proofs still work.)

THEOREM 2.2 (KOSTANT [7]). (1) *$R^\gamma(sv) = R^\gamma(v)$ for $s \in W$, $v \in \mathfrak{a}_{\mathbf{C}}^*$.*

(2) *$R^\gamma(v) = P^\gamma(\bar{v})\, P^\gamma(v)$.*

3. The $A_s^\gamma(v)$. Let $\gamma \in \hat{K}$, then $A_s(v)(v \otimes \lambda) = v \otimes \lambda \cdot A_s^\gamma(v)$ for $v \in V_\gamma$, $\lambda \in V_\gamma^{*M}$, $s \in W$, $v \in \mathfrak{a}_{\mathbf{C}}^*$. In this section we first compute the $A_s^\gamma(v)$.

THEOREM 3.1. *There is a basis $\mu_1, \ldots, \mu_{l(\gamma)}$ of V_γ^{*M} such that $A_s^\gamma(v)$ has the matrix $C_s(v)\, P^\gamma(v)^{-1} P^\gamma(sv)$ where*

$$C_s(v) = \int_{\bar{N} \cap s^{*-1} N s^*} \exp\{-(\varrho + iv)(H(\bar{n}_s))\}\, d\bar{n}_s$$

($H(kan) = \log a$).

PROOF (SKETCH). Let $B_v^\gamma: E_\gamma \to V_\gamma^{*M}$ as follows:

$$B_v^\gamma(\sigma)(v) = (\pi_v(\sigma(v))\, 1)(e) \qquad (1(k) = 1 \text{ for all } k).$$

If $X \in \mathfrak{N}$, $f \in X_\infty$, then

$$(\pi_\nu(X) f)(e) = \frac{d}{dt} f(\exp - tX)\Big|_{t=0} = \frac{d}{dt} f(e)\Big|_{t=0} = 0.$$

Thus if $u \in Uk + \mathfrak{N}U$ then $(\pi_\nu(u)\, 1)(e) = 0$. Thus

$$B_\nu^\gamma(\sigma)(v) = P_{\sigma(v)}(-(\varrho + i\nu)) = P^{\sigma(v)}(\nu).$$

Let $T_\nu : H^* \to X$ be defined by $T_\nu(u) = \pi_\nu(u)\, 1$. Then $T_\nu(v \otimes \sigma) = v \otimes B_\nu^\gamma(\sigma)$ for $v \in V_\gamma$, $\sigma \in E_\gamma$ (we identify H^* with $\sum_{\gamma \in \hat{K}} V_\gamma \otimes E_\gamma$). Now $T_{-i\varrho} : H^* \to X$ is bijective; thus $B_{-i\varrho} : E_\gamma \to V_\gamma^{*M}$ is bijective. It is easily seen that relative to the above basis

$$T_\nu \circ T_{-i\varrho}^{-1}(v \otimes \mu) = v \otimes \mu \cdot P^\gamma(\nu).$$

Now $A_s(\nu) \cdot T_\nu(u) = A_s(\nu)\, \pi_\nu(u)\, 1 = \pi_{s\nu}(u)\, A_s(\nu)\, 1 = C_s(\nu)\, \pi_{s\nu}(u)\, 1 = C_s(\nu)\, T_{s\nu}(u)$. Hence $A_s(\nu) = C_s(\nu)\, T_{s\nu} \circ T_\nu^{-1}$. This proves the result.

We now relate the R^γ with the $A_s(\nu)$. Let $\mathscr{A}_s^\gamma(\nu) = C_s(\nu)^{-1}\, A_s^\gamma(\nu) = P_\gamma(\nu)^{-1}\, P_\gamma(s\nu)$.

THEOREM 3.2. *Let $\nu \in \mathfrak{a}_{\mathbf{C}}^*$. Suppose that $s\nu = \bar{\nu}$. Then $R^\gamma(\nu)$ and $A_s^\gamma(\nu)$ are Hermitian. Furthermore*

$$\mathscr{A}_s^\gamma(\nu)\, R^\gamma(\nu) = R^\gamma(\nu)\, \mathscr{A}_s^\gamma(\nu) = P^\gamma(\bar{\nu})^* P^\gamma(\nu).$$

Hence $R^\gamma(\nu)$ is positive definite if and only if $\mathscr{A}_s^\gamma(\nu)$ is positive definite. Defining $\mathscr{A}_s(\nu) = C_s(\nu)^{-1} A_s(\nu)$, $\mathscr{A}_s(\nu)$ defines a positive definite π_ν invariant inner product on X if and only if $R^\gamma(\nu)$ is positive definite.

PROOF. Suppose that $\det P^\gamma(\nu) \neq 0$. Then

$$R^\gamma(\nu)\, \mathscr{A}_\gamma^s(\nu) = P^\gamma(\bar{\nu})^* P^\gamma(\nu)\, P_\gamma(\nu)^{-1}\, P^\gamma(s\nu) = P^\gamma(\bar{\nu})^* P^\gamma(s\nu).$$

We therefore see that, as meromorphic functions in ν,

(1) $$R^\gamma(\nu)\, \mathscr{A}_\gamma^s(\nu) = P^\gamma(\bar{\nu})^* P^\gamma(s\nu).$$

The right-hand side of (1) depends only on ν and $s\nu$ not on ν and s. We therefore see that since $s\bar{\nu} = \nu$, $s^{-1}\nu = \bar{\nu}$ and

(2) $$R^\gamma(\nu)\, \mathscr{A}_\gamma^{s^{-1}}(\nu) = P^\gamma(\bar{\nu})^* P^\gamma(\bar{\nu}).$$

Now $\mathscr{A}_s^\gamma(\nu)^* = \mathscr{A}_{s^{-1}}^\gamma(s\bar{\nu}) = \mathscr{A}_{s^{-1}}^\gamma(\nu)$. Thus $P^\gamma(\nu)^* P^\gamma(\bar{\nu}) = \mathscr{A}_{s^{-1}}^\gamma(\nu)^* R^\gamma(\nu)^*$. Now $R^\gamma(\nu)^*$

$=P^\gamma(\nu)^* P^\gamma(\bar{\nu})=R^\gamma(\bar{\nu})=R^\gamma(s\nu)=R^\gamma(\nu)$. Hence $P^\gamma(\bar{\nu})^* P^\gamma(\bar{\nu})=\mathscr{A}_s^\gamma(\nu)\, R^\gamma(\nu)$. This proves the result.

We note that the operators $\mathscr{A}_s(\nu)=C_s(\nu)^{-1}A_s(\nu)$ are related to the Kunze-Stein intertwining operators. They have by Theorem 3.2 and results of Kostant [7] the property that (π_ν, X) is unitarizable if and only if $\mathscr{A}_s(\nu)$ is positive definite. $\mathscr{A}_s(\nu)$ is unitary for $\nu\in\mathfrak{a}^*$ and finally $\mathscr{A}_{st}(\nu)=\mathscr{A}_s(t\nu)\,\mathscr{A}_t(\nu)$.

4. Further comments. In Johnson-Wallach [4], the composition series of (π_ν, X) is given in the case $\dim\mathfrak{a}=1$. In the course of our computation we compute explicitly the $\mathscr{A}_s(\nu)$. We determine precisely what parts of the composition series of (π_ν, X) are unitarizable. For example, the normalization of Schiffmann [10], Helgason [3] of the $A_s(\nu)$ gives semidefinite operators on X for all ν such that (π_ν, X) has a nontrivial finite-dimensional subrepresentation.

References

1. Harish-Chandra, *Spherical functions on a semi-simple Lie group.* I, Amer. J. Math. **80** (1958), 241–310. MR **20** #925.

2. S. Helgason, *Differential geometry and symmetric spaces*, Pure and Appl. Math., vol. 12, Academic Press, New York, 1962. MR **26** #2986.

3. ———, *A duality for symmetric spaces with applications to group representations*, Advances in Math. **5** (1970), 1–154. MR **41** #8587.

4. K. Johnson and N. Wallach, *Composition series and intertwining operators for the spherical principal series*, Bull. Amer. Math. Soc. **78** (1972), 1053–1059.

5. A. Knapp and E. Stein, *Intertwining operators for semi-simple groups*, Ann. Math. **93** (1971), 489–578.

6. B. Kostant, *On the existence and irreducibility of certain series of representations*, Bull. Amer. Math. Soc. **75** (1969), 627–642. MR **39** #7031.

7. ———, (to appear).

8. B. Kostant and S. Rallis, (to appear).

9. R. Kunze and E. Stein, *Uniformally bounded representations.* III. *Intertwining operators for the principal series on semisimple groups*, Amer. J. Math. **89** (1967), 385–442. MR **38** #269.

10. G. Schiffmann, *Integrâles d'entralecement et fonctions de Whittaker*, Bull. Soc. Math. France **99** (1971), 3–72.

Rutgers University

ON AN EXCEPTIONAL SERIES OF REPRESENTATIONS

KENNETH JOHNSON

1. Introduction and notation. If we consider the analytic continuation of the spherical principal series for a semisimple Lie group, we naturally arrive at two questions.

(1) Which of these representations are irreducible?

(2) What is the nature of the irreducible subquotients of a representation when it is reducible?

The first question was answered by Kostant [4]. In this article we give a partial answer to the second question by showing that for a certain class of groups, discrete series representations do not appear as subquotients.

Let G be a connected semisimple Lie group with finite center. Let $G=KAN$ be an Iwasawa decomposition of G where K is a maximal compact subgroup, A is a vector subgroup and N is a unipotent subgroup. Let M be the centralizer of A in K. We shall denote the Lie algebras of G, K, A, N and M by $\mathfrak{G}$, $\mathfrak{K}$, $\mathfrak{A}$, $\mathfrak{N}$ and $\mathfrak{M}$ respectively.

For $\alpha \in \mathfrak{A}^*$ set $\mathfrak{N}_\alpha = \{x \in \mathfrak{A} : [HX] = \alpha(H) X \text{ for all } H \in \mathfrak{A}\}$. Set $P = \{\alpha \in \mathfrak{A}^* : \mathfrak{N}_\alpha \neq (0)\}$ and put $m_\alpha = \dim \mathfrak{N}_\alpha$. P is called a positive restricted root system and P determines a Weyl chamber $\mathfrak{A}^+$ in $\mathfrak{A}$ where $\mathfrak{A}^+ = \{H \in \mathfrak{A} : \alpha(H) > 0 \text{ for all } \alpha \in P\}$. If we put $A^+ = \exp \mathfrak{A}^+$ we obtain the polar decomposition of G where $G = K\bar{A}^+K$.

Recall the definition of the spherical principal series. If $\lambda \in \mathfrak{A}_C^*$ we define a character on MAN by sending $x = man \to x^\lambda = \exp\{\lambda(\log a)\}$.

Let H^λ be the set of all measurable functions $f: G \to C$ such that

$$f(gx) = x^{-\lambda} f(g) \quad \text{for } x \in MAN \tag{1}$$

AMS (MOS) subject classifications (1970). Primary 22E45, 31B05; Secondary 22E43, 33A30.

and

$$\int_{K/M} |f(b)|^2 \, db < \infty . \tag{2}$$

Let π_λ denote the action of G on H^λ induced by left translation. If X^λ denotes the K-finite elements of H^λ it is known that π_λ induces an action of the enveloping algebra of $\mathfrak{G}$ on X^λ.

2. Remarks on square-integrability. Suppose $\pi: G \to \mathscr{L}(\mathfrak{H}, \mathfrak{H})$ is an irreducible unitary representation of G on a Hilbert space $\mathfrak{H}$. We wish to determine when π is square-integrable.

Let η be an irreducible representation of K which occurs in π, and let $\varphi_1, \ldots, \varphi_n$ be an orthonormal set of vectors of $\mathfrak{H}$ which spans a vector space V such that $\pi|_K$ transforms V according to η.

Now if $\delta(a) = \prod_{\alpha \in P} (\exp\{\alpha(\log a)\} - \exp\{-\alpha(\log a)\})^{m_\alpha}$ we have that

$$\int_G |\langle \pi(g)\, \varphi_1, \varphi_1 \rangle|^2 \, dg = \int_K \int_{A^+} \int_K \delta(a)\, |\langle \pi(k_1)\, \pi(a)\, \pi(k_2)\, \varphi_1, \varphi_1 \rangle|^2 \, dk_1 \, da \, dk_2$$

$$= n^{-2} \int_{A^+} \delta(a) \sum_{j,k=1}^{n} |\langle \pi(a)\ \varphi_j, \varphi_k \rangle|^2 \, da .$$

We now see that π is square-integrable if and only if

$$\int_{A^+} \exp\{2\varrho(\log a)\}\, |\langle \pi(a)\, \varphi_j, \varphi_k \rangle|^2 \, da < \infty \quad \text{for all } j \text{ and } k$$

where $\varrho = \frac{1}{2} \sum_{\alpha \in P} m_\alpha \alpha$.

3. The classical split rank one groups. In this section $\boldsymbol{F}$ will stand for either $\boldsymbol{R}$, $\boldsymbol{C}$ or $\boldsymbol{H}$ (the quaternions). Consider $\boldsymbol{F}^{n+1}$ as a right vector space over $\boldsymbol{F}$. On $\boldsymbol{F}$ we have the usual norm $|\ |$ and involution. If $\boldsymbol{x} = (x^1, \ldots, x^{n+1}) \in \boldsymbol{F}^{n+1}$ we set $\|\boldsymbol{x}\|^2 = |x^1|^2 + \cdots + |x^n|^2 - |x^{n+1}|^2$.

Let G be the connected component of the group of linear transformations of determinant one which commute with the right action of $\boldsymbol{F}$ and preserve the norm on $\boldsymbol{F}^{n+1}$.

If $\boldsymbol{F} = \boldsymbol{R}$, $\quad G = SO^0(n, 1)$.

If $\boldsymbol{F} = \boldsymbol{C}$, $\quad G = SU(n, 1)$.

If $\boldsymbol{F} = \boldsymbol{H}$, $\quad G = Sp(n, 1)$.

For K we may take all matrices in G of the form

$$\begin{pmatrix} & & 0 \\ & B & \vdots \\ & & 0 \\ 0 \ldots 0 & & b \end{pmatrix}.$$

We may set

$$\mathfrak{A} = \boldsymbol{R}\begin{pmatrix} 0 & 0 \ldots 0 & 1 \\ 0 & & 0 \\ \vdots & \mathbf{0} & \vdots \\ 0 & & 0 \\ 1 & 0 \ldots 0 & 0 \end{pmatrix}.$$

We have $\mathfrak{N}$ equal to all matrices of the form

$$\begin{pmatrix} 0 & x^1 \ldots x^{n-1} & 0 \\ -\bar{x}^1 & & \bar{x}^1 \\ \vdots & & \vdots \\ -\bar{x}^{n-1} & & \bar{x}^{n-1} \\ 0 & x^1 \ldots x^{n-1} & 0 \end{pmatrix}$$

with $x^i \in \boldsymbol{F}$ plus all matrices of the form

$$\begin{pmatrix} u & 0 \ldots 0 & -u \\ 0 & & 0 \\ \vdots & \mathbf{0} & \vdots \\ 0 & & 0 \\ u & 0 \ldots 0 & -u \end{pmatrix}$$

with $u \in \boldsymbol{F}$ and $\operatorname{Re} u = 0$.

If $\boldsymbol{x} = (x^1, \ldots, x^n)$ and $\boldsymbol{y} = (y^1, \ldots, y^n)$ are in $\boldsymbol{F}^n$ we set $\langle \boldsymbol{x}, \boldsymbol{y} \rangle = \bar{x}^1 y^1 + \cdots + \bar{x}^n y^n$. It is easy to see that G acts on the unit ball in $\boldsymbol{F}^n$ (i.e., $g(x^1, \ldots, x^n) = (\zeta^1 (\zeta^{n+1})^{-1}, \ldots, \zeta^n (\zeta^{n+1})^{-1})$ where $g(x^1, \ldots, x^n, 1) = (\zeta^1, \ldots, \zeta^{n+1})$). Moreover, G acts transitively on the interior of the unit ball and on the unit sphere.

4. Principal series for the classical split rank one groups. Since $\dim \mathfrak{A} = 1$, we may identify $\mathfrak{A}_C^*$ with $\boldsymbol{C}$ and parameterize our representations by $\lambda \in \boldsymbol{C}$. After an appropriate normalization we have that

$$(\pi_\lambda(g)\,f)\,(b)=\left(\frac{1-\|g\cdot 0\|^2}{|1-\langle g\cdot 0,\,b\rangle|^2}\right)^{-\lambda}\,f\,(g^{-1}b)$$

where b is on the unit sphere in $\boldsymbol{F}^n$.

We can now state our result.

THEOREM. *When $G=SO^0(n,1)$, $SU(n,1)$ or $Sp(n,1)$ and $n\geqq 3$ and π_λ is reducible, no discrete series representation can occur as a subquotient of H^λ.*

REMARKS. We give the proof when $G=SU(n,1)$ remarking that the other two clases are similar and the case of $G=SO^0(n,1)$ has been done by Takahashi [**5**].

When $G=SU(n,1)$, Kostant's result [**4**] tells us that π_λ is reducible only when $\lambda=-2k$ or $\lambda=2n+2k$ where k is a nonnegative integer. Moreover, since π_{-2k} and π_{2n+2k} are dual representations, we shall only consider the subquotient of π_{-2k}.

As K-vectors the K-finite vectors of H^λ are the same as the K-finite vectors of $L^2(S^{2n-1})$. Thus they are the spherical harmonics.

Let $H^{p,q}$ denote the harmonic polynomials of bidegree (p,q). If $f\in H^{p,q}$, $\Delta f=0$, f is a homogeneous polynomial of total degree $p+q$, and if $z\in\boldsymbol{C}$, $f(zz_1,\ldots,zz_k)=z^p\bar z^q f(z_1,\ldots,z_n)$.

It is known that $H^{p,q}$ is irreducible under the action of K.

Under the action π_{-2k} we have that

(1) $F_k=\sum_{0\leqq p,q\leqq k}H^{p,q}$, $H_k=\sum_{p\geqq 0}\sum_{q\leqq k}H^{p,q}$, $\bar H_k=\sum_{q\geqq 0}\sum_{p\leqq k}H^{p,q}$, $H_k+\bar H_k$, and X^{-2k} are invariant;

(2) F_k, H_k/F_k, $\bar H_k/F_k$ and $X^{-2k}/(H_k+\bar H_k)$ are irreducible; and,

(3) $X^{-2k}/(H_k+\bar H_k)$ is always unitarizable and F_0, H_0/F_0 and $\bar H_0/F_0$ are unitarizable. No others are unitarizable.

PROOF FOR $G=SU(n,1)$. (a) We first show H_0/F_0 and $\bar H_0/F_0$ are not square-integrable. (This proof works for $n=2$ also.)

Let $F(z_1,\ldots,z_n)=z_2$ and let

$$a_t=\exp tA_0\quad\text{when } A_0=\begin{pmatrix}0 & 0\ldots0 & 1\\ 0 & & 0\\ \vdots & \mathbf{0} & \vdots\\ 0 & & 0\\ 1 & 0\ldots0 & 0\end{pmatrix}.$$

Then

$$(\pi_0(a_t)\,F)\,(z_1,\ldots,z_n)=z_2/(\cosh t-(\sinh t)\,z_1)$$

and

$$\langle \pi_0(a_t)\, F, F\rangle = \int\limits_{S^{2n-1}} \frac{|z_2|^2}{\cosh t - (\sinh t)\, z_1}\, d\mu(S^{2n-1}) = \frac{1}{\cosh t} \int\limits_{S^{2n-1}} |z_2|^2\, d\mu(S^{2n-1})$$

by Fourier series. Thus $\langle \pi_0(a_t)\, F, F\rangle = A/\cosh t$. As $\varrho(\log a_t) = nt$ we have that

$$\exp\{2nt\}\ A^2/\cosh^2 t \sim \exp\{(2n-2)\, t\}\ 4A^2$$

for large t. Thus our representation H_0/F_0 is not square-integrable. The case of $\bar{H}_0/F_0$ is now obvious.

(b) Let $G \in H^{k+1,k+1}$ which is independent of z_1 and $\bar{z}_1$. This may be done since $n \geqq 2$. Then

$$(\pi_{-2k}(a_t)\, G)\,(z_1, \ldots, z_n) = \frac{1}{|\cosh t - (\sinh t)\, z_1|^2}\, G(z_1, \ldots, z_n).$$

Then

$$\begin{aligned}
\langle \pi_{-2k}(a_t)\, G, G\rangle &= \int\limits_{S^{2n-1}} \frac{1}{|\cosh t - (\sinh t)\, z_1|^2}\, |G(z_1, \ldots, z_n)|^2\, d\mu(S^{2n-1}) \\
&= \frac{1}{\cosh^2 t} \int\limits_{S^{2n-1}} \frac{1}{|1 - (\tanh t)\, z_1|^2}\, |G(z_2, \ldots, z_n)|^2\, d\mu(S^{2n-1}) \\
&= \frac{1}{\cosh^2 t} \int\limits_{S^{2n-1}} \frac{1}{1 - \tanh^2 t\, |z_1|^2}\, |G(z_2, \ldots, z_n)|^2\, d\mu(S^{2n-1}) \\
&\geqq \frac{1}{\cosh^2 t} \int\limits_{S^{2n-1}} |G(z_2, \ldots, z_n)|^2\, d\mu(S^{2n-1}) = \frac{A}{\cosh^2 t}.
\end{aligned}$$

Now

$$e^{2nt} A^2/\cosh^4 t \sim 16 A^2 e^{(2n-4)\, t}$$

for large t and we see that $X^{2k}/(H_k + \bar{H}_k)$ is not in the discrete series. Q.E.D.

When $G = SU(2, 1)$ or $Sp(2, 1)$ we do not know in general if our representations are square-integrable. However, we have shown that when $G = SU(2, 1)$ some of our representations are square-integrable.

REMARK. The results described in this article are part of a joint work with Nolan Wallach [**3**].

References

1. Harish-Chandra, *Representations of semisimple Lie groups.* II, Trans. Amer. Math. Soc. **76** (1954), 26–65. MR **15**, 398.

2. S. Helgason, *A duality for symmetric spaces with applications to group representations*, Advances in Math. **5** (1970), 1–154. MR **41** #8587.

3. K. Johnson and Nolan Wallach, *Composition series and intertwining operators for the spherical principal series*. I (to appear).

4. B. Kostant, *On the existence and irreducibility of certain series of representations*, Bull. Amer. Math. Soc. **75** (1969), 627–642. MR **39** #7031.

5. R. Takahashi, *Sur les représentations unitaires des groupes de Lorentz généralisés*, Bull. Soc. Math. France **91** (1969), 289–433.

Rutgers University

EXPLICIT FORM OF THE CHARACTERS OF DISCRETE SERIES REPRESENTATIONS OF SEMISIMPLE LIE GROUPS

TAKESHI HIRAI

1. Let G be a connected noncompact real semisimple Lie group with Lie algebra $\mathfrak{g}$. We denote by π an invariant eigendistribution (IED) on G. It is known by Harish-Chandra [**1(b)**] that π is actually a locally summable function on G which is analytic on the open subset G' of G of all regular elements of G. Of course, this function is invariant under (inner automorphisms of) G. In this note, we briefly mention the following three problems:

(1) When does a given G-invariant analytic function on G' define canonically an IED on G? (A necessary and sufficient condition.)

(2) How can we construct all IEDs on G? (Method of construction.)

(3) Determine the explicit formula of the characters of discrete series representations. (Call it "Weyl's formula.")

To talk about problem (3), we must go through (1) and it may be better to mention (2). These problems were also treated in [**2(b)**].

2. Let us state a solution of problem (1). We need several notations and definitions, some of which are not given here but can be found in the cited papers.

Let $C_{ar}(G)$ (or $C_{ar}^0(G)$) be the set of all conjugacy classes of Cartan subgroups (or connected components of Cartan subgroups resp.) of G under inner automorphisms of G. Let $H^1, H^2, \ldots, H^q$ be Cartan subgroups of G which form a complete system of representatives of conjugacy classes. The conjugacy classes containing H^j are denoted by $[H^j]$. Then taking appropriate connected components H_i^j of H^j, H_i^j's form a complete system of representatives of $C_{ar}^0(G)$.

Let $\mathfrak{h}$ be a Cartan subalgebra of $\mathfrak{g}$ and $\mathfrak{h}_c$ its complexification. Let $P^{\mathfrak{h}}$ be the set of all positive roots of $(\mathfrak{g}_c, \mathfrak{h}_c)$ with respect to a lexicographic order. A root α

AMS (MOS) subject classifications (1970). Primary 22E45; Secondary 22E30, 43A65.

of $(\mathfrak{g}_c, \mathfrak{h}_c)$, or simply of $\mathfrak{h}$, is called real (or imaginary) if it takes only real (or pure imaginary) values on $\mathfrak{h}$. Let $P_R^{\mathfrak{h}}$ be the set of all positive real roots of $\mathfrak{h}$. For any root α, choose root vectors X_α, $X_{-\alpha}$ from $\mathfrak{g}_c$ in such a way that $B(X_\alpha, X_{-\alpha})=1$, where B denotes the Killing form of $\mathfrak{g}_c$. Put

$$H_\alpha=[X_\alpha, X_{-\alpha}], \qquad H'_\alpha=2|\alpha|^{-2} H_\alpha, \qquad X'_{\pm\alpha}=2^{1/2}|\alpha|^{-1} X_{\pm\alpha}.$$

Let $H^{\mathfrak{h}}=H$ be the Cartan subgroup of G corresponding to $\mathfrak{h}$. Define for any root α, a homomorphism η_α of H into $\boldsymbol{C}^*$ as

$$\mathrm{Ad}(h)\, X_\alpha=\eta_\alpha(h)\, X_\alpha \qquad (h\in H)$$

and put

$$\Delta'^{\mathfrak{h}}(h)=\prod_{\alpha\in P^{\mathfrak{h}}} (1-\eta_\alpha(h)^{-1}), \qquad \Delta_R'^{\mathfrak{h}}(h)=\prod_{\alpha\in P_R^{\mathfrak{h}}} (1-\eta_\alpha(h)^{-1}),$$

$$\varepsilon_R^{\mathfrak{h}}(h)=\mathrm{sgn}(\Delta_R'^{\mathfrak{h}}(h)) \qquad (h\in H'=H\cap G').$$

Here we assume that G is acceptable (for the definition, see [**1(b)**, p. 484]). Roughly speaking, this means that the homomorphism η_ϱ of $H=H^{\mathfrak{h}}$ into $\boldsymbol{C}^*$ can be canonically defined for any $\mathfrak{h}$, where ϱ is the half-sum of all $\alpha\in P^{\mathfrak{h}}$. This assumption does not hurt the generality. Put

$$\Delta^{\mathfrak{h}}(h)=\eta_\varrho(h)\, \Delta'^{\mathfrak{h}}(h) \qquad (h\in H).$$

Now let π be a given G-invariant analytic function on G'. Put, for $h\in H'$,

$$\tilde{\kappa}^{\mathfrak{h}}(h)=\Delta^{\mathfrak{h}}(h)\, \pi(h), \qquad \kappa^{\mathfrak{h}}(h)=\varepsilon_R^{\mathfrak{h}}(h)\, \tilde{\kappa}^{\mathfrak{h}}(h).$$

Let $H'(R)$ be the subset of H defined by $\Delta_R'^{\mathfrak{h}}(h)\neq 0$. Then by Harish-Chandra [**1(b)**], we know the following:

FUNDAMENTAL THEOREM. *For an IED π on G, the function $\kappa^{\mathfrak{h}}$ (or $\tilde{\kappa}^{\mathfrak{h}}$) on $H'=H\cap G'$ can be analytically extended on $H'(R)$.*

In some point of view, $\kappa^{\mathfrak{h}}$ is more natural than $\tilde{\kappa}^{\mathfrak{h}}$. For instance,

PROPOSITION 1. *The function $\kappa^{\mathfrak{h}}$ corresponding to an IED on G can be continuously extended on the whole Cartan subgroup H.*

For $H=H^j$ or $\mathfrak{h}=\mathfrak{h}^j$ ($\mathfrak{h}^j$ is the Lie algebra of H^j), $\kappa^{\mathfrak{h}}$ is denoted simply by κ^j. Taking into account the order of the roots of $\mathfrak{h}^j$, we may consider that π is completely determined by $\kappa^1, \kappa^2, \dots, \kappa^q$ and that for any $\mathfrak{h}$ such that $H^{\mathfrak{h}}\in[H^j]$, the function $\kappa^{\mathfrak{h}}$ is determined by κ^j.

Let $\mathfrak{h}$ and H be as before and let $W(\mathfrak{h}_c)$ be the Weyl group of $(\mathfrak{g}_c, \mathfrak{h}_c)$ and $W_G(H)$, the set of restrictions on H of all inner automorphisms of G leaving H invariant. For $w \in W_G(H)$, we define $\varepsilon^{\mathfrak{h}}(w, h) = \varepsilon(w, h) = \pm 1$, locally constant in h, as

$$\varepsilon_R^{\mathfrak{h}}(wh)\, \Delta^{\mathfrak{h}}(wh) = \varepsilon(w, h)\, \varepsilon_R^{\mathfrak{h}}(h)\, \Delta^{\mathfrak{h}}(h) \qquad (h \in H').$$

Let $I(\mathfrak{h}_c)$ be the subalgebra of the symmetric algebra $S(\mathfrak{h}_c)$ of $\mathfrak{h}_c$ consisting of all $W(\mathfrak{h}_c)$-invariant elements. Then there exists a unique isomorphism $\gamma_{\mathfrak{h}}$ of the algebra $\mathfrak{Z}$ of all Laplace operators on G onto $I(\mathfrak{h}_c)$ such that, for any $Z \in \mathfrak{Z}$ and G-invariant C^∞-function f on G',

$$(Zf)|_{H'} = [(\Delta^{\mathfrak{h}})^{-1} \circ \gamma_{\mathfrak{h}}(Z) \circ \Delta^{\mathfrak{h}}]\, (f|_{H'}).$$

We denote $\gamma_{\mathfrak{h}}$ for $\mathfrak{h} = \mathfrak{h}^j$ by γ_j.

Now let $\mathfrak{a}$ be a Cartan subalgebra of $\mathfrak{g}$ and take a real root $\alpha \in P_R^{\mathfrak{a}}$. Then we can choose X'_α, $X'_{-\alpha}$, H'_α from $\mathfrak{g}$. Let ν be an automorphism of $\mathfrak{g}_c$ given by

$$\nu = \exp\{-(-1)^{1/2}\, (\pi/4)\, \mathrm{ad}(X'_\alpha + X'_{-\alpha})\}.$$

Put $\mathfrak{b}_c = \nu(\mathfrak{a}_c)$, $\mathfrak{b} = \mathfrak{b}_c \cap \mathfrak{g}_c$. Then $\mathfrak{a}$ and $\mathfrak{b}$ are not conjugate under G, and the root $\beta = \nu\alpha$ of $\mathfrak{b}$ is a singular imaginary root (see [**1 (a)**, p. 561]). Put $A = H^{\mathfrak{a}}$ and $B = H^{\mathfrak{b}}$ and denote the above relation by $A \to B$. Let Σ_α (or Σ'_α) be the subset of A defined by $\eta_\alpha(h) = 1$ (or $\eta_\alpha(h) = 1$, $\eta_\gamma(h) \neq 1$ $(\forall \gamma \neq \pm\alpha, \gamma \in P^{\mathfrak{a}})$ resp.). Then $A \cap B = \Sigma_\alpha = \Sigma_\beta$.

If π is an IED on G and the orders of the roots of $\mathfrak{a}$ and $\mathfrak{b}$ are such that $\nu(P^{\mathfrak{a}}) = P^{\mathfrak{b}}$, then, for any $a \in \Sigma'_\alpha = \Sigma'_\beta$ and $0 \leqq m < \infty$,

$$(*) \qquad (H'_\alpha)^{2m+1}\, \tilde{\kappa}^{\mathfrak{a}}(a) = (H'_\beta)^{2m+1}\, \tilde{\kappa}^{\mathfrak{b}}(a).$$

Here $\tilde{\kappa}^{\mathfrak{a}} = \varepsilon_R^{\mathfrak{a}} \kappa^{\mathfrak{a}}$ and the left-hand side must be considered as a limit value at a.

After these preparations, we can state the solution of problem (1) as follows:

THEOREM 1. *Let π be a G-invariant analytic function on G' and define κ^j (and $\tilde{\kappa}^j$) on $H'^j = H^j \cap G'$ as above for $1 \leqq j \leqq q$. Then a necessary and sufficient condition that π defines canonically an IED on G is given as follows:*

(1) *Every κ^j can be analytically extended on $H'^j(R)$ and satisfies*

$$\kappa^j(wh) = \varepsilon(w, h)\, \kappa^j(h) \qquad (h \in H'^j,\ w \in W_G(H^j)).$$

(This says that κ^j is ε-symmetric.)

(2) *There exists a homomorphism λ of $\mathfrak{Z}$ into $\mathbf{C}$ such that, for every j,*

$$\gamma_j(Z)\, \kappa^j = \lambda(Z)\, \kappa^j \quad on \quad H'^j\ (\forall Z \in \mathfrak{Z}).$$

(3) *For any* $\mathfrak{a}=\mathfrak{h}^j$ (*or* $A=H^j$) *and* $\alpha\in P^{\mathfrak{a}}_{\boldsymbol{R}}$, *the equality* $(*)$ *holds for all* $a\in\Sigma'_\alpha$ *and* $0\leqq m<\infty$. (*It is sufficient to make m run over* $0\leqq m\leqq 4^{-1}$ (dim $G-$rank $G-2$).)

The idea of the proof of this theorem is essentially the same as for the case $G=SU(p, q)$ in [**2(a)**]. The condition (3) expresses the behavior of π near any semiregular element of noncompact type of G (for the definition, see [**1(a)**, p. 554]).

To understand the significance of this theorem, apply it to $G=SL(2, \boldsymbol{R})$. Then we can easily obtain all IEDs on G. And we see that all IEDs are linear combinations of irreducible characters. But this is not the case for $G=SL(n, \boldsymbol{R})$ if $n\geqq 3$ [**2(c)**].

3. From Theorem 1, we obtain some interesting consequences.

If two Cartan subgroups $A=H^{\mathfrak{a}}$, $B=H^{\mathfrak{b}}$ relate to each other as $A\rightarrow B$ (see above), we define $[A]<[B]$. Extending this relation transitively in $C_{ar}(G)$, we get an order $<$ in it. Using Sugiura's result in [**4**], we can write down the diagram of this order for every simple Lie group and see that there always exist unique maximal and minimal elements in $C_{ar}(G)$.

We define also an order $\prec$ in $C^0_{ar}(G)$. Let A and B be as above. Take a connected component A_i of A and a B_j of B. If there exists some $g\in G$ such that $A_i\cap gB_jg^{-1}$ contains a semiregular element (of noncompact type), then we denote this relation as $A_i\rightarrow gB_jg^{-1}$ and define $[A_i]\prec[B_j]$. Extending this relation transitively, we get an order $\prec$ in $C^0_{ar}(G)$. There always exists a unique maximal element in it.

For $G=SU(p, q)$, all Cartan subgroups are connected and the order $<$ is linear. For $G=SL(n, \boldsymbol{R})$, the order $<$ in $C_{ar}(G)$ is linear, but the one $\prec$ in $C^0_{ar}(G)$ is no longer linear. For $G=Sp(n, \boldsymbol{R})$, the order $<$ is already not linear if $n=$rank G $\geqq 2$.

Now, for an IED π on G, let C_π be the subset of $C_{ar}(G)$ consisting of all $[H^j]$ such that $\pi\neq 0$ on H'^j. If $[H^{j_1}],\ldots,[H^{j_\gamma}]$ is the set of all maximal elements in C_π, we say that π is of height $([H^{j_1}],\ldots,[H^{j_\gamma}])$. If C_π has only one maximal element, π is called *extremal*. As a consequence of Theorem 1, we obtain the following:

THEOREM 2. *Suppose that an IED* π *on G is of height* $([H^{j_1}],\ldots,[H^{j_\gamma}])$. *Then the functions* $\kappa^{j_1},\ldots,\kappa^{j_\gamma}$ *can be extended to analytic functions on the whole Cartan subgroups* $H^{j_1},\ldots,H^{j_\gamma}$ *respectively*.

In this case, the functions $\kappa^{j_1},\ldots,\kappa^{j_\gamma}$ are called the highest parts of π. A similar result holds also for the order $\prec$ in $C^0_{ar}(G)$.

4. Considering a converse of Theorem 2, we come to problem (2). The answer is given as follows:

THEOREM 3. *Fix any Cartan subgroup H^l and an order of roots of $\mathfrak{h}^l$. For any ε-symmetric analytic function ξ on the whole H^l which satisfies for some homomorphism λ' of $I(\mathfrak{h}_c^l)$ into $\boldsymbol{C}$,*

$$(**) \qquad Y\xi = \lambda'(Y)\,\xi \qquad (\forall Y \in I(\mathfrak{h}_c^l)),$$

we can construct an extremal IED $\pi = \boldsymbol{T}\xi$ whose highest part is exactly ξ, by a standard method $\boldsymbol{T}$ which constructs π by an induction on the order $\prec$ in $C_{ar}^0(G)$.

The method $\boldsymbol{T}$ is given explicitly in [**2(b)**], and the proof of this theorem depends on the structure theory of Weyl groups due to Iwahori [**3**].

Immediately from Theorem 3, we obtain

COROLLARY. *Every IED on G is a finite sum of extremal ones.*

Thus the above theorem reduces, in a sense, the study of IEDs on G to that on ε-symmetric analytic functions on every Cartan subgroup H^l satisfying the differential equations $(**)$.

PROBLEM. *Are all irreducible characters extremal?*

5. Now we come to problem (3). Naturally we must assume that G has a compact Cartan subgroup, say H^1. Then $[H^1]$ is the unique maximal element in $C_{ar}(G)$ (also in $C_{ar}^0(G)$). On H^1, the explicit form of the characters of discrete series representations (call it "Weyl's formula") is given in [**1(c)**]. Hence we know the highest part ξ of any such character π. Since π is tempered and its infinitesimal character is regular, π is uniquely determined by its highest part ξ.

Unfortunately the method $\boldsymbol{T}$ is not suitable for calculating "Weyl's formula," because the IED $\boldsymbol{T}\xi$ is not tempered. But, modifying $\boldsymbol{T}$ at every step of induction on the order $\prec$ in $C_{ar}^0(G)$, we can obtain "Weyl's formula" for lower rank groups.

For higher rank groups, a heuristic method is useful. For example, for $G = Sp(n, \boldsymbol{R})$, first, the formula is obtained for $n = 2$, 3 and 4 by modifying the method $\boldsymbol{T}$. Next, taking into account the formulas for $n = 2$, 3 and 4 and Theorem 1 (3), we presume "Weyl's formula" for general rank n. Then, we can test by Theorem 1 (3) whether the presumed formula is true or false. Repeating this trial, at last we obtained "Weyl's formula" for $Sp(n, \boldsymbol{R})$. This method is also available for other groups. For $G = SU(p, q)$, "Weyl's formula" has already been given in [**2(a)**].

Here we will give "Weyl's formula" for $G = Sp(2, \boldsymbol{R})$ which already indicates the difference between the holomorphic discrete series and the nonholomorphic one.

The group $G=Sp(2, \boldsymbol{R})$ consists of all 4×4 real matrices g satisfying ${}^t gJg=J$, where

$$J=\begin{bmatrix} 0 & E \\ -E & 0 \end{bmatrix}, \qquad E=\begin{bmatrix} 1 & 0 \\ 0 & 1 \end{bmatrix}.$$

$C_{ar}(G)$ has four elements and a complete system of representatives is given by the following Cartan subgroups $H^{2,0}, H^{1,0}, H^{0,1}, H^{0,0}$:

$$H^{2,0}: \begin{bmatrix} \cos\varphi_1 & 0 & -\sin\varphi_1 & 0 \\ 0 & \cos\varphi_2 & 0 & -\sin\varphi_2 \\ \sin\varphi_1 & 0 & \cos\varphi_1 & 0 \\ 0 & \sin\varphi_2 & 0 & \cos\varphi_2 \end{bmatrix};$$

$$H^{1,0}: \begin{bmatrix} \cos\varphi_1 & 0 & -\sin\varphi_1 & 0 \\ 0 & \varepsilon_2 e^{t_2} & 0 & 0 \\ \sin\varphi_1 & 0 & \cos\varphi_1 & 0 \\ 0 & 0 & 0 & \varepsilon_2 e^{-t_2} \end{bmatrix};$$

$$H^{0,1}: \begin{bmatrix} e^{\tau}u(\theta) & 0 \\ 0 & e^{-\tau}u(\theta) \end{bmatrix}, \qquad u(\theta)=\begin{bmatrix} \cos\theta & -\sin\theta \\ \sin\theta & \cos\theta \end{bmatrix};$$

$$H^{0,0}: \quad d(\varepsilon_1 e^{t_1}, \varepsilon_2 e^{t_2}, \varepsilon_1 e^{-t_1}, \varepsilon_2 e^{-t_2}),$$

where $\varepsilon_i=\pm 1$ and $d(a_1, a_2, a_3, a_4)$ denotes the diagonal matrix with diagonal elements a_1, a_2, a_3, a_4. $H^{1,0}$ has two connected components $H_1^{1,0}$ and $H_0^{1,0}$ corresponding to $\varepsilon_2=+1$ and -1. $H^{0,0}$ has four connected components and let $H_2^{0,0}, H_1^{0,0}$ or $H_0^{0,0}$ be the one corresponding to $\varepsilon_1=\varepsilon_2=+1$, $\varepsilon_1=-\varepsilon_2=+1$ or $\varepsilon_1=\varepsilon_2=-1$ respectively. Then the diagrams of the order $<$ in $C_{ar}(G)$ and the one $\prec$ in $C_{ar}^0(G)$ are as follows (brackets are omitted):

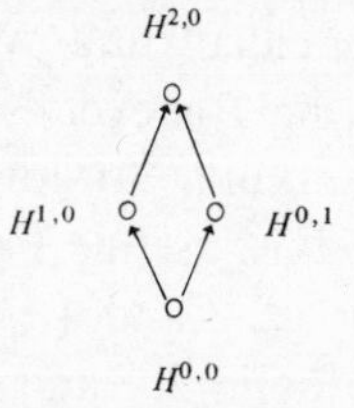

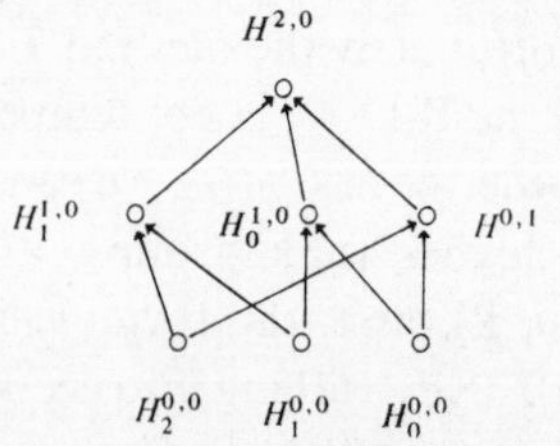

Denote by h the element of $H^{j,k}$ indicated above. For these h, let (δ_1, δ_2) be respectively $(e^{i\varphi_1}, e^{i\varphi_2})$ for $H^{2,0}$; $(e^{i\varphi_1}, \varepsilon_2 e^{t_2})$ for $H^{1,0}$; $(e^z, e^{\bar{z}})$ for $H^{0,1}$, where $z=\tau+i\theta$, $\bar{z}=\tau-i\theta$; $(\varepsilon_1 e^{t_1}, \varepsilon_2 e^{t_2})$ for $H^{0,0}$. The orders of roots are such that the set of $\eta_\alpha(h)$ for $\alpha>0$ is equal to $\{\delta_1\delta_2^{-1}, \delta_1\delta_2, \delta_1^2, \delta_2^2\}$. Let F_+ be the connected component of $H_m^{j,k}\cap H'^{j,k}(R)$ defined by "$\eta_\alpha(h)>1$ for any positive real root α such that $\eta_\alpha(H_m^{j,k})>0$," where $H_m^{2,0}=H^{2,0}$ and $H_m^{0,1}=H^{0,1}$. Then the functions $\kappa_m^{j,k}(h)$ for $h\in F_+\subset H_m^{j,k}$ are given as follows.

Two positive integers $l_1 > l_2 > 0$ determine an infinitesimal character of discrete series representations. Let $\nu_1, \nu_2 = \pm 1$. Then, for any fixed pair $(\nu_1 l_1, \nu_2 l_2)$,

$$\kappa^{2,0}(h) = -\begin{vmatrix} \nu_1 \delta_1^{\nu_1 l_1} & \nu_2 \delta_2^{\nu_2 l_2} \\ \nu_1 \delta_2^{\nu_1 l_1} & \nu_2 \delta_2^{\nu_2 l_2} \end{vmatrix};$$

$$\kappa_m^{1,0}(h) = -\begin{vmatrix} \nu_1 \delta_1^{\nu_1 l_1} & \nu_2 \delta_1^{\nu_2 l_2} \\ -\delta_2^{-l_1} & -\delta_2^{-l_2} \end{vmatrix} \qquad (m=1, 0);$$

$$\kappa^{0,1}(h) = -\begin{vmatrix} -\delta_1^{-l_1} & -\nu_1 \nu_2 \delta_1^{-\nu_1 \nu_2 l_2} \\ -\delta_2^{-l_1} & -\nu_1 \nu_2 \delta_2^{-\nu_1 \nu_2 l_2} \end{vmatrix};$$

$$\kappa_m^{0,0}(h) = -\begin{vmatrix} -\delta_1^{-l_1} & -\delta_1^{-l_2} \\ -\delta_2^{-l_1} & -\nu_1 \nu_2 (\delta_2^{-l_2} + \delta_2^{l_2}) + \delta_2^{l_2} \end{vmatrix} \qquad (m=2, 0);$$

$$\kappa_1^{0,0}(h) = -\begin{vmatrix} -\delta_1^{-l_1} & -\delta_1^{-l_2} \\ -\delta_2^{-l_1} & -\delta_2^{-l_2} \end{vmatrix}.$$

The characters for which $\nu_1 \nu_2 = 1$ correspond to the holomorphic discrete series representations and their contragradient ones. (The last ones are holomorphic with respect to another complex structure.)

References

1. Harish-Chandra,

a) *Some results on an invariant integral on a semisimple Lie algebra*, Ann. of Math. (2) **80** (1964), 551–593. MR **31** #4862b.

b) *Invariant eigendistributions on a semisimple Lie group*, Trans. Amer. Math. Soc. **119** (1965), 457–508. MR **31** #4862d.

c) *Discrete series for semisimple Lie groups.* II. *Explicit determination of the characters*, Acta Math. **116** (1966), 1–111. MR **36** #2745.

2. T. Hirai,

a) *Invariant eigendistributions of Laplace operators on real simple Lie groups.* I. *Case of SU(p, q)*, Japan. J. Math. **40** (1970), 1–68.

b) *Characters of real semi-simple Lie groups and invariant eigendistributions on them*, Sûgaku **23** (1971), 241–260. (Japanese)

c) *Some remarks on invariant eigendistributions on semi-simple Lie groups*, J. of Math. Kyoto Univ. **12** (1972), 393–411.

3. N. Iwahori, *On the structure of a Hecke ring of a Chevalley group over a finite field*, J. Fac. Sci. Univ. Tokyo Sec. I **10** (1964), 215–236. MR **29** #2307.

4. M. Sugiura, *Conjugate classes of Cartan subalgebras in real semisimple Lie algebras*, J. Math. Soc. Japan **11** (1959), 374–434. MR **26** #3827.

KYOTO UNIVERSITY
KYOTO, JAPAN

THE SUBQUOTIENT THEOREM AND SOME APPLICATIONS

J. LEPOWSKY *

Let G be a noncompact connected real semisimple Lie group with finite center. Our purpose is to present some algebraic results on the representations of G and its Lie algebra, and to indicate simplifications of some known results in case G is linear. The main results, together with details and related results, are contained in **[6]** and **[8]**.

Let K be a maximal compact subgroup of G, $\mathfrak{k} \subset \mathfrak{g}$ the corresponding complexified Lie algebras, and $\mathscr{K} \subset \mathscr{G}$ the universal enveloping algebras of $\mathfrak{k}$ and $\mathfrak{g}$, respectively. If V is a $\mathscr{G}$-module and a K-module, we call V a *$(\mathscr{G}, K)$-module* if

$$k \cdot (x \cdot v) = (k \cdot x) \cdot (k \cdot v)$$

for all $k \in K$, $x \in \mathscr{G}$ and $v \in V$. (Module actions are denoted with a dot, and K acts on $\mathscr{G}$ via the natural extension of the adjoint action of K on $\mathfrak{g}$.) We call V a *compatible* $(\mathscr{G}, K)$-module if in addition V is K-finite, and the action of $\mathscr{K}$ on any finite-dimensional K-invariant subspace of V is the action induced by the differential of the action of K. It is well known (see **[3(a)]**) that compatible $(\mathscr{G}, K)$-modules arise in a natural way from Banach space representations of G, and that the study of the irreducible representations of G reduces to a large extent to the study of $\mathscr{G}$-irreducible compatible $(\mathscr{G}, K)$-modules.

Let $G = KAN$ be an Iwasawa decomposition of G adapted to K, and let M be the centralizer of A in K. Denote by $\log: A \to \mathfrak{a}_{\boldsymbol{R}}$ the inverse of the exponential

---

AMS (MOS) subject classifications (1970). Primary 22E45, 17B10, 17B35; Secondary 20G05, 16A64.

Key words and phrases. Real semisimple Lie group, irreducible representation, subquotient, nonunitary principal series, universal enveloping algebra, finite-dimensional representation, composition series, infinitesimal character.

* Partially supported by NSF GP-28323 and NSF GP-33893.

mapping, where $\mathfrak{a}_{\boldsymbol{R}}$ is the Lie algebra of A. Fix a finite-dimensional irreducible M-module Y and an element $\nu \in \mathfrak{a}^*$ (the complex dual of $\mathfrak{a}_{\boldsymbol{R}}$), and define an MAN-module structure on Y by

$$man \cdot y = \exp\{\nu(\log a)\}\, m \cdot y$$

($m \in M$, $a \in A$, $n \in N$ and $y \in Y$). Call this MAN-module $\tilde{Y}$. Let $X^{(Y,\nu)}$ be the space of all C^∞ functions $f: G \to \tilde{Y}$ such that $f(gp) = p^{-1} \cdot f(g)$ for all $g \in G$ and $p \in MAN$. Then $X^{(Y,\nu)}$ is a G-module under the action given by $(g \cdot f)(h) = f(g^{-1}h)$ ($f \in X^{(Y,\nu)}$ and $g, h \in G$). Also, $X^{(Y,\nu)}$ is a $\mathfrak{g}$-module and hence a $\mathscr{G}$-module, by means of the action

$$(x \cdot f)(g) = \frac{d}{dt}((\exp tx) \cdot f)(g)\Big|_{t=0}$$

($f \in X^{(Y,\nu)}$, x in the Lie algebra of G and $g \in G$). Let $X^{Y,\nu}$ be the subspace of K-finite vectors of $X^{(Y,\nu)}$. Then $X^{Y,\nu}$ is $\mathscr{G}$-invariant, and is a compatible $(\mathscr{G}, K)$-module. As Y ranges through a set of representatives of the irreducible representations of M and ν ranges through $\mathfrak{a}^*$, the modules $X^{Y,\nu}$ constitute the *infinitesimal nonunitary principal series* of G. They are the infinitesimal modules (i.e., the compatible $(\mathscr{G}, K)$-modules of K-finite vectors) associated with a family of Hilbert space representations of G called the *nonunitary principal series*.

THEOREM 1 (THE SUBQUOTIENT THEOREM). *Every $\mathscr{G}$-irreducible compatible $(\mathscr{G}, K)$-module is equivalent to a subquotient (i.e., a quotient of submodules) of some infinitesimal nonunitary principal series module.*

The original result of this type is Harish-Chandra's result [**3(b)**, Theorem 4, p. 63], which amounts essentially to Theorem 1 in the special case in which G is linear. Theorem 1 is proved (and [**3(b)**, Theorem 4] is simplified) in [**6**], and a similar result is obtained independently by C. Rader in [**9**]. Before outlining the proof of Theorem 1, we give some applications.

COROLLARY 1. *Let V be a $\mathscr{G}$-irreducible compatible $(\mathscr{G}, K)$-module, and let W be an irreducible K-module. Then the multiplicity with which W occurs in V is at most the maximum of the multiplicities with which irreducible M-modules occur in W.*

REMARK. This estimate is the best possible.

Corollary 1 follows easily from the theorem, but in fact it is much easier, and is proved in the course of the proof of the theorem. The corollary generalizes analogous results of J. Dixmier ([**1(a)**] and [**1(b)**, §15.5]), R. Godement [**2**, §1] and Harish-Chandra ([**3(b)**, Theorem 4 (3), p. 63]; see also [**3(b)**, Theorem 3, p. 36]).

COROLLARY 2. *Every $\mathscr{G}$-irreducible compatible $(\mathscr{G}, K)$-module is the infinitesimal module of some continuous topologically irreducible representation of G on a Hilbert space.*

This follows immediately from the theorem.

If V is a $\mathscr{G}$-module and χ is a complex-valued homomorphism on the center $\mathscr{Z}$ of $\mathscr{G}$, then χ is the *infinitesimal character* of V if $\mathscr{Z}$ acts on V according to the scalars given by χ. A well-known result of Dixmier asserts that every irreducible $\mathscr{G}$-module has an infinitesimal character (see [**6**, Lemma 2.2]).

COROLLARY 3. *There are only finitely many inequivalent $\mathscr{G}$-irreducible compatible $(\mathscr{G}, K)$-modules with a given infinitesimal character.*

This follows from Corollary 2 and Harish-Chandra's deep analytic theorem that there are only finitely many infinitesimally inequivalent irreducible Hilbert space representations of G with a given infinitesimal character. Later we shall indicate how Corollary 3 *and* Harish-Chandra's theorem can be proved quite simply when G is linear, using Theorem 1.

The proofs of Theorem 1 and Corollary 1 are based on the next result, whose statement requires more notation.

Let $\mathscr{G}^K$ denote the centralizer of K in $\mathscr{G}$, and let β be an equivalence class of finite-dimensional irreducible representations of K. Let $\mathscr{I}^\beta$ be the kernel in $\mathscr{K}$ of the representation of $\mathscr{K}$ associated with any member of β. Then $\mathscr{G}^K \cap \mathscr{G}\mathscr{I}^\beta$ is a two-sided ideal of $\mathscr{G}^K$, so that $\mathscr{G}^K/\mathscr{G}^K \cap \mathscr{G}\mathscr{I}^\beta$ is an algebra. This algebra has considerable significance, since its equivalence classes of irreducible representations are in natural 1-1 correspondence with the set of equivalence classes of $\mathscr{G}$-irreducible compatible $(\mathscr{G}, K)$-modules which contain β with positive multiplicity. (This result is proved in a general form in [**7**].) It would be of great interest to understand the full structure of this algebra. The next theorem indicates some of its structure.

Let $\mathfrak{a}$ and $\mathfrak{n}$ be the complexified Lie algebras of A and N, and let $\mathscr{A}$ and $\mathscr{N}$ be their universal enveloping algebras. Then $\mathscr{G} \cong \mathscr{N} \otimes \mathscr{A} \otimes \mathscr{K}$ (linearly), so that $\mathscr{G} = \mathscr{A} \otimes \mathscr{K} \oplus \mathfrak{n}\mathscr{G}$. Let $P: \mathscr{G} \to \mathscr{A} \otimes \mathscr{K}$ be the projection map, and regard $\mathscr{A} \otimes \mathscr{K}$ as an algebra in the natural way. Let $\pi_\beta: \mathscr{K} \to \mathscr{K}/\mathscr{I}^\beta$ be the quotient map, and define

$$P_\beta = (1 \otimes \pi_\beta) \circ P: \mathscr{G} \to \mathscr{A} \otimes \mathscr{K}/\mathscr{I}^\beta.$$

Finally, let $\mathscr{K}^M$ denote the centralizer of M in $\mathscr{K}$.

THEOREM 2. *$P_\beta \mid \mathscr{G}^K$ is an algebra antihomomorphism with kernel precisely $\mathscr{G}^K \cap \mathscr{G}\mathscr{I}^\beta = \mathscr{G}^K \cap \mathscr{I}^\beta\mathscr{G}$, and P_β induces an injection*

$$\mathscr{G}^K/\mathscr{G}^K \cap \mathscr{G}\mathscr{I}^\beta \hookrightarrow \mathscr{A} \otimes (\mathscr{K}^M/\mathscr{K}^M \cap \mathscr{I}^\beta).$$

If we take β to be the trivial one-dimensional class in Theorem 2, we have a well-known theorem of Harish-Chandra. The injectivity statement in Theorem 2 is proved by first proving the corresponding assertion for the Euclidean motion group associated with G, and then reducing to this case by means of the symmetrization mapping. Theorem 2 implies that $\mathcal{G}^K/\mathcal{G}^K \cap \mathcal{G}\mathcal{I}^\beta$ satisfies a polynomial identity, generalizing the well-known commutativity in case β is trivial, and this easily implies Corollary 1. Theorem 1 is proved in **[6]** using Theorem 2 and an adaptation of Harish-Chandra's reasoning in **[3(b)]**. The main idea is to show that every irreducible representation of the image of $\mathcal{G}^K/\mathcal{G}^K \cap \mathcal{G}\mathcal{I}^\beta$ under the above injection factors through a homomorphism of the form $\tau \otimes 1$ of $\mathcal{A} \otimes (\mathcal{K}^M/\mathcal{K}^M \cap \mathcal{I}^\beta)$, where τ is a complex-valued homomorphism of $\mathcal{A}$. This is roughly a kind of noncommutative "going-up" theorem.

The key to extending Theorem 1 and Corollary 1 from the linear case to the nonlinear case is the avoidance of the use of finite-dimensional G-modules which contain a given finite-dimensional irreducible K-module. However, in order to give an elementary proof of Corollary 3 and the related Harish-Chandra theorem, we must rely on finite-dimensional G-modules, and so we assume from now on that G is linear. The following results are contained in **[8]**.

It is well known that the lowest restricted weight space $\gamma(V)$ of any finite-dimensional irreducible G-module V is invariant and irreducible under M. We call an irreducible M-module *extendible* if it is equivalent to $\gamma(V)$ for some V.

THEOREM 3. *Every irreducible M-module is extendible.*

Let $\nu(V) \in \mathfrak{a}^*$ be the lowest restricted weight of an irreducible G-module V. It is well known that $X^{\gamma(V), \nu(V)}$ contains a unique finite-dimensional $\mathcal{G}$-submodule, and this submodule is isomorphic to V (regarded as a $\mathcal{G}$-module). This fact and a partition of unity argument show

THEOREM 4. *Let $\nu \in \mathfrak{a}^*$, and let V be a finite-dimensional irreducible G-module. Then $X^{\gamma(V), \nu+\nu(V)}$ is a $\mathcal{G}$-module quotient of $X^{\mathbf{C}, \nu} \otimes V$. (Here $\mathbf{C}$ is regarded as the trivial M-module.) If the (unique up to scalar) K-fixed vector in $X^{\mathbf{C}, \nu}$ is $\mathcal{G}$-cyclic, then $X^{\gamma(V), \nu+\nu(V)}$ has a cyclic vector for $\mathcal{G}$.*

B. Kostant has found all the modules $X^{\mathbf{C}, \nu}$ with a cyclic K-fixed vector using deep algebraic methods **[5]**, and S. Helgason has obtained a much easier partial result by analytic methods **[4]**. Helgason's result, together with Theorem 4 and a precise version of Theorem 3, yield

THEOREM 5. *Every member of the infinitesimal nonunitary principal series is cyclic, and hence finitely generated, as a $\mathcal{G}$-module.*

THEOREM 6. *Every $\mathcal{G}$-irreducible compatible $(\mathcal{G}, K)$-module is equivalent to a* *subquotient of the tensor product of a module $X^{\mathbf{C},\nu}$ with a cyclic K-fixed vector, and a finite-dimensional irreducible $\mathcal{G}$-module.*

Theorem 6 requires Theorem 1. We now have the following result (which is of course known for all G with finite center).

THEOREM 7. *Every member of the infinitesimal nonunitary principal series has a (finite) composition series.*

This is proved by the following simple algebraic argument due to Kostant: Every member $X^{Y,\nu}$ of the infinitesimal nonunitary principal series satisfies the ascending chain condition, since it is finitely generated (by Theorem 5) over the Noetherian ring $\mathcal{G}$. But it is well known that $X^{Y,\nu}$ is nonsingularly paired with another such module. Hence $X^{Y,\nu}$ also satisfies the descending chain condition, and so has a composition series.

Finally, a straightforward argument using Theorems 1 and 7 implies Corollary 3 and hence also Harish-Chandra's theorem mentioned above (for linear groups G).

REFERENCES

1. J. Dixmier,

a) *Sur les représentations de certaines groupes orthogonaux*, C. R. Acad. Sci. Paris **250** (1960), 3263–3265. MR **22** #5901.

b) *Les C*-algèbres et leurs représentations*, Cahiers Sci., fasc. 29, Gauthier-Villars, Paris, 1964. MR **30** #1404.

2. R. Godement, *A theory of spherical functions*. I, Trans. Amer. Math. Soc. **73** (1952), 496–556. MR **14** #620.

3. Harish-Chandra,

a) *Representations of a semisimple Lie group on a Banach space*. I, Trans. Amer. Math. Soc. **75** (1953), 185–243. MR **15**, 100.

b) *Representations of semisimple Lie groups*. II, Trans. Amer. Math. Soc. **76** (1954), 26–65. MR **15**, 398.

4. S. Helgason, *A duality for symmetric spaces with applications to group representations*, Advances in Math. **5** (1970), 1–154. MR **41** #8587.

5. B. Kostant, *On the existence and irreducibility of certain series of representations*, Publication of 1971 Summer School in Math., edited by I. M. Gel'fand, Bolyai-Janós Math. Soc., Budapest (to appear).

6. J. Lepowsky, *Algebraic results on representations of semisimple Lie groups*, Trans. Amer. Math. Soc. **176** (1973), 1–44.

7. J. Lepowsky and G. W. McCollum, *On the determination of irreducible modules by restriction to a subalgebra*, Trans. Amer. Math. Soc. **176** (1973), 45–57.

8. J. Lepowsky and N. Wallach, *Finite- and infinite-dimensional representations of linear semisimple groups*, Trans. Amer. Math. Soc. (1973) (to appear).

9. C. Rader, *Spherical functions on semisimple Lie groups*, Thesis and unpublished supplements, University of Washington, 1971.

BRANDEIS UNIVERSITY

ON THE CONTINUOUS SPECTRUM FOR A SEMISIMPLE LIE GROUP

P. C. TROMBI

Let G be a connected semisimple Lie group which we assume for simplicity is a noncompact real form of a complex simply connected semisimple Lie group G_c. Let $K \subset G$ be a fixed maximal compact subgroup. It is the purpose of this article to illustrate two things: (i) how each conjugacy class of Cartan subgroups (CSG) contributes to the Plancherel measure of G, and (ii) how the theory of the asymptotics for the Eisenstein integrals (which will be defined shortly) determine this measure. We begin by introducing an appropriate Schwartz space of functions on which these two points can be demonstrated.

Let V be a finite-dimensional complex algebra with involution $v \to v^*$. We assume that V has a Hilbert space structure (compatible with its involution) and that it is a double unitary K-module where $\tau = (\tau_1, \tau_2)$ are unitary representations of K which act on the left and right of V. There are some natural associativity conditions on the product in V and the action of τ which we shall not list here. The space of interest then is the space $\mathscr{C}(G, V, \tau)$ of all C^∞ functions $f: G \to V$ such that

$$f(k_1 x k_2) = \tau_1(k_1) f(x) \tau_2(k_2) \qquad (x \in G,\ k_1,\ k_2 \in K).$$

We shall refer to this by saying that f is *τ-spherical* and if $\mathscr{G}$ denotes the enveloping algebra of the complexification of the Lie algebra $\mathfrak{g}$ of G,

$$\Xi(x) = \int_K \exp\{-\varrho(H(xk))\}\, dk$$

AMS (MOS) subject classifications (1970). Primary 43A80; Secondary 43A90.

denotes the elementary spherical function of weight 0, and $\sigma(x)=\|X\|$ where $x=k\exp X$ is a Cartan decomposition of x, $\mathfrak{g}=\mathfrak{k}+\mathfrak{s}$ the corresponding Cartan decomposition of $\mathfrak{g}$, and $\|\cdot\|$ denotes an $\mathrm{Ad}(K)$ invariant norm on $\mathfrak{s}$ then

$$\sup_{x\in G}(1+\sigma(x))^r\, \Xi^{-1}(x)\|ufv\|_V<\infty$$

for all $r\in \boldsymbol{R}$, and $u, v\in\mathcal{G}$. Since V is an algebra with involution, it is clear that $\mathscr{C}(G, V, \tau)$ is a convolution algebra with involution given by $f\to\tilde{f}$ where $\tilde{f}(x)=f(x^{-1})^*$. As a special case we mention that if $V=C$ (with the usual involution) and $\tau_1=\tau_2=1$ the trivial representation of K, then $\mathscr{C}(G, V, \tau)=\mathscr{I}^2(G)$, the Schwartz space of spherical functions on G.

Harmonic analysis on $\mathscr{I}^2(G)$ begins only after one has distinguished a class of so-called *elementary spherical functions.* They are elementary in the sense that one expects all $f\in\mathscr{I}^2(G)$ to be expanded in terms of them. There are two ways of distinguishing these functions. One way is to require that they be C^∞ spherical functions which satisfy the integral equation

$$\int_K \phi(xky)\,dk=\phi(x)\,\phi(y) \qquad (x, y\in G). \tag{1}$$

The other equivalent way is through differential equations. Namely, if $\mathfrak{q}$ denotes the centralizer of K in $\mathcal{G}$, then the elementary functions ϕ are those for which $\phi(1)=1$ and there exists a homomorphism $\chi:\mathfrak{q}\to C$ such that

$$q\phi=\chi(q)\,\phi \qquad (q\in\mathfrak{q}). \tag{2}$$

Harish-Chandra [1] proved that if $G=KAN$ (resp. $\mathfrak{g}=\mathfrak{k}+\mathfrak{a}+\mathfrak{n}$) is an Iwasawa decomposition of G (resp. $\mathfrak{g}$), $\mathscr{F}=\mathfrak{a}_c^*$ and

$$\phi_\lambda(x)=\int_K \exp\{(i\lambda-\varrho)(H(xk))\}\,dk \qquad (\lambda\in\mathscr{F}) \tag{3}$$

then the $\phi_\lambda(\lambda\in\mathscr{F})$ are precisely all the elementary spherical functions. It turns out that these functions are the spherical matrix elements of the class one principal series.

A variation of (2) (which we will explain presently) can be used to distinguish a class of elementary τ-spherical functions. In fact (2) expresses the point of view that harmonic analysis on $\mathscr{I}^2(G)$ can be thought of as the problem of simultaneously diagonalizing the unbounded operators $q\in\mathfrak{q}$ which are densely defined on $I^2(G)$ (=the square integrable spherical functions). In particular they are defined on $C_c^\infty(G)\cap I^2(G)$ and their restrictions to this subspace form a commutative algebra. If we try to use (2) directly to distinguish a class of τ-spherical functions we run into two problems. One is that the operators in $\mathfrak{q}$ when restricted to $C_c^\infty(G, V, \tau)$ (this space has the obvious definition) stand little chance of forming a commutative algebra. The other problem is that we expect the elementary τ-spherical functions to be matrix functions of irreducible representations. The action of $\mathfrak{q}$ on these functions will be multiplication by a scalar only when τ is irreducible and occurs with multiplicity one in the representation such as the case when $V=C$, $\tau=(1, 1)$ and the representation is a class one principal series. Hence, we must use a smaller subalgebra of $\mathcal{G}$. The obvious choice is $\mathfrak{Z}$ the center of $\mathcal{G}$ as it is already abelian and it acts on matrix elements of irreducible representations as a scalar operator. So we shall say that ϕ is an *elementary τ-spherical function* if there exists a homomorphism $\chi: \mathfrak{Z}\to C$ such that

$$z\phi=\chi(z)\,\phi \qquad (z\in\mathfrak{Z}). \tag{2$'$}$$

We wish to show that we can in fact *construct* a large class of τ-spherical functions satisfying (2)$'$. Let us first give a definition which will be useful in the sequel. Let π be an admissible representation of a reductive group H. We assume that the connected component of the identity, H^0, of H is of finite index in H, the center of H^0 is compact and $\mathrm{Ad}(H)\subset H_c$ where H_c is a complex simply connected semisimple Lie group. Let K denote a maximal compact subgroup of H. For simplicity (as this will be the only case we consider) we shall assume π represents G on a Hilbert space. Then let $\mathcal{A}(\pi)$ denote the complex linear space spanned by the K-finite matrix coefficients of π. If (V_0, τ_0) is a finite-dimensional unitary double K-module we shall write $\mathcal{A}(\pi, V_0, \tau_0)$ to denote the space of all C^∞, τ_0-spherical functions $f: G\to V_0$ such that for all $\lambda\in V^*$ if $g_{f,\lambda}(x)=\lambda(f(x))$ then $g_{f,\lambda}\in\mathcal{A}(\pi)$. We shall refer to functions in $\mathcal{A}(\pi, V_0, \tau_0)$ as *τ_0-spherical matrix functions of π*. Let us remark here that if $P=MAN$ is a parabolic subgroup then M as well as G is a group of the type H described above. Also let us note that if π is an irreducible unitary representation of H and if $\pi\simeq\pi'$ (unitary equivalence) then $\mathcal{A}(\pi)=\mathcal{A}(\pi')$. Hence in this case we shall often write $\mathcal{A}(\mathcal{C}(\pi))$ and $\mathcal{A}(\mathcal{C}(\pi), V_0, \tau_0)$ if $\mathcal{C}(\pi)$ denotes the class of π.

Suppose that G has a compact CSG B. Then it is known that the conjugacy class of B is precisely all compact CSG's of G and associated to that conjugacy class is the discrete series of representations of G which we denote by $\mathcal{E}^2(G)$. Hence, as these are irreducible unitary representations we must have if $\omega_d\in\mathcal{E}^2(G)$ and $\phi\in\mathcal{A}(\omega_d, V, \tau)$ that ϕ satisfies (2)$'$ with χ being the infinitesimal character of ω_d. We also have the following facts (cf. Harish-Chandra [**3**, Lemma 70]): $\dim(\mathcal{A}(\omega_d, V, \tau))<\infty$ for all $\omega_d\in\mathcal{E}^2(G)$ and $\dim(\mathcal{A}(\omega_d, V, \tau))=0$ for all but a finite number of ω_d. Whether G has a discrete series or not we are guided by the principle that discrete series gives eigenfunctions for the center of the enveloping algebra. So we go to certain reductive subgroups of G which have a discrete series and construct from its τ-spherical matrix elements eigenfunctions of $\mathfrak{Z}$. In particular we shall say that a parabolic subgroup $P=MAN$ with Lie algebra $\mathfrak{p}=\mathfrak{m}+\mathfrak{a}+\mathfrak{n}$ is *cuspidal* if there exists a θ-stable CSA $\mathfrak{h}$ (θ is the Cartan involution corresponding to the decomposition $\mathfrak{g}=\mathfrak{k}+\mathfrak{s}$) such that $\mathfrak{h}\cap\mathfrak{s}=\mathfrak{a}$. As $\mathfrak{m}_1=\mathfrak{m}+\mathfrak{a}$ is the centralizer of $\mathfrak{a}$ in $\mathfrak{g}$ it is clear that $\mathfrak{h}\cap\mathfrak{k}\subset\mathfrak{m}$ and hence $\mathfrak{m}$ has a compact Cartan subalgebra. Let $\mathfrak{Z}_{\mathfrak{m}_1}$ denote the center of the enveloping algebra of $\mathfrak{m}_1$. It is known (cf. Trombi and Varadarajan [**1**]) that there exists an injection $\mu'_{\mathfrak{g}/\mathfrak{m}_1}:\mathfrak{Z}\to\mathfrak{Z}_{\mathfrak{m}_1}$ such that $z-\mu'_{\mathfrak{g}/\mathfrak{m}_1}(z)\in\theta(\mathfrak{n})\mathscr{G}\mathfrak{n}$. Hence if $\tau_M=(\tau_1|_{K\cap M},\tau_2|_{K\cap M})$, $\omega\in\mathscr{E}^2(M)$ and $\psi\in\mathscr{A}(\omega, V, \tau_M)$ then we can extend ψ to a function on $G=KMAN$ by putting $\psi(kman)=\tau_1(k)\,\psi(m)$. One can show that, as a function on G, ψ is well defined. For each $x\in G$ we can define a unique element $H(x)\in\mathfrak{a}$ which is determined by the equation $x=km\exp H(x)\,n$. It is clear from this that $H(kxn)=H(x)$ $(x\in G, k\in K, n\in N)$. Consequently, using the above properties of $\mu'_{\mathfrak{g}/\mathfrak{m}_1}$ and $H(x)$ it is obvious that

$$x\to\psi(x)\exp\{(i\nu-\varrho_P)(H(x))\}\qquad(\nu\in\mathfrak{a}_c^*=\mathscr{F})$$

$(\varrho_P(H)=\frac{1}{2}\operatorname{tr}(\operatorname{ad}H\mid\mathfrak{n}))$ is an eigenfunction of $\mathfrak{Z}$ and is left τ_1-spherical. To force this function to be τ-spherical we put

$$(3)'\qquad E(P:\psi:\nu:x)=\int_K\psi(xk)\,\tau_2(k^{-1})\exp\{(i\nu-\varrho_P)(H(xk))\}\,dk.$$

Following Harish-Chandra we shall call such functions *Eisenstein integrals.* (3)′ should be compared with (3) the formula for the elementary spherical functions ϕ_λ. It is clear that $E(P:\psi:\nu)$ is a τ-spherical C^∞ function satisfying (2)′. Moreover, one can show that if $\pi_{P,\omega_\nu}=\operatorname{Ind}_P^G(\sigma_\nu)$ where σ_ν is the representation of $P=MAN$ given by $\sigma_\nu(man)=\exp\{(i\nu+\varrho_P)(\log a)\}\,\sigma(m)$ $(\sigma\in\omega)$ then $E(P:\psi:\nu)\in\mathscr{A}(\pi_{P,\omega_\nu}, V,\tau)$ and the character of $\mathfrak{Z}$ corresponding to $E(P:\psi:\nu)$ is the infinitesmal character of π_{P,ω_ν}. It is also possible to prove that the mapping $\mathscr{A}(\omega, V, \tau_M)\to\mathscr{A}(\pi_{P,\omega_\nu}, V,\tau)$ given by $\psi\to E(P:\psi:\nu)$ is surjective and is bijective if π_{P,ω_ν} is irreducible.

The properties of the representations π_{P,ω_ν} are necessary for the analysis of the asymptotics of the Eisenstein integrals, hence we shall simply list them without giving any proofs.

(a) π_{P,ω_ν} is unitary if $\nu\in\mathscr{F}_R=\{\nu\in\mathscr{F}:\nu(H)\in\mathbf{R},\ \forall H\in\mathfrak{a}\}$.

(b) $\pi_{P,\omega_\nu}(\nu\in\mathscr{F}_R)$ is tempered (i.e., its global character is a tempered distribution) which implies that its matrix coefficients satisfy the weak inequality (cf. Harish-Chandra [**3**]).

(c) If $\mathscr{E}_P(G)$ denotes the equivalence classes of irreducible unitary representations obtained by inducing from $P=MAN$ the representations ω_ν, $\omega\in\mathscr{E}^2(M)$, $\nu\in\mathscr{F}$, and if $P_1=M_1A_1N_1$ is a parabolic subgroup of G for which there exists $x\in G$ such that $A_1^x=A$, then $\mathscr{E}_P(G)=\mathscr{E}_{P_1}(G)$.

(d) By (c) we may restrict our attention to the set $\mathscr{P}(A)$ of all parabolic subgroups of G with split part A. Put $\mathscr{W}(A)=\mathrm{Norm}_G(A)/\mathrm{Cent}_G(A)$ and call $\mathscr{W}(A)$ the *Weyl group of the parabolics of* $\mathscr{P}(A)$. It is clear that $\mathscr{W}(A)$ acts on M and A hence on $\mathscr{E}^2(M)$ and $\mathscr{F}$ in the natural way. If $P_1, P_2\in\mathscr{P}(A)$, $s\in\mathscr{W}(A)$, then $\mathscr{C}(\pi_{P_1,\omega_\nu})=\mathscr{C}(\pi_{P_2,\omega_\nu^s})$ $(\nu\in\mathscr{C}_R)$.

(e) Let $\varpi_P=\prod_{\alpha>0}H_\alpha$ where α runs over all roots of (P, A) and $H_\alpha\in\mathfrak{a}$ is given by $B(H, H_\alpha)=\alpha(H)$ (B is the Killing form of $\mathfrak{g}$). Regarding ϖ_P as a polynomial function on $\mathscr{F}$ we have that π_{P,ω_ν} is irreducible for $\nu\in\mathscr{F}'_R=\{\nu\in\mathscr{F}_R:\varpi_P(\nu)\neq 0\}$.

The representations π_{P,ω_ν} are obviously the generalizations of the principal series of representations associated to a minimal parabolic subgroup (i.e., a parabolic whose split part appears as the vector group in an Iwasawa decomposition of $\mathfrak{g}$). In fact if we apply the definition of π_{P,ω_ν} in the minimal parabolic case then as M is compact and hence all its irreducible unitary representations are square-integrable then we obtain the usual principal series. It is for this reason that we shall say that if $P=MAN$ is a cuspidal parabolic subgroup of G and $\pi_{P,\omega_\nu}=\mathrm{Ind}_P^G(\sigma_\nu)$, $\sigma\in\omega\in\mathscr{E}^2(M)$, $\nu\in\mathscr{F}_R$, that π_{P,ω_ν} belongs to the *principal series of representations associated to the cuspidal parabolic subgroup P*.

What we have shown then is that if $\mathfrak{h}_1,\dots,\mathfrak{h}_s$ is a complete set of representatives of the conjugacy class of CSA's of $\mathfrak{g}$ chosen so that they are θ-stable and $\dim(\mathfrak{h}_i\cap\mathfrak{k})\leqq\dim(\mathfrak{h}_{i+1}\cap\mathfrak{k})$; $A_i=A_{i,I}\cdot A_{i,R}$ the corresponding CSG's where $A_{i,I}=A_i\cap K$, $A_{i,R}=\exp(\mathfrak{h}_i\cap\mathfrak{s})$ then to each A_i there is a series of representations and an associated space of τ-spherical matrix elements $E(P:\psi:\nu)$, $P\in\mathscr{P}(A_{i,R})$, $\psi\in\mathscr{A}(\omega, V, \tau_M)$, $\omega\in\mathscr{E}^2(M)$ and $\nu\in\mathscr{F}$. In general then if $f\in\mathscr{C}(G, V, \tau)$ we must expect that the Fourier expansion of f will involve all of the $E(P:\psi:\nu)$ where P ranges over the parabolics in $\mathscr{P}(A_{i,R})$ $(1\leqq i\leqq s)$ and ψ and ν range over the appropriate spaces. There will of course be exceptions to this. Namely, if $\mathrm{rk}(G)=\mathrm{rk}(K)$ and $f\in{}^\circ\mathscr{C}(G, V, \tau)$, the space of τ-spherical cusp forms, then it is a theorem of Harish-Chandra [3] that f can be expanded in terms of the functions in $\mathscr{A}(\omega_d, V, \tau)$ where ω_d ranges over the discrete series of G. More generally if $P=MAN$ is a fixed parabolic subgroup, put for $f\in\mathscr{C}(G, V, \tau)$, $f^P(x)=\int_N f(xn)\,dn\,(x\in G)$ where dn is the Haar measure on N. Following Harish-Chandra we shall write $f^P\sim 0$ (which is read f^P is negligible) if

$$0=\int_M \operatorname{conj}\phi(m)\cdot f^P(xm)\,dm \qquad (x\in G)$$

for all $\phi\in{}_\circ\mathscr{C}(M, V, \tau_M)$. Let $\mathscr{C}_i(G, V, \tau)$ $(1\leqq i\leqq s)$ be the closed subspace of $\mathscr{C}(G, V, \tau)$ of all functions f such that $f^P\sim 0$ unless A is conjugate to $A_{i,R}$ under K.

Then it is a theorem of Harish-Chandra [4] that $\mathscr{C}(G, V, \tau)=\sum_{1\leqq i\leqq s}\mathscr{C}_i(G, V, \tau)$, the sum being direct and smooth. Also the expansion of $f\in\mathscr{C}_i(G, V, \tau)$ will involve only the functions $E(P:\psi:\nu)$, $P\in\mathscr{P}(A_{i,R})$, $P=MA_{i,R}N$, $\psi\in\mathscr{A}(\omega, V, \tau_M)$, $\omega\in\mathscr{E}^2(M)$ and $\nu\in\mathscr{F}_R$. Note that if $\mathrm{rk}(G)=\mathrm{rk}(K)$ then $\mathscr{C}_1(G, V, \tau)={}^\circ\mathscr{C}(G, V, \tau)$. We are interested in getting the contribution to the Plancherel measure coming from π_{P,ω_ν} for a fixed P and ω with ν varying over $\mathscr{F}_R$. For this reason we shall now make the following definitions. Let $P=MAN$ be an arbitrary cuspidal parabolic subgroup.

DEFINITION. Let $\omega\in\mathscr{E}^2(M)$ be fixed. For $\alpha\in C_c^\infty(\mathfrak{F}_R)$, $\psi\in\mathscr{A}(\omega, V, \tau_M)$ and $\gamma_\omega(\nu)$ a positive Borel measure on $\mathfrak{F}_R$ we define a function ϕ_α on G by

$$\phi_\alpha(x)=\int_R \alpha(\nu)\, E(P:\psi:\nu:x)\, d\gamma_\omega(\nu) \qquad (x\in G). \tag{4}$$

We shall call ϕ_α a *wave packet associated with* $\alpha d\gamma_\omega(\nu)$.

Suppose that $\phi_\alpha\in\mathscr{C}(G, V, \tau)$ (note that there are measures for which ϕ_α will belong to $\mathscr{C}(G, V, \tau)$ for instance $|\omega_P(\nu)|^r\, d\nu$ for r sufficiently large (cf. Harish-Chandra [5] and [2])). One can show that if $P'=M'A'N'$ is another cuspidal parabolic subgroup of G and if we define $\pi_{P',\omega'_{\nu'}}$ as usual, $\Theta_{\omega'_{\nu'}}$ denoting its global character, then if $\Theta_{\omega'_{\nu'}}\neq\Theta_{\omega_\nu}$ $(\forall\nu\in\mathscr{F}_R)$ we have $\Theta_{\omega'_{\nu'}}(\phi_\alpha)=0$. It is natural therefore to define the *Fourier transform* of $f\in\mathscr{C}(G, V, \tau)$ for which $\Theta_{\omega'_{\nu'}}(f)=0$ for all characters $\Theta_{\omega'_{\nu'}}\neq\Theta_{\omega_\nu}$ $(\forall\nu\in\mathscr{F}_R)$ to be

$$\hat{f}(\omega, \nu)=\int_G \mathrm{conj}(\Theta_{\omega_\nu}(x))\, f(x)\, dx. \tag{5}$$

We shall say that $d\gamma_\omega(\nu)$ is the *Plancherel measure corresponding to wave packets formed with Eisenstein integrals of the type* $E(P:\psi:\nu)$ $(\psi\in\mathscr{A}(\omega, V, \tau_M))$ if

$$\phi_\alpha(1)=\int_{\mathscr{F}_R} \hat{\phi}_\alpha(\omega, \nu)\, d\gamma_\omega(\nu). \tag{6}$$

Since $\mathscr{C}(G, V, \tau)$ is a convolution algebra, (6) would imply the usual Plancherel theorem that $\|\phi_\alpha\|_2=\|\hat{\phi}_\alpha\|_2$. It is the measure $\gamma_\omega(\nu)$ we wish to determine.

In order to show that a wave packet associated to $\alpha d\gamma_\omega(\nu)$ belongs to Schwartz space it is clear that we must investigate the behavior at infinity on G as well as $\mathscr{F}_R$ of the Eisenstein integrals $E(P:\psi:\nu)$ and relate the asymptotics to the measure $\gamma_\omega(\nu)$. This is done by considering the differential equations (2)$'$.

From (2)$'$ we get for any $P_1, P_2\in(A)$ that there are linear maps

$$c_{P_1|P_2}(s:\nu):\mathscr{A}(\omega, V, \tau_M)\to\mathscr{A}(\omega^s, V, \tau_M) \qquad (s\in\mathscr{W}(A), \nu\in\mathscr{F}'_R)$$

such that (cf. Harish-Chandra [4] for notation)

$$\lim_{a\to\infty;\,P_1}\Big\{\exp\{\varrho_{P_1}(\log a)\}\, E(P_2:\psi:\nu:ma) - \sum_{s\in\mathscr{W}(A)} (c_{P_1|P_2}(s:\nu)\,\psi)(m)\exp\{is\nu(\log a)\}\Big\}=0 \qquad (ma\in MA) \tag{7}$$

where $P_2=MAN_2$, $\psi\in\mathscr{A}(\omega, V, \tau_M)$ and $\nu\in\mathscr{F}'_R$. We call the finite sum in (7) the *constant term of* $E(P_2:\psi:\nu)$ *along* P_1. The linear transformations $c_{P_1|P_2}(s:\nu)$ will be called *c-functions* in analogy with the spherical functions. A careful analysis of this finite Fourier series must now be made. In fact we can be more precise. The idea behind showing a wave packet lands in $\mathscr{C}(G, V, \tau)$ is to replace $E(P:\psi:\nu)$ in the definition of ϕ_α by its constant term and to show the result for the new integral (the error term is easily controlled). One can show that the c-functions are meromorphic on $\mathscr{F}_R$. Hence we must take care that the Plancherel measure has zeros precisely where the c-functions have poles (on $\mathscr{F}_R$) and that the orders match. This is essentially the role of the Maass-Selberg relations which we shall now describe.

Let χ_{τ_j} be the character of τ_j $(j=1, 2)$ and let $\mathscr{H}^\omega$ denote the representation space of π_{P,ω_ν} (the definition of the representation space of π_{P,ω_ν} does not change as ν varies over $\mathscr{F}$). Put $\mathscr{H}^\omega_\tau=\{h\in\mathscr{H}^\omega \mid \chi_{\tau_1} *_K h *_K \chi_{\tau_2}=h\}$. Then we can define a bilinear mapping of End $\mathscr{H}^\omega_\tau\times V$ into $\mathscr{A}(\omega, V, \tau_M)$. For $T\in\operatorname{End}\mathscr{H}^\omega_\tau$ and $v\in V$ we shall denote the image of (T, v) under this mapping by $\psi_{T\otimes v}$. One can then show that the mapping $(T, v)\to\psi_{T\otimes v}$ extends to a bijection of $\operatorname{End}\mathscr{H}^\omega_\tau\otimes V$ with $\mathscr{A}(\omega, V, \tau_M)$. It is a consequence of what Harish-Chandra calls the Maass-Selberg relations that the following holds: Let $P_1, P_2\in\mathscr{P}(A)$, $\nu\in\mathscr{F}'_R$, $T\in\operatorname{End}\mathscr{H}^\omega_\tau$, $v\in V$ and $s\in\mathscr{W}(A)$; then there exists a constant $c>0$ *not* depending on P_1, P_2 or s such that

$$\|c_{P_1|P_2}(s:\nu)\,\psi_{T\otimes v}\|_2^2=c\,\|T\|_2^2\,\|v\|^2. \tag{8}$$

Here the norm on the left of (8) is the L^2-norm and $\|T\|_2$ denotes the Hilbert-Schmidt norm of T. We now must introduce the Plancherel measure and show how (8) implies that its zeros match the poles of the c-functions.

Recall that in the case of $\mathscr{I}^2(G)$ the Plancherel measure turns out to be $|c(\lambda)|^{-2}\,d\lambda$. We wish to show that essentially the same result holds for $\mathscr{C}(G, V, \tau)$. The existence of the Plancherel measure comes from the irreducibility of the representations π_{P,ω_ν} for $\nu\in\mathscr{F}'_R$. In particular one shows that the c-functions

$c_{P|P}(1:\nu)$ and $c_{\bar{P}|P}(1:\nu)$ have integral formulae (which look like generalizations of the integral formula for $c(\lambda)$). The integral formulae give two important results:

(i) If $\nu\in\mathscr{F}$ then there exist unique $\nu_I, \nu_R\in\mathscr{F}_R$ such that $\nu=\nu_R+i\nu_I$. Let $\mathscr{F}(P)=\{\nu\in\mathscr{F}\mid \nu_I(H_\alpha)>0, \alpha>0 \text{ a root of } (P,A)\}$. Then $c_{P|P}(1:-\nu)$ and $c_{P|\bar{P}}(1:\nu)$ have holomorphic continuations to $\mathscr{F}(P)$.

(ii) There exists unique $c_P=c_P(\omega_\nu)$ in End $\mathscr{H}_\tau^\omega$ such that, for all $\nu\in\mathscr{F}'_R$,

$$c_{P|P}(1:\nu)\,\psi_{T\otimes v}=\psi_{c_PT\otimes v} \qquad (T\in\operatorname{End}\mathscr{H}_\tau^\omega, v\in V). \tag{9}$$

The importance of (9) is the following:

LEMMA. *For all* $\nu\in\mathscr{F}'_R$, $c_{P|P}=c_{P|P}(1:\nu)$, $c_{\bar{P}|P}=c_{\bar{P}|P}(1:\nu)$ *and* $c_P=c_P(\omega_\nu)$,
(1) $c_{\bar{P}|P}\psi_T=\psi_{Tc_{\bar{P}}^*\otimes v}$,
(2) $c_P\pi_{P,\omega_\nu}|_{\mathscr{H}_\tau^\omega}=\pi_{\bar{P},\omega_\nu}|_{\mathscr{H}_\tau^\omega}c_P$,
(3) $c_{\bar{P}}=(c_P)^*$.

It is easy to deduce from (2) of the lemma, using the unitarity of $\pi_{P,\omega_\nu}(\nu\in\mathscr{F}_R)$ that, for all $\nu\in\mathscr{F}'_R$,

$$c_P^*c_P\pi_{P,\omega_\nu}|_{\mathscr{H}_\tau^\omega}=\pi_{P,\omega_\nu}|_{\mathscr{H}_\tau^\omega}c_P^*c_P. \tag{10}$$

That is, $c_P^*c_P$ is an intertwining operator of $\pi_{P,\omega_\nu}|_{\mathscr{H}_\tau^\omega}$. Since π_{P,ω_ν} is irreducible for $\nu\in\mathscr{F}'_R$, we must have that there exists a function $\mu_\omega(\nu)$ defined on $\mathscr{F}'_R$ such that

$$c_P^*c_P=\mu_\omega(\nu)^{-1}I \qquad (I=\text{identity operator}). \tag{11}$$

It is a theorem of Harish-Chandra [5] and also follows from a recent result of Wallach [1] that the c-functions have meromorphic continuations to $\mathscr{F}$ (Wallach shows they are quotients of gamma functions). Hence $\mu_\omega(\nu)$ has a meromorphic continuation to $\mathscr{F}$. In fact the next lemma shows that $\mu_\omega(\nu)$ defines a slowly increasing, positive Borel measure on $\mathscr{F}_R$ (note we have removed the prime on $\mathscr{F}'_R$).

LEMMA. (1) $\mu_\omega(\nu)$ *is holomorphic on* $\mathscr{F}_R$;
(2) $\mu_\omega(\nu)>0$ *on* $\mathscr{F}'_R$ *and* $\mu_\omega(\nu)\geqq 0$ *on* $\mathscr{F}_R$;
(3) $\mu_{\omega s}(s\nu)=\mu_\omega(\nu)$ $(s\in\mathscr{W}(A))$;
(4) *there exists* $c>0$ *and* $k\geqq 0$ *such that* $\mu_\omega(\nu)\leqq c(1+\|\nu\|^2)^k$ $(\nu\in\mathscr{F}_R)$.

Now we observe that (11) together with (8) gives

$$\begin{aligned}\|c_{P|P}(1:\nu)\,\psi_{T\otimes v}\|_2^2=\|\psi_{c_PT\otimes v}\|_2^2&=C\|c_PT\|_2^2\,\|v\|^2\\ &=C\mu_\omega(\nu)^{-1}\,\|T\|_2^2\,\|v\|^2\end{aligned}$$

which implies that

$$\mu_\omega(\nu) \, \|c_{P_1|P_2}(s:\nu)\,\psi\|_2^2 = \|\psi\|_2^2$$

for all $P_1, P_2 \in \mathscr{P}(A)$, $s \in \mathscr{W}(A)$, $\nu \in \mathscr{F}'_R$ and $\psi \in A(\omega, V, \gamma_M)$. This shows that $\mu_\omega(\nu)\, c_{P_1|P_2}(s:\nu)$ remains bounded on $\mathscr{F}'_R$ hence has an analytic continuation to $\mathscr{F}_R$. Finally we come to the principal result of this article.

THEOREM. *The measure $d(\omega)\,\mu_\omega(\nu)\,d\nu$ ($d(\omega)$ = formal degree of ω, $d\nu$ = Lebesgue measure) is the Plancherel measure corresponding to wave packets formed from Eisenstein integrals of the form $E(P:\psi:\nu)$, $P \in \mathscr{P}(A)$, $\psi \in \mathscr{A}(\omega, V, \tau_M)$.*

REFERENCES

Harish-Chandra

1. *Spherical functions on a semisimple Lie group*. I, Amer. J. Math. **80** (1958), 241–310. MR **20** #925.
2. *Spherical functions on a semisimple Lie group*. II, Amer. J. Math. **80** (1958), 553–613. MR **21** #92.
3. *Discrete series for semisimple Lie groups*. II. *Explicit determination of the characters*, Acta Math. **116** (1966), 1–111. MR **36** #2745.
4. *Harmonic analysis on semisimple Lie groups*, Bull. Amer. Math. Soc. **76** (1970), 529–551. MR **41** #1933.
5. *On the theory of the Eisenstein integral*, Conference on Harmonic Analysis, Lecture Notes in Math., vol. 266, Springer-Verlag, Berlin and New York, 1971.

P. C. Trombi and V. S. Varadarajan

1. *Asymptotic behaviour of eigenfunctions on a semisimple Lie group: the discrete spectrum*, Acta Math. **129** (1972), 237–279.

N. Wallach

1. *On Harish-Chandra's generalized c-functions*, Amer. J. Math. (submitted).

INSTITUTE FOR ADVANCED STUDY

THE SPECTRUM OF A REDUCTIVE LIE GROUP

JOSEPH A. WOLF

Harish-Chandra's constructions of various series of representations, and his Plancherel formula, apply (roughly speaking) to those reductive Lie groups G such that the analytic subgroup for the derived algebra $[\mathfrak{g}, \mathfrak{g}]$ has finite center. See Peter Trombi's summary just preceding. Here I want to indicate the extension of that work to a class of reductive groups which includes all semisimple groups and is stable under passage to the reductive part of a cuspidal parabolic subgroup. The extension is definitive for construction of the various series. However, it is provisional for the Plancherel theorem; when the details of Harish-Chandra's work become available his method should extend to give a sharper result with less effort.

1. Relative discrete series

1.1. Notion of relative discrete series. Let G be a unimodular locally compact group and Z a closed normal abelian subgroup. Given a unitary character $\zeta \in \hat{Z}$ we have the representation space

$$L_2(G/Z, \zeta) = \left\{ f: G \to \mathbf{C}: f(gz) = \zeta(z)^{-1} f(g), \forall z \in Z, g \in G \text{ and } \int_{G/Z} |f(g)|^2 \, d(gZ) < \infty \right\}$$

for $l_\zeta = \mathrm{Ind}_{Z \uparrow G}(\zeta)$. Evidently $L_2(G) = \int_{\hat{Z}} L_2(G/Z, \zeta) \, d\zeta$ and G has left regular representation $\int_{\hat{Z}} l_\zeta \, d\zeta$.

$\hat{G}$ is the set of equivalence classes of irreducible unitary representations of G. If $\zeta \in \hat{Z}$ denote $\hat{G}_\zeta = \{[\pi] \in \hat{G} : \zeta \text{ is a summand of } \pi|_Z\}$. A class $[\pi] \in \hat{G}$ is *ζ-discrete* if π is equivalent to a subrepresentation of l_ζ. The ζ-discrete classes form the

---

AMS (MOS) subject classifications (1970). Primary 43A80; Secondary 22E45, 43A85.

ζ-discrete series $\hat{G}_{\zeta\text{-disc}} \subset \hat{G}_\zeta \subset \hat{G}$. The *relative* (to Z) *discrete series* is $\hat{G}_{\text{disc}} = \bigcup_{\zeta \in \hat{Z}} \hat{G}_{\zeta\text{-disc}}$.

Suppose Z central in G. If $[\pi] \in \hat{G}_\zeta$ the following are equivalent:

(1) There exist nonzero φ, ψ in the representation space H_π such that $\langle \varphi, \pi(\cdot)\, \psi \rangle \in L_2(G/Z, \zeta)$.

(2) If $\varphi, \psi \in H_\pi$ then $\langle \varphi, \pi(\cdot)\, \psi \rangle \in L_2(G/Z, \zeta)$.

(3) $[\pi] \in \hat{G}_{\zeta\text{-disc}}$.

Under those conditions there is a number $d_\pi > 0$ such that

$$\int_{G/Z} \langle \varphi_1, \pi(g)\, \psi_1 \rangle\, \overline{\langle \varphi_2, \pi(g)\, \psi_2 \rangle}\, d(gZ) = d_\pi^{-1} \langle \varphi_1, \varphi_2 \rangle\, \overline{\langle \psi_1, \psi_2 \rangle}$$

for all $\varphi_i, \psi_i \in H_\pi$. The number d_π is the *formal degree* of π.

1.2. Exact working hypotheses. From now on, G is reductive Lie group, i.e. its Lie algebra $\mathfrak{g} = \mathfrak{c} \oplus \mathfrak{g}_1$ with $\mathfrak{c}$ central and $\mathfrak{g}_1 = [\mathfrak{g}, \mathfrak{g}]$ semisimple. We suppose

(1.2.1) if $g \in G$ then $\operatorname{ad}(g)$ is an inner automorphism on $\mathfrak{g}_C$.

We also suppose that the closed normal abelian subgroup $Z \subset G$ has the following properties:

(1.2.2a) Z centralizes the identity component G^0 and $|G/ZG^0| < \infty$.

(1.2.2b) $Z \cap G^0$ is cocompact in the center Z_{G^0} of G^0.

Two comments. If $|G/G^0| < \infty$ then Z_{G^0} satisfies (1.2.2). And $\hat{G}_{\text{disc}}$ is independent of choice of subgroup $Z \subset G$ that satisfies (1.2.2).

Without comment we use the notation

(1.2.3a) $G^\dagger = \{g \in G\colon \operatorname{ad}(g) \text{ is an inner automorphism on } G^0\}$.

Then evidently

(1.2.3b) $G^\dagger = Z_G(G^0)\, G^0$ where $Z_G(G^0)$ is the G-centralizer of G^0.

Note $Z \subset Z_G(G^0)$ with $Z_G(G^0)/Z$ compact. So $ZG^0 \subset G^\dagger$.

1.3. Discrete series for connected groups with compact center. The Harish-Chandra analysis of discrete series for connected reductive acceptable groups extends without change to the groups G^0 of §1.2 for which Z_{G^0} is compact. We state the result.

If G^0 has no compact Cartan subgroup then $(G^0)\hat{}_{\text{disc}}$ is empty.

Let $H^0 \subset G^0$ compact Cartan subgroup. Denote $L=\{\lambda \in i\mathfrak{h}^*: e^\lambda$ is well-defined on $H^0\}$. Choose a positive root system Σ^+ and make the usual definitions:

$$(1.3.1) \quad \varrho=\tfrac{1}{2} \sum_{\varphi\in\Sigma^+} \varphi, \qquad \varpi(\lambda)=\prod_{\varphi\in\Sigma^+} \langle\varphi, \lambda\rangle, \qquad \Delta=\prod_{\varphi\in\Sigma^+} (e^{\varphi/2}-e^{-\varphi/2}).$$

We arrange $\varrho\in L$ by passing to a 2-sheeted "cover" of G if necessary; then Δ is well defined on H^0. Let $L'=\{\lambda\in L: \varpi(\lambda)\neq 0\}$, the regular set in L. If $\lambda\in L'$ then

$$q(\lambda)=|\{\varphi\in\Sigma^+ \text{ compact}: \langle\varphi, \lambda\rangle<0\}|+|\{\varphi\in\Sigma^+ \text{ noncompact}: \langle\varphi, \lambda\rangle>0\}|.$$

Suppose $\lambda\in L'$ and $\xi=e^{\lambda-\varrho}|_{Z_{G^0}}$. Then there is a unique class $[\pi_\lambda]=\omega(\lambda) \in (G^0)\hat{}_{\xi\text{-disc}}$ whose distribution character has restriction to the regular elliptic set given by

$$(1.3.2) \qquad \Theta_{\pi_\lambda}|_{H^0\cap G'}=(-1)^{q(\lambda)}\, \Delta^{-1} \sum_{W(G^0, H^0)} \det(w)\, e^{w\lambda}.$$

Every class in $(G^0)\hat{}_{\text{disc}}$ is one of these $[\pi_\lambda]$, and $[\pi_\lambda]=[\pi_{\lambda'}]$ precisely when λ' is in the Weyl group orbit $W(G^0, H^0)(\lambda)$. Dual class $[\pi_\lambda^*]=[\pi_{-\lambda}]$. The infinitesimal character of $[\pi_\lambda]$ is χ_λ, so the Casimir element goes to $\|\lambda\|^2-\|\varrho\|^2$. Finally, for appropriate normalization of Haar measure, $[\pi_\lambda]$ has formal degree $|\varpi(\lambda)|$.

1.4. Relative discrete series for connected groups. In §1.4 we suppose Z central in G. In particular, our considerations apply to $Z\cap G^0$ in G^0.

Let $S=\{s\in C: |s|=1\}$, the circle group. $1\in\hat{S}$ is defined by $1(s)=s$. Given $\zeta\in\hat{Z}$ we have the quotient group

$$(1.4.1) \qquad G[\zeta]=\{S\times G\}/\{(\zeta(z)^{-1}, z): z\in Z\}.$$

It is the Mackey central extension $1\to S\to G[\zeta]\to G/Z\to 1$ for $\delta\zeta\in Z^2(G/Z; S)$. Anyway, $G[\zeta]$ is a reductive Lie group with Lie algebra $\mathfrak{s}\oplus(\mathfrak{g}/\mathfrak{z})$, with identity component of compact center, and with $|G[\zeta]/G[\zeta]^0|<\infty$. Projection $S\times G \to G[\zeta]$ restricts to a homomorphism

$$(1.4.2) \quad p: G\to G[\zeta] \quad \text{where } f\to f\cdot p \text{ maps } L_2(G[\zeta]/S, 1)\cong L_2(G/Z, \zeta).$$

1.4.3. PROPOSITION. *$\varepsilon[\psi]=[\psi\cdot p]$ defines a bijection $\varepsilon: G[\zeta]\hat{}_1\to\hat{G}_\zeta$ that carries Plancherel measure to Plancherel measure and maps $G[\zeta]\hat{}_{1\text{-disc}}$ onto $\hat{G}_{\zeta\text{-disc}}$. Distribution characters satisfy $\Theta_{\varepsilon[\psi]}=\Theta_\psi\cdot p$.*

We know $G[\zeta]\hat{}_{1\text{-disc}}$ (for connected G) from §1.3. Apply Proposition 1.4.3. Then $(G^0)\hat{}_{\text{disc}}$ is given as follows:

If $G^0/Z\cap G^0$ has no compact Cartan subgroup then $(G^0)^{\widehat{}}_{\text{disc}}$ is empty.

Let $H^0/Z\cap G^0$ be a compact Cartan subgroup of $G^0/Z\cap G^0$. Define L, ϱ, Δ, ϖ, L' and q as in §1.3. Replace G by a 2-sheeted cover if necessary, Z by a subgroup of index 2 if necessary, so that e^{ϱ} is well defined on $H^0/Z\cap G^0$. If $\lambda\in L'$ and $\xi=e^{\lambda-\varrho}|_{Z_{G^0}}$, then there is a unique class $[\pi_\lambda]\in(G^0)^{\widehat{}}_{\text{disc}}$ whose distribution character

$$\Theta_{\pi_\lambda}|_{H^0\cap G'}=(-1)^{q(\lambda)}\,\Delta^{-1}\sum_{W(G^0,H^0)}\det(w)\,e^{w\lambda}. \tag{1.4.4}$$

Every class in $(G^0)^{\widehat{}}_{\text{disc}}$ is one of those $[\pi_\lambda]$. Classes $[\pi_\lambda]=[\pi_{\lambda'}]$ just when $\lambda'\in W(G^0,H^0)(\lambda)$. $[\pi_\lambda^*]=[\pi_{-\lambda}]$. The infinitesimal character of $[\pi_\lambda]$ is χ_λ and the formal degree $d_{\pi_\lambda}=|\varpi(\lambda)|$.

1.5. Relative discrete series in general. One passes from $(G^0)^{\widehat{}}_{\text{disc}}$ to $(G^\dagger)^{\widehat{}}_{\text{disc}}$ by (1.2.3b) and a $\otimes$ construction, then up to $\hat{G}_{\text{disc}}$ by (1.2.1), (1.2.2) and $\text{Ind}_{G^\dagger\uparrow G}$.

Suppose that G/Z has a compact Cartan subgroup H/Z. Let $\lambda\in L'$, $\xi=e^{\lambda-\varrho}|_{Z_{G^0}}$ and $[\chi]\in Z_G(G^0)^{\widehat{}}_{\xi}$. Note $[\chi\otimes\pi_\lambda]\in(G^\dagger)^{\widehat{}}_{\zeta\text{-disc}}$ where $\zeta\in\hat{Z}$ is a summand of $\xi|_Z$. Then

$$[\pi_{\chi,\lambda}]=[\text{Ind}_{G^\dagger\uparrow G}(\chi\otimes\pi_\lambda)] \text{ is in } \hat{G}_{\zeta\text{-disc}}. \tag{1.5.1}$$

Further, every element of $\hat{G}_{\zeta\text{-disc}}$ is one of these $[\pi_{\chi,\lambda}]$.

Choose $\{x_1,\ldots,x_r\}$ representatives of G modulo $G^\dagger$ with $\text{ad}(x_i)\,H=H$. Let $w_i\in W(\mathfrak{g}_C,\mathfrak{h}_C)$ be the element specified (using (1.2.1)) by x_i. Then the distribution character $\Theta_{\pi_{\chi,\lambda}}$ has support in $G^\dagger$, where it is given by

$$\Theta_{\pi_{\chi,\lambda}}(xg)=\sum_{1\le i\le r}\{\text{trace }\chi(x_i^{-1}xx_i)\}\;\Theta_{\pi_{w_i(\lambda)}}(g) \tag{1.5.2}$$

for $x\in Z_G(G^0)$ and $g\in(G^0)'$.

Classes $[\pi_{\chi,\lambda}]=[\pi_{\chi',\lambda'}]$ just when there is an x_i with $[\chi']=[\chi\cdot\text{ad}(x_i)^{-1}]$ and $\lambda'\in W(G^0,H^0)(w_i\lambda)$. Also $[\pi_{\chi,\lambda}]$ has dual $[\pi_{\chi^*,-\lambda}]$, and infinitesimal character χ_λ.

2. The nondegenerate series

2.1. Cuspidal parabolic subgroups. Let K/Z be a maximal compact subgroup of G/Z. In other words, K is the fixed point set of a Cartan involution θ of G. Now choose

$$\{H_1,\ldots,H_l\}:\theta\text{-stable Cartan subgroups of } G \tag{2.1.1}$$

such that every Cartan subgroup is conjugate to just one of the H_i. Stability under θ gives splittings

$$\mathfrak{h}_j=\mathfrak{t}_j\oplus\mathfrak{a}_j \quad\text{and}\quad H_j=T_j\times A_j \tag{2.1.2}$$

where $T_j = H_j \cap K$, $\mathfrak{a}_j = \{x \in \mathfrak{h}_j : \theta x = -x\}$ and $A_j = \exp(\mathfrak{a}_j)$.

The $\mathfrak{a}_j$-roots of $\mathfrak{g}$ are the nonzero real linear functionals φ on $\mathfrak{a}_j$ such that

$$\mathfrak{g}^\varphi = \{x \in \mathfrak{g} : [\alpha, x] = \varphi(\alpha)\, x \text{ for all } \alpha \in \mathfrak{a}_j\} \neq 0.$$

Let $\Sigma_{\mathfrak{a}_j}$ be the $\mathfrak{a}_j$-root system and choose a positive subsystem $\Sigma^+_{\mathfrak{a}_j}$. That specifies

$$\mathfrak{n}_j = \sum_{\varphi \in \Sigma^+_{\mathfrak{a}_j}} \mathfrak{g}^\varphi \quad \text{and} \quad N_j = \exp_G(\mathfrak{n}_j), \tag{2.1.3}$$

and

$$P_j = \{g \in G : \mathrm{ad}(g)\, N_j = N_j\}. \tag{2.1.4}$$

Then P_j is a (real) parabolic subgroup of G with unipotent radical $P^u_j = N_j$. Also $P_j = P^r_j \cdot P^u_j$ (semidirect) $= M_j A_j N_j$ where

$$P^r_j = \{g \in G : \mathrm{ad}(g)\, \alpha = \alpha \text{ all } \alpha \in \mathfrak{a}_j\} = M_j \times A_j. \tag{2.1.5}$$

The P_j are *cuspidal parabolic subgroups* of G. They are characterized by the fact that M_j/Z has a compact Cartan subgroup T_j/Z.

2.1.6. LEMMA. *M_j inherits* (1.2.1) *and* (1.2.2) *from G: Every* $\mathrm{ad}(m)$ *is inner on* $\mathfrak{m}_{jC}$, *Z centralizes M^0_j and $|M_j/ZM^0_j| < \infty$, and $Z \cap M^0_j$ is cocompact in the center of M^0_j.*

2.2. The series for a Cartan subgroup. The relative discrete series of M_j is given as in §1.5. Denote $L_j = \{\nu \in i\mathfrak{t}^*_j : e^\nu \text{ well defined on } T^0_j\}$. Choose a positive $\mathfrak{t}_{jC}$-root system $\Sigma^+_{\mathfrak{t}_j}$ on $\mathfrak{m}_{jC}$. Define $\varrho_{\mathfrak{t}_j}$, $\varpi_{\mathfrak{t}_j}(\nu)$ and $\Delta_{\mathfrak{t}_j}$ as in (1.3.1). We may assume $\varrho_{\mathfrak{t}_j} \in L_j$, thus is in its $\mathfrak{m}_j$-regular set $L''_j = \{\nu \in L_j : \varpi_{\mathfrak{t}_j}(\nu) \neq 0\}$. Let $\nu \in L''_j$, $\xi = \exp(\nu - \varrho_{\mathfrak{t}_j})\,|_{\text{center of } M^0_j}$ and $[\chi] \in Z_{M_j}(M^0_j)\hat{}_\xi$. That gives the relative discrete classes $[\eta_\nu]$ of M^0_j, $[\chi \otimes \eta_\nu]$ of $M^\dagger_j = Z_{M_j}(M^0_j)\, M^0_j$, and $[\eta_{\chi,\nu}] = [\mathrm{Ind}_{M^\dagger_j \uparrow M_j}(\chi \otimes \eta_\nu)]$ of M_j.

$(P^r_j)\hat{}_{\mathrm{disc}}$ consists of the $[\eta_{\chi,\nu} \otimes e^{i\sigma}]$, χ and ν as above and $\sigma \in \mathfrak{a}^*_j$. Extend $\eta_{\chi,\nu} \otimes e^{i\sigma}$ to $P_j = M_j A_j N_j$ by $(\eta_{\chi,\nu} \otimes e^{i\sigma})\,(man) = \eta_{\chi,\nu}(m) \cdot e^{i\sigma}(a)$. Then we have the (unitarily) induced representations

$$\pi_{\chi,\nu,\sigma} = \mathrm{Ind}_{P_j \uparrow G}(\eta_{\chi,\nu} \otimes e^{i\sigma}). \tag{2.2.1}$$

By *H_j-series* of G we mean the set of all unitary equivalence classes of representations (2.2.1). The H_j-series depends only on the conjugacy class of H_j. The various H_j-series are the *nondegenerate series*. Two cases are the relative discrete series (H_j/Z compact) and the principal series (P_j minimal parabolic).

2.3. Nondegenerate series characters. Here are the basic facts. Let $[\pi_{\chi,\nu,\sigma}]$ be an H_j-series class. Then $\pi_{\chi,\nu,\sigma}$ is a finite sum of irreducible classes, and is irreducible itself whenever $\langle\sigma,\varphi\rangle\neq 0$ for every $\mathfrak{h}_{jC}$-root φ of $\mathfrak{g}_C$ such that $\varphi|_{\mathfrak{a}_j}\neq 0$. Every irreducible summand of $\pi_{\chi,\nu,\sigma}$ is in $\hat{G}_\zeta$ where $\zeta\in\hat{Z}$ and $[\chi]\in Z_{M_j}(M_j^0)\hat{}_\zeta$. The class $[\pi_{\chi,\nu,\sigma}]$ has infinitesimal character $\chi_{\nu+i\sigma}$ relative to $\mathfrak{h}_j$. The distribution character $\Theta_{\pi_{\chi,\nu,\sigma}}$ exists and is a locally integrable function with support in the closure of

$$\bigcup_{g\in G^\dagger}\ \bigcup_{H\subset M_jA_j} gHg^{-1}\subset G^\dagger \qquad (H \text{ is any Cartan subgroup of } M_jA_j). \tag{2.3.1}$$

Finally that character is given on $H_j\cap G'$ by

$$\Theta_{\pi_{\chi,\nu,\sigma}}(ta)=|\Delta_{\mathfrak{t}_j}(t)/\Delta_{\mathfrak{h}_j}(ta)|\sum|N_{M_j}(T_j)(w(t))|^{-1}\Psi_{\eta_{\chi,\nu}}(w(t))\,e^{i\sigma}(w(a)) \tag{2.3.2}$$

where $t\in T_j$ and $a\in A_j$. Here $N_G(H_j)$ is the G-normalizer of H_j, the sum runs over the finite set of all $w(ta)$ in $N_G(H_j)(ta)$, $N_{M_j}(T_j)$ is defined similarly, and $\Psi_{\eta_{\chi,\nu}}$ is the character of $\eta_{\chi,\nu}$.

3. Plancherel measure

3.1. Statement of result. Fix $\zeta\in\hat{Z}$. Define

$$L_{j,\zeta}=\{\nu\in L_j: e^\nu\in(T_j^0)\hat{}_\zeta\} \quad\text{and}\quad L''_{j,\zeta}=L_{j,\zeta}\cap L''_j. \tag{3.1.1}$$

Given $\nu\in L''_{j,\zeta}$ and $\sigma\in\mathfrak{a}_j^*$, the corresponding H-series classes that transform by ζ are the $[\pi_{\chi,\nu,\sigma}]$ with $[\chi]\in Z_{M_j}(M_j^0)\hat{}_\zeta$. They give us discrete sums

$$\pi_{j,\zeta,\lambda}=\sum(\dim\chi)\,\pi_{\chi,\nu,\sigma} \quad\text{and}\quad \Theta_{j,\zeta,\lambda}=\Theta_{\pi_{j,\zeta,\lambda}}=\sum(\dim\chi)\,\Theta_{\pi_{\chi,\nu,\sigma}} \tag{3.1.2}$$

where $\lambda=\nu+i\sigma$ and $[\chi]$ runs over the appropriate subset of $Z_{M_j}(M_j^0)\hat{}_\zeta$. Here is our extension of a weak form of Harish-Chandra's Plancherel formula.

3.1.3. THEOREM. *There are unique measurable functions $m_{j,\zeta,\nu}$ on $\mathfrak{a}_j^*$, $\nu\in L''_{j,\zeta}$, with these properties.*

1. *$m_{j,\zeta,\nu}$ is invariant by the Weyl group $W(G,A_j)$.*
2. *If $f\in L_2(G/Z,\zeta)$ is C^∞ with support compact modulo Z, then*

$$\sum_{1\leqq j\leqq l}\ \sum_{\nu\in L''_{j,\zeta}}|\varpi_{\mathfrak{t}_j}(\nu)|\int_{\mathfrak{a}_j^*}|\Theta_{j,\zeta,\nu+i\sigma}(f)\,m_{j,\zeta,\nu}(\sigma)|\,d\sigma<\infty \tag{3.1.4a}$$

and

(3.1.4b) $$f(1)=\sum_{1\leqq j\leqq l}\ \sum_{\nu\in L''_{j,\zeta}} |\varpi_{\mathfrak{t}_j}(\nu)| \int_{\mathfrak{a}_j^*} \Theta_{j,\zeta,\nu+i\sigma}(f)\, m_{j,\zeta,\nu}(\sigma)\, d\sigma .$$

3.2. Reduction from G to ZG^0. Denote $G^1=ZG^0$, $M_j^1=M_j\cap G^1$, etc. Define $\pi^1_{j,\zeta,\lambda}$ as in (3.1.2) on G^1. Then $\pi_{j,\zeta,\lambda}=\mathrm{Ind}_{G^1\uparrow G}(\pi^1_{j,\zeta,\lambda})$. If Theorem 3.1.3 holds for G^1 with functions $m^1_{j,\zeta,\nu}$ now it holds for G with functions $m_{j,\zeta,\nu}=|G/G^1|^{-1}m^1_{j,\zeta,\nu}$.

3.3. Reduction from ZG^0 to $(ZG^0)[\zeta]$. For simplicity now let $G=ZG^0$. Enlarge Z to that $|Z_G/Z|<\infty$. The considerations of §1.4 apply. Define $b: L_j[\zeta]_1 \to L_{j,\zeta}$ by $e^\nu\cdot q=\zeta\otimes e^{b(\nu)}$. If $\lambda=\nu+i\sigma$ let $b(\lambda)=b(\nu)+i\sigma$. In Proposition 1.4.3 we have $[\pi_{j,1,\lambda}\cdot p]=[\pi_{j,\zeta,b(\lambda)}]$ and $\Theta_{j,1,\lambda}\cdot p=\Theta_{j,\zeta,b(\lambda)}$. If Theorem 3.1.3 holds for $(G[\zeta], S, 1)$ with functions $m_{j,1,\nu}$ then it holds for (G, Z, ζ) with functions $m_{j,\zeta,b(\nu)}=m_{j,1,\nu}\cdot p_*$.

3.4. The function E. As seen above, the proof of Theorem 3.1.3 reduces to the case where G is connected, Z_G is a finite extension of the circle group S, and $\zeta=1\in\hat{S}$. We may also assume $K=Z_K^0\times[K,K]$. Then one can construct a class function $E: G\to S$, analytic on the regular set, with the following properties. $E(g)=E(g_{ss})$ where g_{ss} is the semisimple part. If $s\in S$ and $g\in G$ then $E(sg)=sE(g)$. Each $E|_{H_j}\in\hat{H}_j$ with A_j in its kernel. And $E|_K\in\hat{K}$. In effect S is a direct factor of K, $E|_K$ is projection of K to S, and then E is specified by the other properties.

3.5. Reduction from $(ZG^0)[\zeta]$ to $G^0/Z\cap G^0$. We take $(G,Z,\zeta)=(G,S,1)$ as in §3.4. Denote $L_{j,n}=\{\nu\in L_j: e^\nu(s)=s^n \text{ for } s\in S\}$ and $L''_{j,n}=L_{j,n}\cap L''_j$. Define $\varepsilon_j\in L_{j,1}$ by $\exp(\varepsilon_j)=E|_{T_j^0}$, so $L_{j,0}=\{\nu-\varepsilon_j: \nu\in L_{j,1}\}$. Arguing from Harish-Chandra's Plancherel formula, and from the explicit form of G' and the $\Theta_{j,\zeta,\lambda}$, one can prove

3.5.1. PROPOSITION. *Let $f\in L_2(G/S,1)$ be continuous at 1, C^∞ on G', and bounded by a rapidly decreasing function. Let $B_{j,\nu}\subset\mathfrak{a}_j^*$ be sets of Lebesgue measure zero. Suppose $\Theta_{j,1,\lambda}(f)=0$ whenever $1\leqq j\leqq l$, $\lambda=\nu+i\sigma$ with $\nu\in L''_{j,1}$, and $\sigma\in\mathfrak{a}_j^*-B_j$. Then $\Theta_{j,0,\lambda}(Ef)=0$ whenever $1\leqq j\leqq l$ and $\lambda\in L''_{j,0}+i\mathfrak{a}_j^*$.*

Proposition 3.5.1 and Harish-Chandra's formula on G/S give absolutely continuous Borel measures $\mu_{j,1,\nu}$ on $\mathfrak{a}_j^*$ such that

$$f(1)=\sum_{1\leqq j\leqq l}\ \sum_{\nu\in L''_{j,1}} |\varpi(\nu)| \int_{\mathfrak{a}_j^*} \Theta_{j,1,\nu+i\sigma}(f)\, d\mu_{j,1,\nu}(\sigma)$$

in $C_c^\infty(G)\cap L_2(G/S,1)$. Theorem 3.1.3 follows for $(ZG^0)[\zeta]$, then finally for (G, Z, ζ).

3.6. Two consequences. For realization of nondegenerate series representations on partially holomorphic cohomology spaces one needs

3.6.1. COROLLARY. *Suppose that* $\hat{G}_{\text{disc}}$ *is not empty. If* $[\pi] \in \hat{G}$ *let* T_π *be the distribution* $f \mapsto \text{trace} \int_K f(k)\, \pi(k)\, dk$ *on* K. *If* $\zeta \in \hat{Z}$ *then* $\{[\pi] \in \hat{G}_\zeta - \hat{G}_{\zeta\text{-disc}} : T_\pi|_{K \cap G'} \neq 0\}$ *has Plancherel measure zero in* $\hat{G}_\zeta$.

For realization of nondegenerate series representations on spaces of partially-harmonic-spinors one needs

3.6.2. COROLLARY. *Let* $\Omega \in \mathfrak{G}$ *be the Casimir element. If* c *is a number and* $\zeta \in \hat{Z}$ *then* $\{[\pi] \in \hat{G}_\zeta - \hat{G}_{\zeta\text{-disc}} : \chi_\pi(\Omega) = c\}$ *has Plancherel measure zero in* $\hat{G}_\zeta$.

UNIVERSITY OF CALIFORNIA, BERKELEY

GEOMETRIC REALIZATIONS OF REPRESENTATIONS OF REDUCTIVE LIE GROUPS

JOSEPH A. WOLF

1. General idea. Let G be a reductive Lie group, $H=T_H\times A_H$ a Cartan subgroup, and $P_H=MAN$ a cuspidal parabolic subgroup associated to H. We find complex manifolds X on which G acts, and certain orbits $Y_H=G(x_H)\subset X$, such that P_H is the G-stabilizer of the maximal complex analytic piece $S_{[x_H]}$ of Y_H that passes through x_H. This is done so that the isotropy subgroup of G at x_H is UAN with $T\subset U\subset M$, and a certain quotient U/Z is compact. If $[\mu]\in\hat{U}$ and $e^{i\sigma}\in\hat{A}$ then $[\mu\otimes e^{i\sigma}]\in(UAN)\hat{}$ defines a G-homogeneous Hermitian vector bundle $\mathscr{V}_{\mu,\sigma}\to Y_H$ that is holomorphic over the complex analytic pieces. Then G acts on the space $H_2^{0,q}(\mathscr{V}_{\mu,\sigma})$ of L_2 partially harmonic $(0, q)$-forms with values in $\mathscr{V}_{\mu,\sigma}$, by a unitary representation $\pi^q_{\mu,\sigma}$. Roughly speaking we realize every H-series representation of G by the $\pi^q_{\mu,\sigma}$. The relative discrete series, which is an interesting special case, plays a key role.

2. The flag manifold orbits. We work under the following fixed hypotheses. G is a reductive Lie group, i.e., its Lie algebra $\mathfrak{g}=\mathfrak{c}\oplus\mathfrak{g}_1$ with $\mathfrak{c}$ central and $\mathfrak{g}_1$ semisimple. Further

(2.1) if $g\in G$ then $\mathrm{ad}(g)$ is an inner automorphism on $\mathfrak{g}_C$.

Finally, G has a closed normal abelian subgroup Z such that

(2.2a) Z centralizes the identity component G^0 and $|G/ZG^0|<\infty$,

(2.2b) $Z\cap G^0$ is cocompact in the center Z_{G^0} of G^0.

Let $\bar{G}=G^0/Z_{G^0}$ and $\bar{G}_C$ its complexification. If $\bar{P}$ is a parabolic subgroup of $\bar{G}_C$ then by (2.1), G acts on the complex flag manifold $X=\bar{G}_C/\bar{P}$ by: $g(\bar{x}\bar{P})$ is the point at which $\bar{G}_C$ has isotropy group $\mathrm{ad}(g)\,\mathrm{ad}(\bar{x})\,\bar{P}$. This action is holomorphic.

Fix a Cartan subgroup $H\subset G$. One can construct pairs (X, x_H) such that

---

AMS (MOS) subject classifications (1970). Primary 43A80; Secondary 22E45, 43A85.

$x_H \in X$ complex flag, with the following properties: The G-normalizer $N_{[x_H]}$ of the holomorphic arc component (maximal complex analytic piece) $S_{[x_H]}$ of $G(x_H)$ through x_H has the same Lie algebra as a cuspidal parabolic subgroup $P_H = MAN$ associated to H. Further $S_{[x_H]}$ has an $N_{[x_H]}$-invariant positive Radon measure. Finally G has isotropy group UAN at x_H with $T \subset U \subset M$ and U/Z compact. For example, one could take $\bar{P}$ to be a Borel subgroup of $\bar{G}_C$.

We remark that G permutes the holomorphic arc components of $G(x_H)$, and that the component through gx_H (which is $S_{[gx_H]} = gS_{[x_H]}$) has G-normalizer $\mathrm{ad}(g)\, N_{[x_H]}$.

3. Partially harmonic L_2 forms. Let $[\mu] \in \hat{U}$ and $\sigma \in \mathfrak{a}^*$. Denote $\varrho(\alpha) = \frac{1}{2}\mathrm{trace}_n(\mathrm{ad}\,\alpha)$. Define a representation of UAN on the space V_μ of $[\mu]$ by $\gamma_{\mu,\sigma}(uan) = e^{i\sigma+\varrho}(a)\,\mu(u)$. Then we have

$$p: \mathscr{V}_{\mu,\sigma} \to G/UAN = G(x_H) \text{ associated complex vector bundle.} \tag{3.1}$$

There is a unique assignment of complex structures to the pieces $p^{-1}S_{[gx_H]}$, stable under G, such that $\mathscr{V}_{\mu,\sigma}|_{S_{[gx_H]}}$ is a holomorphic vector bundle.

Let $\mathscr{T} \to G(x_H)$ be the complex G-homogeneous bundle such that each $\mathscr{T}|_{S_{[gx_H]}}$ is the holomorphic tangent bundle there. By *partially smooth* (p, q)*-form* with values in $\mathscr{V}_{\mu,\sigma}$ we mean a measurable section of $\mathscr{V}_{\mu,\sigma} \otimes \Lambda^p \mathscr{T}^* \otimes \Lambda^q \overline{\mathscr{T}}^*$ that is C^∞ over each holomorphic arc component. Let $A^{p,q}(\mathscr{V}_{\mu,\sigma})$ denote the space of all such forms. The Dolbeault operator of X specifies operators $\bar{\partial}: A^{p,q}(\mathscr{V}_{\mu,\sigma}) \to A^{p,q+1}(\mathscr{V}_{\mu,\sigma})$. Using K-invariant metrics, where K is the fixed point set of a Cartan involution that stabilizes H, we get Hodge-Kodaira maps

$$A^{p,q}(\mathscr{V}_{\mu,\sigma}) \xrightarrow{\#} A^{n-p,n-q}(\mathscr{V}^*_{\mu,\sigma}) \xrightarrow{\tilde{\#}} A^{p,q}(\mathscr{V}_{\mu,\sigma})$$

where $n = \dim_C S_{[x_H]}$. That specifies a pre-Hilbert space

$$A_2^{p,q}(\mathscr{V}_{\mu,\sigma}) = \left\{ \omega \in A^{p,q}(\mathscr{V}_{\mu,\sigma}) : \int_{K/Z} \left(\int_{S[kx_H]} \omega \bar{\wedge} \# \omega \right) d(kZ) < \infty \right\}.$$

$L_2^{p,q}(\mathscr{V}_{\mu,\sigma})$ is the Hilbert space completion. The partial Hodge-Kodaira-Laplace operator,

$$\square = (\bar{\partial} + \bar{\partial}^*)^2 = \bar{\partial}\bar{\partial}^* + \bar{\partial}^*\bar{\partial}, \qquad \bar{\partial}^* = -\tilde{\#}\bar{\partial}\#, \tag{3.2}$$

is essentially selfadjoint from domain $\{\omega \in A^{p,q}(\mathscr{V}_{\mu,\sigma}) : \mathrm{supp}(\omega) \text{ compact}\}$. Its kernel

$$H_2^{p,q}(\mathscr{V}_{\mu,\sigma}) = \{\omega \in L_2^{p,q}(\mathscr{V}_{\mu,\sigma}) : \square^*(\omega) = 0\} \tag{3.3}$$

is the space of *square-integrable partially harmonic (p, q)-forms* with values in $\mathscr{V}_{\mu,\sigma}$. G acts there by a unitary representation $\pi^{p,q}_{\mu,\sigma}$. We write $\pi^q_{\mu,\sigma}$ for $\pi^{0,q}_{\mu,\sigma}$.

4. Main theorem. Following the notation used in the preceding article, † denotes the elements that give rise to inner automorphisms: $G^\dagger = Z_G(G^0)\,G^0$, $M^\dagger = Z_M(M^0)\,M^0$ and $U^\dagger = Z_U(U^0)\,U^0$. Let $L''_{\mathfrak{t}}$ denote $\{\nu \in i\mathfrak{t}^* : e^\nu$ defined on T^0 and $\mathfrak{m}$-regular$\}$. We are interested in the classes of the $\mu_{\chi,\nu} = \mathrm{Ind}_{U^\dagger\uparrow U}(\chi \otimes \mu_\nu)$ where $\mu_\nu \in \hat{U}^0$ has highest weight ν and $\chi \in Z_U(U^0)^{\wedge}_{\xi}$, $\xi = e^\nu|_{\mathrm{center}(U^0)}$. Note that $\mu_{\chi,\nu}$ is irreducible if $\nu \in L''_{\mathfrak{t}}$.

4.1. THEOREM. *Let $[\mu_{\chi,\nu}] \in \hat{U}$ as above where $\nu + \varrho_{\mathfrak{t}} \in L''_{\mathfrak{t}}$. Let $\sigma \in \mathfrak{a}^*$ and $\pi^q_{\chi,\nu,\sigma}$ be the representation of G on $H^{0,q}_2(\mathscr{V}_{\mu_{\chi,\nu},\sigma})$.*

1. *The irreducible subrepresentations of $\pi^q_{\chi,\nu,\sigma}$ are just its constituents equivalent to irreducible subrepresentations of H-series representations of G. Let $\Theta^H_{\chi,\nu,\sigma,q}$ denote the sum of their distribution characters. Then, in the notation of the preceding article,*

$$\sum_{q \geq 0} (-1)^q\, \Theta^H_{\chi,\nu,\sigma,q} = (-1)^{n+q_H(\nu+\varrho_{\mathfrak{t}})}\, \Theta_{\pi_{\chi,\nu+\varrho_{\mathfrak{t}},\sigma}}. \tag{4.2}$$

2. *There is a constant $b_H \geq 0$ dependent only on $[\mathfrak{m}, \mathfrak{m}]$ such that if $|\langle \nu + \varrho_{\mathfrak{t}}, \psi\rangle| > b_H$ for all $\psi \in \Sigma^+_{\mathfrak{t}}$, and if $q \neq q_H(\nu + \varrho_{\mathfrak{t}})$, then $H^{0,q}_2(\mathscr{V}_{\mu_{\chi,\nu},\sigma}) = 0$.*
3. *If q_0 is an integer such that $q \neq q_0$ implies $H^{0,q}_2(\mathscr{V}_{\mu_{\chi,\nu},\sigma}) = 0$, then $[\pi^{q_0}_{\chi,\nu,\sigma}]$ is the H-series class $[\pi_{\chi,\nu+\varrho_{\mathfrak{t}},\sigma}]$.*

The rest of this article is a brief sketch of the idea of proof of Theorem 4.1.

5. Reduction to discrete series. Let $\eta^q_{\chi,\nu}$ denote the (unitary) representation of M on $H^{0,q}_2(\mathscr{V}_{\mu_{\chi,\nu}})$ where $\mathscr{V}_{\mu_{\chi,\nu}} = \mathscr{V}_{\mu_{\chi,\nu},\sigma}\,|\,_{M(x_H)}$. One can prove

$$\pi^q_{\chi,\nu,\sigma} = \mathrm{Ind}_{P_H\uparrow G}(\eta^q_{\chi,\nu} \otimes e^{i\sigma}). \tag{5.1}$$

The Plancherel theorem (3.1.3) of the preceding article combines with (5.1) to prove the assertion on the irreducible constituents of $\pi^q_{\chi,\nu,\sigma}$ in Theorem 4.1. If one knows the corresponding discrete series result for the $\eta^q_{\chi,\nu}$, then Theorem 4.1 follows by standard H-series considerations.

6. Idea of proof for discrete series. Considerations are reduced to the case where H/Z is compact. Thus $G(x_H)$ is an open submanifold of X with a G-invariant Hermitian metric, and $\pi^q_{\chi,\nu,\sigma}$ is properly written $\pi^q_{\chi,\nu}$.

One checks that $\pi^q_{\chi,\nu}$ is induced from the corresponding representation of $G^\dagger$.

This reduces Theorem 4.1 from G to $G^{\dagger}$. There $\pi^q_{\chi,\nu}=\chi\otimes\pi^q_\nu$ where π^q_ν is the corresponding representation of G^0. In summary we may assume G connected and examine its action π^q_ν on $H_2^{0,q}(\mathscr{V}_{\mu_\nu})$.

We may assume $x_H=1\cdot\bar{P}\in\bar{G}_C/\bar{P}=X$. Root orderings give a Borel subgroup $\bar{B}\subset\bar{P}$ of $\bar{G}_C$. Let $y_H=1\cdot\bar{B}\in\bar{G}_C/\bar{B}=Y$. The holomorphic fibration $Y\to X$ gives a proper holomorphic fibration $G(y_H)\to G(x_H)$. Let $\mathscr{L}_\nu\to G(y_H)$ be the holomorphic line bundle for $e^\nu\in\hat{H}$. An L_2-version of the Leray spectral sequence, using the Borel-Weil theorem extended to U/H, gives $H_2^{0,q}(\mathscr{L}_\nu)\cong H_2^{0,q}(\mathscr{V}_{\mu_\nu})$ unitary equivalence. These reduce Theorem 4.1 further to the case $X=\bar{G}_C/\bar{B}$ and $U=H$.

In the case to which we are reduced, cohomology consisting of the elements of $H_2^{0,q}(\mathscr{L}_\nu)$ can be compared with Lie algebra cohomology. The alternating sum formula (4.2) can then be extracted.

The vanishing theorem (part 2 of Theorem 4.1) is a Lie algebra computation of Griffiths and Schmid.

In the case considered (after our reductions) in part 3 of Theorem 4.1, the alternating sum formula shows that $\pi^{q_0}_\nu$ has relative discrete series component $\pi_{\nu+\varrho}$. A consequence of the Plancherel Theorem (see Corollary 3.6.1 of the preceding article) eliminates other constituents from the direct integral expression of $\pi^{q_0}_\nu$. Thus $\pi^{q_0}_\nu=\pi_{\nu+\varrho}$.

7. Remark on harmonic spinors. One can also follow these considerations with L_2 spinors killed by Dirac operators. The vanishing theorem (Parthasarathy) is better, but the result is not so geometric.

UNIVERSITY OF CALIFORNIA, BERKELEY

THE USE OF PARTIAL DIFFERENTIAL EQUATIONS FOR THE STUDY OF GROUP REPRESENTATIONS

LEON EHRENPREIS*

I. The group $SL(2, \boldsymbol{R})$. Let G denote the group $SL(2, \boldsymbol{R})$. We shall study the three-dimensional representation ϱ which we shall regard as the symmetric square of the fundamental two-dimensional representation. We can regard ϱ as the isomorphism of $G/\pm I$ with the group $SO(1, 2)$ of linear transformations preserving the form $t^2-x^2-y^2$. We shall usually write $g\cdot(t, x, y)$ for $\varrho(g)\cdot(t, x, y)$. We write (u, v, w) for dual variables to (t, x, y).

Now, G leaves invariant the wave operator

$$\Box=\frac{\partial^2}{\partial t^2}-\frac{\partial^2}{\partial x^2}-\frac{\partial^2}{\partial y^2}.$$

Thus, we have a representation of G on the space of solutions W_λ of $\Box f=\lambda f$ for any λ. The classical theory of separation of variables suggests that, to find irreducible subspaces of W_λ, we should try to solve

$$\Box f=\lambda f, \tag{1}$$

$$f(r, \theta)=f_1(r)f_2(\theta). \tag{2}$$

Here r, θ are a form of polar coordinates, that is, $r^2=\pm(t^2-x^2-y^2)$, the $\pm$ sign taken so as to make $r^2\geqq 0$ and θ is a coordinate on the hyperboloids $r^2=\text{constant}$.

We can look at things from another point of view. By the fundamental principle (see [1]) solutions of $\Box f=\lambda f$ can be represented as Fourier integrals of suitable measures on the hyperboloids $-u^2+v^2+w^2=\lambda$. (The Fourier transform is defined using the inner product $(t, x, y)\cdot(u, v, w)=tu-xv-yw$.) These hyperboloids fall naturally into three classes:

AMS (MOS) subject classifications (1970). Primary 22E24, 35C15, 35L05.

* Work supported by NSF grant GP-28499.

(a) $\lambda=0$, *light cone,*

(b) $\lambda<0$, *hyperboloid of two sheets,*

(c) $\lambda>0$, *hyperboloid of one sheet.*

(Actually, if λ is not real there are other possibilities which we shall not discuss here.)

Now, the Fourier transform or rather the bilinear form $tu-xv-yw$ gives a way of letting G act on the light cone and the hyperboids. In terms of this action we see that:

(A) The positive light cone, $\lambda=0$, $u>0$, is of the form G/MN.

(B) The positive hyperboloid of two sheets, $\lambda<0$, $u>0$, is of the form G/K.

(C) The hyperboloid of one sheet $\lambda>0$ is of the form G/A.

Here we are using the usual notation K, M, A, N for subgroups of G. Thus the method of separation of variables, in view of the fact that the representation of G on W_λ corresponds to the representation on functions on $-u^2+v^2+w^2=\lambda$ by the fundamental principle, attempts to decompose the representation of G on functions on G/MN, G/A, and G/K.

We wish to point out some important ideas which fit nicely into the above picture:

1. *Cauchy and Dirichlet problems.* A solution of the homogeneous wave equation $\Box f=0$ in the forward light cone, that is, $t^2 \geqq x^2+y^2$, $t \geqq 0$, can be determined by either its *Cauchy data* on G/K, that is, the restriction of f and its normal derivative to $t^2-x^2-y^2=1$, $t>0$, or by its Dirichlet data on $t^2=x^2+y^2$, $t \geqq 0$. The explicit formulas are due to Hadamard (see [**2**]). In fact, this correspondence sets up an isomorphism between $L^2(G/K) \oplus L^2(G/K)$ and $L^2(G/MN)$. This isomorphism is clearly intertwining for G and thus explains why only the class 1 principal series arises in $L^2(G/K)$. (Since, by definition, only the class 1 principal series appears in the decomposition of $L^2(G/MN)$.)

2. *Eisenstein series and Eisenstein integrals.* The fundamental principle suggests that we write solutions of $\Box f=0$ as Fourier integrals on G/MN (actually, the whole light cone which is two copies of G/MN, but this is not important). The decomposition of $L^2(G/MN)$ is accomplished by right action of A on G/MN, which is meaningful since A normalizes MN. In accordance with (1) and (2), we take the Fourier transform of functions on the light cone which are homogeneous. In particular, the Fourier transform of the K-invariant homogeneous function on the light cone, when restricted to G/K, is a constant times the elementary spherical function which is readily computed:

$$\begin{aligned} \varphi_s(a) &= \frac{1}{\Gamma(s)} \int \exp\{ir[(a^2+a^{-2})/2+((a^2-a^{-2})/2)\cos \pi\theta]\}\, r^s\, dr \\ &= \int [(a^2+a^{-2})/2+((a^2-a^{-2})/2)\cos 2\pi\theta]^{-s}\, d\theta . \end{aligned} \tag{3}$$

The first integral expresses φ_s in the form that Harish-Chandra terms "Eisenstein integral."

If, instead of trying to find the K-invariant homogeneous function, we sought a function which was invariant under the modular group $\Gamma = SL(2, \mathbf{Z})$, then we could try to start with the sum of δ functions at an invariant lattice on G/MN, take its Mellin transform, and then take the Fourier transform. When restricted to G/K we obtain in this way the Eisenstein series.

It should be noted that Fourier transform, on the one hand, and the relation between the Dirichlet and Cauchy problems, on the other hand, are two intertwinings between G/K and G/MN. This accounts for the functional equations for the Eisenstein series and Eisenstein integrals.

3. *Fundamental solutions and the Euler formula.* We are now going to make use in full of the fact that $\Box$ is *hyperbolic*. This means that we can find, uniquely, two fundamental solutions $e^{\pm}$ with singularity at (1, 0, 0) such that

(4) $$\Box e^{\pm} = \delta_{(1,0,0)},$$

(5) $$\text{Support } e^{\pm} \subset (1, 0, 0) + \{t^2 \geqq x^2 + y^2,\ t \geqq 0 \text{ (resp. } t \leqq 0)\}.$$

Relation (2) suggests that we take the Mellin transform of $e^{\pm}$, since, for $\lambda = 0$, the only choice for $f_2(r)$ is r^s. If we do this, we obtain K-invariant homogeneous functions $E_s^{\pm}(r, \theta)$ in the forward light cones. From (4) and (5) we easily deduce that

(6) $$[\Delta + s(1-s)]\, E_s^{\pm}(1, \theta) = \delta_{(1,0,0)}.$$

Here Δ is the Laplacian on G/K.

(7) $E_s^{\pm}(1, \theta)$ is, for $\theta \neq (1, 0, 0)$, a function of s which is holomorphic and bounded in $\mathrm{Re}(s) \geqq -1$ $(\mathrm{Re}(s) \leqq 0)$.

Using the explicit form for $e^{\pm}$ we see that $E_s^{\pm}(1, \theta)$, which is K-invariant and hence is a function of a, is just the associated Legendre function $Q_{-s}(a)$ [resp. $Q_{s-1}(a)$]. Notice that $\Box(e^+ - e^-) = 0$ so $[\Delta + s(1-s)]\,(E_s^+ - E_s^-) = 0$. This suggests that $E_s^+ - E_s^-$ is a multiple of φ_s which is the Legendre function $P_{-s}(a)$. On using the explicit formula for $e^{\pm}$ it is easy to deduce the classical formula of Hobson

(8) $$\pi P_s(a) = \tan \pi s\, [Q_s(a) + Q_{-s-1}(a)].$$

We refer to (8) as the *Euler formula*.

In view of (7) it is easy to invert (8) to obtain, for $\mathrm{Re}(s) > \frac{1}{2}$,

(9) $$Q_s(a) = \int_{RZ = -1/2} \frac{P_Z(a)\,(Z + \frac{1}{2}) \coth \pi Z}{Z(1-Z) - s(1-s)}\, dZ.$$

Using (6) we now obtain the Plancherel formula

$$\delta_{a=1} = \int_{RZ=-1/2} P_Z(a)\,(Z+\tfrac{1}{2})\coth \pi Z\, dZ. \tag{10}$$

4. *Discrete series.* We remark here only that the space of all discrete series can be identified with the space of solutions $\Box f=0$ which vanish inside the light cone.[1] When we express such solutions as Fourier integrals over the light cone, we obtain the usual way of writing the discrete series.

II. General semisimple Lie groups. We cannot go into the details of the extension of the above ideas to general semisimple Lie groups. Actually, we have thus far carried out the method only for the groups $SL(n, \boldsymbol{R})$ and $Sp(n, \boldsymbol{R})$. There does not seem to be any essential obstacle to carrying out our program for general semisimple groups in normal form (Chevalley groups over $\boldsymbol{R}$).

The first problem is to find the correct analog for a semisimple group of the three-dimensional representation of $SL(2, \boldsymbol{R})$. We want a representation ϱ of the semisimple group G with the following properties:

(a) There is an orbit which is G/MN.

(b) Right action of A on G/MN is the restriction to the "cone" G/MN of global linear transformations commuting with G. (We say that the action of A is contained in the scalar group of the representation.)

(c) There is an orbit which is G/K.

Following a suggestion of Bert Kostant, the correct representation to take is the direct sum of the symmetric squares of the fundamental representations. For this representation we can carry out 1, 2, 3 above, that is, we can set up Cauchy and Dirichlet problems and the corresponding intertwining operator; we can construct Eisenstein series and integrals; we can find suitable fundamental solutions, the Euler formula (which now expresses φ_s as a sum over the Weyl group), and the Plancherel formula for G/K. However, we have up to now succeeded in constructing only some discrete series in a suitable manner.

The details will appear in a book now under preparation.

References

1. L. Ehrenpreis, *Fourier analysis in several complex variables*, Pure and Appl. Math., vol. 17, Interscience, New York, 1970. MR **44** #3066.

2. M. Riesz, *L'intégrale de Riemann-Liouville et le problème de Cauchy*, Acta Math. **81** (1949), 1–223. MR **10**, 713.

Yeshiva University

[1] This result was observed independently by Robert Strichartz.

SOME REMARKS ON BOUNDARY BEHAVIOR OF ANALYTIC FUNCTIONS

CHARLES FEFFERMAN

In the early days of harmonic analysis most real-variable questions could be attacked only by using complex function theory. Over the years, real-variable theory grew in scope and power, until today it provides us with tools to handle the problems of complex function theory. This note mentions some of the tools, and sketches some old and some new work on the use of real variables to study boundary behavior of analytic functions.

We begin with an analytic function $F(y, t)$ defined on the upper half-plane $R^2_+ = \{(y, t) \mid t > 0\}$. To avoid technicalities, suppose $F \to 0$ at infinity. Two basic quantities connected with the boundary behavior of F are the *maximal function* $F^*(x) = \sup_{|y-x|<t} |F(y, t)|$, and the *Lusin area integral*

$$S(F)(x) = \left(\iint_{|y-x|<t} |F'(y, t)|^2 \, dy \, dt\right)^{1/2},$$

both defined on R^1.

$F^*(x)$ measures the size of F near x, and so is of obvious interest for us. $S(F)$ is more profound, for it measures the manner in which F becomes large, by looking at the size of its derivative. The following standard results (see [**BGS**], [**S**], [**Z**]) illustrate the uses of F^* and $S(F)$.

THEOREM 1. *Fix p $(0 < p < \infty)$. The following are equivalent.*

(a) *F has "boundary values" in $L^p(R^1)$, i.e., $F \in H^p$.*

(b) *F^* belongs to $L^p(R^1)$.*

(c) *$S(F)$ belongs to $L^p(R^1)$.*

THEOREM 2. *At almost every point $x \in R^1$, the following are equivalent.*

AMS (MOS) subject classifications (1970). Primary 30A78, 31B25; Secondary 32A30.

(a) *$F(y, t)$ converges to a limit as (y, t) tends "nontangentially" to x.*
(b) $F^*(x) < \infty$.
(c) $S(F)(x) < \infty$.

Actually, Theorem 2 and (b)⇔(c) of Theorem 1 also hold for harmonic functions u if we define

$$u^*(x) = \sup_{|y-x|<t} |u(y, t)| \quad \text{and} \quad S(u)(x) = \left(\iint_{|y-x|<t} |\nabla u(y, t)|^2 \, dy \, dt\right)^{1/2}.$$

Now the key point in this discussion is that the area integral provides a link between the real and imaginary parts of the analytic function $F = u + iv$. For we know from the Cauchy-Riemann equations that $|\nabla u| = |\nabla v| = |F'|$ everywhere, so that trivially $S(u) = S(F)$. Theorems 1 and 2 and their harmonic analogues show that the basic properties of F and F^* already show up in the size of u^*; for instance $F \in H^p \Leftrightarrow S(F) \in L^p \Leftrightarrow S(u) \in L^p \Leftrightarrow u^* \in L^p$. Thus we may redefine H^p – it consists of all harmonic functions u satisfying $u^* \in H^p$.

So far, we have achieved part of our goal of translating H^p and similar complex-variable ideas into real-variable terms. Starting with analytic functions $u + iv$, we have been able to get rid of v, and reduce our problems to u; now we have to get rid of u. For simplicity, we first restrict attention to H^1.

Any reasonable harmonic function u arises as the Poisson integral of some distribution $u = P_t * f$; and since we are studying $u \in H^1$, it follows that $f \in L^1$. Let us investigate some functions $f \in L^1(R^1)$ to see whether their Poisson integrals belong to H^1. Two typical L^1 functions are $f_\varepsilon(x) = \varepsilon^{-1} f(x/\varepsilon)$ and $g_\varepsilon(x) = \varepsilon^{-1} g(x/\varepsilon)$, where $f, g \in C_0^\infty(R^1)$ and, say, $f \geqq 0$ while g has total integral zero. A few moments' thought shows that if u_ε is the Poisson integral of f_ε, then $U_\varepsilon^*(x) \sim 1/(\varepsilon + |x|) \notin L^1$, so that $u_\varepsilon \notin H^1$: However, the Poisson integral W_ε of g_ε is much smaller than u_ε, because of the cancellation in $W_\varepsilon(x, t) = \int_{R^1} P_t(y) \, g_\varepsilon(x-y) \, dy$. Thus $W_\varepsilon^* \sim \varepsilon/(\varepsilon^2 + x^2) \in L^1$, and W_ε belongs to H^1. Our two examples illustrate the heuristic meaning of H^1. Poisson integrals of L^1 functions f are in H^1 if and only if f has a great deal of cancellation. The essential idea is now that "cancellation" of a distribution f is a natural intuitive property of f, having nothing to do with harmonic functions. We may define it intrinsically as follows: Let f be a distribution and $\{\varphi_\varepsilon\}$ a reasonable, smooth approximate identity on R^1. Form the maximal function $f^*(x) = \sup_{\varepsilon > 0} |\varphi_\varepsilon * f(x)|$, and say that f is in H^p if f^* belongs to L^p. If $\{\varphi_\varepsilon\}$ is the Poisson kernel, this is consistent with our old definition of H^p. However, as suggested by our examples f_ε and g_ε, the assertion $f^* \in L^p$ is entirely a property of f, and does not depend on the choice of $\{\varphi_\varepsilon\}$. Hence, we have achieved our goal of transferring the theory of H^p-spaces into a real-variable setting. Although there is not enough space here to explain, several of the hardest H^p-theorems of classical Fourier analysis can be simplified and sharpened by using the real-variable definition (see [**FS**]).

We conclude with a few words on functions of several variables. The ideas sketched above go over essentially intact to the Cauchy-Riemann systems in [**S**] and also to the study of analytic functions on the unit ball in $\boldsymbol{C}^n$. (Here, one defines H^p-spaces of distributions on the Heisenberg group, using the natural family of dilations to form approximate identities.) However, the polydisc presents extremely difficult new problems, about which very little is known so far. We forego a detailed discussion of the (somewhat confusing) state of affairs but only point out that some of the less difficult problems involve curious variants of the classical area integral, such as

$$\tilde{S}(F)(y_1, y_2) = \left(\iint_{|y_1 - x_1| < t_2;} \iint_{|y_2 - x_2| < t_2} \left| \frac{\partial F}{\partial x_1}(x_1 + it_1, x_2 + it_2) \right|^2 \cdot \left| \frac{\partial F}{\partial x_2}(x_1 + it_1, x_2 + it_2) \right|^2 dx_1\, dx_2\, dt_1\, dt_2 \right)^{1/4}.$$

References

[**BGS**] D. L. Burkholder, R. F. Gundy and M. L. Silverstein, *A maximal function characterization of the class H^p*, Trans. Amer. Math. Soc. **157** (1971), 137–153. MR **43** #527.

[**FS**] C. Fefferman and E. M. Stein, *H^p-spaces of several variables*, Acta Math. **129** (1972), 137–193.

[**S**] E. M. Stein, *Singular integrals and differentiability properties of functions*, Princeton Math. Series, no. 30, Princeton Univ. Press, Princeton, N.J., 1970. MR **44** #7280.

[**Z**] A. Zygmund, *Trigonometrical series*, Vols. I, II, 2nd rev. ed., Cambridge Univ. Press, New York, 1959. MR **21** #6498.

University of Chicago

BOUNDARY BEHAVIOR OF POISSON INTEGRALS

LARS-ÅKE LINDAHL

Let G be a connected semisimple Lie group with finite center, let $G=KAN$ be an Iwasawa decomposition and write each element $g \in G$ as $g=k(g)(\exp H(g))\,n(g)$ with $H(g) \in \mathfrak{a}$, the Lie algebra of A. We shall say that $a \in A$ tends to ∞ if $\lambda(H(a)) \to +\infty$ for every positive restricted root λ. Let 2ϱ be the sum of the positive restricted roots. Finally, let M be the centralizer of A in K and put $\bar{N}=\theta N$, where θ is the Cartan involution.

The spaces K/M and $\bar{N}$ can be considered as boundaries of the symmetric space G/K, and to each of these boundaries there is associated a Poisson kernel and a Poisson integral. The two kernels are related by the map $\bar{n} \to k(\bar{n})\,M$ which is a diffeomorphism of $\bar{N}$ onto an open set in K/M, whose complement has lower dimension. Therefore, it is enough to consider the Poisson kernel relative to $\bar{N}$ which is defined by

$$P(gK, \bar{n}) = \exp\{-2\varrho(H(g^{-1}\bar{n}))\}.$$

The Poisson integral F of a function $f \in L^p(\bar{N})$, $1 \leqq p \leqq \infty$, is a function on G/K given by

$$F(gK) = \int_{\bar{N}} f(\bar{n})\, P(gK, \bar{n})\, d\bar{n}.$$

This makes sense, since $P(gK, \cdot) \in L^p(\bar{N})$ for every $p \geqq 1$. Putting $\psi(\bar{n}) = P(K, \bar{n})$ we obtain

$$F(\bar{n}_0 aK) = \int_{\bar{N}} f(\bar{n}_0 a \bar{n} a^{-1})\, \psi(\bar{n})\, d\bar{n}. \tag{1}$$

AMS (MOS) subject classifications (1970). Primary 43A85; Secondary 22E30.

We shall assume that the Haar measure $d\bar{n}$ is normalized so that $\int \psi(\bar{n})\, d\bar{n}=1$. Since $a\bar{n}a^{-1}\to e$ as $a\to\infty$ we obtain the following:

PROPOSITION 1. *If* $f\in L^\infty(\bar{N})$ *is continuous at* $\bar{n}_0$ *then* $F(\bar{n}_0 aK)\to f(\bar{n}_0)$ *as* $a\to\infty$.

It follows from this that $P(aK,\cdot)$ is an approximate identity. Standard arguments now give

PROPOSITION 2. *Let* $F_a(\bar{n})=F(\bar{n}aK)$. *As* $a\to\infty$, $F_a\to f$
(i) *in* L^p *if* $f\in L^p$ *and* $1\leqq p<\infty$,
(ii) *weak-* against* L^1 *if* $f\in L^\infty$,
(iii) *uniformly if* f *is bounded and uniformly continuous on* $\bar{N}$.

The reader is referred to [4] for details.

The question of convergence a.e. is of course harder to settle. The first Fatou type theorem was proved by Helgason and Korányi [2], and it can be stated as follows:

THEOREM 1. *Let* $f\in L^\infty(\bar{N})$ *and fix* H *in the positive Weyl chamber* $\mathfrak{a}^+$. *Then* $F(k(\bar{n})(\exp tH)K)\to f(\bar{n})$ *a.e. as* $t\to+\infty$.

For rank one spaces G/K, Knapp [3] proved that L^∞ could be replaced by L^1 (and finite signed measures) in the above theorem.

Following Korányi [5] we now define the type of convergence that generalizes the classical nontangential convergence. The function h on G/K is said to converge to the function f on $\bar{N}$ at $\bar{n}_0$ *admissibly and unrestrictedly* if, for each nonempty compact set $U\subset\bar{N}$,

$$\lim_{a\to\infty} h(\bar{n}_0 a\bar{n}K)=f(\bar{n}_0)$$

uniformly for $\bar{n}\in U$.

In [4], [5] and [6], Theorem 1 was extended to admissible unrestricted convergence a.e. for L^∞-functions (and for L^1-functions in the rank one case). For the polydisc which corresponds to the case when G is a direct sum of copies of $SL(2;\boldsymbol{R})$, it is a classical result of Marcinkiewicz and Zygmund that F converges to f admissibly and unrestrictedly a.e. if $f\in L^p$ and $p>1$, and that there are examples of $f\in L^1$ such that admissible unrestricted convergence fails (see e.g. the last chapter of [8]). It seems reasonable to conjecture that in any symmetric space the Poisson integral of an L^p-function, $p>1$, should converge admissibly and unrestrictedly a.e. The best general result in that direction seems to be the following theorem [7]; similar results have also been obtained by A. W. Knapp and E. Stein (unpublished).

THEOREM 2. *Let $H_0 \in \mathfrak{a}^+$ be the element such that $\lambda(H_0)=1$ for all simple restricted roots λ and put $p_0 = 2\varrho(H_0)-1$. If $p>p_0$ and $f \in L^p(\bar{N})$ then F converges to f admissibly and unrestrictedly a.e. When $G=SL(n;\boldsymbol{R})$, p_0 can be replaced by $n-2$.*

The proof of Theorem 2 follows classical lines. Let, for each compact set $U \subset \bar{N}$,

$$M_U f(\bar{n}) = \sup\{|F(\bar{n}a\bar{n}_1 K)|;\, a \in A,\, \bar{n}_1 \in U\}.$$

Since the theorem holds for bounded continuous functions f, it suffices to prove that there is a constant C_p (depending on U) such that

$$\|M_U f\|_p \leqq C_p \|f\|_p, \qquad p>p_0. \tag{2}$$

Since $P(\bar{n}_0 a\bar{n}_1 K, \bar{n}) \leqq \text{const}\, P(\bar{n}_0 aK, \bar{n})$ for $\bar{n}_1 \in U$, it is enough to prove (2) when $U = \{e\}$, and this is done by comparing $Mf = M_{\{e\}}f$ with a suitable Hardy-Littlewood maximal function on $\bar{N}$. Introduce a basis $X_1, \ldots, X_\nu$ in $\bar{\mathfrak{n}}$ such that (i) each X_j belongs to some root space $\mathfrak{g}_{-\lambda_j}$ say, and (ii) $[X_i, X_j]$ belongs to the span of $X_1, \ldots, X_{j-1}$ for all i, j. Then the map

$$\varphi: (x_1, \ldots, x_\nu) \to (\exp x_1 X_1)\cdots(\exp x_\nu X_\nu)$$

is a diffeomorphism of R^ν onto $\bar{N}$. Define the following maximal function

$$f^*(\bar{n}_0) = \sup (\operatorname{meas}(\omega))^{-1} \int_{\omega=\varphi(I_1 \times \cdots \times I_\nu)} |f(\bar{n}_0\bar{n})|\, d\bar{n},$$

where the I_j range over all symmetric open intervals around 0.

Using induction it can be shown that $\|f^*\|_p \leqq C_p \|f\|_p$, $p>1$ ([**4**], [**6**]). Hence (2) will follow from the estimate

$$\|Mf\|_p \leqq C_p \|f^*\|_p, \quad p>p_0. \tag{3}$$

For this, some information is needed about the behavior of $\psi(\bar{n})$ at infinity. Put, for $\bar{n} = \varphi(x_1, \ldots, x_\nu)$,

$$|\bar{n}| = \max_{1 \leqq j \leqq \nu} |x_j|^{1/\lambda_j(H_0)}.$$

Then

$$\int_{\bar{N}} \psi(\bar{n})^{1/2+\varepsilon}\, d\bar{n} < \infty \quad \text{if } \varepsilon>0, \tag{4}$$

$$(5) \qquad \psi(\bar n) \leqq \operatorname{const} |\bar n|^{-2}.$$

(4) is well known and (5) follows from the inequality

$$1+\exp\{-2t+2\varrho(H(\bar n))\} \geqq \exp 2\varrho(H(\bar n^{\exp tH_0})), \quad t>0$$

[**1**, p. 290] and the fact that $\psi \in C_0(\bar N)$.

In order to estimate Mf we now split the integral (1) into a sum of integrals over the sets where $2^{-j} < \psi(\bar n) \leqq 2^{-j+1}$, $j=1, 2, \ldots$, then apply Hölder's inequality and use (4) and (5). With $B_j = \{\bar n ; |\bar n| < 2^{j/2}\}$ we obtain finally

$$|F(\bar n a K)| \leqq \operatorname{const} ((f^p)^*(\bar n))^{1/p} \cdot \sum_{j=1}^{\infty} (\operatorname{meas} B_j)^{1/p} ((2^{-j})^{q-1/2-\varepsilon})^{1/q}.$$

When $p > p_0$ the sum turns out to be finite. Hence

$$Mf(\bar n) \leqq C_p((f^p)^*(\bar n))^{1/p}, \quad p > p_0,$$

from which (3) follows.

For details and results concerning other boundaries than the maximal one we refer to [**7**].

References

1. Harish-Chandra, *Spherical functions on a semisimple Lie group*. I, Amer. J. Math. **80** (1958), 241–310. MR **20** #925.

2. S. Helgason and A. Korányi, *A Fatou-type theorem for harmonic functions on symmetric spaces*, Bull. Amer. Math. Soc. **74** (1968), 258–263. MR **37** #4753.

3. A. W. Knapp, *Fatou's theorem for symmetric spaces*. I, Ann. of Math. (2) **88** (1968), 106–127. MR **37** #1528.

4. A. W. Knapp and R. E. Williamson, *Poisson integrals and semisimple groups*, J. Analyse Math. **24** (1971), 53–76.

5. A. Korányi, *Boundary behavior of Poisson integrals on symmetric spaces*, Trans. Amer. Math. Soc. **140** (1969), 393–409. MR **39** #7132.

6. ———, *Harmonic functions on symmetric spaces*, Symmetric Spaces, Dekker, New York, 1972.

7. L.-Å. Lindahl, *Fatou's theorem for symmetric spaces*, Ark. Mat. **10** (1972), 33–47.

8. A. Zygmund, *Trigonometrical series*, 2nd ed., Cambridge Univ. Press, New York, 1959. MR **21** #6498.

Institut Mittag-Leffler
Djursholm, Sweden

GENERALIZED POISSON INTEGRALS AND THEIR BOUNDARY BEHAVIOR

H. LEE MICHELSON

1. Characterization of the spaces $H_\lambda^p(X)$. Let X be a Riemannian symmetric space homeomorphic to a Euclidean space. $X = G/K$, where G is the direct product of $\boldsymbol{R}^n$ and a real semisimple Lie group G' with finite center, and K is a maximal compact subgroup of G' (and hence of G). We write an Iwasawa decomposition $G = KAN$, where $A = \boldsymbol{R}^n \times A'$ and $G' = KA'N$ is an Iwasawa decomposition of G'. For $g \in G$ we write $k(g) \in K$, $H(g) \in \mathfrak{A}$, the Lie algebra of A, such that $g \in k(g) \exp H(g)\, N$. Let $\mathfrak{A}^*$ be the dual of $\mathfrak{A}$ and $\mathfrak{A}_+^*$ be the positive Weyl chamber in $\mathfrak{A}^*$. Let P^+ be the set of positive restricted roots, ϱ be their half-sum with multiplicities. Let M be the centralizer of $\mathfrak{A}$ in K, dk_M be the K-invariant probability measure on K/M. For $\lambda \in \mathfrak{A}^* - i\bar{\mathfrak{A}}_+^*$ let

$$\phi_\lambda(gK) = \int_{K/M} \exp\{-(i\lambda + \varrho)\,(H(g^{-1}k))\}\, dk_M,$$

$$P_\lambda(gK, kM) = [\phi_\lambda(gK)]^{-1} \exp\{-(i\lambda + \varrho)\,(H(g^{-1}k))\},$$

$$F_\lambda(x, kM) = \int_{K/M} P_\lambda(x, kM)\, f(kM)\, dk_M,$$

for f a function, measure, distribution, etc. on K/M. We call f the "boundary values" of F_λ. Then $\phi_\lambda F_\lambda$ is an eigenfunction of all G-invariant differential operators on X. For rank $X = 1$ Helgason has shown that all such eigenfunctions are of this form for f some "functional" on K/M; for the non-Euclidean disc the functionals which occur are precisely the analytic functionals. For X of arbitrary

AMS (MOS) subject classifications (1970). Primary 53C35, 43A85.

rank, Karpelevič has shown that nonnegative eigenfunctions exist precisely for member $\lambda \in -i\mathfrak{A}_+^*$ and that in that case all such nonnegative eigenfunctions are of the form F_λ with f a finite regular positive measure on K/M. Furstenberg and the author have shown the following:

THEOREM 1. *For $\lambda \in -i\mathfrak{A}_+^*$, let $H_\lambda^p(X)$ be the set of $F: X \to \boldsymbol{C}$ with the following properties:*

(1) $F \cdot \phi_\lambda$ is an eigenfunction of the Laplace-Beltrami operator Δ with $\Delta(F \cdot \phi_\lambda)/(F \cdot \phi_\lambda) = \Delta\phi_\lambda/\phi_\lambda$.

(2) $\|F\|_{H_\lambda^p(X)} =_{\text{def}} \sup_{x \in X} \|F^x\|_p < \infty$, where $F^x: K \to X$ is defined by $F^x(k) = F(k \cdot x)$.

Then $H_\lambda^p(X)$ consists of the set of functions F_λ for $f \in L^p(K/M)$, $1 < p \leqq \infty$, and for f a finite regular signed measure on K/M, $p = 1$. The correspondence $f \leftrightarrow F_\lambda$ is an isometric bijection.

COROLLARY 1. *If $F \in H_\lambda^p(X)$, $F \cdot \phi_\lambda$ is an eigenfunction of every G-invariant differential operator D on X, with $D(F \cdot \phi_\lambda)/(F \cdot \phi_\lambda) = D\phi_\lambda/\phi_\lambda$.*

COROLLARY 2. *If $X = \boldsymbol{R}^n$, $H_\lambda^p(X)$ consists precisely of the constants.*

REMARK ON THE PROOF OF THEOREM 1. For $p = \infty$ the proof is the same as that of Karpelevič for the harmonic case ($\lambda = -i\varrho$). For $1 \leqq p < \infty$, one considers convolutions of F with functions from $L^q(K)$ which approximate the identity in $L^1(K)$ and passes to the limit.

An analogue of Theorem 1 for other $\lambda \in \mathfrak{A}^* - i\mathfrak{A}_+^*$ does not appear to be a reasonable conjecture because the kernel P_λ is no longer everywhere nonnegative.

2. Fatou theorems for F_λ. We are interested in Fatou-type theorems for convergence of $F_\lambda^{a \cdot x}$ to $\tilde{f}$ $[\tilde{f}(k) = f(kM)]$ as $a \to +\infty$ $[(\forall \alpha \in P^+)\, \alpha(\log a) \to +\infty]$. Throughout this section, except as otherwise stated, the conditions on λ will be $\lambda \in \mathfrak{A}^* - i\mathfrak{A}_+^*$ or rank $X = 1$ and $\lambda = 0$. Asymptotic behavior of ϕ_λ at $\lambda \in -i\partial\mathfrak{A}_+^*$ is not well known for rank $X > 1$, and asymptotic behavior of ϕ_λ for $\lambda \in \mathfrak{A}^*$ yields no Fatou-type results even for rank $X = 1$. The asymptotic behavior of ϕ_λ for λ which we consider is given by

$$
\begin{aligned}
\phi_\lambda(kaK) &\sim c(\lambda)\exp\{(i\lambda - \varrho)(\log a)\} && \text{as } a \to +\infty,\ \lambda \in \mathfrak{A}^* - i\mathfrak{A}_+^*,\\
&\sim \text{const} \cdot \varrho(\log a)\exp\{-\varrho(\log a)\} && \text{as } a \to +\infty,\ \lambda = 0,\ \text{rank } X = 1.
\end{aligned}
$$

The following "soft" Fatou-type theorem now follows easily from elementary properties of P_λ.

THEOREM 1. $\lim_{a\to+\infty} F_\lambda^{a\cdot x}=\tilde{f}$

(i) *uniformly*, $f\in C(K/M)$,

(ii) *in* $L^p(K)$, $f\in L^p(K/M)$, $1\leqq p<\infty$,

(iii) *weak* against* $L^1(K)$, $f\in L^\infty(K/M)$,

(iv) *weak* against* $C(K)$, *f a finite regular signed measure on* K/M,

(v) *weak* against* $C^\infty(K)$, *f a distribution on* K/M.

We are interested, however, in pointwise results. We shall consider pointwise results of two kinds:

(1) on a Lebesgue point set of f (defined below),

(2) on an unspecified set of full measure in K/M.

We shall see that all local results known for the harmonic case ($\lambda=-i\varrho$) have been generalized to our situation, while generalization of global results involves some essential difficulties and has been accomplished in only a few special cases.

The type of pointwise convergence that we consider is unrestricted admissible convergence as defined by Korányi. We say that a function F on X converges admissibly and unrestrictedly to a number L at $kM\in K/M$ if $\lim_{a\to+\infty} F(ka\cdot x)=L$, uniformly for x in any compact subset of X.

We let $\bar{N}$ be the Cartan involute of N and say that a point $k_0M\in K/M$ is a Lebesgue point of $f\in L'(K/M)$ if $k_0M=k(\bar{n}_0)M$, $\bar{n}_0\in\bar{N}$, and, for some (hence any) compact neighborhood V of e in $\bar{N}$,

$$\lim_{a\to+\infty}\frac{1}{\operatorname{meas}(aVa^{-1})}\int_{aVa^{-1}}|f(k(\bar{n}_0\bar{n})M)-f(k_0M)|\,d\bar{n}=0.$$

For f a signed measure on K/M we say that k_0M is a Lebesgue point of f and that $f(k_0M)=L$ if this condition holds with L substituted for $f(k_0M)$. This notion of Lebesgue point was first considered by Korányi. A theorem first published by Knapp and Williamson states that, for $f\in L^p(K/M)$, $1<p\leqq\infty$, or for rank $X=1$ and f a finite regular signed measure on K/M, almost all points of K/M are Lebesgue points of f.

The following local Fatou-type theorem is best possible:

THEOREM 2. *Let* $f\in L^\infty(K/M)$ *or let* rank $X=1$ *and f be a finite regular signed measure on* K/M. *Then* F_λ *converges admissibly and unrestrictedly to f at Lebesgue points of f, in particular, almost everywhere.*

REMARKS ON THE PROOF. (We point out only the new features that do not appear in the harmonic case.)

This theorem for radial convergence at points of continuity is due to Helgason. The basic estimate used is the absolute convergence of the integral $c(\lambda)=$

$\int_{\bar{N}} \exp\{-(i\lambda+\varrho)(H(\bar{n}))\}\, d\bar{n}$, for $\lambda\in\mathfrak{A}^*-i\mathfrak{A}^*_+$. For rank $X=1$, $\lambda=0$, a more concrete estimate is used.

I have improved the result of Helgason to that stated here. The principal difficulty is in comparing $\phi_\lambda(\bar{n}aK)$ with $\phi_\lambda(aK)$. I accomplish this comparison by using a lemma of Harish-Chandra that, where $\bar{n}a\in Ka'K$, $a'\in\exp\bar{\mathfrak{A}}_+$ (the closed positive Weyl chamber in $\mathfrak{A}$),

$$\lim_{a\to+\infty} (\log a'-\log a)=H(\bar{n}),$$

and from the fact, which follows easily from Harish-Chandra's proof, that the convergence is uniform on compact subsets of $\bar{N}$. (*Note.* For rank $X=1$ the convergence is actually uniform on $\bar{N}$. It may be worth investigating whether the same is true in general and whether such a result may be useful in proving global Fatou theorems.)

Global results are more difficult to prove even in the harmonic case because the global behavior of the Poisson kernel is less well known than its local behavior and are more difficult to generalize because $\phi_\lambda(\bar{n}aK)$ is not controllable on all of $\bar{N}$. The following special cases are known. The first result follows *a fortiori* from that of Lindahl for $\lambda=-i\varrho$.

THEOREM 3. *Let $H_0\in\mathfrak{A}_+$ be such that $\alpha(H_0)=1$ for each simple root α. Let $\lambda\in-i\varrho+\mathfrak{A}^*+i\bar{\mathfrak{A}}^*_+$, $f\in L^p(K/M)\cap L^1(K/M)$, $p>2\varrho(H_0)-1$. Then F_λ converges admissibly and unrestrictedly to f almost everywhere. For $X=SL(n,\mathbf{R})/SO(n)$ this results holds for $p>n-2$.*

Some global results are known for products of rank one spaces, where a maximal theorem of Knapp on K/M is available. We have the following:

THEOREM 4. *Let $X=\prod_i X_i$, rank $X_i=1$, $\lambda=\prod_i\lambda_i$, where $\lambda_i\in\mathfrak{A}^*_i-i\mathfrak{A}^*_{i+}$ or $\lambda_i=0$. Let $f\in L^p(K/M)$, $1<p\leqq\infty$. Then F_λ converges admissibly and unrestrictedly to f almost everywhere.*

For $H\in\mathfrak{A}_+$ we say that F converges admissibly and restrictedly with respect to H to L at $kM\in K/M$ if $\lim_{t\to+\infty} F(k\exp tH\cdot x)=L$, uniformly for x in any compact subset of X.

THEOREM 5. *Let $X=\prod_i X_i$, $\lambda\in\mathfrak{A}^*-i\mathfrak{A}^*_+$, $H\in\mathfrak{A}_+$, f be a finite regular signed measure on K/M. Then F_λ convergences admissibly and restrictedly with respect to H almost everywhere to the Radon-Nikodym derivative of f.*

REMARKS. Theorem 5 can be localized for certain values of λ. Specifically if $\lambda=\prod_i\lambda_i$, $\operatorname{Re} i\lambda_i(H_0)\geqq\sum_{j\neq i}\varrho_j(H_0)$, the theorem can be localized for $f\in L^1(K/M)$, and if the inequalities are strict it can be localized even for measures. The localization is to a "restricted Lebesgue point" set with respect to H, which has full measure.

Theorem 5 is unknown for the case $\lambda=0$. The type of weak (1, 1) estimate used to prove Theorem 4 in this case does not iterate to products.

Theorem 5 cannot be improved to unrestricted admissible convergence, which, for $\lambda\in\mathfrak{A}^*-i\mathfrak{A}^*_+$, implies a strong differentiation property. This negative result is also unknown for $\lambda=0$.

Finally, the following special case is known for all $\lambda\in\mathfrak{A}^*-i\mathfrak{A}^*_+$.

THEOREM 6. *Let X be of type* A_2. *Let $\gamma_\lambda(\varrho+i\lambda)-\varrho\in\mathfrak{A}^*_+ +i\mathfrak{A}^*$. Let $f\in L^p(K/M)$, $p>(\varrho(H_0)-\gamma_\lambda)/(1-\gamma_\lambda)$. Then F_λ converges to f admissibly and restrictedly with respect to H_0 almost everywhere. If $X=SL(3,\boldsymbol{R})/SO(3)$, this result holds for $p>1$.*

REFERENCE

Theorems 1–5 of §2 appear in the author's paper, *Fatou theorems for eigenfunctions of the invariant differential operators on symmetric spaces*, Trans. Amer. Math. Soc. **177** (1973), 257–274. Complete proofs of Theorems 1–4 are given there. The proof of Theorem 5 is only sketched there but is given in full in the author's doctoral dissertation at Yeshiva University.

YESHIVA UNIVERSITY

MASSACHUSETTS INSTITUTE OF TECHNOLOGY

REFINEMENTS OF ABEL SUMMABILITY FOR JACOBI SERIES

RICHARD ASKEY *

Abstract. Since the spherical harmonics on rank one symmetric spaces are known, it is possible to solve some of the more detailed questions of harmonic analysis on these spaces. As an illustration a wide class of positive measures is constructed on spheres, projective spaces and Euclidean spaces from their expansions in spherical functions.

When dealing with multiplier problems in harmonic analysis there are three types of problems: those that correspond to singular integrals, which are usually difficult; those which come from L^1 functions or bounded measures, which are easier; and a special case of this problem when the measure is positive. For a given multiplier sequence it usually is hard to prove the nonnegativity of the associated measure, but if it can be proven there are usually interesting consequences. To make this last statement concrete consider Jacobi polynomial expansions,

$$f(x) \sim \sum_{n=0}^{\infty} a_n h_n R_n^{(\alpha,\beta)}(x), \tag{1}$$

where

$$R_n^{(\alpha,\beta)}(x) = {}_2F_1(-n, n+\alpha+\beta+1; \alpha+1; (1-x)/2),$$

$$h_n^{-1} = \int_{-1}^{1} [R_n^{(\alpha,\beta)}(x)]^2 (1-x)^\alpha (1+x)^\beta \, dx, \qquad \alpha, \beta > -1,$$

AMS (MOS) subject classifications (1970). Primary 43A55, 43A35; Secondary 43A22.

* Supported in part by NSF grant GP-33897X.

$$a_n = \int_{-1}^{1} f(x)\, R_n^{(\alpha,\beta)}(x)\,(1-x)^\alpha (1+x)^\beta\, dx.$$

For certain values of (α, β), Jacobi polynomials are the spherical functions on the compact, two point homogeneous spaces, i.e., spheres and projective spaces ([**1**], [**6**]). The Poisson kernel for Jacobi series (1) is

$$P_r(x, y) = \sum_{n=0}^{\infty} r^n h_n R_n^{(\alpha,\beta)}(x)\, R_n^{(\alpha,\beta)}(y), \tag{2}$$

and the Abel sum of (1) is

$$f_r(x) = \sum_{n=0}^{\infty} r^n h_n R_n^{(\alpha,\beta)}(x) = \int_{-1}^{1} f(y)\, P_r(x, y)\,(1-y)^\alpha (1+y)^\beta\, dy. \tag{3}$$

Bailey [**4**] proved that (2) is positive for $-1 \leqq x, y \leqq 1, 0 \leqq r < 1, \alpha, \beta > -1$ by summing the series. It is a routine argument to show that $f_r(x) \to f(x)$ uniformly when $f(x)$ is continuous. In the case of Fourier series on the circle it is a well-known fact that the Poisson kernel is an approximate identity for all continuous functions, not only for even functions. The same is true for spheres [**9**] and projective spaces.

The Poisson kernel can be used to construct a large class of nonnegative series. If

$$a_n = \int_0^1 r^n\, d\mu(r), \qquad d\mu(r) \geqq 0, \tag{4}$$

then $\sum_{n=0}^{\infty} a_n h_n R_n^{(\alpha,\beta)}(x)$ is the Jacobi series expansion of a nonnegative measure. Recall Hausdorff's theorem that (4) holds if and only if

$$\Delta^k a_n = \sum_{j=0}^{k} (-1)^j \binom{k}{j} a_{n+j} \geqq 0, \qquad k, n = 0, 1, \ldots. \tag{5}$$

When $\alpha = \beta = -\frac{1}{2}$, Fejér [**5**] improved this result by dropping all of the inequalities (5) for $k > 2$. This suggests that a similar refinement can be found for all Jacobi series. Fejér's theorem can be proven by summation by parts and his classical result on the nonnegativity of the $(C, 1)$ means of $\frac{1}{2} + \sum_{n=1}^{\infty} \cos n\theta$. The corresponding theorem for Jacobi series should be the following:

CONJECTURE 1. *Let* $\alpha \geqq \beta \geqq -\frac{1}{2}$, $-1 \leqq x \leqq 1$. *Then*

$$\sum_{k=0}^{n} \frac{\Gamma(n-k+\alpha+\beta+3)}{\Gamma(n-k+1)} \cdot \frac{(2k+\alpha+\beta+1)\,\Gamma(k+\alpha+\beta+1)}{\Gamma(k+\beta+1)} P_k^{(\alpha,\beta)}(x) \geqq 0. \tag{6}$$

(6) is a positive multiple of the $(C, \alpha+\beta+2)$ means of (2) when $r=1$, $y=1$. This conjecture has been proven for $\alpha=\beta\geqq-\frac{1}{2}$ ([**7**], [**3**]) and $|\beta|\leqq\alpha\leqq\beta+1$, $3-\beta\leqq\alpha\leqq\beta+2$ [**2**]. It is best possible in the sense that $\Gamma(n-k+\gamma)$ cannot be used for any $\gamma<\alpha+\beta+3$ and have the nonnegativity of the series (6) for all x, $-1\leqq x\leqq 1$.

A summation by parts (possibly a fractional number of times) can be used to show the following theorem.

THEOREM 1. *If Conjecture* 1 *holds for* (α, β), $-1\leqq x\leqq 1$, *then the series* (1) *is the Jacobi expansion of a nonnegative measure when*

$$a_n \to c, \qquad c \geqq 0, \qquad \Delta^{\alpha+\beta+3} a_n \geqq 0. \tag{7}$$

The operator $\Delta^{\gamma} a_n$ is defined by

$$\Delta^{\gamma} a_n = \sum_{k=0}^{\infty} (-\gamma)_k \, a_{n+k}/k!$$

where $(\alpha)_k = \Gamma(k+\alpha)/\Gamma(\alpha) = (\alpha)(\alpha+1)\cdots(\alpha+k-1)$.

The symmetric spaces for which best possible results of this type are known are the spheres of all dimensions, the real projective space of dimension three, the complex projective space of real dimension four, and the quaternionic projective space of real dimension eight.

If one is willing to settle for less and have nonnegativity only in a fixed interval containing $x=1$, then there is a further conjecture which lowers the order of Cesàro summability (or the order of the differences (7)) for all $\beta>-\frac{1}{2}$, which means that for all symmetric spaces except the real projective spaces there should be a refinement which gives positivity in a neighborhood of the north pole for spheres and near the corresponding point for projective spaces.

CONJECTURE 2. *Let* $\alpha\geqq\beta\geqq-\frac{1}{2}$. *Then there is an* $x_0<1$ *so that*

$$\sum_{k=0}^{n} \frac{\Gamma(n-k+\alpha+\frac{5}{2})}{(n-k)!} \cdot \frac{(2k+\alpha+\beta+1)\,\Gamma(k+\alpha+\beta+1)}{\Gamma(k+\beta+1)} P_k^{(\alpha,\beta)}(x) \geqq 0, \tag{8}$$

$x_0\leqq x\leqq 1$, $n=0, 1, \ldots$.

Some incomplete calculations suggest that if $x_0(\alpha, \beta)$ is the smallest x_0 which works for all n then $x_0(\alpha, \beta) \leqq 0$, $\alpha \geqq \beta \geqq -\frac{1}{2}$, and $x_0(\alpha, \beta)$ is an increasing function of β for fixed α. This is tied up with the geometry of spheres. Euclidean spaces can be obtained as the limit of a patch around the north pole as the diameter of the sphere increases. The extra difference conditions which are necessary in Theorem 1 for spheres (one higher difference for each added dimension) as opposed to projective spaces (one half difference for each added dimension) comes from the closing of spheres at the south pole.

Either conjecture leads to the following conjecture:

CONJECTURE 3. *If $J_\alpha(t)$ denotes the Bessel function of the first kind then*

$$\int_0^x (x-t)^{\alpha+3/2} t^{\alpha+1} J_\alpha(t)\, dt \geqq 0, \qquad \alpha \geqq -\tfrac{1}{2},\ x \geqq 0.$$

Conjecture 3 has been proven for $\alpha = -\frac{1}{2} + k$, $k = 0, 1, \ldots$, which correspond to odd-dimensional spaces, for other values of α only weaker results are known at present [**2**]. One corollary of these results is the following theorem, which generalizes a well-known result of Pólya [**8**] in one dimension.

THEOREM 2. *Let $f(x) = f(x_1, \ldots, x_n) = g((x_1^2 + \cdots + x_n^2)^{1/2})$ be a radial function on $\boldsymbol{R}^n$. If $g(t) \to 0$ as $t \to \infty$, and $(-1)^{[n/2]} g^{[n/2]}(t)$ is convex for $t \geqq 0$ then $f(x)$ is positive definite, or equivalently it is the Fourier transform of a positive measure.*

REFERENCES

1. R. Askey, *Jacobi polynomial expansions with positive coefficients and imbeddings of projective spaces*, Bull. Amer. Math. Soc. **74** (1968), 301–304. MR **36** #4039.

2. ———, *Summability of Jacobi series*, Trans. Amer. Math. Soc. **179** (1973), 71–84.

3. R. Askey and H. Pollard, *Some absolutely monotonic and completely monotonic functions*, SIAM J. Math. Anal. **5** (1974).

4. W. N. Bailey, *The generating function of Jacobi polynomials*, J. London Math. Soc. **13** (1938), 8–12.

5. L. Fejér, *Trigonometrische Reihen und Potenzreihen mit mehrfach monotoner Koeffizientenfolge*, Trans. Amer. Math. Soc. **39** (1936), 18–59.

6. R. Gangolli, *Positive definite kernels on homogeneous spaces and certain stochastic processes related to Levy's Brownian motion of several parameters*, Ann. Inst. H. Poincaré Sect. B **3** (1967), 121–226. MR **35** #6172.

7. E. Kogbetliantz, *Recherches sur la sommabilité des séries ultersphériques par la méthode des moyennes arithmétiques*, J. Math. Pures Appl. (9) **3** (1924), 107–187.

8. G. Pólya, *Remarks on characteristic functions*, Proc. Berkeley Sympos. on Statist. and Probability, Univ. of California Press, Berkeley, Calif., 1949, pp. 115–123. MR **10**, 463.

9. I. J. Schoenberg, *Positive definite functions on spheres*, Duke Math. J. **9** (1942), 96–108. MR **3**, 232.

UNIVERSITY OF WISCONSIN, MADISON

SPHERICAL FUNCTIONS ON RANK ONE SYMMETRIC SPACES AND GENERALIZATIONS

M. FLENSTED-JENSEN

1. Fourier analysis on a noncompact rank one symmetric space G/K. Let $G=K\bar{A}^+K$ and $G=KAN$ be the Cartan- and Iwasawa-decompositions of G. Let $C^\infty(K; G/K)$ be the C^∞, K-invariant functions on G/K. The radial part of the Laplacian on G/K is

(1) $$\Delta = d^2/dt^2 + ((2\alpha+1)\coth t + (2\beta+1)\tanh t)\, d/dt$$

where $t \in [0, \infty]$ and α and β are certain half-integers. The *spherical functions* φ_λ on G/K are the solutions to $\Delta\varphi + (\lambda^2+\varrho^2)\varphi = 0$ with $\varphi(0)=1$ and $\varrho = \alpha+\beta+1$. φ_λ is given by

(2) $$\varphi_\lambda(t) = \varphi_\lambda^{(\alpha,\beta)}(t) = {}_2F_1(\tfrac{1}{2}(\varrho+i\lambda), \tfrac{1}{2}(\varrho-i\lambda); \alpha+1; -(\mathrm{sh}\,t)^2).$$

For $f \in C_0^\infty(K; G/K)$ the *Fourier-transform* and its *inverse* are

(3) $$\tilde{f}(\lambda) = \int_{G/K} f(x)\,\varphi_\lambda(x)\,dx,$$

(4) $$f(x) = \int_0^\infty \tilde{f}(\lambda)\,\varphi_\lambda(x)\,|c(\lambda)|^{-2}\,d\lambda.$$

Here $c(\lambda)$ is a certain function depending on α and β.

(5) There is a *Paley-Wiener theorem.*

$L^1(K; G/K)$ has a convolution-structure such that

(6) $$(f_1 * f_2)^\sim(\lambda) = \tilde{f}_1(\lambda)\,\tilde{f}_2(\lambda).$$

AMS (MOS) subject classifications (1970). Primary 43A85, 33A75, 22E30.

Related to this, there is the *product formula*:

(7) $$\varphi_\lambda(t_1)\,\varphi_\lambda(t_2)=\int_K \varphi_\lambda(t_1 k t_2)\,dk.$$

There is an *integral-formula* for φ_λ:

(8) $$\varphi_\lambda(t)=\int_K \exp\{(i\lambda-\varrho)\,(H(tk))\}\,dk;$$

a *Radon-transform* for $f\in C_0(K;G/K)$:

(9) $$F_f(t)=e^{\varrho(t)}\int_N f(tn)\,dn,$$

and the relation:

(10) $$\tilde{f}(\lambda)=\int_0^\infty F_f(t)\cos\lambda t\,dt.$$

The interesting thing is that all the formulas and statements (1)–(10) keep their validity for all $\alpha\geqq\beta\geqq-\frac{1}{2}$, in the sense that all the formulas can be brought to an analytic form, where the group-structure is not involved ([**1**], [**2**] and [**3**]). Then for all $\alpha\geqq\beta\geqq-\frac{1}{2}$, (1)–(10) are statements in the harmonic analysis with respect to Jacobi function expansions. Even in most of the cases of half-integers α and β, there are no corresponding groups. Professor Helgason called the above "analysis without a space to do it on." The space was found by T. Koornwinder, who communicated the following result to me:

THEOREM 1. *Let* $\alpha\geqq\beta\geqq-\frac{1}{2}$, $\alpha,\beta\in\frac{1}{2}\mathbf{Z}$, *and let* $G=SO(2\alpha+2,2\beta+2)$, $H=SO(2\alpha+2,2\beta+1)$ *and* $K=SO(2\alpha+2)\times SO(2\beta+2)$. *Let* Δ *be the Laplacian on* G/H. *Then* $\varphi_\lambda^{(\alpha,\beta)}(t)$ *is characterized by*

(i) $\varphi_\lambda^{(\alpha,\beta)}\in C^\infty(K;G/H)$ *and* $\varphi_\lambda^{(\alpha,\beta)}(e)=1$,

(ii) $\Delta\varphi_\lambda^{(\alpha,\beta)}+(\lambda^2+\varrho^2)\,\varphi_\lambda^{(\alpha,\beta)}=0.$

It turned out to be possible to do the analysis indicated by (1)–(10) on these pseudo-Riemannian symmetric spaces G/H.

2. Analysis on pseudo-Riemannian symmetric spaces. Let G be semisimple, noncompact, with finite center. Let τ be an involutive automorphism with fix-point group H. Then there exists a Cartan-involution σ, with compact fix-point group

K, and such that $\tau\sigma=\sigma\tau$. Corresponding to σ and τ we have the eigenspace decompositions of the Lie algebra $g=k+p=h+q$.

Let b be a maximal abelian subalgebra in $p\cap q$, extend b to a maximal abelian subalgebra $a\subset p$. There is a Weyl-group on B, and we have the decomposition

$$G=K\bar{B}^+H \qquad (\bar{B}^+\text{-element unique!}).$$

Now I want to propose two definitions:

The *"rank" of* G/H is the dimension of B.

The *"spherical" functions on* G/H are the functions satisfying:

(i) $\varphi\in C^\infty(K;\, G/H)$ and $\varphi(e)=1$,

(ii) φ is an eigenfunction for each $D\in D(G/H)$.

CONJECTURE. (i)+(ii) *is equivalent to* (i)+(ii)′, *where*

(ii)′ *φ is an eigenfunction for each $D\in D(K\backslash G)$.*

In the following we take (i)+(ii)′ as definition.

EXAMPLE 1. H is compact; then $H=K$ and we are back in the Riemannian case.

EXAMPLE 2. Let G_1/K_1 be a symmetric space (not necessarily rank one), let $G=G_1\times G_1$, $K=K_1\times K_1$ and $H=$diagonal in G. Identify G/H and G_1 by $\pi(g_1, g_2)=g_1g_2^{-1}$. The "rank" of G/H is the rank of G_1/K_1 in this case. The "spherical" functions on G/H are the usual spherical functions on G_1.

EXAMPLE 3. See Theorem 1.

DEFINITION. The triple (G, H, K) is said to *satisfy condition* (A_0) if there exists a subalgebra $a_0\subset a$ such that

(i) $\dim a_0=\dim b$ and $a_0\cap h=\{0\}$,

(ii) for all $a\in A_0$ and $k\in K$, $ak\in KaH$.

Then it follows that $G=K\bar{A}_0^+H$.

THEOREM 2. *Let (G, H, K) satisfy condition (A_0). Let φ be a "spherical" function on G/H. Then $\psi(g)=\int_K\varphi(gk)\,dk$ is a spherical function on G/K, $\varphi|_{A_0}=\psi|_{A_0}$ and*

$$\text{(11)} \qquad \textit{for all } a_1, a_2\in A_0,\quad \varphi(a_1)\,\varphi(a_2)=\int_K\varphi(a_1ka_2)\,dk.$$

This product formula gives a convolution-structure.

DEFINITION. Let $f_1, f_2\in C_0(K;\, G/H)$. Let, for $y\in G/H$, $A(y)\in\bar{A}_0^+$ be determined by $yH\in KA(y)\,H$; then

$$\text{(12)} \qquad f_1*f_2(x)=\int_{G/H}f_1(y)\int_K f_2(A(y)^{-1}\,kx)\,dk\,dy.$$

For a "spherical" function φ we get

$$\int_{G/H} f_1 * f_2(x)\,\varphi(x)\,dx = \int_{G/H} f_1(x)\,\varphi(x)\,dx \cdot \int_{G/H} f_2(x)\,\varphi(x)\,dx;$$

thus φ is a character of $(C_0(K; G/H), *)$.

EXAMPLE 1. Take $A_0 = A$; then (12) is the convolution on G.

EXAMPLE 2. Take $A = A_1 \times A_1$ and $A_0 = A_1 \times e$; then (12) gives the convolution on G_1.

EXAMPLE 3. Take A_0 to be the following subgroup of $SO(2\alpha+2, 2\beta+2)$:

$$A_0 = \left\{ \begin{pmatrix} I_{2(\alpha-\beta)} & 0 & 0 \\ 0 & \operatorname{cht} I_{2\beta+2} & \operatorname{sht} I_{2\beta+2} \\ 0 & \operatorname{sht} I_{2\beta+2} & \operatorname{cht} I_{2\beta+2} \end{pmatrix} \middle| \, t \in \boldsymbol{R} \right\}.$$

Here I_γ means the $(\gamma \times \gamma)$-identity matrix. In this case (11) and (12) are an interpretation of the convolution-structure for Jacobi functions of parameters α, β; see (6) and (7). (12) does not give the usual convolution on G, except when $\beta = -\frac{1}{2}$.

3. Generalized Iwasawa decomposition. Let (G, H, K) satisfy condition (A_0). Let Δ^+ be the set of positive roots on a. Define $P^+ = \{\alpha \in \Delta^+ \mid \alpha | a_0 \neq 0\}$ and $P^- = \{\alpha \in \Delta^+ \mid \alpha | a_0 = 0\}$, and $n_0 = \sum_{\alpha \in P^+} g_\alpha$, $k_0 = \{x + \sigma x \mid x \in g_\alpha, \alpha \in P^-\}$ then n_0 is a nilpotent and k_0 a compact Lie algebra.

In Examples 1, 2 and 3, the following hold:

(i) $g = k_0 + a_0 + n_0 + h$ and $G = K_0 A_0 N_0 H$.

(ii) For $f \in C_0(G/H)$: $\int_{G/H} f(x)\,dx = \int_{K_0 \times A_0 \times N_0/N_0 \cap H} f(kan)\,dk\,da\,dn$.

The *Radon-transform* is defined in these cases by

$$F_f(a) = e^{\varrho(a)} \int_{N_0/N_0 \cap H} f(an)\,dn.$$

In Example 3 this leads to an interpretation and a proof of (4), (5), (8), (9) and (10) for Jacobi functions with parameters $\alpha, \beta \in \frac{1}{2}\boldsymbol{Z}$, $\alpha \geqq \beta \geqq \frac{1}{2}$.

More details and proofs will appear elsewhere.

REFERENCES

1. M. Flensted-Jensen, *Paley-Wiener type theorems for a differential operator connected with symmetric spaces*, Ark. Mat. **10** (1972), 143–162.

2. M. Flensted-Jensen and T. H. Koornwinder, *The convolution structure for Jacobi function expansions*, Ark. Mat. (to appear).

3. T. H. Koornwinder, *The addition formula for Jacobi polynomials*, Indag. Math. **34** (1972), 188–191.

KØBENHAVNS UNIVERSITETS MATEMATISKE INSTITUT
KØBENHAVN, DANMARK

BESSEL FUNCTIONS, REPRESENTATION THEORY, AND AUTOMORPHIC FUNCTIONS

STEPHEN GELBART

We shall describe the relation between two types of operator-valued Bessel functions of matrix arguments and indicate some connections with representation theory and automorphic functions. By $M_{n,m}$ we denote the $n \times m$ matrix space and by P_m the cone of $m \times m$ positive definite matrices. We let ω denote an irreducible unitary representation of $SO(n)$ which is class 1 with respect to $SO(n-m)$ and r_ω its number of $SO(n-m)$-fixed vectors. Then we define $J_\omega(R, P)$ on $P_m \times P_m$ by

$$(1) \qquad J_\omega^{n,m}(R, P) = \int_{SO(n)} \omega^*(u) \exp\{2\pi i \operatorname{tr}({}^t\bar{I} u \bar{I} R^{1/2} P^{1/2})\} \, du,$$

where $\omega^*(u)$ is the upper left-hand $r_\omega \times r_\omega$ submatrix of $\omega(u)$, and $\bar{I} = \binom{I_m}{0} \in M_{n,m}$. Such matrix-valued Bessel kernels arise naturally in decomposing the Fourier operator on $M_{n,m}$ with respect to the action of $SO(n)$ given by left matrix multiplication (cf. [**2**]). If $m=1$, and ω has highest weight $(k, 0, \ldots, 0)$, then $J_\omega^{n,1}(R, P) = c(RP)^{(n-2)/2} J_{k+(n-2)/2}(2\pi RP)$.

A second type of Bessel kernel is defined in the context of the Siegel upper half-space $\Sigma_m = \{\text{symmetric } Z = X + iY,\ Y \in P_m\}$. Recall that the automorphism group of Σ_m is the real symplectic group $Sp(m, \boldsymbol{R})$. If ϱ is a finite-dimensional irreducible polynomial representation of $GL(m, \boldsymbol{C})$, we set

$$(2) \quad J_\varrho^m(R, P) = \int_{Z = X + iY_0} \varrho(Z)^{-1} \exp\{-2\pi i \operatorname{tr}(RZ^{-1})\} \exp\{-2\pi i \operatorname{tr}(PZ)\} \, dZ$$

with Y_0 arbitrary in P_m. For one-dimensional ϱ this is the Bessel function intro-

---

AMS (MOS) subject classifications (1970). Primary 33A75, 10D05; Secondary 22E45, 22E55.

duced by Bochner in [1] and used by him to extend to Σ_m a summation formula of the type

$$\sum a_n \varphi(n^{1/2}) = \sum a_n (U_k \varphi)(n^{1/2}).$$

Here φ is a suitable radial function, U_k is the Bessel transform of order k, and $\Phi(y) = \sum a_n \exp\{-2\pi n y\}$ is a modular form of degree k. For general ϱ these functions arise in computing Fourier coefficients of the Poincaré-Eisenstein series

$$E_{S,\varrho}(Z) = \sum \varrho(CZ+D)^{-1} \exp\{2\pi i \operatorname{tr}(SMZ)\},$$

where S is a positive definite symmetric integral matrix, and the summation is extended over the coset representatives $M = \left(\begin{smallmatrix} A & B \\ C & D \end{smallmatrix}\right)$ for $Sp(n, \boldsymbol{Z})$ modulo the translations of Σ_m (cf. [4]). Growth estimates for these J_ϱ^m would be of definite interest.

How are the kernels (1) and (2) related? When $m=1$,

$$J_\varrho^1(R, P) = c(R^{-1/2} P^{1/2})^{k-1} J_{k-1}(4\pi R^{1/2} P^{1/2}), \tag{3}$$

if $\varrho(A) = A^k$. For $m > 1$, we have the following generalization of (3).

THEOREM 1 [2]. *Suppose $n \geqq 2m$. Fix ω of highest weight $(m_1, \ldots, m_n, 0, \ldots, 0)$, and let ϱ_ω denote the unique finite-dimensional representation of $GL(m, \boldsymbol{C})$ which is polynomial, irreducible, and of highest weight $(m_1, \ldots, m_n)$. Then, for some choice of matrix representatives for ω and ϱ_ω,*

$$\begin{aligned} J_\omega^{n,m}(R, P)\, \varrho_\omega(R^{1/2}) (\det R)^\nu = \frac{\pi^{mn/2} \varrho_\omega(iP^{1/2})}{(2\pi i)^{m(m+1)/2}} \int_{Z = A_0 + iB} & (\det Z)^{-n/2} \varrho_\omega(\pi^{-1} Z)^{-1} \\ & \cdot \exp(\operatorname{tr}(RZ) - \operatorname{tr}(\pi^2 P Z^{-1}))\, dZ \end{aligned} \tag{4}$$

where $\gamma = (n-m-1)/2$.

For $r_\omega = 1$, this formula was known (Herz [5]). The proof of the general case involves a generalization of the classical theory of spherical harmonics to the setting of matrix space. In particular, one must prove that r_ω equals the dimension of ϱ_ω and that ω is realizable on a space of $\boldsymbol{C}^{r_\omega}$-valued $SO(r)$-harmonic polynomials on $M_{n,m}$ satisfying $p(XA) = p(X)\, \varrho_\omega(A)$, for $X \in M_{n,m}$, $A \in GL(n, \boldsymbol{R})$.

Our application of (4) to representation theory involves a new construction of the holomorphic discrete series representations of $Sp(m, \boldsymbol{R})$. If we imbed this group in the abstract symplectic group associated with $M_{n,m}$ (cf. [6]) then by Weil there is a natural projective representation $W(g)$ of $Sp(m, \boldsymbol{R})$ on $L^2(M_{n,m})$

which for $n=2m$ is an ordinary representation and may be described on the generators of $Sp(m, \mathbf{R})$ as follows:

$$W\begin{pmatrix} I_m & B \\ 0 & I_m \end{pmatrix} F(X) = \exp\{\pi i \operatorname{tr}(XBX^t)\} F(X), \text{ and } W\begin{pmatrix} 0 & -I_m \\ I_m & 0 \end{pmatrix} F(X) = i^{-m^2} \hat{F}(X).$$

Since $W(g)$ clearly commutes with the action of $SO(2m)$, it respects the decomposition of $L^2(M_{2m,m})$ into subspaces belonging to the irreducible representations ω of $SO(2m)$.

THEOREM 2 [3]. (a) *$W(g)=\Sigma_\omega \oplus W_\omega$ contains all the holomorphic discrete series representations of $Sp(m, \mathbf{R})$*; (b) *if $\omega=(m_1, \ldots, m_n)$ with $m_m \neq 0$, then W_ω is isomorphic to r_ω copies of a discrete series representation.*

K. Gross and R. Kunze have given a similar construction of certain holomorphic discrete series for $U(n, n)$ in [7].

REFERENCES

1. S. Bochner, *Bessel functions and modular relations of higher type and hyperbolic differential equations*, Comm. Sém. Math. Univ. Lund, Tome Supplémentaire, 1952, 12–20. MR **15**, 422.

2. S. Gelbart, *Harmonics on Stiefel manifolds and generalized Hankel transforms*, Bull. Amer. Math. Soc. **78** (1972), 451–455; see also, *A theory of Stiefel harmonics*, Trans. Amer. Math. Soc. (to appear).

3. ———, *Holomorphic discrete series for the real symplectic group*, Invent. Math. **19** (1973), 49–58.

4. R. Godement, *Séminaire Cartan*, 1957/58. Exposé 9.

5. C. Herz, *Bessel functions of matrix argument*, Ann. of Math. (2) **61** (1955), 474–523. MR **16**, 1107.

6. A. Weil, *Sur certains groupes d'opérateurs unitaires*, Acta Math. **111** (1964), 143–211. MR **29** #2324.

7. K. Gross and R. Kunze, *Bessel transforms and discrete series*, Maryland Harmonic Analysis Conference Proceedings, Lecture Notes in Math., vol. 276, Springer-Verlag, Berlin and New York, 1972.

INSTITUTE FOR ADVANCED STUDY

CORNELL UNIVERSITY

GENERALIZED BESSEL TRANSFORMS AND UNITARY REPRESENTATIONS

KENNETH I. GROSS AND RAY A. KUNZE*

In our paper [2], the first in a series devoted to generalizations of classical Bessel functions and their interconnections with discrete series representations of certain classical groups, we described a collection of Bessel functions J_λ defined on complex matrix space $M=C^{n\times n}$ with values in $d_\lambda \times d_\lambda$ matrices. Here, λ runs through the dual $\tilde{U}$ of the unitary subgroup $U=U(n)$ of the general linear group $A=GL(n, C)$, and d_λ denotes the degree of λ. We then applied these Bessel functions to the representation theory of the classical group $G=U(n, n)$; that is, the group of matrices $g\in M^{2\times 2}$ such that $gpg^*=p$, where p is the "Weyl reflection" $p=\left(\begin{smallmatrix} 0 & I \\ -I & 0\end{smallmatrix}\right)\in M^{2\times 2}$.

The results in [2] may be summarized roughly as follows:

(i) $L^2(A)$ decomposes under the left action of U into a discrete direct sum

$$L^2(A)\cong \sum_{\lambda\in\tilde{U}} \oplus d_\lambda L^2(A, \lambda)$$

where $L^2(A, \lambda)$ is the space of λ-covariant square-integrable functions from A to $C^{d_\lambda\times 1}$.

(ii) The "Weil representation" R of G has a corresponding decomposition

$$R\cong \sum_{\lambda\in\tilde{U}} \oplus d_\lambda R(\cdot, \lambda)$$

where $R(\cdot, \lambda)$ is a unitary representation of G in $L^2(A, \lambda)$.

(iii) Let $L=VC$ be the maximal parabolic subgroup of G of elements $l=v(s)\,c(a)$, where $v(s)=\left(\begin{smallmatrix} I & 0 \\ s & I\end{smallmatrix}\right)$ and $c(a)=\left(\begin{smallmatrix} a^{*-1} & 0 \\ 0 & a\end{smallmatrix}\right)$, with $a\in A$ and $s\in S$, the vector

---

AMS (MOS) subject classifications (1970). Primary 22E30, 22E45, 33A75; Secondary 32A07, 32M15, 33A40, 43A75, 43A80, 43A85, 46E20, 57E25, 81A78.

* This research was partially supported by the National Science Foundation under grant no. GP-22795 and grant no. GP-30061 respectively.

group of Hermitian $n \times n$ matrices. Then $R(\cdot, \lambda)$ is already irreducible when restricted to L. Indeed, L is isomorphic to the semidirect product of S with A, and on L the representation $R(\cdot, \lambda)$ is given by the standard Mackey formulation for semidirect products.

(iv) For the matrix p, which is in G, $R(p, \lambda)$ is the Hankel transform associated with the Bessel function $J_{\delta^{-1}\lambda}$, δ being the nth power of the determinant. In particular, since $R(p)$ is essentially the Fourier transform on $L^2(M)$ (modified to act in $L^2(A)$), the decomposition

$$R(p) \cong \sum_{\lambda \in \hat{U}} \oplus d_\lambda R(p, \lambda)$$

gives a direct generalization of the classical decomposition of the complex Fourier transform into integer-order Hankel transforms.

(v) With the highest weight of λ suitably constrained, $R(\cdot, \lambda)$ is a square-integrable representation of G, and hence lies in the discrete series. In fact, $R(\cdot, \lambda)$ is unitarily equivalent via a Laplace transformation to a representation $T(\cdot, \lambda)$ which acts in a Hilbert space of holomorphic functions on the generalized upper half-plane $H = S + iP$, where P is the cone in S of positive-definite matrices. Thus, these representations are in the holomorphic discrete series.

S. Gelbart has recently obtained analogous results for $Sp(n, \boldsymbol{R})$ by different methods [**1**].

In what follows we give a brief outline of an extension of part of the above results which is appropriate for treating the remaining representations in the holomorphic discrete series of G. The important differences between this case and that of [**2**] center around the fact that the Weil representation does not appear and the Hankel transforms are not related to the Fourier transform.

Let λ be an irreducible unitary representation of $U \times U$ and denote by the same symbol its holomorphic extension to $A \times A$. Set $\varrho_\lambda(a) = \lambda(a, a^{-1})$ for $a \in A$, and define the *Bessel function* $J_\lambda : A \to \boldsymbol{C}^{d_\lambda \times d_\lambda}$ by

$$J_\lambda(a) = \varrho_\lambda(i) \int_U \exp(-i \operatorname{Re} \operatorname{tr} au)\, \lambda(u^{-1}, aua^{*-1})\, du$$

where du is normalized Haar measure on U. Notice that J_λ transforms covariantly on both left and right by the unitary representation $u \mapsto \lambda(u, u)$ of U.

Now assume that ϱ_λ has a holomorphic extension from A to M, and let $\mathscr{H}_\lambda$ denote the Hilbert space (possibly zero) of holomorphic functions F on the half-space H with values in $\boldsymbol{C}^{d_\lambda \times 1}$ such that

$$\|F\|_\lambda^2 = \int_H F(x+iy)^*\, \varrho_\lambda(y)\, F(x+iy)\, \delta(y)^{-2}\, dx\, dy < \infty$$

where $z=x+iy$ with $x\in S$, $y\in P$. Recall that the measure $\delta(y)^{-2}\,dx\,dy$ on H is invariant under the action $(z, g)\mapsto z\tilde{g}$ of G on H by linear fractional transformations. Specifically, if $g=\left(\begin{smallmatrix} a & b \\ c & d\end{smallmatrix}\right)\in G$, then $z\tilde{g}=(zb+d)^{-1}(za+c)$ for $z\in H$. For $F\in\mathscr{H}_\lambda$ and $g\in G$ set

$$(T(g,\lambda)\,F)(z)=\lambda((z^*b+d)^{*-1}, zb+d)\,F(z\tilde{g}).$$

If $\mathscr{H}_\lambda\neq\{0\}$, it can be seen that the representation $g\mapsto T(g,\lambda)$ is a member of the holomorphic discrete series for G [**3**].

We next describe a new realization $R(\cdot,\lambda)$ of these representations, and in the process show that the nontriviality of $\mathscr{H}_\lambda$ is dependent upon the convergence of the *generalized gamma integral*

$$\gamma(\lambda)=\int_P \exp(-2\,\mathrm{tr}\,y)\,\varrho_\lambda(y)\,\delta(y)^{-1}\,dy.$$

Indeed, $\mathscr{H}_\lambda\neq\{0\}$ if and only if $\gamma(\delta^{-1}\lambda)$ exists as an absolutely convergent integral, $\delta^{-1}\lambda$ being the representation $(a_1, a_2)\mapsto\delta(a_1)^{-1}\,\lambda(a_1, a_2)$ of $A\times A$.

Let $L^2(A,\lambda)$ denote the Hilbert space of measurable functions $f\colon A\mapsto \boldsymbol{C}^{d_\lambda\times 1}$ satisfying the covariance condition $f(ua)=\lambda(u,u)\,f(a)$ for $(u,a)\in U\times A$, and such that

$$\|f\|_\lambda^2=\int_A f(a)^*\,\gamma(\delta^{-1}\lambda)\,f(a)\,da<\infty.$$

We define a Laplace transform $\mathscr{L}\colon f\mapsto F$ on $L^2(A,\lambda)$ by

$$F(z)=c\int_A \exp(i\,\mathrm{tr}\,za^*a)\,\lambda(a^*, a^{-1})\,f(a)\,da$$

for $z\in H$, c being a suitably chosen constant (independent of λ).

THEOREM. *$\mathscr{L}\colon f\mapsto F$ is a unitary map of $L^2(A,\lambda)$ onto $\mathscr{H}_\lambda$. Hence, the equation $R(g,\lambda)=\mathscr{L}^{-1}\,T(g,\lambda)\,\mathscr{L}$, $g\in G$, defines a unitary representation of G in the space $L^2(A,\lambda)$. The operator $R(p,\lambda)$ on $L^2(A,\lambda)$ is the Hankel transform associated with the Bessel function $J_{\delta^{-1}\lambda}$, and defined by*

$$(1)\qquad (R(p,\lambda)\,f)(a)=\pi^{-n^2}\varrho_\lambda(-i)\int_A J_{\delta^{-1}\lambda}(2ab^*)\,\delta(ab^*)\,f(b)\,db.$$

In this realization of the holomorphic discrete series of G, we can reduce square-integrability to a condition on the single operator $R(p,\lambda)$; more specifically, to the property

$$\int_A \|J_{\delta^{-1}\lambda}(a)\|^2\, da < \infty. \tag{2}$$

The key to the proof of (2), as well as to the derivation of (1), is the following Fourier transform formula (which for the case $n=1$ is a well-known classical result):

$$\exp(-i\,\mathrm{tr}\, z^{-1}a^*a)\,\delta(z^{-1})\,\varrho_\lambda(az^{-1}) = (i\pi)^{n^2}\int_M \exp(-2i\,\mathrm{Re}\,\mathrm{tr}\, aw^*)\exp(i\,\mathrm{tr}\, zw^*w)\,\varrho_\lambda(w)\,dw$$

valid for $z \in H$ and $a \in A$.

Finally, let us make the connection with the results of [2]. We may write λ in the form $\lambda(u_1, u_2) = \lambda_1(u_1) \otimes \lambda_2(u_2)$ where $\lambda_1, \lambda_2 \in \tilde{U}$. For the special case in which λ_2 is the identity character of U, the results outlined above yield the theory described in [2].

REFERENCES

1. S. Gelbart, *Holomorphic discrete series for the real symplectic groups*, Invent. Math. **19** (1973), 49–58.

2. K. Gross and R. Kunze, *Fourier-Bessel transforms and holomorphic discrete series*, Conference in Harmonic Analysis, Lecture Notes in Math., vol. 266, Springer-Verlag, Berlin and New York, 1971, pp. 79–122.

3. Harish-Chandra, *Representations of semi-simple Lie groups*. V, Amer. J. Math. **78** (1956), 1–41. MR **18**, 490.

DARTMOUTH COLLEGE

UNIVERSITY OF CALIFORNIA, IRVINE

EXPLICIT FORMULAS FOR SPECIAL FUNCTIONS RELATED TO SYMMETRIC SPACES

TOM KOORNWINDER

It is our purpose to derive new explicit formulas for special functions by group theoretic interpretation. These formulas should be significant in the context of harmonic analysis for the classical expansions. Here we sketch a proof for the addition formula for Jacobi polynomials (see [**8**], [**9**]).

The radial part of the Laplace-Beltrami operator on a compact symmetric space of rank one G/K is

$$\omega_{p,q} = d^2/d\theta^2 + (p\cot\theta + 2q\cot 2\theta)\, d/d\theta, \tag{1}$$

where $0<\theta<\pi/2$ and p, q are certain nonnegative integers. The eigenfunctions of $\omega_{p,q}$, regular in $\theta=0$ and $\pi/2$, are Jacobi polynomials $P_n^{(\alpha,\beta)}(\cos 2\theta)$ $(\alpha=\frac{1}{2}(p+q-1),\ \beta=\frac{1}{2}(q-1))$. It was pointed out by Gangolli [**4**] that the convolution for radial functions on G/K implies a positive convolution structure for Jacobi series for certain values of α and β.

The radial convolution structure on G/K is closely connected with the product formula

$$\varphi(x)\,\varphi(y) = \int_K \varphi(xky)\, dk \tag{2}$$

for spherical functions φ. Let

$$\varphi(xky) = \sum_{\delta\in\hat{K}} \varphi_\delta(x,k,y) \tag{3}$$

AMS (MOS) subject classifications (1970). Primary 33A65, 33A75, 43A90; Secondary 33A45, 42A56.

be the corresponding expansion of $\varphi(xky)$ as a function of $k \in K$. We call this expansion the addition formula for φ. In the case of $SO(q)/SO(q-1)$, formulas (2) and (3), when written in analytic form, are well-known results for Gegenbauer polynomials [**2**, §3.15, (19) and (20)].

We obtain the addition formula for Jacobi polynomials by generalizing the method used in the case $SO(q)/SO(q-1)$. Let G/K be of rank one, let V be a subspace of $L^2(G/K)$, irreducible under G, and let $\varphi \in V$ be a spherical function. When $e \in G/K$ is the K-invariant point then the function $\varphi(\xi)$ only depends on the distance $d(\xi, e)$, and we write $\varphi(\xi) = p(d(\xi, e))$ for $\xi \in G/K$. Let the functions $f_k(\xi)$ $(k = 1, \ldots, N)$ form an arbitrary orthonormal base of V. Then

$$\sum_{k=1}^{N} f_k(\xi)\, \overline{f_k(\eta)} = \text{const}\, p(d(\xi, \eta)), \tag{4}$$

$p(d(\xi, \eta))$ is the kernel function for V. Next, we decompose V with respect to K, and we have $V = \sum_{\delta \in \hat{K}} V_\delta$. Let the functions $f_{k_\delta}(\xi)$ $(k_\delta = 1, \ldots, N_\delta)$ form an orthonormal base of V_δ and write

$$\psi_\delta(\xi, \eta) = \sum_{k_\delta = 1}^{N_\delta} f_{k_\delta}(\xi)\, \overline{f_{k_\delta}(\eta)}. \tag{5}$$

Then the sum

$$\sum_{\delta \in \hat{K}} \psi_\delta(\xi, \eta) = \text{const}\, p(d(\xi, \eta)) \tag{6}$$

is the required addition formula. The problem is to choose suitable coordinates and to express the functions $\psi_\delta(\xi, \eta)$ in terms of special functions.

We have solved this problem for the complex projective space $SU(q)/U(q-1)$ with spherical functions $P_n^{(q-2,0)}$. Observe that the homogeneous space $U(q)/U(q-1)$ is the unit sphere in the complex vector space $\boldsymbol{C}^q$. The functions on this sphere which are invariant under scalar multiplication by $e^{i\psi}$ are precisely the functions on $SU(q)/U(q-1)$. Ikeda and Kayama ([**6**], [**7**]) developed a theory for functions on $U(q)/U(q-1)$ analogous to the classical theory of spherical harmonics. By using their results we obtained the explicit addition formula

$$\begin{aligned} &P_n^{(q-2,0)}(2|\cos\theta_1 \cos\theta_2 + \sin\theta_1 \sin\theta_2 r e^{i\varphi}|^2 - 1) \\ &\quad = \sum_{k=0}^{n} \sum_{l=0}^{k} c_{n,k,l}^{(q-2,0)}\, f_{n,k,l}^{(q-2,0)}(\cos 2\theta_1)\, f_{n,k,l}^{(q-2,0)}(\cos 2\theta_2)\, P_{k,l}^{(q-2,0)}(r e^{i\varphi}), \end{aligned} \tag{7}$$

where $c_{n,k,l}^{(q-2,0)}$ are constants,

$$f_{n,k,l}^{(q-2,0)}(\cos 2\theta)=(\sin\theta)^{k+l}(\cos\theta)^{k-l}\,P_{n-k}^{(q-2+k+l,\,k-l)}(\cos 2\theta)$$

and

$$P_{k,l}^{(q-2,0)}(re^{i\varphi})=P_l^{(q-3,\,k-l)}(2r^2-1)\,r^{k-l}\cos(k-l)\,\varphi.$$

The functions $P_{k,l}^{(q-2,0)}(x+iy)$ are polynomials in x and y, orthogonal in the unit disk with respect to the measure $(1-x^2-y^2)^{q-3}\,dx\,dy$. Formula (7) is an orthogonal expansion in terms of these polynomials.

Let $G=KAK$ be the Cartan decomposition associated with the symmetric space G/K. Suppose that the expansion (3) can be written as

$$\varphi(a_1ka_2)=\sum_{\delta\in\hat{K}} f_\delta(a_1)\,f_\delta(a_2)\,p_\delta(k)\qquad(a_1,a_2\in A,\,k\in K).\tag{8}$$

This is the case for $SU(q)/U(q-1)$. Let M be the centralizer of A in K. Then the function $p_\delta(k)$ is a spherical function on the homogeneous space K/M. In our example we have $K/M=U(q-1)/U(q-2)$, which is not a symmetric space. The $U(q-2)$-orbits in $U(q-1)/U(q-2)$ depend on two parameters. This explains the double expansion in (7).

The addition formula for Jacobi polynomials $P_n^{(\alpha,\beta)}(x)$ with general α and β can be obtained by repeated differentiation of both sides of (7) with respect to φ and by doing analytic continuation with respect to α and β. When $\alpha\geqq\beta\geqq-\frac{1}{2}$, the addition formula implies a product formula and the product formula immediately gives the positivity of the convolution structure for Jacobi series. This result was earlier obtained by Gasper [**5**]. He obtained the kernel K in

$$\frac{P_n^{(\alpha,\beta)}(x)\,P_n^{(\alpha,\beta)}(y)}{P_n^{(\alpha,\beta)}(1)}=\int_{-1}^{+1}K(x,y,z)\,P_n^{(\alpha,\beta)}(z)\,(1-z)^\alpha(1+z)^\beta\,dz\tag{9}$$

explicitly as a nonnegative function. Formula (9) also follows from our product formula (see [**11**]).

There is an interpretation of Jacobi polynomials $P_n^{(p/2-1,\,q/2-1)}$ as spherical harmonics of degree $2n$ in $q+p$ dimensions which are invariant under the rotation group $SO(q)\times SO(p)$ (see Zernike and Brinkman [**12**], Braaksma and Meulenbeld [**1**]). In this interpretation another group theoretic proof can be given for our formulas (Flensted-Jensen [**3**], Koornwinder [**10**]).

References

1. B. L. J. Braaksma and B. Meulenbeld, *Jacobi polynomials as spherical harmonics*, Nederl. Akad. Wetensch. Proc. Ser. A **71** = Indag. Math. **30** (1968), 384–389. MR **38** #3481.

2. A. Erdélyi, et al., *Higher transcendental functions.* Vol. I. *The hypergeometric function, Legendre functions*, McGraw-Hill, New York, 1953. MR **15**, 419.

3. M. Flensted-Jensen, *Spherical functions on rank one symmetric spaces and generalizations*, Proc. Sympos. Pure Math., vol. 26, Amer. Math. Soc., Providence, R.I., 1974, pp. 339–342.

4. R. Gangolli, *Positive definite kernels on homogeneous spaces and certain stochastic processes related to Levy's Brownian motion of several parameters*, Ann. Inst. H. Poincaré Sect. B **3** (1967), 121–226. MR **35** #6172.

5. G. Gasper, *Positivity and the convolution structure for Jacobi series*, Ann. of Math. (2) **93** (1971), 112–118. MR **44** #1852.

6. M. Ikeda, *On spherical functions for the unitary group.* I, II, III, Mem. Fac. Engrg. Hiroshima Univ. **3** (1967), no. 1, 17–75. MR **36** #4042; 4043; 4044.

7. M. Ikeda and T. Kayama, *On spherical functions for the unitary group.* IV. *The case of higher dimensions*, Mem. Fac. Engrg. Hiroshima Univ. **3** (1967), 77–100. MR **36** #4045.

8. T. H. Koornwinder, *The addition formula for Jacobi polynomials.* I, *Summary of results*, Indag. Math. **36** (1972), 188–191.

9. ———, *The addition formula for Jacobi polynomials.* II, III, Math. Centrum Afd. Toegepaste Wisk. Reports TW 133, 135 (1972).

10. ———, *The addition formula for Jacobi polynomials and spherical harmonics*, SIAM J. Appl. Math. **25** (1973).

11. ———, *Jacobi polynomials.* II, *An analytic proof of the product formula*, SIAM J. Math. Anal. **5** (1974).

12. F. Zernike and H. C. Brinkman, *Hypersphärische Funktionen und die in sphärischen Bereichen orthogonalen Polynome*, Nederl. Akad. Wetensch. Proc. **38** (1935), 161–173.

MATHEMATISCH CENTRUM
AMSTERDAM, THE NETHERLANDS

LIE ALGEBRAS AND GENERALIZATIONS OF HYPERGEOMETRIC FUNCTIONS

WILLARD MILLER, JR.

We show how Lie algebras can be employed to study ${}_2F_1$ and its generalizations. We use the differential recurrence relations obeyed by a family of hypergeometric functions to generate a Lie algebra whose action determines basic properties of the corresponding functions.

For the ${}_pF_q$ we introduce functions and operators

$$f_{\alpha_j\beta_k}(t_j, u_k, x) = {}_pF_q\left(\begin{smallmatrix}\alpha_j\\ \beta_k\end{smallmatrix} \mid x\right) t_1^{\alpha_1}\cdots t_p^{\alpha_p} u_1^{\beta_1}\cdots u_q^{\beta_q},$$

$$E_{\alpha_l} = t_l(x\partial_x + t_l\partial_{t_l}), \qquad E_{-\beta_s} = u_s^{-1}(x\partial_x + u_s\partial_{u_s} - 1), \qquad 1 \leqq l \leqq p,$$

$$E_{\alpha_1\cdots\beta_q} = t_1\cdots u_q\partial_x, \qquad T_l = t_l\partial_{t_l}, \qquad U_s = u_s\partial_{u_s}, \qquad 1 \leqq s \leqq q,$$

obtained from the recurrence formulas for ${}_pF_q$. The operators generate a Lie algebra $\mathcal{G}_{p,q}$ of dimension $2(p+q)+1$ and the $f_{\alpha_j\beta_k}$ form bases for $\mathcal{G}_{p,q}$-representations. Let

$$L_{p,q} = E_{\alpha_1}\cdots E_{\alpha_p} - E_{\alpha_1\cdots\beta_q}E_{-\beta_1}\cdots E_{-\beta_q}.$$

THEOREM. *If* (1) $L_{p,q}f = 0$, (2) $T_lf = \alpha_l f$, $1 \leqq l \leqq p$, (3) $U_sf = \beta_s f$, $1 \leqq s \leqq q$, *and* (4) f *analytic at* $x=0$, *then* $f = cf_{\alpha_j\beta_k}$, c *constant.*

THEOREM. *The null space of* $L_{p,q}$ *is invariant under* $\mathcal{G}_{p,q}$.

WEISNER'S PRINCIPLE. *If* (1) $L_{p,q}f = 0$, (2) $f = \sum_{\alpha_j\beta_k} h_{\alpha_j\beta_k}(x)\, t_1^{\alpha_1}\cdots u_q^{\beta_q}$, (3) f *analytic at* $x=0$, *and* (4) $L_{p,q}$ *can be applied term-by-term to the sum, then* $h_{\alpha_j\beta_k}(x) = c_{\alpha_j\beta_k} f_{\alpha_j\beta_k}$, c *a constant.*

AMS (MOS) subject classifications (1970). Primary 33A75; Secondary 22E70.

We can consider any analytic solution f of $L_{p,q}f=0$ as a generating function for the ${}_pF_q$ and use these theorems to determine the expansion coefficients. In practice f is characterized as a simultaneous eigenfunction of $p+q$ operators in the enveloping algebra of $\mathcal{G}_{p,q}$ [**1**].

By a simple transformation and change of variable we obtain $E_{\alpha_j}=\partial_{z_j}$, $E_{\beta_k}=\partial_{w_k}$, $E_{\alpha_1\cdots\beta_q}=\partial_{w_{q+1}}$,

$$(*)\qquad L_{p,q}f=(\partial_{z_1}\cdots\partial_{z_p}-\partial_{w_1}\cdots\partial_{w_{q+1}})\,f=0.$$

In addition to the $\mathcal{G}_{p,q}$ symmetries, permutation symmetries of equation (*) are now apparent.

THEOREM. *If* $L_{p,q}f=0$, $L_{p',q'}f'=0$ *then* $L_{p+p',q+q'}(ff')=0$.

In special cases the symmetry algebra is larger:

	function	$Lf=0$	*algebra*	*dimension*	*reference*
1.	${}_2F_1$	$\Delta_4 f=0$	$sl(4)\cong o(6)$	15	[**2**], [**3**]
2.	${}_1F_1$	$\Delta_2 f=\partial_t f$		9	[**2**]
3.	D_ν	$\Delta_1 f=\partial_t f$		6	[**4**]
4.	${}_2F_1\left(\begin{smallmatrix}-\alpha,\,\beta\\ 1-\alpha-\beta\end{smallmatrix}\middle\vert\, x\right)$	$\Delta_3 f=0$	$o(5)$	10	[**3**].

Analogous results for Lauricella functions are ([**2**], [**5**]):

	function	$L_k f=0,\ 1\leqq k\leqq n$	*algebra*	*dimension*
5.	F_A	$(\partial u\partial u_k-\partial v_k\partial w_k)\,f=0$		$6n+2$
6.	F_B	$(\partial u_k\partial v_k-\partial w_k\partial w)\,f=0$		$6n+2$
7.	F_C	$(\partial u\partial v-\partial u_k\partial w_k)\,f=0$		$3n+4$
8.	F_D	$(\partial u\partial u_k-\partial v_k\partial v)\,f=0$	$sl(n+3)$	$(n+3)^2-1$

REFERENCES

1. W. Miller, Jr., *Lie theory and generalized hypergeometric functions*, SIAM J. Math. Anal. **3** (1972), 31–44.

2. ———, *Lie theory and generalizations of the hypergeometric functions*, SIAM J. Appl. Math. (to appear).

3. ———, *Symmetries of differential equations: The Euler-Darboux and hypergeometric equations*, SIAM J. Math. Anal. **4** (1972), 314–328.

4. L. Weisner, *Generating functions for hermite functions*, Canad. J. Math. **11** (1959), 141–147. MR **22** #786.

5. W. Miller, Jr., *Lie theory and the Lauricella functions* F_D, J. Math. Phys. **13** (1972), 1393–1399.

UNIVERSITY OF MINNESOTA

A NONCOMPACT ANALOGUE OF SPHERICAL HARMONICS

R. STRICHARTZ

The theory of spherical harmonics states that the space of spherical harmonics of degree k on $S^{n-1} \subseteq \boldsymbol{R}^n$ can be characterized in the following ways:

(1) the irreducible subspace of $L^2(S^{n-1})$ corresponding to the representation of $SO(n)$ of highest weight $(k, 0, \ldots, 0)$;

(2) the restriction to S^{n-1} of homogeneous polynomials of degree k on $\boldsymbol{R}^n$ satisfying $\Delta u=0$;

(3) the eigenspace of the spherical Laplacian (the tangential part of Δ) with eigenvalue $-k(n-2+k)$;

(4) the linear span of $(x_1+ix_2)^k$ under the action of $SO(n)$.

We replace $SO(n)$ by $O(n, N)$ acting on $\boldsymbol{R}^{n+N}$ preserving the quadratic form $Q=-x_1^2-\cdots-x_n^2+t_1^2+\cdots+t_N^2$. The group acts transitively on the hyperboloids $\{Q=c\neq 0\}$ and the cone $C=\{Q=0\}\backslash\{0\}$. The hyperboloid $H=\{Q=1\}$ plays the role of the unit sphere in our theory. If we consider $\square=-\Delta_x+\Delta_t$ in place of the Laplacian we have a complete analogue of (1)–(4) above; roughly speaking the following are equivalent:

(1′) irreducible subspaces of $L^2(C)$ or $L^2(H)$;

(2′) restrictions to C or H of homogeneous functions of degree σ (of fixed parity) on $\boldsymbol{R}^{n+N}$ satisfying $\square u=0$;

(3′) eigenspaces of the tangential part of $\square$ on H with eigenvalue $-\sigma\cdot(n+N-2+\sigma)$ (of fixed parity);

(4′) the span of $|x_1-t_1|^\sigma$ or $|x_1-t_1|^\sigma \operatorname{sgn}(x_1-t_1)$ under the action of $O(n, N)$.

In addition we describe which values of σ occur and how they are put together to form $L^2(C)$ and $L^2(H)$.

$L^2(C)$: The values $\sigma=-q+i\varrho$ for $q=(n+N-2)/2$ and $-\infty<\varrho<+\infty$ occur.

AMS (MOS) subject classifications (1970). Primary 43A85, 22E43, 22E45.

The representations corresponding to $q \pm i\varrho$ are equivalent but distinct. A typical function $F \in L^2(C)$ can be written

$$(*)\quad F(x,t) = \frac{1}{2\pi} \int_{-\infty}^{\infty} \int_{S^{n-1}} \int_{S^{N-1}} |x\cdot\zeta' - t\cdot\tau'|^{-q+i\varrho} \cdot (\psi_0(\zeta',\tau',\varrho)\, a(\varrho)^2 + \operatorname{sgn}(x\cdot\zeta' - t\cdot\tau')\, \psi_1(\zeta',\tau',\varrho)\, b(\varrho)^2)\, d\zeta'\, d\tau'\, d\varrho$$

where

$$a(\varrho) = b(\varrho) = 2(2\pi)^{-q}\, \Gamma(q+i\varrho)/\Gamma(i\varrho) \quad \text{if } n \text{ or } N \text{ is odd},$$

$$\begin{Bmatrix} a(\varrho) \\ b(\varrho) \end{Bmatrix} = \tfrac{1}{2}(2\pi)^{-q} \frac{\Gamma(q+i\varrho)}{\Gamma(i\varrho)} \begin{Bmatrix} \left|\tanh \dfrac{\pi\varrho}{2}\right| \\ \left|\coth \dfrac{\pi\varrho}{2}\right| \end{Bmatrix} \quad \text{if } n, N \text{ is even}.$$

$L^2(H)$: There is a decomposition $L^2(H) = \mathscr{H}_+ \oplus \mathscr{H}_-$ into the discrete and continuous spectrum. The values of σ entering into $\mathscr{H}_+$ are all integers $> -q$, and the corresponding solutions of $\Box u = 0$ vanish on one side of the cone. $\mathscr{H}_- \oplus \mathscr{H}_-$ is unitarily equivalent to $L^2(C)$, so a typical function $F \in \mathscr{H}_-$ can be written in the form $(*)$ with the ϱ-integration extended from 0 to ∞.

Complete details are published under the title *Harmonic analysis on hyperboloids* in the Journal of Functional Analysis **12** (1973), 341–383. The special cases $O(3,1)$ and $O(1,3)$ may be found in *Generalized functions*, Vol. 5 by I. M. Gel'fand, M. I. Graev and N. Ja. Vilenkin, and $O(n,1)$ may be found in *Special functions and the theory of group representations* by N. Ja. Vilenkin. The case $O(2,2)$ is equivalent to the Plancherel formula for $SL(2, \boldsymbol{R})$.

Ideas from the theory of partial differential equations come from *L'intégral de Riemann-Liouville* by M. Riesz, Acta Math. 1949, and *The stationary observer problem for $\Box u = Mu$* by the author in Journal of Differential Equations, 1971.

For a different approach to the same problem, see N. Limić, J. Niederle and R. Raczka, Journal of Mathematical Physics **8** (1967). For a similar approach to related problems, see L. Ehrenpreis' article in these PROCEEDINGS.

CORNELL UNIVERSITY

SUBELLIPTIC OPERATORS ON THE HEISENBERG GROUP

G. B. FOLLAND

A formally selfadjoint second order differential operator $\mathscr{L}$ on a Riemannian manifold M is called *subelliptic of order* ε $(0<\varepsilon<1)$ if for every compact $K\subset M$ there exists $c>0$ such that, for all $u\in C_0^\infty(K)$,

$$\|u\|_\varepsilon^2 \leqq c(|(\mathscr{L}u, u)| + \|u\|^2). \tag{1}$$

Here $\| \ \|_\varepsilon$ is the Sobolev norm of order ε and $\| \ \|$ is the L^2 norm. The basic property of subelliptic operators is the following:

THEOREM A (KOHN-NIRENBERG [**5**]). *If $\mathscr{L}$ is subelliptic of order ε then $\mathscr{L}$ is hypoelliptic, and for each compact $K\subset M$ and $s\geqq 0$ there exists $c>0$ such that, for all $u\in C_0^\infty(K)$,*

$$\|u\|_{s+2\varepsilon}^2 \leqq c(\|\mathscr{L}u\|_s^2 + \|u\|^2). \tag{2}$$

The basic tool for proving subellipticity is the following theorem (cf. Kohn [**4**]):

THEOREM B (HÖRMANDER-KOHN-RADKEVIČ). *Let $K\subset M$ be compact and $X_1, \ldots, X_k$ complex vector fields on K whose linear span is closed under complex conjugation and such that $\{X_j, [X_{j_1}, X_{j_2}], \ldots, [X_{j_1}, [X_{j_2}, \ldots [X_{j_{p-1}}, X_{j_p}]\ldots]]\}$ spans all vector fields on K. Then there exists $c>0$ such that, for all $u\in C_0^\infty(K)$,*

$$\|u\|_{2^{1-p}} \leqq c\left(\sum_1^k \|X_j u\|^2 + \|u\|^2\right).$$

AMS (MOS) subject classifications (1970). Primary 35C05, 35H05, 43A80; Secondary 35B45, 44A25.

Subelliptic operators of order $\frac{1}{2}$ arise from the theory of several complex variables in the study of the tangential Cauchy-Riemann operators on the boundary of a complex manifold (cf. [**2**]).

We now construct a subelliptic operator on the Heisenberg group N, which is the nilpotent Lie group with underlying manifold $\boldsymbol{C}^n \times \boldsymbol{R}$ and group law $(z, t)(z', t') = (z+z', t+t'+2\,\mathrm{Im}(z, z'))$. Letting $z = x+iy$, $x_1, \ldots, x_n, y_1, \ldots, y_n, t$ are coordinates on N, and the vector fields

$$X_j = \frac{\partial}{\partial x_j} + 2y_j \frac{\partial}{\partial t}, \qquad Y_j = \frac{\partial}{\partial y_j} - 2y_j \frac{\partial}{\partial t}, \qquad T = \frac{\partial}{\partial t},$$

are a basis for the Lie algebra of N. We set $\mathscr{L} = \sum_1^n (X_j^2 + Y_j^2)$.

THEOREM 1. *$\mathscr{L}$ is subelliptic of order $\frac{1}{2}$.*

Indeed, since X_j and Y_j are skew-adjoint (with respect to a left-invariant metric on N), we have $|(\mathscr{L}u, u)| = \sum_1^n (\|X_j u\|^2 + \|Y_j u\|^2)$, which by Theorem B implies the estimate (1) with $\varepsilon = \frac{1}{2}$, since $[Y_j, X_j] = 4T$.

Using the natural dilations and norm function on N, we can construct a fundamental solution for $\mathscr{L}$. For $0 < r < \infty$, we define the dilation $\delta_r(z, t) = (rz, r^2 t)$, and we define the norm $\varrho(z, t) = (|z|^4 + t^2)^{1/4}$.

THEOREM 2. *$\mathscr{L}\varrho^{-2n} = c_n \mu_0$, where μ_0 is the point mass at the origin and c_n is a nonzero constant.*

COROLLARY 1. *If f is a "reasonable" function, $u = f * (c_n^{-1} \varrho^{-2n})$ satisfies $\mathscr{L}u = f$, where $*$ denotes convolution on the group N.*

Using the theory of singular integrals on nilpotent groups (cf. [**1**] and [**3**]) we easily deduce the following regularity theorem for $\mathscr{L}$.

THEOREM 3. *Let f and u be as in Corollary 1. Then the mappings taking f to $X_j X_k u$, $X_j Y_k u$, $Y_k Y_j u$, and Tu are bounded on L^p, $1 < p < \infty$, and are weak-type $(1, 1)$. This is not true of $X_j Tu$, $Y_j Tu$, or $T^2 u$.*

We can also estimate higher derivatives of u in terms of derivatives of f, and this provides a very precise interpretation of the estimate (2). These facts indicate strongly that nilpotent singular integrals should be useful in studying more general subelliptic operators.

REFERENCES

1. R. Coifman and G. Weiss, *Analyse harmonique non-commutative sur certaines espaces homogènes*, Lecture Notes in Math., vol. 242, Springer-Verlag, Berlin and New York, 1971.

2. G. B. Folland and J. J. Kohn, *The Neumann problem for the Cauchy-Riemann complex*, Ann. of Math. Studies, no. 75, Princeton Univ. Press, Princeton, N.J., 1972.

3. A. W. Knapp and E. M. Stein, *Intertwining operators for semi-simple groups*, Ann. of Math. (2) **93** (1971), 489–578.

4. J. J. Kohn, *Pseudo-differential operators and non-elliptic problems*, Pseudo-differential operators (C.I.M.E. Stresa, 1968), Edizioni Cremonese, Rome, 1969, pp. 157–165. MR **41** #3972.

5. J. J. Kohn and L. Nirenberg, *Non-coercive boundary value problems*, Comm. Pure Appl. Math. **18** (1965), 443–492. MR **31** #6041.

COURANT INSTITUTE OF MATHEMATICAL SCIENCES, NEW YORK UNIVERSITY

SINGULAR INTEGRALS RELATED TO NILPOTENT GROUPS AND $\bar{\partial}$ ESTIMATES*

E. M. STEIN

Our purpose is to give a general discussion, together with some motivation, of a new direction in the study of singular integral operators. The kind of extension of singular integral operators, and the types of estimates to be made for them arise in the following areas:[1]

(1) In the context of nilpotent groups, motivated partly by the study of intertwining operators. See [**5**], [**8**], and [**1**], and §5 below.

(2) The singular integrals that occur in the solution of the $\bar{\partial}u = f$ problem, as given by Grauert and Lieb [**2**], Henkin [**3**], and Kerzman [**4**].

(3) Boundary behavior of holomorphic functions of several complex variables. See [**12**] and §3 below.

(4) Singular integrals related to the Bergman kernel. See §4.

From the qualitative point of view what is common to these areas (and distinguishes them from the more classical integrals and estimates) is the splitting of directions at each point, together with the nonisotropic way that singularities behave and estimates are made. The nonisotropy is not always apparent for L^p estimates, but its role becomes clear when calculating with the appropriate Lipschitz (or Hölder) inequalities. It is with the latter type of estimates that we are principally concerned here. Detailed proofs and further results will be given at another occasion.

1. New Lipschitz spaces. For each Riemannian space $\mathscr{R}$ we can speak of the standard Lipschitz spaces $\Lambda_\alpha(\mathscr{R})$, $0<\alpha<\infty$ (see e.g. [**11**] for the case $\mathscr{R} = \boldsymbol{R}^n$).

We now modify the above definitions so as to take into account the nonisotropy

AMS (MOS) subject classifications (1970). Primary 44A25, 33N15, 26A16.

* A more expanded form of this note appears in Bull. Amer. Math. Soc. **79** (1973), 439–445. §5 contains the corrected version of results that were inadvertently stated incorrectly in the other note.

[1] Further background for this approach may be found in [**10**].

which is crucial in what follows. For each $x \in \mathscr{R}$ let T_x denote the tangent space at x. Assume that there is given a direct sum decomposition $T_x = T_x^1 \oplus T_x^2$, with the property that the mappings $x \to T_x^i$ are smooth. We give two basic examples of this. $\mathscr{D}$ is an open subdomain of $\boldsymbol{C}^n$ with smooth boundary.

EXAMPLE 1. $\mathscr{R} = \partial\mathscr{D}$. For each $x \in \partial\mathscr{D}$, let v_x be the unit normal vector at x. Set $T_x^2 = \{\boldsymbol{R} i v_x\}$, and T_x^1 the orthogonal complement of T_x^2 in T_x.

EXAMPLE 2. $\mathscr{R} = \mathscr{D}$. Define a smooth mapping $x \to \bar{x}$ of $\mathscr{D}$ to $\partial\mathscr{D}$ with the property that whenever x is close to $\partial\mathscr{D}$ then $\bar{x}$ is the normal projection of x to $\partial\mathscr{D}$. Set $T_x^2 = \{\boldsymbol{C} v_{\bar{x}}\}$, and T_x^1 the orthogonal complement of T_x^2 in T_x. For these examples see also [**12**, pp. 34 and 55].

Return to the general case. For $0 < \alpha < \infty$ we define the space $\Gamma_\alpha(\mathscr{R})$ to consist of those functions u which are in $\Lambda_\alpha(\mathscr{R})$, but satisfy the additional property that when restricted to curves whose tangents lie in T_x^1, these functions are in $\Lambda_{2\alpha}$. More precisely, let l be the smallest integer $> 2\alpha$. Then for the norm we take

$$\|u\|_{\Gamma_\alpha(\mathscr{R})} = \|u\|_{\Lambda_\alpha(\mathscr{R})} + \sup\{\|u(x(\cdot))\|_{\Lambda_{2\alpha}(I)}\},$$

where the "sup" is taken over all curves $x(\cdot)$ which have C^l norm bounded by 1 and have the property that $x'(t) \in T^1_{x(t)}$, $t \in I$.

These definitions can be considerably generalized, but we shall not pursue this point here. We make some remarks about the spaces $\Lambda_1(\mathscr{R})$ and $\Gamma_{1/2}(\mathscr{R})$. It is well known that whenever $f \in \Lambda_1(I)$, $|f(t+h) - f(t)| \leqq A\,|h| \log 1/|h|$, as $h \to 0$, but no better estimate involving the first difference can be made for this class.[2] With this in mind we define $\tilde{\Lambda}_1(\mathscr{R})$ to be the class of bounded u so that $|u(x) - u(y)| \leqq A\,|x-y| \log 1/|x-y|$, whenever $|x-y| \leqq \frac{1}{2}$; we give $\tilde{\Lambda}_1(\mathscr{R})$ an obvious norm. Clearly then $\Lambda_1(\mathscr{R}) \subseteq \tilde{\Lambda}_1(\mathscr{R})$. There is a similar situation for $\Gamma_{1/2}(\mathscr{R})$. We define $\tilde{\Gamma}_{1/2}(\mathscr{R})$ to be the class of all bounded functions u so that

(a) $|u(x) - u(y)| \leqq A\,|x-y|^{1/2}$, all $x, y \in \mathscr{R}$, and

(b) $|u(x(t_1)) - u(x(t_2))| \leqq A\,|t_1 - t_2| \log 1/|t_1 - t_2|$ whenever $|t_1 - t_2| \leqq \frac{1}{2}$ and $t \to x(t)$ is a curve in the set $\mathscr{C}^2$ whose tangents, $x'(t)$, lie in $T^1_{x(t)}$. Again, $\Gamma_{1/2}(\mathscr{R}) \subset \tilde{\Gamma}_{1/2}(\mathscr{R})$.

2. Estimates for $\bar{\partial} u = f$. Now let $\mathscr{D}$ be a smooth bounded domain in $\boldsymbol{C}^n$ which is strictly pseudoconvex. Let f be a once continuously differentiable (0, 1) form defined in $\mathscr{D}$, which satisfies the integrability condition $\bar{\partial} f = 0$. The problem is to estimate the smoothness of solutions of $\bar{\partial} u = f$. We consider the splitting described in Example 2 of § 1 and the resulting $\Gamma(\mathscr{D})$ spaces.

Recent integral representations for u and estimates are due to Grauert and Lieb [**2**], Henkin [**3**], and Kerzman [**4**]. Basing ourselves on these integral representations, then for the resulting linear mapping $f \to u$ we can prove

[2] See Zygmund [**13**, p. 44]. Note, however, that there the Λ_1 spaces are designated as Λ_*.

THEOREM 1. *Suppose f is bounded on $\mathscr{D}$ and $\bar{\partial} f=0$. Then the solution $u \in \tilde{\Gamma}_{1/2}(\mathscr{D})$; also $\|u\|_{\Gamma_{1/2}(\mathscr{D})} \leqq A\,\|f\|_\infty$.*

An immediate corollary is the following:

COROLLARY.[3] *$u \in \Lambda_{1/2}(\mathscr{D})$, and $\|u\|_{\Lambda_{1/2}(\mathscr{D})} \leqq A\,\|f\|_\infty$.*

3. Holomorphic functions. Using reasoning similar to the above, we can prove the following theorem for holomorphic functions. Here $\mathscr{D}$ is any bounded smooth domain in $\boldsymbol{C}^n$, not necessarily pseudoconvex.

THEOREM 2. *Suppose u is homomorphic in $\mathscr{D}$ and continuous in $\bar{\mathscr{D}}$. Then*

$$u \in \Gamma_\alpha(\mathscr{D}) \Leftrightarrow u \in \Lambda_\alpha(\mathscr{D}) \Leftrightarrow u \in \Lambda_\alpha(\partial\mathscr{D}) \Leftrightarrow u \in \Gamma_\alpha(\partial\mathscr{D}), \qquad 0<\alpha<\infty .$$

4. Bergman kernel operator for the unit ball. Here $\mathscr{R}=\mathscr{D}=$ unit ball in $\boldsymbol{C}^n$, and we consider the projection operator P given by the Bergman kernel of $\mathscr{D}$, namely,

$$(Pu)(z)=\int_{|\zeta|\leqq 1} K(z,\zeta)\,u(\zeta)\,d\sigma(\zeta),$$

where $K(z,\zeta)=(n!/\pi^n)\,(1-z\cdot\bar{\zeta})^{-n-1}$.

THEOREM 3. *The mapping $u \to P(u)$*
(1) *takes $\Gamma_\alpha(\mathscr{D})$ continously to $\Gamma_\alpha(\mathscr{D})$. $0<\alpha<\infty$,*
(2) *takes $L^p(\mathscr{D})$ continuously to $L^p(\mathscr{D})$, $1<p<\infty$.*

The proof of part (1) uses the characterization given in Theorem 2. An application is as follows. Let u be any solution of $\bar{\partial} u=f$ in the unit ball. Then Kohn's solution (in [**6**]) is given by $u_0=u-P(u)$, since it is characterized by being orthogonal to holomorphic functions. The theorem then shows that any Γ_α or L^p estimate made for u (such as in §4 above) holds also for Kohn's solution, in the case of the unit ball.

5. Nilpotent groups. As we have already indicated, there are several closely related models for the kind of singular integrals we are dealing with. Besides those already discussed there are the "fractional integrals" which we now describe.

Let X be a simply-connected nilpotent Lie group, and let $x \to |x|$ be a "norm function" on it in the sense of [**5**]. Thus $|x|>0$ iff $x \neq$ identity, $|x|$ is C^∞ in the com-

---

[3] For the corollary see Romanov and Henkin [**9**].

plement of the identity, $|x|=|x^{-1}|$, and $dx/|x|$ is invariant under an appropriate one-parameter group of automorphisms of X (dilations). For α positive and sufficiently small it is natural to define the space $*\Gamma_\alpha(X)$ to consist of those functions f on X for which

$$\sup_{x\in X}|f(xy)-f(x)|\leqq A|y|^\alpha .$$

A similar, but not equivalent definition is obtained if left translation is used in place of right translations.

Let Ω be a function defined on $X-\{\text{identity}\}$ which is C^∞ there, and homogeneous of degree 0, i.e., is invariant under the dilations. We shall consider the fractional integrals I_α given by

$$I_\alpha(f)(x)=\int_X \frac{\Omega(y)}{|y|^{1-\alpha}} f(xy^{-1})\,dy,\quad 0<\alpha<1. \tag{1}$$

This can be rewritten as $I_\alpha(f)=\int K(y^{-1}x)f(y)\,dy$, with $K(x)=\Omega(x)/|x|^{1-\alpha}$. In order to be able to deal with bounded f (for which (1) may not converge), we modify I_α by adding a constant, and define

$$\tilde I_\alpha(f)(x)=\int_X [K(y^{-1}x)-K(y^{-1})]\,f(y)\,dy.$$

We let d be the largest positive exponent satisfying the condition that $\|x\|\leqq A|x|^d$, when $\|x\|\leqq 1$, where $\|\cdot\|$ is a standard Euclidean norm on X. (For the determination of d, see [**5**, p. 498].)

THEOREM 4. (1) *If f is bounded, then $\tilde I_\alpha(f)\in *\Gamma_\alpha(X)$, whenever $0<\alpha<d$.*
(2) *If f is bounded and $f\in *\Gamma_\beta(X)$, then $\tilde I_\alpha(f)\in *\Gamma_{\alpha+\beta}(X)$, whenever $0<\alpha+\beta<d$.*
(3) *If $f\in L^p(X)$, then $I_\alpha(f)\in L^q(X)$, whenever $1/q=(1/p)-\alpha$, and $1<p<q<\infty$.*

REMARKS. 1. When $X=\boldsymbol{R}^n$, we can take $|x|=\|x\|^n$. Then $d=1/n$, and $*\Gamma_\alpha(X)$ is essentially $\Lambda_{n\alpha}(\boldsymbol{R}^n)$.

2. Closer to the results above we take $X=\{(z,w)\}=\boldsymbol{C}^{n-1}\times\boldsymbol{R}^1$, with $(z_1,w_1)\cdot(z_2,w_2)=(z_1+z_2,\,w_1+w_2+2\operatorname{Im} z_1\bar z_2)$. Here we can take $|(z,w)|=(|z|^4+w^2)^{n/2}$, and hence $d=\frac{1}{2}n$. (See also [**10**, p. 179].) $*\Gamma_\alpha(X)$ is essentially $\Gamma_{n\alpha}(X)$, $0<\alpha<d$.

3. Of further interest is the fact, recently obtained by G. Folland (see his note in these Proceedings), that a fundamental solution of the real part at the $\Box_b$ Laplacian of Kohn [7], corresponding to the unit ball, can be realized explicitly as an integral of the form (1).

REFERENCES

1. R. Coifman and G. Weiss, *Analyse harmonique non-commutative sur certains espaces homogenes*, Lecture Notes in Math., Springer-Verlag, Berlin, 1971.

2. H. Grauert and I. Lieb, *Das Ramirezsche Integral und die Lösung der Gleichung $\bar{\partial}f=\alpha$ im Bereich der beschränkten Formen*, Complex Analysis (Proc. Conf., Rice Univ., Houston, Tex., 1969), Rice Univ. Studies **56** (1970), no. 2, 29–50. MR **42** #7938.

3. G. Henkin, *Integral representations of holomorphic functions in strongly pseudo-convex domains and applications to the $\bar{\partial}$-problem*, Mat. Sb. **82 (124)** (1970), 300–308 = Math. USSR Sb. **11** (1970), 273–281. MR **42** #534.

4. N. Kerzman, *Hölder and L^p estimates for solutions of $\bar{\partial}u=f$ in strongly pseudo-convex domains*, Comm. Pure Appl. Math. **24** (1971), 301–379. MR **43** #7658.

5. A. W. Knapp and E. M. Stein, *Intertwining operators for semi-simple groups*, Ann. of Math. (2) **93** (1971), 489–578.

6. J. J. Kohn, *Harmonic integrals in pseudo-convex domains.* I, II, Ann. of Math. (2) **78** (1963), 112–148; **79** (1964), 450–472. MR **27** #2999; **34** #8010.

7. ———, *Boundaries of complex manifolds*, Proc. Conf. Complex Analysis (Minneapolis, 1964), Springer, Berlin, 1965, pp. 81–94. MR **30** #5334.

8. A. Korányi and I. Vagi, *Singular integrals on homogeneous spaces and some problems of classical analysis*, Ann. Scuola Norm. Pisa, 1971.

9. A. V. Romanov and G. M. Henkin, *Exact Hölder estimates for the solutions of the $\bar{\partial}$-equation*, Izv. Akad. Nauk SSSR Ser. Mat. **35** (1971), 1171–1183 = Math. USSR Izv. **5** (1971), 1180–1191. MR **45** #220.

10. E. M. Stein, *Some problems in harmonic analysis suggested by symmetric spaces and semi-simple groups*, Proc. Internat. Congress Math. (Nice, 1970), vol. 1, Gauthier-Villars, Paris, 1971, pp. 173–189.

11. ———, *Singular integrals and differentiability properties of functions*, Princeton Math. Series, no. 30, Princeton Univ. Press, Princeton, N. J., 1970. MR **44** #7280.

12. ———, *Boundary behavior of holomorphic functions of several complex variables*, Math. Notes, Princeton, N. J., 1972.

13. A. Zygmund, *Trigonometrical series*, Vol. I, 2nd ed., Cambridge Univ. Press, New York, 1959. MR **21** #6498.

PRINCETON UNIVERSITY

OPERATORS TRANSFERRED BY REPRESENTATIONS OF AN AMENABLE GROUP

RONALD R. COIFMAN AND GUIDO WEISS

Our purpose is to illustrate a general method for obtaining norm and weak-type inequalities, as well as maximal inequalities, for operators associated with spaces on which act representations of amenable groups. Suppose $k \in L^1(G)$ has compact support, where G is an amenable group. Then the operator obtained by forming the (right or left) convolution with k, $f \to k * f$, is bounded on $L^p(G)$, $1 \leqq p \leqq \infty$. Let $N_p(k)$ denote its norm. If $\mathfrak{M}$ is a measure space and R a uniformly bounded representation of G acting on $L^p(\mathfrak{M})$ we can define the operator $\hat{k}_R$ mapping $L^p(\mathfrak{M})$ into itself by letting

$$(\hat{k}_R F)(x) = \int_G k(u)(R_{u^{-1}} F)(x)\, dx$$

for $F \in L^p(\mathfrak{M})$ and $x \in \mathfrak{M}$. It follows immediately from this definition that $\hat{k}_R$ is bounded with norm $\|k\|_1$. The basic feature of the method we shall illustrate is the sharper estimate:

(1) *The norm of the operator $\hat{k}_R$ on $L^p(\mathfrak{M})$ does not exceed $c^2 N_p(k)$, where c is the supremum of the norms of the transformations R_u, $u \in G$.*

An important class of representations of G arises from measure preserving transformations. That is, suppose S is a representation of G consisting of measure preserving transformations on $\mathfrak{M}$. If we define R by letting $(R_u F)(x) = F(S_{u^{-1}} x)$ we obtain a representation of G consisting of isometries of $L^p(\mathfrak{M})$ for $1 \leqq p \leqq \infty$. The convolution operator defined by k, being bounded on $L^p(G)$, is certainly of weak-type (p, p). Let $W_p(k)$ denote its weak-type norm.

AMS (MOS) subject classifications (1970). Primary 43A15, 43A55.

(2) *If R arises from measure preserving transformations then the weak-type norm of the operator $\hat{k}_R$ on $L^p(\mathfrak{M})$ does not exceed $W_p(k)$.*

Suppose $k^{(n)} \in L^1(G)$ has compact support for $n=1, 2, 3, \ldots$, and the operator K^* is defined by

$$(K^*f)(v) = \sup_{1 \leq n < \infty} \left| \int_G k^{(n)}(u) f(vu^{-1})\, du \right| .$$

Let $\hat{K}_R^*$ then be defined by

$$(\hat{K}_R^* F)(x) = \sup_{1 \leq n < \infty} |(\hat{k}_R^{(n)} F)(x)| .$$

(3) *If $\|K^*f\|_p \leq M_p \|f\|_p$ for all $f \in L^p(G)$, $1 \leq p \leq \infty$, then*
$$\|\hat{K}_R^* F\|_p \leq c^2 M_p \|F\|_p .$$

These inequalities also extend to certain operators that are not of convolution type and generalize the commutator operators of Calderón (see [**1**]). Moreover, one can often transfer convergence results on G (say $G = T$, the torus, and $k^{(n)}$ is the Dirichlet kernel) to convergence results on $\mathfrak{M}$. Special cases of these observations and inequalities have been described by Calderón [**2**] and Herz [**7**]; a more complete presentation, including some proofs, will appear in [**3**].

A basic operator is obtained by considering an ergodic flow S on $\mathfrak{M}$ and the Hilbert transform on the real line $\boldsymbol{R}$ (see Cotlar [**4**]). That is, the transformation defined by the integral

$$\text{(4)} \qquad \text{P.V.} \int_{-\infty}^{\infty} F(S_u x)\, du/u$$

for $x \in \mathfrak{M}$. It is well known that the Calderón-Zygmund singular integrals on $\boldsymbol{R}^n$ with odd kernels can be realized as averages of such operators (where the averaging is done over all one-parameter subgroups of $\boldsymbol{R}^n$). Expression (4) allows us to form analogous singular integral operators for every Lie group:

$$\text{(5)} \qquad \int_{|y|=1} \Omega(y) \left\{ \text{P.V.} \int_{-\infty}^{\infty} f(u \exp ty)\, dt/t \right\} dy,$$

where u lies in the group and y is a unit vector in the Lie algebra.

Expression (5), however, does not always yield the most natural realization of singular integrals. Let us consider the case when G is a compact connected Lie group of dimension N. Let T^n be a maximal torus which we identify with the cube $Q^n=\{\theta=(\theta_1, \theta_2, \ldots, \theta_n)\in \boldsymbol{R}^n: -\pi\leqq\theta_j<\pi, j=1, 2, \ldots, n\}$. If R is a finite-dimensional unitary representation of G we define the operator K_R by

$$(K_R F)(v)=\text{P.V.}\int_G \frac{R(u)-I}{[p(u)]^{N+1}} f(vu^{-1})\,du,$$

where $v\in G$ and p is a nonnegative central function such that, for $t=t(\theta)\in T^n$,

$$p(t(\theta))=\left(\sum_{j=1}^{n} \theta_j^2\right)^{1/2}$$

when $\theta\in Q^n$ belongs to a neighborhood of 0 and p is bounded away from zero outside this neighborhood. Consider the coefficients $\{r_{ij}(u)\}$ of $R(u)$ with respect to a basis that diagonalizes $R(u)$ when $u\in T^n$. They can be expressed as averages, over the conjugates of T^n, of singular integrals of Calderón-Zygmund type on T^n. An application of (1) and Minkowski's integral inequality gives us:

K_R is a bounded operator on $L^p(G)$, $1<p<\infty$; moreover,

$$\text{P.V.}\int_G \frac{r_{ij}(u)-\delta_{ij}}{[p(u)]^{N+1}} f(vu^{-1})\,du, \tag{6}$$

as an operator on $L^p(G)$, has norm not exceeding $l_R A_p$, where A_p depends only on G and p and $l_R=\sup_{u\in G}\{\|R(u)-I\|/p(u)\}$.

Our method can be used to transfer results of convergence or summability of Fourier series on a compact Lie group G. For example, if $G=SU(2)$ and χ^l, $l=0, \frac{1}{2}, 1, \frac{3}{2}, \ldots$, is the character of the irreducible representation of weight l, then it follows from (1) and a classical result of M. Riesz that the operators with kernels $\chi^l(u)/[p(u)]^2$ are uniformly bounded on $L^p(SU(2))$, $1<p<\infty$. From this we can deduce corresponding inequalities for Riesz summability of order 1 on $SU(2)$. Moreover, convergence theorems can also be transferred from those associated with T^n or $\boldsymbol{R}^n$ (in this connection, see also Calderón [**2**]).

Another type of result obtainable from (1) and (2) is the following generalization of a theorem of de Leeuw [**5**] (see also Fife [**6**] and Herz [**8**]):

(7) *Suppose G is a locally compact abelian group and m is a normalized*[1] *function on $\hat{G}$ such that the operator $f \to (m\hat{f})\check{}$ is bounded on $L^p(G)$ with operator norm $N_p(m)$ for some $p \in [1, \infty]$. Assume that R is a representation of G arising from measure preserving transformations on a measure space $\mathfrak{M}$ and E is the spectral measure on $\hat{G}$ for which $R_u = \int_{\hat{G}} \langle \chi, u \rangle \, dE(\chi)$. Then the operator $\int_{\hat{G}} m(\chi) \, dE(\chi)$ is bounded on $L^p(\mathfrak{M})$ with norm not exceeding $N_p(m)$.*

These are some of the applications of this method. There are many more. In particular, we were recently told that Lohoué and Peyriere [9] have transferred theorems from the group of motions of $\boldsymbol{R}^2$ to the circle by making use of a special case of (1).

References

1. A. P. Calderón, *Commutators of singular integral operators*, Proc. Nat. Acad. Sci. U.S.A. **53** (1965), 1092–1099. MR **31** #1575.

2. ———, *Ergodic theory and translation invariant operators*, Proc. Nat. Acad. Sci. U.S.A. **59** (1968), 349–353. MR **37** #2939.

3. R. R. Coifman and Guido Weiss, *Operators associated with representations of amenable groups, singular integrals induced by ergodic flows, the rotation method and multipliers*, Studia Math. **47** (1973), 285–303.

4. M. Cotlar, *A unified theory of Hilbert transforms and ergodic theorems*, Rev. Mat. Cuyana **1** (1955), 105–167. MR **18**, 893.

5. K. de Leeuw, *On L^p multipliers*, Ann. of Math. (2) **81** (1965), 364–379. MR **30** #5127.

6. D. Fife, *Spectral decomposition of ergodic flows on L^p*, Bull. Amer. Math. Soc. **76** (1970), 138–141.

7. C. Herz, *The theory of p-spaces with application to convolution operators*, Trans. Amer. Math. Soc. **154** (1971), 69–82. MR **42** #7833.

8. ———, *Problems of extrapolation and spectral synthesis on groups*, Conf. on Harmonic Analysis, Lecture Notes in Math., Springer-Verlag, Berlin, pp. 157–166.

9. N. Lohoué and J. Peyriere, *Analyse harmonique sur le groupe de déplacements du Plan*, Ann. Inst. Fourier (Grenoble) (to appear).

Washington University

[1] One can construct a sequence $\{\varphi_n\}$ of functions on G which acts as an approximation of the identity; m is said to be *normalized* (with respect to $\{\hat{\varphi}_n\}$) if $(\varphi * m)(\chi)$ converges to $m(\chi)$ for all $\chi \in \hat{G}$ (for details see [**3**]).

INVARIANT MEANS AND UNITARY REPRESENTATIONS

PIERRE EYMARD

This is an attempt to build a theory of amenability for any unitary representation π of a locally compact group G, in the same spirit that the classical theory of amenable groups concerns the regular representation ϱ of the group in $L^2(G)$.

I. Let us recall first that G is an amenable group if

(M) *on the vector space $\mathscr{CB}(G)$ of the continuous bounded functions on G, there exist a mean (i.e., a positive normalized linear form) which is G-invariant.*

This property is easily proved to be equivalent with the fixed point property, in a more general appearance:

(PF) *for every compact convex nonvoid set Q, if G acts continuously and affinely on Q, then there exists in Q a fixed point under G.*

For instance, the compact extensions of solvable groups are amenable, but the noncompact connected semisimple Lie groups are not amenable.

In 1965, H. Reiter and A. Hulanicki proved the following criterion, in terms of harmonic analysis, namely G is amenable iff

(F) *the trivial one-dimensional representation i_G of G is weakly contained (in the sense of Fell) in the regular representation ϱ of G.*

AMS (MOS) subject classifications (1970). Primary 22-02, 22D10, 43A07; Secondary 22D15, 22D30, 43A35.

My purpose is to extend this equivalence (M)$\Leftrightarrow$(F) to representations other than ϱ. In order to show the difficulties of the general case, it is useful to describe *grosso modo* the chain of arguments in the classical case of ϱ. For $p=1$ or 2, one introduces the following property:

(P_p) *for every compact set K in G, for every $\varepsilon>0$, there exists a function $f\in L^p(G)$, $f\geqq 0$ and $\|f\|_p=1$, such that, for every $x\in K$, $\|{}_xf-f\|_p\leqq\varepsilon$.*

One proves easily that (F)$\Leftrightarrow$(P_2) and that (P_2)$\Leftrightarrow$(P_1), but this last equivalence is important, because it is the turning point between the Hilbertian properties (unitary representations) and the mean property. Actually those normalized $f\in L^1_+(G)$, which are given by (P_1), are nothing but means on $\mathscr{CB}(G)$, and these means are "almost" G-invariant. Then a compactness argument allows us to deduce (M) from (P_1). The converse (M)$\Rightarrow$(F) is more intricate and I have no time for details here, but one can prove that (P_1) is a consequence of the fixed point property.

How can we generalize, replacing in (F) the regular representation ϱ by any unitary representation π? A difficulty occurs immediately in the general case. In the classical case, the representation space was $L^2(G)$, which contains many *positive* functions, and that was important in obtaining in (P_1) these almost invariant means. In the general case the representation space $\mathscr{H}_\pi$ has no "positive" elements, therefore we can only work in particular cases.

II. Let H be a closed subgroup of G, and let π be the *quasi-regular* representation of G in $L^2(G/H)$. In this case one can give a sense to each property defined above and prove their equivalence. Limiting ourselves to the most important results, we have the following theorem, partly obtained by F. P. Greenleaf and partly by myself.

THEOREM 1. *The properties* (M_H), (F_H), (PF_H) *and* (RG) *are equivalent, where*
(M_H) *there exists on $\mathscr{CB}(G/H)$ a G-invariant mean;*
(F_H) *i_G is weakly contained in the quasi-regular representation π;*
(PF_H) *for every convex compact set Q, if G acts continuously and affinely on Q in such a manner that there exists in Q a fixed point under H, then there exists in Q a fixed point under G.*

(RG) is a universal property, of which Reiter and Glicksberg gave particular cases. To explain it, let me define some notations. Let E and F be two Banach spaces and $\tau: E\to F$ a Banach-morphism. For every $x\in G$, let $A_x\in\mathscr{L}(E)$ such that: $|||A_x|||=1$; $x\mapsto A_x$ is strongly continuous; $A_e=I$; $A_{xy}=A_xA_y$, $x\in G$, $y\in G$; and, for every $h\in H$, one has $\tau\circ A_h=\tau$. Let J be the vector subspace of E spanned by the

$A_x g - g$, $g \in E$, $x \in E$. For $f \in G$, let C_f be the convex set generated in F by the $\tau(A_x f)$, $x \in G$. Then the property (RG) is

(RG) *Every time we are in such a situation, we have the inequality*

$$\mathrm{dist}_F(0, C_f) \leqq \mathrm{dist}_E(f, J).$$

We call a homogeneous space G/H with the properties of Theorem 1 *G-amenable*. Observe that if G is an amenable group, then every G/H is a G-amenable homogeneous space; and if H is an amenable group, then G/H is G-amenable iff G is an amenable group. Therefore the only interesting new cases are those where neither G nor H is an amenable group, but where nevertheless G/H is amenable.

If the group G has the Kajdan property, i.e., if $\{i_G\}$ is an open set in $\hat{G}$ for the Fell topology, for instance if $G = SL(3, \boldsymbol{R})$ or $SO(2, 3)$, then every amenable homogeneous space G/H, with H unimodular, is of finite volume, hence trivially amenable. But take $G = SL(2, \boldsymbol{R})$, $\Gamma = SL(2, \boldsymbol{Z})$ and H = the second derived group of Γ, then G/H is an amenable homogeneous space which is not of finite volume.

With the criterion (F_H) one can decide whether the imaginary Lobatschevskian space $SL(2, \boldsymbol{C})/SL(2, \boldsymbol{R})$ is amenable or not; this can be read on the explicit decomposition of π into irreducible representations, given by Gelfand and Graev, and we find that this space is *not* amenable.

III. I considered in §II the case of $\pi = \mathrm{ind}(i_H \uparrow G)$. Now let us work with other induced representations.

Let $G = KN$ be a locally compact group, a semidirect topological product of two closed subgroups K and N, where N is a *normal* subgroup of G.

Let σ be a *one-dimensional* unitary representation of N. If $k \in K$, we put $\sigma_k(n) = \sigma(k^{-1}nk)$, where $n \in N$, and we denote by $O_K(\sigma) = \{\sigma_k \mid k \in K\}$ the orbit of σ under K in $\hat{N}$. Let π_σ be the representation of G in $L^2(K)$, which is *induced* from σ, and defined for $n \in N$, $k \in K$, $\varphi \in L^2(K)$ by the formula

$$[\pi_\sigma(nk)\,\varphi]\,(t) = \sigma(t^{-1}nt)\,\varphi(k^{-1}t).$$

THEOREM 2. *The following properties are equivalent:*

(F^σ) *i_G is weakly contained in π_σ.*

(F^σ_o) *K is an amenable group, and i_N is in the closure of $O_K(\sigma)$ in $\hat{N}$.*

(M^σ) *On $\mathscr{CB}(K)$ there exists a K-invariant mean M, with the following supplementary stability property that, for every probability measure μ on N and every $\varphi \in \mathscr{CB}(K)$, one has*

$$M_t[\hat{\mu}(\sigma_t)\,\varphi(t)] = M(\varphi).$$

There are a number of other equivalent properties (cf. [**2**]). I only indicate here that, in this situation, we have the following *precised* fixed point property for K:

(PF^σ) *For every convex set Q, if K acts affinely and continuously on Q, then for every $b \in Q$ and for every* σ-essential *set $S \subset K$, there exists in Sb a point fixed under K.*

Here S is said to be *σ-essential* if every $\varphi \in \mathscr{CB}(K)$ which vanishes identically on S belongs to the closed ideal of $\mathscr{CB}(K)$ generated by the functions $k \mapsto 1 - \sigma_k(n)$, where $n \in N$.

The criterion (F_o^σ) is easy to verify in particular cases. For instance it is true for the affine group of $\boldsymbol{R}$, when σ is any nontrivial character of the translation subgroup, and for other solvable groups.

References

1. P. Eymard, *Sur les moyennes invariantes et les représentations unitaires*, C. R. Acad. Sci. Paris Sér. A-B **272** (1971), A1649–A1652. MR **43** #6740.

2. ———, *Moyennes invariantes et représentations unitaires*, Lecture Notes in Math., no. 300, Springer-Verlag, Berlin and New York, 1972.

3. F. P. Greenleaf, *Amenable actions of locally compact groups*, J. Functional Analysis **4** (1969), 295–315. MR **40** #268.

4. A. Hulanicki, *Means and Følner conditions on locally compact groups*, Studia Math. **27** (1966), 87–104. MR **33** #4178.

5. H. Reiter, *On some properties of locally compact groups*, Nederl. Akad. Wetensch. Proc. Ser. A **68** = Indag. Math. **27** (1965), 697–701. MR **33** #3114.

Université de Nancy
Nancy, France

TWO CONJECTURES ABOUT REDUCTIVE p-ADIC GROUPS

ROGER HOWE

1. I want to discuss two conjectures which, if true, seem to be important for harmonic analysis on reductive p-adic groups. Although each conjecture is interesting in itself, it is their interaction which is most important. They are in a sense dual to one another. Hence, after a brief introduction of each conjecture, I will discuss them together. For simplicity, I will restrict my attention to the case of Gl_n. Our unexplained terminology will conform to that of Harish-Chandra [1].

2. The first conjecture is illustrated and motivated by the following very pretty result of Allan Silberger.

THEOREM. *Let π_1 and π_2 be two infinite-dimensional irreducible representations of PGl_2. Then $\theta_{\pi_1} - \theta_{\pi_2}$, the difference of the characters of the π_i, is a locally constant function on PGl_2.*

COROLLARY. *θ_{π_1} is a locally integrable function on PGl_2, locally constant on the regular set, and bounded on the regular elliptic set.*

PROOF OF THE COROLLARY. Take π_2 to be a principal series representation.

One cannot expect such an extremely pleasant situation in general, but notice the qualitative import of Silberger's theorem. It says there is essentially only one kind of singularity that a character, as a function on PGl_2, can have. The case of PGl_2 is particularly simple because the only singular points of PGl_2 have trivial semisimple part. In general, one will expect characters to have singularities along the whole singular set, and these singularities need not be coordinated. However, one might hope that around a given point a character could exhibit only a finite number of types of singular behavior. For a general point, I do not now know how to be very precise about this, but for the identity I can make a concrete suggestion.

AMS (MOS) subject classifications (1970). Primary 22E50.

Fix a Borel subgroup B of $G=Gl_n$, and let $\{P_i\}_{i=1}^{l}$ denote the set of parabolics containing B and which correspond to flags of the kernels of powers of nilpotent elements. For each P_i let $\varrho_i=\mathrm{ind}_{P_i}^{G}\, 1$.

CONJECTURE 1. *Let π be an arbitrary admissible irreducible representation of G. Then there are constants $a_i(\pi)$ such that the distribution $\theta_\pi-\sum_{i=1}^{l} a_i(\pi)\,\theta_{\varrho_i}$ vanishes in a neighborhood of the identity.*

REMARKS. (a) One direct consequence of this conjecture is that θ_π would be a locally integrable function in a neighborhood of the identity. Presumably, a suitably extended conjecture, covering all points, would imply local integrability everywhere.

(b) The conjecture also shows θ_π cannot vanish on the regular set. For then all the $a_i(\pi)$ would be zero, which is impossible.

(c) One may show the $a_i(\pi)$ are necessarily integers. This may have some implications for the formal degree of supercuspidal representations.

3. The second conjecture is related to Harish-Chandra's map F_f. Recall that an invariant distribution on G is a distribution left unchanged by inner automorphisms. For a compact set $X\subseteq G$, let X^G be the invariant set generated by X. Let $I(X)$ be the space of invariant distributions supported on X^G. Let $H\subseteq G$ be an arbitrary open compact subgroup. Let $I(X, H)$ be the space of all linear functions on $C_c^\infty(G/H)$ which are restrictions of distributions in $I(X)$.

CONJECTURE 2. *The dimension of $I(X, H)$ is finite.*

Let $A\subseteq G$ be a Cartan subgroup, and let $\{\mathcal{O}_i\}_{i=1}^{l}$ be the unipotent classes in G. Each of the $\mathcal{O}_i$ carries an essentially unique measure $d\mathcal{O}_i$ invariant under inner automorphisms. As Shalika [3] has shown, there are functions Γ_i^A such that, for any $f\in C_c^\infty(G)$, $F_f^A-\sum_{i=1}^{l} a_i(f)\,\Gamma_i^A$ vanishes in a neighborhood of the identity, where

$$a_i(f)=\int_{\mathcal{O}_i} f\, d\mathcal{O}_i.$$

A similar asymptotic expansion for F_f^A will in fact hold around any point of A. The Γ_i^A, which Shalika calls "germs", are probably key objects in the harmonic analysis of G. However, very little is known to date about them, although there are several conjectures. Conjecture 2 would imply that the asymptotic expansion given by the Γ_i^A would hold in a fixed neighborhood of the identity for arbitrary $f\in C_c^\infty(G/H)$. The knowledge of such uniformity in the asymptotic expansion would allow Harish-Chandra to establish all he wants to know about F_f^A on the Schwartz space of G.

Another application of Conjecture 2 is the observation that it plus local integrability of supercuspidal characters would prove that all irreducible unitary representations are admissible. For in [2], Harish-Chandra showed admissibility would follow if the space ${}^{\circ}C(G//H, \chi)$ of supercusp forms bi-invariant by H and transforming according to χ by the center of G was of finite dimension. But the elliptic set of G is contained (modulo the center of G) in X^G for some compact X. And if supercuspidal characters are integrable, one sees that the left translates of $I(X)$ separate points of ${}^{\circ}C(G//H, X)$, which would then be finite dimensional by Conjecture 2. We note also that since a suitable strengthening of Conjecture 1 would imply local integrability of characters, the two conjectures do tend to complement each other. We will see more examples of this shortly.

Finally, we remark that Conjecture 2 may be verified without too much difficulty for Gl_2 and Gl_3. For Gl_2, one does not need integrability of characters, or one has it by a slight extension of Silberger's theorem. Hence, one obtains a quick proof of admissibility in this case.

4. Now I want to show how the two conjectures interact. First let me formulate an analogue of Conjecture 2 for the Lie algebra. The Lie algebra of $G = Gl_n$ is M_n, the $n \times n$ matrices. We call it $\mathfrak{A}$. Again, G acts on $\mathfrak{A}$ by the adjoint action. We refer to a distribution on $\mathfrak{A}$ as invariant if it is left unchanged by the adjoint action. Let $L \subseteq \mathfrak{A}$ be an open compact subgroup. Then $C_c^\infty(\mathfrak{A}/L)$ is the space of compactly supported functions on $\mathfrak{A}$ constant on cosets of L. Let $X \subseteq \mathfrak{A}$ be a compact set, let X^G be the set swept out by X under the adjoint action, and let $J(X)$ be the space of invariant distributions supported on X^G. Let $J(X, L)$ be the space of restrictions of elements of $J(X)$ to $C_c^\infty(\mathfrak{A}/L)$.

THEOREM. *$J(X, L)$ has finite dimension.*

This analogue of Conjecture 2 relates to Conjecture 1 via Fourier transform. To define Fourier transform, consider the bilinear form $B(x, y) = \operatorname{tr}(xy)$ for $x, y \in \mathfrak{A}$. The product xy is the ordinary associative product in M_n. The form B is symmetric, nondegenerate, and invariant under the adjoint action. Pick a character χ of the base field and define Fourier transform by

$$\hat{f}(x) = \int_{\mathfrak{A}} f(y)\, \chi(B(x, y))\, dy,$$

where $f \in C_c^\infty(\mathfrak{A})$ and dy is Haar measure on $\mathfrak{A}$. For a set $X \in \mathfrak{A}$, define $X' = \{y \in \mathfrak{A}, \chi(B(y, X)) = 1\}$. Then, if $L \subseteq \mathfrak{A}$ is an open compact subgroup, so is L', and we see that $C_c^\infty(\mathfrak{A}/L)$ is precisely the Fourier transform of $C_c^\infty(L')$.

We now commit a slight atrocity (which does not, however, make the succeeding discussion as special as it might seem to do). Consider the map $\lambda: G \to \mathfrak{A}$ given by $\lambda(g)=g-1$. Then λ is an injection of G onto an open dense set of $\mathfrak{A}$. Also, λ is equivariant with respect to inner automorphisms and the adjoint action, and it establishes a bijection between the unipotent classes in G and the nilpotent classes in $\mathfrak{A}$. For a given unipotent class $\mathcal{O}_i$, let $\mathcal{Q}_i=\lambda(\mathcal{O}_i)$ be the corresponding nilpotent class. We will indiscriminately let $\mathcal{Q}_i$ denote either the orbit itself or the distribution on $\mathfrak{A}$ defined by the invariant measure on $\mathcal{Q}_i$.

Given a distribution D on $\mathfrak{A}$ let $\lambda^{-1}(D)$ be the distribution on G obtained by taking $f\in C_c^\infty(G)$, translating by λ, then applying D. Also, $\hat{D}$ denotes the Fourier transform of D.

There is a well-known correspondence between the parabolic subgroups $\{P_i\}_{i=1}^l$ and the unipotent classes. Specifically, if P_i is a given parabolic, let $\mathcal{O}_i$ be the unipotent class whose intersection with the unipotent radical of P_i is dense. Then $P_i \Leftrightarrow \mathcal{O}_i$ is the correspondence in question.

PROPOSITION. *For each i, $\lambda^{-1}(\hat{\mathcal{Q}}_i)$ is, up to scalar, equal to θ_{ϱ_i} near 1 where ϱ_i is as in Conjecture* 1.

PROPOSITION. *Suppose $D\in J(X)$ for compact $X\subseteq G$. Then there are numbers $a_i(D)$ such that $\hat{D}-\sum_{i=1}^l a_i(D)\,\hat{\mathcal{Q}}_i$ vanishes near zero.*

Combining these two propositions, one obtains the following criterion for Conjecture 1 to hold.

CRITERION. Let σ be an irreducible admissible representation of G. Then Conjecture 1 holds for σ if and only if there exists $D\in J(X)$, for compact $X\subseteq\mathfrak{A}$, such that $\theta_\sigma-\lambda^{-1}(\hat{D})$ vanishes near the identity.

It may be verified that supercuspidal representations satisfy this criterion. Hence, all representations induced irreducibly from supercuspidals satisfy Conjecture 1. According to Jacquet's theorem [1], this is in a sense "almost all" admissible representations. The typical representation not of this sort is the special representation; but the special representation satisfies Conjecture 1 by the formula of Borel-Serre. Hence, Conjecture 1 seems very likely.

Now using the above information, one may return to F_f and derive information about Shalika's Γ_i^A. For example, on each elliptic A, the germ corresponding to the identity is a constant times $|D_G|^{1/2}$.

REFERENCES

1. Harish-Chandra, *Harmonic analysis on reductive p-adic groups*, Proc. Sympos. Pure Math., vol. 26, Amer. Math. Soc., Providence, R.I., 1974, pp. 167–192.

2. ———, *Harmonic analysis on reductive p-adic groups* (notes by G. van Dijk), Lecture Notes in Math., vol. 162, Springer-Verlag, New York, 1970.

3. J. Shalika, *A theorem on semisimple p-adic groups*, Ann. of Math. (2) **95** (1972), 226–242.

SUNY AT STONY BROOK

ZETA FUNCTIONS OF SIMPLE ALGEBRAS (LOCAL THEORY)

HERVÉ JACQUET

This is a brief account of a joint work with R. Godement which is a generalization of Tate's work (cf. [**1**] and [**2**]).

1. Notations. Let F be a commutative, local field. To simplify, we take F to be nonarchimedean. We denote by $|x|$ the absolute value of an x in F and by q the cardinality of the residual field of F.

Let G be a reductive group, defined over F. We use the notions introduced in Harish-Chandra's lectures: If π is an admissible representation of G_F on a complex vector space V, we have a representation π' of G_F on the algebraic dual V' of V and an admissible representation $\tilde{\pi}$ of G_F on the subspace $\tilde{V}$ of smooth vectors in V'. We may also consider the algebraic dual $(\tilde{V})'$ of $\tilde{V}$ on which we have a representation $(\tilde{\pi})'$. Of course V may be regarded as the space of smooth vectors in $(\tilde{V})'$ and then π is just the representation $(\tilde{\pi})^{\sim}$.

Let X be a locally compact, totally disconnected space. We shall denote by $\boldsymbol{D}(X)$ the space of locally constant, compactly supported functions on X. A V-valued distribution will be just a linear map from $\boldsymbol{D}(X)$ to V.

Suppose that G_F operates on X and that π is an admissible representation of G_F on V. A V-valued distribution μ on X will be said to be π-invariant if it satisfies the following condition:

$$\int f(g^{-1}x)\, d\mu(x) = \pi(g) \int f(x)\, d\mu(x).$$

If $X = G_F/H$ where H is a closed, unimodular subgroup of G_F and if μ is a π-invariant distribution on X there is in $(\tilde{V})'$ a vector v invariant under H such that

AMS (MOS) subject classifications (1970). Primary 12B35, Secondary 22E50.

$$\left\langle \int f(x)\, d\mu(x), \tilde{v} \right\rangle = \int_{G_F/H} f(g) \langle v, \tilde{\pi}(g^{-1}) \tilde{v} \rangle \, dg, \quad \text{for all } \tilde{v} \in \tilde{V}.$$

In particular, if H contains the unipotent radical of a parabolic subgroup and if π (and therefore $\tilde{\pi}$) is supercuspidal, we must have $v=0$ and also $\mu=0$.

2. The main result. From now on G will be the group $GL(n)$ and $X=M(n, F)$. Choose a nontrivial additive character ψ of F. If $\Phi \in \boldsymbol{D}(X)$ we define the Fourier transform $\hat{\Phi}(x)$ by

$$\hat{\Phi}(x) = \int_X \Phi(y)\, \psi(\operatorname{Tr} xy)\, dy.$$

Of course $\hat{\Phi} \in \boldsymbol{D}(X)$. We normalize the Haar measure dy in such a way that $(\hat{\Phi})\hat{}\ (x) = \Phi(-x)$. Let π be an irreducible admissible representation of G_F on V. We want to study the integrals

$$Z(s, f, \Phi) = \int_{G_F} \Phi(g)\, f(g)\, |\det g|^s \, d^\times g \tag{2.1}$$

where $d^\times g$ is a Haar measure on G_F, Φ belongs to $\boldsymbol{D}(X)$, s to $\boldsymbol{C}$ and f is a coefficient of π, that is a function of the form $f(g) = \langle \pi(g)\, v, \tilde{v} \rangle$ where v is in V and $\tilde{v}$ in $\tilde{V}$. The function $\check{f}$ defined by $\check{f}(g) = f(g^{-1})$ is then a coefficient of $\tilde{\pi}$. This leads us to consider also the integrals

$$Z(s, \check{f}, \Phi) = \int_{G_F} \Phi(g)\, f(g^{-1})\, |\det g|^s \, d^\times g. \tag{2.2}$$

THEOREM (2.3). (1) *For* $\operatorname{Re} s$ *large enough the integrals* (2.1) *and* (2.2) *are absolutely convergent.*

(2) *They are rational functions of* q^{-s}.

(3) *As such, they satisfy the functional equation*

$$Z(n-s, \check{f}, \hat{\Phi}) = c(s, \pi, \psi)\, Z(s, f, \Phi)$$

where $c(s, \pi, \psi)$ *is a rational function of* q^{-s} *which depends only on* π *and* ψ.

Observe that in this statement π and $\tilde{\pi}$ play the same role. It may happen that Theorem (2.3) is true for a pair $(\pi, \tilde{\pi})$ even though π is not irreducible. Suppose this is the case. Then it is easily seen that one can sharpen the results as follows. First let $I(\pi)$ be the vector space generated by the integrals $Z(s+\frac{1}{2}(n-1) f, \Phi)$ where

f is a coefficient of π. Then $I(\pi)$ is in fact a fractional ideal of the ring $\mathbf{C}[q^{-s}, q^s]$ with a unique generator of the form

$$L(s, \pi) = 1/P(q^{-s})$$

where P is a polynomial such that $P(0)=1$. There is a similar factor $L(s, \tilde{\pi})$ and the functional equation reads:

$$Z(1-s+\tfrac{1}{2}(n-1), \check{f}, \hat{\Phi})/L(1-s, \tilde{\pi}) = \varepsilon(s, \pi, \psi)\, Z(s+\tfrac{1}{2}(n-1), f, \Phi)/L(s, \pi)$$

where the factor $\varepsilon(s, \pi, \psi)$ is, as a function of s, just a constant times a power of q^{-s}.

We pass now to the demonstration of Theorem (2.3).

3. Reduction to the supercuspidal case. The first step is trivial.

PROPOSITION (3.1). *Suppose Theorem* (2.3) *is true for the pair* $(\pi, \tilde{\pi})$. *Let* σ *be a component of* π *(i.e., a subquotient). Then the theorem is true for the pair* $(\sigma, \tilde{\sigma})$. *More precisely*

$$c(s, \sigma, \psi) = c(s, \pi, \psi), \qquad L(s, \sigma)/L(s, \pi) \in \mathbf{C}[q^{-s}].$$

For the second step consider the following situation. Let P be a parabolic subgroup of G and U its unipotent radical. Then

$$P/U \simeq \prod G_i, \quad G_i = GL(n_i), \quad \sum n_i = n.$$

Let σ_i be an irreducible admissible representation of G_{iF}. Then $\sigma = \times_i \sigma_i$ is an admissible irreducible representation of P_F/U_F that is a representation of P_F trivial on U_F. We can form the representation

$$\pi = \mathrm{Ind}(G, P, \delta_P^{1/2}\sigma) \tag{3.2}$$

where δ_P is the module of P (cf. Harish-Chandra's lectures). The coefficients of π can be computed in terms of the coefficients of σ. It is then not too difficult to prove the following.

PROPOSITION (3.3). *Suppose that Theorem* (2.3) *is true for each pair* $(\sigma_i, \tilde{\sigma}_i)$. *Then it is true for* $(\pi, \tilde{\pi})$. *More precisely*

$$L(s, \pi) = \prod_i L(s, \sigma_i), \quad L(s, \tilde{\pi}) = \prod_i L(s, \tilde{\sigma}_i), \quad \varepsilon(s, \pi, \psi) = \prod_i \varepsilon(s, \sigma_i, \psi).$$

Since each irreducible admissible representation of G_F is contained in a representation of the form (3.2) where the σ_i are supercuspidal, it is clear that in order to prove Theorem (2.3) it is enough to prove it for a pair $(\sigma, \tilde{\sigma})$ where σ, and therefore $\tilde{\sigma}$, is supercuspidal. This is taken up in the next section.

4. The supercuspidal case. Suppose that σ is irreducible and supercuspidal. If $n=1$, G_F is just the group $F^\times$ and σ a quasi-character of $F^\times$. The result is then already in Tate's work.

Suppose now that $n>1$. Let ω be the central exponent of σ so that ω is a quasi character of $Z_F \simeq F^\times$ (Z being the center of G). Then every coefficient f of σ satisfies

$$f(ga)=f(g)\,\omega(a) \quad \text{for } g\in G_F,\ a\in F^\times,$$

and is compactly supported mod Z_F. Moreover, given f, there is a compact subgroup H of G_F such that

$$\int_H f(gh)\,dh=0 \quad \text{for all } g \text{ in } G_F. \tag{4.1}$$

(Take for H a large enough compact open subgroup in the unipotent radical of a parabolic.) It is clear that the convergence of $Z(s, f, \Phi)$ depends on the convergence of the integrals

$$\int_{F^\times} \varphi(a)\,|a|^{ns}\,\omega(a)\,d^\times a \quad \text{where } \varphi(a)=\Phi(ga) \text{ and } g\in G_F. \tag{4.2}$$

Since φ belongs to $\boldsymbol{D}(F)$, we see that (4.2) is absolutely convergent for Res large enough. The same is therefore true of $Z(s, f, \Phi)$. Now suppose that $\Phi(0)=0$. Then, in (4.2), $\varphi(0)=0$, and (4.2) is always convergent. Again the same is true of $Z(s, f, \Phi)$. In general if Φ is given in $\boldsymbol{D}(X)$ set

$$\Phi_1(x)=\Phi(x)-(\operatorname{vol} H)^{-1}\int_H \Phi(xh)\,dh.$$

Then $\Phi_1(0)=0$ so that $Z(s, f, \Phi_1)$ is always convergent. On the other hand, by (4.1), $Z(s, f, \Phi)=Z(s, f, \Phi_1)$. Hence $Z(s, f, \Phi)$ is an entire function of s, i.e., a polynomial in q^{-s}, q^s. The same is true of $Z(s, \check{f}, \Phi)$ so that we have proved the two first assertions of Theorem (2.3) and even that $L(s, \pi)=L(s, \tilde{\pi})=1$.

To prove the functional equation it is enough to show that $Z(0, f, \Phi)$ and $Z(n, \check{f}, \hat{\Phi})$ are proportional (with a proportionality coefficient which depends only on σ and ψ). Let $L(V)$ denote the space of linear operators on V, the space of σ. Given Φ in $\boldsymbol{D}(X)$ there are $\mu_1(\Phi)$ and $\mu_2(\Phi)$ in $L(V)$ such that, for every v in V

and $\tilde{v}$ *in* $\tilde{V}$,

$$Z(0, f, \Phi) = \langle \mu_1(\Phi)\, v, \tilde{v} \rangle, \qquad Z(n, \check{f}, \hat{\Phi}) = \langle \mu_2(\Phi)\, v, \tilde{v} \rangle,$$

where $f(g) = \langle \pi(g)\, v, \tilde{v} \rangle$. Clearly μ_1 and μ_2 are $L(V)$-valued distributions on X. If $\mu = \mu_1$ or μ_2, μ satisfies the following condition:

$$\int \Phi(gxh)\, d\mu(x) = \sigma^{-1}(g) \int \Phi(x)\, d\mu(x)\, \sigma^{-1}(h).$$

Let μ be any distribution satisfying the above condition. Then it is easily seen that μ takes its values in $V \otimes \tilde{V}$ (identified in the usual way with a subspace of $L(V)$). Moreover $G \times G$ operates on X by $(g, h) \cdot x = gxh^{-1}$ and the above condition amounts to saying that μ is τ-invariant where τ is the representation $\sigma \times \tilde{\sigma}$ of $G \times G$. Note that τ is a supercuspidal representation. Clearly it will be enough to prove the following:

PROPOSITION (4.3). *Suppose that μ is a τ-invariant distribution on X. Then μ is proportional to μ_1.*

For $\Phi \in \boldsymbol{D}(G) \subset \boldsymbol{D}(X)$, $\mu_1(\Phi)$ is simply the integral $\int \Phi(g)\, \pi(g)\, d^\times g$. It easily follows that the restriction of μ to G_F is proportional to the restriction of μ_1. Subtracting from μ a suitable multiple of μ_1 we may assume that the restriction of μ to G_F is zero. What we have to show then is that μ is zero.

Let X_i be the set of matrices whose rank is $\geqq i$. Then X_i is open, $X_0 = X$, $X_n = G_F$ and $X_i \supset X_{i+1}$. Moreover, for $i \geqq 1$, $Y_i = X_{i-1} - X_i$ is a single orbit of $G \times G$, closed in X_{i-1}. Let ν_i be the restriction of μ to X_i. We have to show that $\nu_i = 0$ implies that $\nu_{i-1} = 0$. So suppose that ν_i vanishes. Then the support of ν_{i-1} is contained in Y_i such that ν_{i-1} can be regarded as a distribution on $Y_i = G \times G / H$. Of course, this distribution is τ-invariant. Since τ is supercuspidal and H contains the unipotent radical of a parabolic subgroup of $G \times G$, this implies that ν_{i-1} vanishes. This completes the proof of Proposition (4.3) and also Theorem (2.3).

REMARK. The proof given here in the supercuspidal case is due to A. Weil.

REFERENCES

1. R. Godement and H. Jacquet, *Zeta functions of simple algebras*, Lecture Notes in Math., vol. 260, Springer-Verlag, Berlin and New York, 1972.

2. J. Tate, *Fourier analysis in number fields, and Hecke's zeta -functions*, Doctoral Dissertation, Princeton Univ., Princeton, N. J., 1950; reprinted in Algebraic number theory (Proc. Instructional Conf., Brighton, 1965), Thompson, Washington, D.C., 1967, pp. 305–347. MR **36** #121.

THE GRADUATE SCHOOL OF THE
CITY UNIVERSITY OF NEW YORK

ON WORK OF MACDONALD AND $L^2(G/B)$ FOR A p-ADIC GROUP

ALLAN J. SILBERGER

Let $\boldsymbol{G}$ be a connected, simply-connected, almost simple linear algebraic group defined over a nonarchimedean local field Ω. Let G denote the set of Ω-rational points of $\boldsymbol{G}$. Then G is a totally disconnected unimodular group. Let $\boldsymbol{A}$ denote a maximal Ω-split torus of $\boldsymbol{G}$, so $\operatorname{rank}_\Omega \boldsymbol{G} = \dim \boldsymbol{A} = l$ (say). According to the theory of Bruhat and Tits [2] there are exactly $l+1$ conjugacy classes of maximal compact subgroups in G. An Iwahori subgroup is an intersection of $l+1$ properly chosen nonconjugate maximal compact subgroups of G (see §1 for a more precise description).

For K an open compact subgroup of G let $C(G/K)$ [respectively, $C(G//K)$] denote the functions on G which are right [respectively, two-sided] invariant under translation by K. The subspace of compactly supported functions $C_c(G//K) \subset C(G//K)$ has the structure of a convolution algebra. As G/K is discrete, $C(G/K)$ consists of continuous functions. The space $C_c(G//K)$ is dense in $L^2(G//K)$.

In [**5a**] Macdonald chooses a maximal compact subgroup K which has a simple description in terms of Bruhat and Tits's structure theory. He computes zonal spherical functions and determines an explicit Plancherel formula for $L^2(G//K)$. Analogues of the c-functions of Harish-Chandra occur, and Macdonald's results have a natural formulation in terms of these new functions.

Here, generalizing Macdonald's work, we give analogous formulas for the zonal spherical functions on G associated to certain subalgebras $C_c^\sigma(G//B) \subset C_c(G//B)$, where B is an Iwahori subgroup and $C_c^\sigma(G//B)$ corresponds to an arbitrary one-dimensional representation σ of the finite Weyl group W_0 of G. For each σ, Fourier transformation with respect to the class one or spherical principal series defines an isomorphism of $C_c^\sigma(G//B)$ onto an algebra of polyno-

AMS (MOS) subject classifications (1970). Primary 22E35.

mials which satisfy an invariance condition for W_0. This had been proved for $C_c(G//K)$ by Satake [**6**].

We shall show that the Fourier analysis of $C_c(G//B)$ or $L^2(G/B)$ relates closely to the problem of determining the full support for the Plancherel measure in the analytic continuation of the principal series. I believe that the two problems are essentially equivalent, although I can, at the moment, prove only that knowledge of the Fourier decomposition of $L^2(G/B)$ implies knowledge of the full Plancherel's measure for the spherical principal series.

1. Affine root structures on a *p*-adic group. Macdonald's approach to the problem of computing spherical functions for G involves the structure theory of Bruhat and Tits. Even to formulate results we need to summarize parts of this theory.

Let V be an l-dimensional real vector space. Let Σ_0 be a reduced irreducible root system in the dual space V^*. Let W_0 be the Weyl group of Σ_0. For each root $a \in \Sigma_0$ and each integer r define an affine linear function $x \mapsto a(x)+r$ on V. Write Σ for the set of all such functions, termed affine roots. Associated to Σ there is an infinite "affine Weyl group" W generated by the reflections w_α in the hyperplanes $\alpha^{-1}(0)$, $\alpha \in \Sigma$. The group W is a semidirect product of W_0 with a free group T of rank l which consists of all the translations in W. The mapping $t_a = w_{a+1} \circ w_a \mapsto t_a(0) = a'$ sends Σ_0 onto the set of dual roots in V. The functions t_a form a base for T and the correspondence we have defined provides a convenient description for the action of W_0 on T in W.

Let $\{a_1, \ldots, a_l\}$ be a set of simple roots in Σ_0, Σ_0^+ (Σ_0^-) the associated positive (negative) roots. Set $a_0 = 1 - b$, where b is the highest root in Σ_0^+. Then the cone $C_0 = \{x \in V \mid a_i(x) > 0,\ i = 1, \ldots, l\}$ is a chamber for W_0 and the open simplex $C = \{x \in V \mid a_i(x) > 0,\ i = 0, \ldots, l\}$ is a chamber for W. The reflections w_{α_i} $(i = 0, \ldots, l)$ generate W.

The group G has an affine root structure which we now describe. Let $M = \mathscr{Z}(A)$, the centralizer of A in G. Let $\mathscr{N} = \mathscr{N}(A)$, the normalizer of A in G. There exist compact unipotent subgroups, called root subgroups, $\{N_\alpha : \alpha \in \Sigma)$ and an epimorphism $\gamma : x \mapsto \bar{x}$ of $\mathscr{N}$ to W with the following properties (let $\langle A, B \rangle$ denote the group generated by the sets A and B; set $H = \ker \gamma$):

(1) $x N_\alpha x^{-1} = N_{\bar{x}(\alpha)}$ $(x \in \mathscr{N},\ \alpha \in \Sigma)$.

(2) $N_{\alpha+1} \subsetneq N_\alpha$, $(N_\alpha : N_{\alpha+1}) < \infty$, and $\bigcap_{r \in \boldsymbol{Z}} N_{\alpha+r} = \{1\}$.

(3) If $\beta = -\alpha + r$ $(0 \neq r \in \boldsymbol{Z})$, then $\langle N_\alpha, N_\beta, H \rangle = N_\alpha H N_\beta$.

(4) $\langle N_\alpha, N_{-\alpha}, H \rangle = N_\alpha \gamma^{-1}(w_\alpha) N_\alpha \cup N_\alpha H N_{-\alpha+1}$.

(5) If $\beta \neq \pm\alpha + r$ for every $r \in \boldsymbol{Z}$, then the commutator group $[N_\alpha, N_\beta] \subset \langle N_{r\alpha+s\beta} \mid r\alpha + s\beta \in \Sigma,\ r > 0,\ s > 0 \rangle$.

(6) For each $a \in \Sigma_0$ let $N_{(a)} = \bigcup_{r \in \boldsymbol{Z}} N_{a+r}$ and let N $(\bar{N}) = \langle N_{(a)} \mid a \in \Sigma_0^+$ $(\Sigma_0^-) \rangle$. Then $N \cap M\bar{N} = \{1\}$.

(7) $G=\langle \mathcal{N}, N_\alpha : \alpha\in\Sigma\rangle$.

The group $K=\langle H, N_\alpha \mid \alpha(0)\geqq 0\rangle$ is the maximal compact subgroup for which Macdonald has formulated his results. The group K is "good" in the sense that $G=PK$ for any parabolic subgroup P of G. This is a necessary condition in order that $C_c(G//K)$ be a commutative algebra and, according to Satake [6], it is also sufficient.

The group $B=\langle H, N_a, N_{-a+1} : a\in\Sigma_0^+\rangle$ is an Iwahori subgroup.[1] To the pair $(\mathcal{N}, B)$ there corresponds a Tits system for G with Weyl group W and parabolic subgroups, the so-called parahoric subgroups – proper open subgroups which contain a conjugate of B. The maximal parahoric subgroups containing a given Iwahori subgroup represent the conjugacy classes of maximal compact subgroups in G. If S is a face of the closed simplex $\bar{C}$, then the group $\langle H, N_\alpha \mid \alpha(S)\geqq 0\rangle$ is a parahoric subgroup. The maximal parahoric subgroups which contain B correspond to the vertices of $\bar{C}$ (K to the origin $=$ unique fixed point of W_0) and "good" maximal compact subgroups correspond to vertices at which root hyperplanes, one from each family of parallel hyperplanes, intersect.

Let T^+ (respectively, T^{++}) be the set of all translations $t\in T$ such that $t(0)\in C_0$ (respectively, $t(0)\in\bar{C}_0$). Let $M^+=\gamma^{-1}(T^+)$ and $M^{++}=\gamma^{-1}(T^{++})$. We have the Cartan decomposition

$$K\backslash G/K \overset{1\text{-}1}{\longleftrightarrow} T^{++}$$

and the Iwasawa decomposition

$$N\backslash G/K \overset{1\text{-}1}{\longleftrightarrow} T.$$

We define $q_\alpha=(N_{\alpha-1}:N_\alpha)$. Then $q_{w\alpha}=q_\alpha$ for all $\alpha\in\Sigma$ and $w\in W$, so $q_\alpha=q_{\alpha+2}$ for all $\alpha\in\Sigma$; however, it is not necessarily true that $q_\alpha=q_{\alpha+1}$. We enlarge the root system Σ_0 by adjoining $a/2$ ($a\in\Sigma_0$) precisely when $q_a\neq q_{a+1}$ and discover that the resulting set of vectors is again a root system (though, of course, no longer in general reduced). Set $q_{a/2}=q_{a+1}/q_a$, so that $q_{a/2}\neq 1$ if and only if $a/2\in\Sigma_1$.

To define the Poincaré polynomial $Q(\xi)$ for the root system Σ_1, associate to each $b\in\Sigma_1$ a symbol ξ_b such that $\xi_{wb}=\xi_b$ for $w\in W_0$. Let $\xi(w)=\prod \xi_b$, the product taken over all positive $b\in\Sigma_1$ such that $w^{-1}b$ is negative. Set $Q(\xi)=\sum_{w\in W_0}\xi(w)$. Then $Q(\xi)$ is a polynomial in no more than three variables, depending on the number of distinct root lengths in Σ_1. For $w\in W_0$ let $q(w)=(B:B\cap wBw^{-1})$. Then $q(w)=\xi(w)$ evaluated at $\xi_b=q_b$. We write $Q(q^{-1})=Q(\xi)$ evaluated at $\xi_b=q_b^{-1}$.

2. Spherical principal series for G and σ-spherical functions. We use notation

[1] The class of groups conjugate to B equals the class of Iwahori subgroups, i.e., a minimal parahoric subgroup is an Iwahori subgroup.

introduced in [3, especially §7]. Set $A = A \cap G$. Write P for the semidirect product MN. Then (P, A) is a minimal p-pair of G with MN its Levi decomposition. Let $\chi_\nu(m) = q^{i\langle\nu, H(m)\rangle} = s_\nu(\gamma(m))$ ($m \in M$ and $\nu \in \mathfrak{A}^*_C = V \otimes C$) and consider the induced representation $\pi_\nu = \mathrm{Ind}_P^G(\delta_P^{1/2}\chi_\nu)$. Recall that π_ν acts on the space $\mathscr{H}_\nu$ of all smooth complex valued functions ψ on G which satisfy the relation $\psi(pg) = \delta_P(p)^{1/2}\chi_\nu(p)\psi(g)$ ($p \in P$, $g \in G$). As elements of $\mathscr{H}_\nu$ are determined by their values on any set of representatives for $P\backslash G$, we may identify $\mathscr{H}_\nu$ with either $C^\infty(P \cap K\backslash K)$ or a subspace of $C^\infty(\bar{N})$.

From the explicit form of the representation one sees immediately that $\mathscr{H}_\nu$ transforms under $\pi_\nu \mid K$ as $\mathrm{Ind}_{P\cap K}^K 1$. Taking $P \cap K \subset B$, we note that there is a subspace $U = U_\nu \subset \mathscr{H}_\nu$ which transforms under $\pi_\nu \mid K$ as $\mathrm{Ind}_B^P 1$; moreover U_ν contains all the B-invariant vectors in $\mathscr{H}_\nu$. From a result of Iwahori and Tits (cf. [4] for an indication of the proof) we know that $C(K//B)$ is isomorphic to $C(W_0)$, the group algebra of the finite Weyl group. It follows that there are exactly $|W_0|$ B-invariant vectors in $\mathscr{H}_\nu$. It also follows that to any linear character σ of W_0 there corresponds a representation $u^\sigma \in \mathrm{Ind}_B^K 1$ which occurs simply.

The space $\mathscr{H}_\nu$ has an $L^2(K)$ scalar product (,) defined on it. Let ψ_ν^σ be a normalized B-invariant vector in the subspace of U_ν which transforms under $\pi_\nu \mid K$ as u^σ. Then the function

$$\omega_\nu^\sigma(g) = (\pi_\nu(g)\,\psi_\nu^\sigma, \psi_\nu^\sigma)$$

is a zonal spherical function for a subalgebra $C_c^\sigma(G//B)$ of $C_c(G//B)$. Sufficient that ω_ν^σ be a positive definite function on G is the condition that $\nu \in \mathfrak{A}^*$. This condition is known to be unnecessary even for the case $\sigma = 1$. In the case $\sigma = 1$ the dual measure, given explicitly by Macdonald, usually has support in $\mathfrak{A}^*$. For $\sigma \neq 1$ this is always only a proper subset of the full support of the measure.

3. Explicit formulas for $\omega_\nu^\sigma(g)$. Since W_0 is generated by reflections, $\sigma(W_0) \subset \{\pm 1\}$. Define

$$c^\sigma(s_\nu) = \prod_{b \in \Sigma_1^+} \frac{1 - q_{b/2}^{-1/2} q_b^{-1} s_\nu(\sigma(w_b)\, t_b)^{-1}}{1 - q_{b/2}^{-1/2} s_\nu(\sigma(w_b)\, t_b)^{-1}}.$$

The product is over all positive roots in Σ_1; t_b is the translation which satisfies $t_b(0) = b'$, the dual root associated to b. The factor $q_{b/2} = 1$ if $b/2 \notin \Sigma_1$. If b and $b/2$ both are roots, then one of the factors in the denominator of $c^\sigma(s_\nu)$ cancels into the numerator, so $c^\sigma(s_\nu)$ may be defined for all ν such that $s_\nu(t_b) \neq 1$ for all $b \in \Sigma_1$.

The spherical function ω_ν^σ corresponds to a representation of K, so, because there is a Cartan decomposition for G, it suffices, in order to specify the function, to give its values at a set of representatives for $K\backslash G/K$, i.e., for representatives in M of $\gamma^{-1}(T^{++}) = M^{++}$. Since the function is B-invariant, its values do not depend upon the representative.

THEOREM 1. *Let $\sigma: W_0 \to \{\pm 1\}$ be a homomorphism. Let ω_ν^σ be the corresponding function defined above. Let $\nu \in \mathfrak{A}_C^*$ and $m \in \gamma^{-1}(T^{++})$. Then*

$$\omega_\nu^\sigma(m) = \frac{\delta_P(m)^{-1/2}}{Q(q^{-1})} \sum_{w \in W_0} c^\sigma(ws_\nu^{-1})(ws_\nu^{-1})(\gamma(m)).$$

The calculation involves careful inspection of the technique developed by Macdonald for the case $\sigma = 1$. At the moment we do not have an inversion formula for $C_c^\sigma(G//B)$. However, Theorem 1 combined with Macdonald's results implies the following weaker result.

Let $C_c(T)^{W_0}$ denote the subalgebra of the group algebra of T which consists of W_0-invariant elements. For $f \in C_c^\sigma(G//B)$ define $\hat{\omega}_\nu^\sigma(f) = f * \omega_\nu^\sigma(1)$, the $*$ denoting convolution.

THEOREM 2. *The Fourier transform $f \mapsto \hat{\omega}_\nu^\sigma(f)$ maps $C_c^\sigma(G//B)$ isomorphically onto the algebra $C_c(T)^{W_0}$. There exists a measure d_{μ_ν} on the space $\Lambda^{\sigma^+} \subset \mathfrak{A}_C^*$ which indexes positive definite functions ω_ν^σ such that*

$$f(1) = \int_{\Lambda^{\sigma+}} \hat{\omega}_\nu^\sigma(f)\, d_{\mu_\nu}$$

for any $f \in C_c^\sigma(G//B)$.

4. $L^2(G/B)$ and the spherical principal series. We shall prove the following result:

THEOREM 3. *Any subquotient of π_ν contains a B-invariant vector.*

PROOF. It suffices to consider an irreducible subquotient π of π_ν. We observe first that Theorem 14 (1) of [3] (cf. notation too) actually implies $f_{P'} \sim 0$ for $P' \notin \mathscr{P}(A)$ and all $f \in \mathscr{A}(\pi, \tau)$. This follows from the fact that, for $P' = M'N'$ and $\varphi \in C_c(M', \tau_{M'})$, the scalar product $\int_M (f_{P'}, \varphi)\, dm$ may be regarded as a value of an analytic function. Harish-Chandra's theorem implies that the analytic function vanishes on a set large enough so that it must be identically zero. Therefore, by Lemma 9 of [3], $f_{P'} \not\sim 0$ for some $P' \in \mathscr{P}(A)$, so $\mathscr{X}_\pi(P', A) \neq \emptyset$. It then follows from Theorems 10 and 11 of [3] that $\pi \subset \operatorname{Ind}_{P'}^G \delta_{P'}^{1/2} \chi$ for some $\chi \in \mathscr{X}(M)$ such that $\chi \mid H = 1$. Since Iwahori subgroups are conjugate as are minimal parabolic subgroups, it suffices to consider $\pi \subset \pi_\nu$. Let $\mathscr{H}$ denote the subspace of $\mathscr{H}_\nu$ which transforms as π.

Identify $\mathscr{H}_\nu$ with a subspace of $C^\infty(\bar{N})$. Let E_B denote the characteristic function of B and notice that convolution from the left with E_B projects $\mathscr{H}_\nu$ on the subspace

U_v; moreover, for $\psi \in \mathscr{H}_v$, $E_B * \psi(1) = \int_X \psi(\bar{n})\, d\bar{n}$, integration being over an open compact subgroup $X \subset \bar{N}$. Clearly, there exists $\psi \in \mathscr{H}$ such that $\psi(1) \neq 0$; the function ψ is constant on some open compact subgroup $Y \subset \bar{N}$. Transforming ψ by $\pi(m)$, $m \in M^+ = \gamma^{-1}(T^+)$, we have $\pi(m)\, \psi(\bar{n}) = \chi(m)\, \delta_P(m)^{1/2} \psi(m^{-1}\bar{n}m^{+1})$. By (1) of the structure theory, we may choose m so that $\pi(m)\, \psi$ is constant on X, which implies that $E_B * (\pi(m)\, \psi)\, (1) \neq 0$. It follows that $\mathscr{H}$ contains a B-invariant vector.

COROLLARY 4. *π_v is algebraically irreducible if and only if the corresponding $|W_0|$-dimensional representation of $C_c(G//B)$ is irreducible. A composition series for π_v has length at most equal to the sum of the degrees of the irreducible representations of W_0.*

The last statement of Corollary 4 has also been proved by W. Casselman.

COROLLARY 5. *Every irreducible subquotient of the spherical principal series which lies in the support of the Plancherel measure for $L^2(G)$ also lies in the support of the Plancherel measure for $L^2(G/B)$.*

CONJECTURE 1. *Conversely, Fourier transformation of $C_c(G//B)$ with respect to the spherical principal series is an isomorphism. The inversion formula for $C_c(G//B)$ involves only the spherical principal series.*

CONJECTURE 2. *Every discrete irreducible representation in $L^2(G/B)$ corresponds to a one-dimensional representation of $C_c(G//B)$. Every unitary representation in the support of the Plancherel measure for $L^2(G/B)$ contains a representation u^σ in its restriction to K.*

CONJECTURE 3. *$C_c(G//B) \simeq C_c(W)$, the group algebra of the affine Weyl group.*

REFERENCES

1. N. Bourbaki, *Eléments de mathématique.* Fasc. XXXIV. *Groupes et algèbres de Lie.* Chap. IV: *Groupes de Coxeter et système de Tits.*. Chap. V: *Groupes engendrés par des réflexions.* Chap. VI: *Systèmes de racines*, Actualités Sci. Indust., no. 1337, Hermann, Paris, 1968. MR **39** #1590.

2. F. Bruhat and J. Tits

a) BN-*paires de type affine et données radicielles*, C. R. Acad. Sci. Paris Sér. A-B **263** (1966), A598-A601. MR **39** #4160.

b) *Groupes simples résiduellement déployés sur un corps local*, C. R. Acad. Sci. Paris Sér. A-B **263** (1966), A766-A768. MR **39** #4161.

c) *Groupes algébriques simples sur un corps local*, C. R. Acad. Sci. Paris Sér. A-B **263** (1966), A822-A825. MR **39** #4162.

d) *Groupes algébriques simples sur un corps local: Cohomologie galoisienne, décompositions d'Iwasawa et de Cartan*, C. R. Acad. Sci. Paris Sér. A-B **263** (1966), A867-A869. MR **39** #4163.

3. Harish-Chandra, *Harmonic analysis on reductive p-adic groups*, Proc. Sympos. Pure Math., vol. 26, Amer. Math. Soc., Providence, R. I., 1974, pp. 167–192.

4. N. Iwahori, *Generalized Tits systems (Bruhat decomposition) on p-adic semisimple groups*, Proc. Sympos. Pure Math., vol. 9, Amer. Math. Soc., Providence, R.I., 1966, pp. 71–83. MR **35** #6693.

5. I. G. Macdonald

a) *Spherical functions on groups of p-adic type*, Publications of the Ramanujan Institute for Advanced Study No. 2, Ramanujan Institue for Advanced Study, Madras, India, 1972.

b) Same title, Mimeographed notes for lectures at Nancy, 1971.

c) *Harmonic analysis on semisimple groups*, Proc. Internat. Congress Math. (Nice, 1970), vol. 2, Gauthier-Villars, Paris, 1971, pp. 331–335.

d) *Spherical functions on a p-adic Chevalley group*, Bull. Amer. Math. Soc. **74** (1968), 520–525. MR **36** #5141.

6. I. Satake, *Theory of spherical functions on reductive algebraic groups over p-adic fields*, Inst. Hautes Études Sci. Publ. Math. No. **18** (1963), 5–69. MR **35** #4059.

INSTITUTE FOR ADVANCED STUDY

CHARACTER FORMULAS FOR SL_2

P. J. SALLY, JR.*

One of the central problems in the representation theory of p-adic groups is the determination of the supercuspidal representations and the calculation of the characters of these representations. Some of the general properties of supercuspidal representations and their characters have been worked out by Harish-Chandra, Howe and Jacquet (see [4(a)], [4(b)], [5(b)], [6]). One general method for constructing supercuspidal representations is that of inducing certain representations from a maximal compact subgroup. This method was originally discovered by Mautner [7], and has been carried out for particular groups by several authors ([3], [5(a)], [9], [10], [11]).

Let Π be a supercuspidal representation of a semisimple p-adic group G. Suppose that $\Pi = \mathrm{ind}_{K\uparrow G}\,\pi$ where K is some compact open subgroup of G and π is an irreducible, unitary representation of K with character χ_π. Let dx be a Haar measure on G which, when restricted to K, gives a Haar measure $d\kappa$ on K satisfying $\int_K d\kappa = 1$. If $f \in C_c^\infty(G)$, the space of locally constant, compactly supported functions on G, then $\Pi(f) = \int_G f(x)\,\Pi(x)\,dx$ is of finite rank and hence of trace class. It is easy to see that (see, for example, [10])

$$(1) \qquad \mathrm{trace}\,\Pi(f) = \Theta_\Pi(f) = \int_G \int_K f(x\kappa x^{-1})\,\chi_\pi(\kappa)\,d\kappa\,dx.$$

This is simply the Frobenius formula for induced characters.

This effectiveness of (1) as a tool for computing character formulas hinges on two points. First, one must have a detailed knowledge of the character χ_π. Second, if we define

---

AMS (MOS) subject classifications (1970). Primary 22E35, 22E50; Secondary 20G25.

* Research supported in part by the NSF.

(2) $$\tilde{\chi}_\pi(x)=\chi_\pi(x), \qquad x\in K, \\ =0, \qquad x\notin K,$$

and set

(3) $$\Theta_\Pi(y)=\int_{G/K} \tilde{\chi}_\pi(x^{-1}yx)\,d\dot{x}, \quad d\dot{x} \text{ an invariant measure on } G/K,$$

we would like to apply Fubini's theorem and write

(4) $$\Theta_\Pi(f)=\int_G f(x)\,\Theta_\Pi(x)\,dx.$$

In the cases mentioned above, complete knowledge of the characters χ_π is available only for the two by two groups (some information about χ_π in a more general setting is given in [**3**]). We shall discuss this below. Moreover, as one sees already in the case of SL_2, it is possible to employ (3) and (4) only when f is supported on the elliptic set, or, equivalently, when y is a regular element in a compact Cartan subgroup. To obtain the formula for Θ_Π on the noncompact Cartan subgroup in SL_2, it is necessary to use some other method such as summing matrix coefficients (see [**8(a)**]).

We now turn to the specific case of $SL_2(k)$, k a nonarchimedean local field. Let $\mathcal{O}$ be the ring of integers in k, $\mathcal{P}$ the maximal ideal in $\mathcal{O}$, τ a prime element in $\mathcal{P}$, U the units in $\mathcal{O}$, and ε a primitive $(q-1)$st root of unity in U, where $q=[\mathcal{O}/\mathcal{P}]$. We let $|\cdot|$ denote an absolute value on k such that $|\tau|=q^{-1}$. We assume that q is odd. Up to isomorphism any quadratic extension of k can be written in the form $V_\theta=k(\theta^{1/2})$, $\theta=\tau, \varepsilon\tau, \varepsilon$. Write $V=V_\theta$ and denote by $N_{V/k}$ the norm from V to k. Let C_θ be the kernel of $N_{V/k}$, and let sgn_θ be the character of order two on $k^\times$ whose kernel is $N_{V/k}(V^\times)$. If Φ is a nontrivial character on k^+ and $b\in k$, we set $\Phi_b(x)=\Phi(bx)$. The map $b\mapsto\Phi_b$ is a topological isomorphism of k^+ and $\hat{k}^+$.

On V, we have the nondegenerate bilinear form $B(z, w)=\mathrm{trace}(z\bar{w})$. For a nontrivial $\Phi\in\hat{k}^+$, the Fourier transform on V (relative to Φ) is defined by

(5) $$\hat{f}(z)=\int_V f(w)\,\Phi(B(z, w))\,d_\Phi w,$$

where $f\in L^1(V)$ and $d_\Phi w$ is an additive Haar measure on V normalized so that $\hat{f}\hat{\ }(z)=f(-z)$.

We set

(6) $$\kappa(\Phi, V)=\mathrm{P.V.}\int_V \Phi(N_{V/k}(z))\,d_\Phi z \quad \text{(principal value integral).}$$

$\kappa(\Phi, V)$ is a fourth root of unity ([**8**(**a**)], [**8**(**c**)]).

The *big representation* or *Weil representation* of $G=SL_2(k)$ was constructed by Shalika [**9**] and Tanaka [**12**]. For Φ a nontrivial character on k^+, $f\in\mathscr{S}(V)$ (the Schwartz-Bruhat space on V), and $g=\left(\begin{smallmatrix}\alpha & \beta\\ \gamma & \delta\end{smallmatrix}\right)\in G$, we set

$$\Pi(\Phi, V)(g)\, f(z)=\kappa(\Phi, V)\frac{\operatorname{sgn}_\theta(-\gamma)}{|\gamma|}\int_V \Phi\left[\frac{\alpha N(z)+\delta N(w)-B(z, w)}{\gamma}\right] f(w)\, d_\Phi w, \quad \gamma\neq 0, \tag{7}$$

$$=|\alpha|\operatorname{sgn}_\theta(\alpha)\,\Phi(\alpha\beta N(z))\, f(\alpha z), \qquad \gamma=0,$$

where $N=N_{V/k}$. In particular, for $p=\left(\begin{smallmatrix}0 & 1\\ -1 & 0\end{smallmatrix}\right)$, $\Pi(\Phi, V)(p)\, f(z)=\kappa(\Phi, V)\,\hat{f}(z)$. The map $g\mapsto\Pi(\Phi, V)(g)$ defines a representation of G on $\mathscr{S}(V)$ and extends to a continuous unitary representation of G on $L^2(V, d_\Phi z)$ which we also denote by $\Pi(\Phi, V)$.

The supercuspidal representations of G are obtained by decomposing the Weil representation relative to the representations of the orthogonal group of the form $B(\cdot, \cdot)$, that is, relative to the characters of C_θ. For $\psi\in\hat{C}_\theta$, we set

$$\mathscr{H}_{\psi, V}=\{f\in L^2(V): f(tz)=\psi(t)\, f(z),\ t\in C_\theta,\ \text{a.e. } z\in V\}. \tag{8}$$

Then $\Pi(\Phi, V)$ acts on $\mathscr{H}_{\psi, V}$ and determines a continuous unitary representation of G on $\mathscr{H}_{\psi, V}$ which we denote by $\Pi(\Phi, \psi, V)$.

The various equivalences and other properties relating to the representations $\Pi(\Phi, \psi, V)$ are listed in [**8**(**a**)]. We remark here that $\Pi(\Phi, \psi, V)$ is irreducible unless $\psi=\psi_0$, the character of order two on C_θ, in which case $\Pi(\Phi, \psi_0, V)$ splits as the direct sum of two irreducible representations. If ψ is the trivial character, then the representations $\Pi(\Phi, 1, V)$ are equivalent to the irreducible components of the reducible representations in the principal series for G. If $\psi\neq 1$, then $\Pi(\Phi, \psi, V)$ (or its irreducible components when $\psi=\psi_0$) is a supercuspidal representation.

REMARKS. (i) The representations $\Pi(\Phi, \psi, V)$ were first constructed by Gel'fand and Graev [**2**] in a somewhat different realization without the use of the Weil representation.

(ii) The results cited above have also been proved by Casselman [**1**] to include the case of q even.

Now let

$$K=SL_2(\mathscr{O}) \quad\text{and}\quad K^*=\begin{pmatrix}\tau & 0\\ 0 & 1\end{pmatrix}K\begin{pmatrix}\tau^{-1} & 0\\ 0 & 1\end{pmatrix}.$$

The pair $\{K, K^*\}$ is a complete set (up to conjugacy) of maximal compact subgroups of G. One of the major results of Shalika [**9**] is that the representations $\Pi(\Phi, \psi, V)$, $\psi^2\neq 1$, and the irreducible components of $\Pi(\Phi, \psi_0, V)$ are all induced from irreducible representations of either K or K^*.

For purposes of illustration, we shall henceforth restrict our considerations to the unramified discrete series, that is, the case $\theta=\varepsilon$, and V will always denote V_ε. Let $\mathcal{O}_\varepsilon$ denote the ring of integers in V, and $\mathcal{P}_\varepsilon$ the maximal ideal in $\mathcal{O}_\varepsilon$. We define a filtration in C_ε by setting $C_\varepsilon^{(h)}=C_\varepsilon\cap(1+\mathcal{P}_\varepsilon^h)$, $h=1,2,\ldots$.

Take $\psi\in\hat{C}_\varepsilon$ such that $\operatorname{cond}\psi=C_\varepsilon^{(h)}$, $h\geqq 1$ (cond = conductor), and assume that $\psi\neq\psi_0$. Take $\Phi\in\hat{k}^+$ such that $\operatorname{cond}\Phi=\mathcal{P}^h$, and define

$$H(\Phi,\psi,V)=\{f\in\mathcal{S}(V): f \text{ is supported on } \mathcal{O}_\varepsilon,\ f \text{ is constant on cosets of } \mathcal{P}_\varepsilon^h, \text{ and } f(tz)=\psi(t)f(z),\ t\in C_\varepsilon,\ z\in V\}. \tag{9}$$

One checks easily that the functions in $H(\Phi,\psi,V)$ are actually supported on U_ε, the units in $\mathcal{O}_\varepsilon$, and that

$$\dim H(\Phi,\psi,V)=(q-1)\,q^{h-1}. \tag{10}$$

Furthermore, $\Pi(\Phi,\psi,V)$ restricted to K acts on $H(\Phi,\psi,V)$ and we write $\pi(\Phi,\psi,V)$ for the representation thus obtained. The action of $\pi(\Phi,\psi,V)$ on $H(\Phi,\psi,V)$ is, of course, described by (7).

PROPOSITION 1 [**9**]. *Suppose* $\psi\in\hat{C}_\varepsilon$, $\operatorname{cond}\psi=C_\varepsilon^{(h)}$, $h\geqq 1$ *and* $\psi\neq\psi_0$. *Then* $\pi(\Phi,\psi,V)$ *is an irreducible representation of* K *of degree* $(q-1)\,q^{h-1}$, *and* $\Pi(\Phi,\psi,V)$ *is equivalent to* $\operatorname{ind}_{K\uparrow G}\pi(\Phi,\psi,V)$.

With the above information, it is a straightforward matter to determine the characters of the representations $\pi(\Phi,\psi,V)$. Simply take the obvious orthonormal basis for $H(\Phi,\psi,V)$, compute the matrix coefficients and add. The final form of the characters is simplified by the use of the following notation. We let Δ_0 denote the characteristic function of $\mathcal{O}$ (or $\mathcal{O}_\varepsilon$) and Δ_h, $h\geqq 1$, the characteristic function of $\mathcal{P}^h$ (or $\mathcal{P}_\varepsilon^h$).

PROPOSITION 2. *Let* $\pi=\pi(\Phi,\psi,V)$ $\operatorname{cond}\Phi=\mathcal{P}^h$, $\operatorname{cond}\psi=C_\varepsilon^{(h)}$, $h\geqq 1$, $\psi\neq\psi_0$. *Then, for* $g=\left(\begin{smallmatrix}\alpha & \beta\\ \gamma & \delta\end{smallmatrix}\right)\in K$, *we have, for* $|\gamma|>q^{-h}$,

$$\operatorname{tr}\pi(g)=\kappa(\Phi,V)\,\frac{\operatorname{sgn}_\varepsilon(\gamma)}{|\gamma|}\,[(q+1)/q]\times$$

$$\int_{C_\varepsilon}\psi(t)\,\Delta_0\left(\frac{\alpha-t}{\gamma}\right)\Delta_0\left(\frac{\delta-t^{-1}}{\gamma}\right)\cdot\left[q^h\Delta_h\left(\frac{\alpha+\delta-\operatorname{tr}(t)}{\gamma}\right)-q^{h-1}\Delta_{h-1}\left(\frac{\alpha+\delta-\operatorname{tr}(t)}{\gamma}\right)\right]dt$$

where $\operatorname{tr}(t)=t+t^{-1}$ *and* dt *is normalized so that* $\int_{C_\varepsilon}dt=1$. *If* $|\gamma|\leqq q^{-h}$, *it is sufficient to take*

$$g=\begin{pmatrix}\alpha & \beta\\ 0 & \alpha^{-1}\end{pmatrix},\qquad \alpha\in U,\ \beta\in\mathcal{O}.$$

We then have

$$\operatorname{tr}\pi(g)=[(q+1)/q]\,[q^h\Delta_h(\beta)-q^{h-1}\Delta_{h-1}(\beta)]\,q^h\int_{C_\varepsilon}(\psi(t))^-\,\Delta_h(\alpha t-1)\,dt.$$

With the formulas given in Proposition 2, one may utilise (3) to compute the character of $\Pi(\Phi, \psi, V)$ on the compact Cartan subgroups of G. Although direct, the process is somewhat lengthy and technical. The results are given in [**8(a)**]. An entirely analogous procedure is valid for the ramified discrete series ($\theta=\tau$, $\varepsilon\tau$).

We note that the formal degree of the representation $\Pi(\Phi, \psi, V)$ is just the degree of the inducing representation, that is, the degree of $\pi(\Phi, \psi, V)$. From Proposition 1, it follows that, for Φ fixed, $\operatorname{cond}\Phi=\mathscr{P}^h$, and any $\psi\in\hat{C}_\varepsilon$ such that $\operatorname{cond}\psi=C_\varepsilon^{(h)}$, $\deg\pi(\Phi, \psi, V)=(q-1)\,q^{h-1}$. Thus, for purposes of the Plancherel theorem, it would be sufficient to compute

$$\Theta_{\Pi(\Phi,h,V)}=\sum_{\psi\in\hat{C}_\varepsilon;\operatorname{cond}\psi=C_\varepsilon^{(h)}}\Theta_{\Pi(\Phi,\psi,V)},\tag{11}$$

and $\Theta_{\Pi(\Phi,h,V)}$ can be computed (using (3)) from the character of

$$\sum_{\psi\in\hat{C}_\varepsilon;\,\operatorname{cond}\psi=C_\varepsilon^{(h)}}\oplus\,\pi(\Phi, \psi, V).$$

This latter character can be obtained from Proposition 2, and the computation of the induced character is considerably simplified in this case.

References

1. W. Casselman, *On the representations of $SL_2(k)$ related to binary quadratic forms*, Amer. J. Math. **94** (1972), 810–834.

2. I. M. Gel'fand and M. I. Graev, *Representations of the group of second-order matrices with elements in a locally compact field and special functions on locally compact fields*, Usephi Mat. Nauk **18** (1963), no. 4 (112) 29–99 = Russian Math. Surveys **18** (1963), no. 4, 29–99. MR **27** #5864.

3. P. Gerardin, *Sur les représentations des groupes de Chevalley p-adiques* (to appear).

4. Harish-Chandra, a) *Harmonic analysis on reductive p-adic groups*, Lecture Notes in Math. vol. 162, Springer-Verlag, Berlin and New York, 1970.

b) *Harmonic analysis on reductive p-adic groups*, Proc. Sympos. Pure Math., vol. 26, Amer. Math. Soc., Providence, R.I., 1974, pp. 167–192.

5. R. Howe, a) *Tamely ramified supercuspidal representations of GL_n* (to appear).

b) *Some qualitative results on the representation theory of GL_n over a p-adic field* (to appear).

6. H. Jacquet, *Représentations des groups linéaires p-adiques*, Theory of Group Representations and Fourier Analysis (C.I.M.E., II Ciclo Montecatini Terme, 1970), Edizioni Cremonese, Rome, 1970, pp. 119–220. MR **45** #453.

7. F. Mautner, *Spherical functions over p-adic fields*. II, Amer J. Math. **86** (1964), 171–200. MR **29** #3582.

8. P. J. Sally, Jr. and J. A. Shalika, a) *Characters of the discrete series of representations of SL(2) over a local field*, Proc. Nat. Acad. Sci. U.S.A. **61** (1968), 1231–1237. MR **38** #5994.

b) *The Plancherel formula for SL(2) over a local field*, Proc. Nat. Acad. Sci. U.S.A. **63** (1969), 661–667.

c) *The Fourier transform on SL_2 over a non-archimedean local field* (to appear).

9. J. A. Shalika, *Representations of the two by two unimodular group over local fields*, Thesis, Johns Hopkins University, Baltimore, Md., 1966.

10. T. Shintani, *On certain square integrable irreducible unitary representations of some p-adic linear groups*, J. Math. Soc. Japan **20** (1968), 522–565. MR **38** #2252.

11. A. J. Silberger, *PGL_2 over the p-adics: its representations, spherical functions, and Fourier analysis*, Lecture Notes in Math., vol. 166, Springer-Verlag, Berlin and New York, 1970. MR **44** #2891.

12. S. Tanaka, *On irreducible unitary representations of some special linear groups of the second order*. I, Osaka J. Math. **3** (1966), 217–227. MR **36** #6541.

UNIVERSITY OF CHICAGO

CHARACTERS OF FINITE CHEVALLEY GROUPS

T. A. SPRINGER

This contribution contains a brief discussion of some problems and results in the character theory of finite Chevalley groups.

1. The groups. Let k be a finite field with q elements. Let $\boldsymbol{G}$ be a linear algebraic group defined over k, let $G=\boldsymbol{G}(k)$ be its group of k-rational points. If, which we assume from now on, $\boldsymbol{G}$ is Zariski-connected and reductive then we shall say that G is a finite *Chevalley group*. The class of these groups is somewhat larger than the class of groups originally studied by Chevalley, but it is the appropriate one for the problems to be discussed here.

EXAMPLES. (i) Moreover let $\boldsymbol{G}$ be semisimple, adjoint and k-split. Then G is a Chevalley group in the usual sense.

(ii) $G=\boldsymbol{GL}_n(k)$.

This is the problem with which we are concerned here: *Find the irreducible characters of* G. After stating it, it should be said immediately that one is very far from a complete solution. Such a solution ought to give, in the first place, a parametrization of the irreducible characters. One would hope such a parametrization to be dual, in some way, to one of the conjugacy classes of G. This is indeed the case in the special case that $G=\boldsymbol{GL}_n(k)$. There the irreducible characters have been determined by J. A. Green. His results give a description of the irreducible characters which is dual to that of the conjugacy classes of $\boldsymbol{GL}_n(k)$ via Jordan normal forms; see [**1**, p. 126 and pp. 153–155]. But in the general case the situation seems to be less favourable. For one thing, in an arbitrary Chevalley group one does not know a satisfactory description of all conjugacy classes. (For information about the conjugacy classes of Chevalley groups see [**1**, part E]).

AMS (MOS) subject classifications (1970). Primary 20C30, 20G40.

However, certain conjugacy classes in G, for example the semisimple ones, can be handled quite well. This leads one to hope that, similarly, one may be able to deal with certain irreducible characters, "dual" to the semisimple classes of G.

Secondly, assuming that one has obtained, in some way, an irreducible character θ of G, one would like to have a "character formula," enabling one to find the value of θ on all elements of G. In the representation theory of compact Lie groups, Weyl's character formula, which gives the value of an irreducible character on a regular element, determines by continuity the value on all elements. In our case there is no obvious way in which the values of θ on the "good" elements of G (say the regular semisimple ones) determine the value of θ on the "bad" elements (for example, the unipotent ones). It seems that, in order to deal with the values of θ on unipotent elements, the finite Lie algebra defined by $\boldsymbol{G}$ has to be used (see the conjecture in §5).

Finally, one can ask for an explicit realization of the irreducible representation defined by a given irreducible character.

Below we shall discuss some methods of obtaining irreducible characters of G, all of them using induction. After that, something will be said about the problem of finding character formulas. Little is known about this problem. The same is true for the construction of explicit irreducible representations, which will not be discussed here.

2. Induction from parabolic subgroups. $\boldsymbol{G}$ is always assumed to be connected and reductive. Let $\boldsymbol{P}$ be a parabolic k-subgroup of $\boldsymbol{G}$. Let $\boldsymbol{U}=U(\boldsymbol{P})$ be its unipotent radical. Denote by $\boldsymbol{L}$ a Levi subgroup of $\boldsymbol{P}$, so that $\boldsymbol{L}$ is a connected reductive k-subgroup of $\boldsymbol{P}$, such that $\boldsymbol{P}$ is the semidirect product $\boldsymbol{P}=\boldsymbol{L}\cdot\boldsymbol{U}$. $P=\boldsymbol{P}(k)$ is called a parabolic subgroup of G and $U=\boldsymbol{U}(k)$ is called its unipotent radical, which we also write $U(P)$. P (resp. U) determines $\boldsymbol{P}$ completely.

Two parabolic k-groups $\boldsymbol{P}$ and $\boldsymbol{Q}$ of $\boldsymbol{G}$ are said to be *associated*, if they have k-conjugate Levi subgroups. We then also say that the corresponding subgroups P and Q of G are associated. This defines an equivalence relation on the set of parabolic subgroups of G. A complex valued function f on G is called a *cusp form* on G if for all $x\in G$ and all parabolic subgroups $P\neq G$ we have $\sum_{y\in U(P)} f(xy)=0$. An irreducible representation π of G is called *cuspidal* or *parabolic*, if its matrix elements are cusp forms. This is equivalent to its character being a cusp form. The character of π is said to be a cuspidal or parabolic character.

Denote by $\mathscr{E}(G)$ the set of irreducible characters of G. Let $\boldsymbol{P}=\boldsymbol{L}\cdot\boldsymbol{U}$ be a parabolic subgroup as above, let $P=L\cdot U$ be the corresponding parabolic subgroup of G. If $\theta\in\mathscr{E}(L)$, extend θ to a character θ' of P by

$$\theta'(lu)=\theta(l) \qquad (l\in L,\, u\in U),$$

and let

$$\Theta = i_{P \to G}\theta'$$

be the character of G induced by θ'. Let $^\circ\mathscr{E}(P)$ be the set of irreducible characters of G which occur in some Θ, for θ a *cuspidal* character of L. The following theorem is due to Harish-Chandra [**5**]. For a proof, in a slightly more general situation, see [**10**, no. 5].

THEOREM 1. (i) *Let P and Q be parabolic subgroups of G. If P and Q are associated then $^\circ\mathscr{E}(P) = {}^\circ\mathscr{E}(Q)$. If they are not associated then $^\circ\mathscr{E}(P) \cap {}^\circ\mathscr{E}(Q) = \emptyset$.*

(ii) *$\mathscr{E}(G)$ is the union of all $^\circ\mathscr{E}(P)$, P running through the parabolic subgroups of G.*

Theorem 1 reduces, to some extent, the determination of all irreducible characters of G to that of the cuspidal ones of G and certain of its subgroups. This situation is quite similar to that for infinite-dimensional irreducible representations of real and p-adic Lie groups, where similar results have been proved by Harish-Chandra.

In connection with Theorem 1, the problem arises of decomposing Θ into irreducible characters (θ being cuspidal). Harish-Chandra has determined the dimension of the commuting algebra of Θ (see [**10**, 5.5]); this permits one to prove irreducibility of Θ in certain cases. Beyond this, not much is known in general. The special case that $\boldsymbol{P}$ is a Borel subgroup and θ is the trivial character has been studied quite extensively, and much more is known there. For recent results see [**2**] and [**3**]. In that case Θ contains the very interesting "Steinberg-character" of G. For its properties see [**12**] and [**7**]; see also [**11**].

3. Whittaker models. Let $\boldsymbol{B}$ be a k-Borel subgroup of $\boldsymbol{G}$, with unipotent radical $\boldsymbol{U}$. Let $\boldsymbol{T}$ be a maximal k-torus in $\boldsymbol{G}$ contained in $\boldsymbol{B}$. Assume, for simplicity, that $\boldsymbol{G}$ is semisimple and k-split, so that $\boldsymbol{T}$ is a k-split torus. Let R be the root system of $\boldsymbol{G}$ with respect to $\boldsymbol{T}$. $\boldsymbol{B}$ determines an order on R, let S be the basis of R (i.e., the system of simple roots) defined by the order. Denote by $\boldsymbol{X}_r \subset \boldsymbol{G}$ the 1-parameter additive subgroup of $\boldsymbol{G}$ associated with $r \in R$ and let X_r be the corresponding subgroup of G. If $r > 0$, then $X_r \subset U$.

A character ϕ of U is called *regular*, if the restriction $\phi \mid X_r$ of ϕ to X_r is nontrivial if and only if $r \in S$. We then have the following theorem, first announced by Gel'fand and Graev in [**4**]. For a proof (in a more general situation) see [**13**, p. 258].

THEOREM 2. *Let ϕ be a regular character of U. Then the induced character $\Phi = i_{U \to G}\phi$ is multiplicity free.*

If an irreducible character χ occurs in Φ (with multiplicity 1, by Theorem 2) we say that χ has a *Whittaker model*. This terminology is suggested by that used in the representation theory of $\boldsymbol{PGL}_2(F)$ over a local field F (see [**6**, p. 60]). In fact, it follows at once from Theorem 1 that if χ has a Whittaker model, the irreducible representation defined by χ can be realized by left translations in a unique space of complex valued functions f on G such that $f(xu)=\phi(u)f(x)$ $(x\in G, u\in U)$.

Not every irreducible character occurs in Φ (for example the trivial character does not). Nor does every cuspidal character have a Whittaker model (see [**1**, p. 165]). However it is easily proved that (under a mild assumption) a cuspidal irreducible character whose degree is prime to the characteristic of the underlying field k does have a Whittaker model; see [**11**].

It was proved by F. Rodier (see his contribution in this volume or [**11**]) that, θ and Θ being as in §2 (in the paragraph before Theorem 1), the fact that θ has a Whittaker model implies that there exists exactly one irreducible character χ of G occurs in Θ with multiplicity 1 and has a Whittaker model.

4. The use of maximal tori. For simplicity, assume again $\boldsymbol{G}$ to be semisimple and k-split. Fix a k-split maximal k-torus $\boldsymbol{T}_0$ of $\boldsymbol{G}$. Let $r=\dim \boldsymbol{T}_0$ be the rank of G. Denote by $\boldsymbol{N}_0$ the normalizer of $\boldsymbol{T}_0$ and let $W=\boldsymbol{N}_0/\boldsymbol{T}_0$ be the Weyl group. Let $\boldsymbol{T}$ be any maximal k-torus of $\boldsymbol{G}$, with normalizer $\boldsymbol{N}$. Put $W(\boldsymbol{T})=\boldsymbol{N}/\boldsymbol{T}$.

It is known that there is a bijection of the set of k-conjugacy classes of maximal k-tori onto the set of conjugacy classes of W. Moreover if $\boldsymbol{T}$ corresponds to the class of $w\in W$, then (a) $W(\boldsymbol{T})$ is isomorphic to the centralizer of w in W, (b) $\boldsymbol{T}$ is anisotropic if and only if w has no eigenvalue 1 (in the natural representation of W). For these facts see [**1**, p. 189 and p. 191].

$W(\boldsymbol{T})$ acts on $T=\boldsymbol{T}(k)$ and on the character group $\hat{T}$ of T. We say that $t\in T$ (or $\psi\in\hat{T}$) is *regular* if the trivial element is the only element of $W(\boldsymbol{T})$ fixing t (respectively ψ).

The number $|T|$ of elements of T equals $f(q)$, where f is a polynomial of degree r, with leading coefficient 1. Moreover there exists a constant c such that we have, $|T_{\mathrm{reg}}|$ denoting the number of regular elements of T,

$$\big|\,|T_{\mathrm{reg}}|-q^r\big|\leqq cq^{r-1},$$

which we write as

$$|T_{\mathrm{reg}}|=q^r+O(q^{r-1}).$$

A similar result holds for the number of regular elements of $\hat{T}$ (see [**10**, 6.15]).

THEOREM 3. *Let $\boldsymbol{T}$ be a maximal k-torus in $\boldsymbol{G}$. Assume that the nonregular elements of T are contained in a subgroup of T of order $O(q^{r-1})$.*

(i) *Then there exist $q^r+O(q^{r-1})$ regular characters ψ of T to which one can* *associate an irreducible character* $\chi_{T,\psi}$ *of* G; *the number of distinct irreducible characters so obtained is* $W(T)^{-1}q^r+O(q^{r-1})$.

(ii) *If, moreover,* T *is anisotropic then all except* $O(q^{r-1})$ *of the* $\chi_{T,\psi}$ *are cuspidal.*

The proof uses some ideas from the theory of exceptional characters of finite groups (see [**10**, 7.2]).

5. Values of characters. Except for some obvious cases and the case of the Steinberg character, there are no general results about the explicit determination of the values of irreducible characters. In order to show what such an explicit determination ought to look like, we shall briefly discuss some conjectural statements about the characters $\chi_{T,\psi}$ of Theorem 3.

A conjecture of Macdonald (see [**1**, p. 117]) leads one to expect that we have, if T is anisotropic and if t is a regular element of T,

$$\chi_{T,\psi}(t)=(-1)^r \sum_{w\in W(T)} \psi(w\cdot t)$$

(the notations being as in §4). Moreover, the degree of $\chi_{T,\psi}$ should be $|G|\,|T|^{-1}|U|^{-1}$ (where U is as in §3).

To deal with the values of $\chi_{T,\psi}$ on the unipotent elements of G, it seems that the Lie algebra of $\boldsymbol{G}$ has to be used. Assume now also that $\boldsymbol{G}$ is semisimple and k-split, with an irreducible root system. Moreover we make the following hypothesis on the characteristic p of k: p does not divide a highest root coefficient of R and if R is type A_l, then p does not divide $l+1$.

Let $\mathfrak{g}$ be the finite Lie algebra over k formed by the k-rational left-invariant vector fields on $\boldsymbol{G}$. G acts on $\mathfrak{g}$ via the adjoint action Ad.

Under our assumptions on p, there exists a bijection c of the set of unipotent elements of G into the set of nilpotent elements of $\mathfrak{g}$, such that

$$c(xux^{-1})=\mathrm{Ad}(x)\,c(u) \qquad (x\in G)$$

(see [**8**]).

Moreover, our hypothesis on p also implies (see [**1**, p. 184]) that there exists a nondegenerate symmetric bilinear form $B(\ ,\ )$ on $\mathfrak{g}\times\mathfrak{g}$, which is $\mathrm{Ad}(G)$-invariant. Fix a nontrivial complex character τ of k and put

$$\langle X, Y\rangle=\tau(B(X, Y)) \qquad (X, Y\in\mathfrak{g}).$$

If f is a complex valued function on $\mathfrak{g}$ define its Fourier transform $\hat{f}$ by

$$\hat{f}(X)=q^{-d/2}\sum_{Y\in\mathfrak{g}}\langle X, Y\rangle f(Y),$$

where $d=\dim \boldsymbol{G}$.

Now let $\boldsymbol{T}$ be as in Theorem 3, let $\mathfrak{t}$ be the subalgebra of $\mathfrak{g}$ defined by $\boldsymbol{T}$. Fix

$A \in \mathfrak{t}$ and let $O = \mathrm{Ad}(G)\, A$ be its orbit in $\mathfrak{g}$. Let μ_O be the characteristic function of O in $\mathfrak{g}$. Let ψ be a regular character of T such that $\chi_{T,\psi}$ exists.

CONJECTURE. *If q is sufficiently large, there exists $A \in \mathfrak{t}$ such that*

$$\chi_{T,\psi}(u) = q^{-r/2}\, \hat{\mu}_O(cu),$$

for any unipotent $u \in G$.

We have imposed some restrictions on G, which are not fulfilled if $G = GL_n$. But it is not difficult to formulate a similar statement in that case. This statement can be proved (see [9]).

REFERENCES

1. A. Borel, et al., *Seminar on algebraic groups and related finite groups*, Lecture Notes in Math., no. 131, Springer-Verlag, Berlin, 1970.

2. C. T. Benson and C. W. Curtis, *On the degrees and rationality of certain characters of finite Chevalley groups*, Trans. Amer. Math. Soc. **165** (1972), 251–273.

3. C. W. Curtis, N. Iwahori and R. Kilmoyer, *Hecke algebras and characters of parabolic type of finite groups, with (B, N)-pairs*, Inst. Hautes Études Sci. Publ. Math. no. 40 (1971), 81–116.

4. I. M. Gel'fand and M. I. Graev, *Construction of irreducible representations of simple algebraic groups over a finite field*, Dokl. Akad. Nauk SSSR **147** (1962), 529–532 = Soviet Math. Dokl. **3** (1962), 1646–1649. MR **26** #6271.

5. Harish-Chandra, *Eisenstein series over finite fields*, Functional Analysis and Related Fields, Springer-Verlag, Berlin and New York, 1970, pp. 76–88.

6. H. Jacquet and R. P. Langlands, *Automorphic forms on GL(2)*, Lecture Notes in Math., vol. 114, Springer-Verlag, Berlin, 1970.

7. L. Solomon, *The Steinberg character of a finite group with BN-pair*, Theory of Finite Groups (Sympos., Harvard Univ., Cambridge, Mass., 1968), Benjamin, New York, 1969, pp. 213–221. MR **40** #220.

8. T. A. Springer, *The unipotent variety of a semisimple group*, Algebraic Geometry (Internat. Colloq., Tata Inst. Fund. Res., Bombay, 1968), Oxford Univ. Press, London, 1969, pp. 373–391. MR **41** #8429.

9. ———, *Generalization of Green's polynomials*, Proc. Sympos. Pure Math., vol. 21, Amer. Math. Soc., Providence, R.I., 1971, pp. 149–153.

10. ———, *On the characters of certain finite groups*, Proc. Summer School on Theory of Group Representations Budapest (to appear).

11. ———, *Caractères de groupes de Chevalley finis*, Séminar N. Bourbaki, no. 429, 1972/73.

12. R. Steinberg, *Endomorphisms of linear algebraic groups*, Mem. Amer. Math. Soc. No. 80 (1968). MR **37** #6288.

13. ———, *Lectures on Chevalley groups*, Yale University, New Haven, Conn., 1969.

MATHEMATISCH INSTITUUT DER RIJKSUNIVERSITEIT,
UTRECHT, NETHERLANDS

ON THE DISCRETE SERIES FOR CHEVALLEY GROUPS

PAUL GÉRARDIN

1. Weil's method for representations of metaplectic groups [**13**], associated with Mackey's theorems on induced representations, give supercuspidal representations [**5**] of Chevalley groups over p-adic fields ([**7**] and [**11**] for SL_2, [**9**] for PGL_2, [**8**] for SL_n, [**6**] for GL_n). Here we give an explicit correspondence between the strongly regular characteristics of some unramified compact Cartan subgroup of a simply connected Chevalley group over a p-adic field, and supercuspidal representations. A complete proof will appear elsewhere [**4**].

2. Let k be a local p-adic field of residual characteristic p; let $\mathcal{O}$ be its ring of integers, $\mathfrak{p}$ its maximal ideal, $\bar{k}=\mathcal{O}/\mathfrak{p}$ its residual field of order q. We shall index by $n \leqq \infty$ the corresponding objects in the unramified extension of degree n over k. We denote by F the Frobenius substitution of k_∞ over k, and also by F the corresponding automorphism of $\bar{k}_\infty$ over $\bar{k}$. When x is an element of the quadratic extension k_{2n} of k_n, we write $\bar{x}$ for x^{F^n}. Finally τ is a character of k which puts $\mathcal{O}$ in duality with itself.

3. Two groups.

3.1. Let $\mathfrak{G}=\mathfrak{G}_r$ the Lie algebra sl_2 with the positive root r and its Chevalley basis (H_r, X_r, X_{-r}); we denote by $\mathfrak{A}$ the Cartan subalgebra with basis H_r, and $\mathfrak{M}$ the supplementary subspace generated by X_r and X_{-r}.

3.2. LEMMA. *Let $\mathfrak{q}\subset\mathfrak{p}$ be an ideal of $\mathcal{O}$, and $H(\mathfrak{q})$ be the set of matrices defined as*

$$\begin{pmatrix}1+t & u\\ v & 1+w\end{pmatrix},\qquad u, v\in\mathfrak{q}/\mathfrak{p}\mathfrak{q},\ t, w\in\mathfrak{q}^2/\mathfrak{p}\mathfrak{q}^2,\ t+w=uv.$$

Then

AMS (MOS) *subject classifications* (1970). Primary 20G05, 20G25.

(i) *$H(\mathfrak{q})$ is a Heisenberg group*

$$1 \to A(\mathfrak{q}^2/\mathfrak{p}\mathfrak{q}^2) \to H(\mathfrak{q}) \to \mathfrak{M}(\mathfrak{q}/\mathfrak{p}\mathfrak{q}) \to 1,$$

(1)

$$\begin{pmatrix} 1+t & u \\ v & 1+w \end{pmatrix} \mapsto uX_r + vX_{-r},$$

where the center $A(\mathfrak{q}^2/\mathfrak{p}\mathfrak{q}^2)$ is the set of $\left(\begin{smallmatrix} 1+t & 0 \\ 0 & 1-t \end{smallmatrix}\right)$, $t \in \mathfrak{q}^2/\mathfrak{p}\mathfrak{q}^2$.

(ii) *By conjugation we have the automorphisms of $H(\mathfrak{q})$ defined by the matrices $\left(\begin{smallmatrix} t & 0 \\ 0 & 1 \end{smallmatrix}\right)$ where $t \in \mathcal{O}^*$, the group of units of $\mathcal{O}$.*

3.3. Let $i(r)$ be a sign, and $\sigma(r)$ one of the two automorphisms of $SL_2(k_\infty)$, resp. $sl_2(k_\infty)$ defined by

split case
$$\begin{pmatrix} a & b \\ c & d \end{pmatrix}^{\sigma(r)} = \begin{pmatrix} a^F & i(r)\, b^F \\ i(r)\, c^F & d^F \end{pmatrix},$$

twisted case
$$\begin{pmatrix} a & b \\ c & d \end{pmatrix}^{\sigma(r)} = \begin{pmatrix} d^F & -i(r)\, c^F \\ -i(r)\, b^F & a^F \end{pmatrix}.$$

3.4. LEMMA. *In Lemma 3.2, let $\mathfrak{q}$ be the unramified quadratic extension of an ideal $\mathfrak{f} \subset \mathfrak{p}^2$, and σ as in 3.3. Then we can take the σ-invariants in (1). Moreover, let $A'(\sigma, \mathcal{O})$ be the group of elements $\left(\begin{smallmatrix} t & 0 \\ 0 & 1 \end{smallmatrix}\right)$ with $t \in \mathcal{O}^*$ in the split case (resp., $t \in \mathcal{O}_2^*$ and $t\bar{t} = 1$ in the twisted case), then $A'(\sigma, \mathcal{O})$ acts on $H(\sigma, \mathfrak{f})$, the set of σ-invariants of $H(\mathfrak{f}_2)$.*

4. Chevalley groups.

4.1. Let R be a reduced irreducible root system, and l, N, W, $P(R)$ as in [**10**]. We denote by $\ddot{W}$ the extended Weyl group of R [**12**].

4.2. Let k be as in §2, and suppose moreover that p is good [**1**, p. 178] and does not divide the discriminant of the Killing form B of a Chevalley system $\mathfrak{G}_\mathbf{Z}$ over R ([**1**, p. 180], [**3**, p. 318]). Let $\mathfrak{A}_\mathbf{Z}$ be the Cartan subalgebra generated by the H_r and $\mathfrak{M}_\mathbf{Z}$ the supplementary subspace generated by the X_r, for $r \in R$.

4.3. Let G be the simply connected split semisimple scheme over $\mathbf{Z}$ defined by R and $\mathfrak{G}_\mathbf{Z}$. We write $G = G(k)$, $K = G(\mathcal{O})$ the standard maximal compact subgroup, $A(k)$ the standard maximal split torus; for an ideal $\mathfrak{f}$ of $\mathcal{O}$, we write $G(\mathfrak{f})$ for the kernel of reduction modulo $\mathfrak{f}$ on $G(\mathcal{O})$, and similarly $G(\mathfrak{f}/\mathfrak{f}')$ if $\mathfrak{f}' \subset \mathfrak{f}$. Those groups are given by generators and relations, as in [**10**].

4.4. LEMMA. (i) *If $\mathfrak{f}^2 \subset \mathfrak{f}' \subset \mathfrak{f} \subset \mathfrak{p}$ between ideals of $\mathcal{O}$, then the mapping $\exp: \mathfrak{G}(\mathfrak{f}/\mathfrak{f}') \to G(\mathfrak{f}/\mathfrak{f}')$ is an isomorphism.*

(ii) *If $\mathfrak{f} \subset \mathfrak{p}$, then the commutator of two elements of $G(\mathfrak{f}/\mathfrak{p}\mathfrak{f}^2)$ is given by the bracket of their corresponding projections in $\mathfrak{G}(\mathfrak{f}/\mathfrak{p}\mathfrak{f})$, via the isomorphism $\exp: \mathfrak{G}(\mathfrak{f}^2/\mathfrak{p}\mathfrak{f}^2) \to G(\mathfrak{f}^2/\mathfrak{p}\mathfrak{f}^2)$.*

5. Some unramified Cartan subgroups.

5.1. Let $s \in W$, $\ddot{s}$ above s in $\ddot{W}$, $h(s)$ the order of s in W, S the cyclic group generated by s in W, $W(s)$ the centraliser of s. We define an automorphism σ of $G(k_\infty)$ (resp., $\mathfrak{G}(k_\infty)$) by $\sigma = F \circ \operatorname{Int} \ddot{s}$ (resp., $\sigma = F \circ \operatorname{Ad} \ddot{s}$). We denote by $G(s, k)$ (resp., $\mathfrak{G}(s, k)$) for the σ-fixed points. Similarly we have $G(s, \mathcal{O})$, $A(s, k)$, $G(s, \mathfrak{f})$, $\mathfrak{G}(s, \mathcal{O}), \ldots$.

5.2. LEMMA. *Let σ be as above; there exists $g \in G(\mathcal{O}_{2h(s)})$ such that* $\operatorname{Int} g$ *conjugates $G(s, k)$ with $G(k)$, and similarly for the Lie algebras with* $\operatorname{Ad} g$.

5.3. LEMMA. *Suppose there exists a regular point of $A(\bar{k}_\infty)$ which is fixed by σ. Then $A(s, k)$ is a Cartan subgroup of $G(s, k)$. Moreover $A(s, k)$ is compact – and is equal to $A(s, \mathcal{O})$ – if and only if* $\det(s-1) \neq 0$.

5.4. DEFINITION. Let λ be a character of $A(s, k)$, and $\mathfrak{f}$ the ideal of $\mathcal{O}$ which is its conductor. We shall say that λ is strongly regular if $\mathfrak{f} \subset \mathfrak{p}^2$ and if the element $\lambda^0 \in \mathfrak{A}(s, \bar{k})$ defined by $\lambda(\exp(H)) = \tau(\pi^{-\operatorname{val} \mathfrak{f}} B(\lambda^0, H))$ for $H \in \mathfrak{A}(s, \mathfrak{p}^{-1}\mathfrak{f}/\mathfrak{f})$ is regular; $\prod_{r \in R} (\lambda^0, r) \neq 0$, i.e., $w \cdot \lambda^0 = \lambda^0$ for a w in W implies $w = 1$.

6. The representations W_λ.

6.1. Let λ be a strong character of $A(s, k)$ and f its conductor. We define an ideal $\mathfrak{f}'$ by $\mathfrak{f} = \mathfrak{f}'^2$ if $\mathfrak{f}$ is even and $\mathfrak{f} = \mathfrak{p}\mathfrak{f}'^2$ if $\mathfrak{f}$ is odd, and we write $D(s, \mathfrak{f})$ for the subgroup of $G(s, \mathcal{O})$ defined by $D(s, \mathfrak{f}) = A(s, \mathcal{O})\, G(s, \mathfrak{f}')$.

6.2. LEMMA. *If $\mathfrak{f} = \mathfrak{f}'^2$, then λ extends canonically as a representation W_λ of degree* 1 *of $D(s, \mathfrak{f})$. In fact, we have the exact sequence*

$$1 \to \mathfrak{M}(s, \mathfrak{f}'/\mathfrak{f}) \to D(s, \mathfrak{f})/G(s, \mathfrak{f}) \to A(s, \mathcal{O}/\mathfrak{f}) \to 1.$$

6.3. LEMMA. *If $\mathfrak{f} = \mathfrak{p}\mathfrak{f}'^2$, the cokernel of $\mathfrak{M}(s, \mathfrak{p}\mathfrak{f}'/\mathfrak{f}) \to D(s, \mathfrak{f})/G(s, \mathfrak{f})$ is an extension by $A(s, \mathcal{O}/\mathfrak{f}'^2)$ of a product of Heisenberg groups, each associated with an orbit of the group $\pm S$ in R; if $r \in R$, let $n(r)$ be the order of $S \cdot r$.*

(i) *If $S \cdot r \neq -S \cdot r$, define $i(r)$ by $\operatorname{Ad} \ddot{s}^{n(r)} X_r = i(r) X_r$; then the Heisenberg group associated to r is $H(\sigma_r, \mathfrak{f}_{n(r)})$* (3.3 *and* 3.4).

(ii) *If $Sr = -Sr$, then $n(r) = 2n'(r)$ is even,* $\operatorname{Ad} \ddot{s}^{n'(r)} X_r = -i(r) X_{-r}$, *and the Heisenberg group associated to r is $H(\sigma_r, \mathfrak{f}_{n'(r)})$.*

6.4. LEMMA. *If $\mathfrak{f} = \mathfrak{p}\mathfrak{f}'^2$, then λ defines canonically an irreducible representation of $D(s, \mathfrak{f})$ of degree q^N, that we denote W_λ.*

7. THEOREM. *Let R, k, p satisfy all the above conditions, and $s \in W$ which does*

not have 1 *as an eigenvalue; we suppose p prime to* $h(s)$ (5.1), *and that there exists a regular point of* $A(\bar{k}_\infty)$ *which is fixed by* σ; *let* W_λ *be as in* 6.2 *or* 6.3 *for a strongly regular character* λ *of* $A(s, k)$. *Then*

(i) *The induced representation* $\mathrm{Ind}_{D(s,\mathfrak{f})}^{G(s,k)} W_\lambda$ *defines an irreducible supercuspidal admissible* [5] *representation* ω_λ *of* $G(k)$: $\omega_\lambda \in {}^\circ\mathscr{E}(G)$.

(ii) *The mapping* $\lambda \mapsto \omega_\lambda$ *is an injection of the* $W(s)$*-orbits of the strongly regular characters of* $A(s, k)$ *into* ${}^\circ\mathscr{E}(G)$.

(iii) *Let* σ_λ *be the representation of K defined by* $\mathrm{Ind}_{D(s,\mathfrak{f})}^{G(s,\sigma)} W_\lambda$; *then* σ_λ *is an irreducible cuspidal representation of order* 1 *of K, i.e.,*

$$\int_{N(\mathfrak{p}^{-1}\mathfrak{f})} \sigma_\lambda(u)\, du = 0$$

for every proper horocyclic part N of R.

(iv) *Normalising Haar measure on G by the condition* $\int_K dx = 1$, *the formal degree* $d(\omega_\lambda)$ *of* ω_λ *is the degree of* σ_λ, *say,*

$$d(\omega_\lambda) = q^{N(\mathrm{val}\,\mathfrak{f}-1)} Q(q)\,(q-1)^l / X_s(q)$$

where Q is the Poincaré polynomial of R and X_s *the characteristic polynomial of the transformation s on* $P(R)$.

(v) *Let* $t \in A(s, k)$ *such that its image in* $A(s, \bar{k})$ *is a regular element; then (g is as in Lemma* 5.2*)*

$$\mathrm{tr}\ \omega_\lambda(gtg^{-1}) = \sum_{W(s)} (w\lambda)(t) \qquad \text{if the conductor } \mathfrak{f} \text{ of } \lambda \text{ is even},$$

$$\mathrm{tr}\ \omega_\lambda(gtg^{-1}) = (-1)^l \sum_{W(s)} (w\lambda)(t) \quad \text{if } \mathfrak{f} \text{ is odd}.$$

8. Remarks.

8.1. If s has 1 as an eigenvalue, this method gives irreducible representations of K.

8.2. If s has 1 as an eigenvalue, a similar method gives a principal series of representations of G associated with a standard parabolic subgroup defined by s.

References

1. A. Borel, et al., *Seminar on algebraic groups and related finite groups*, Lecture Notes in Math., vol. 131, Springer-Verlag, Berlin, 1970.

2. N. Bourbaki, *Eléments de mathématique*. Fasc. XXXIV. *Groupes et algèbres de Lie*. Chap. IV: *Groupes de Coxeter et systèmes de Tits*. Chap. V: *Groupes engendrés par des réflexions*. Chap. VI: *Systèmes de racines*, Actualités Sci. Indust., no. 1337, Hermann, Paris, 1968. MR **39** #1590.

3. M. Demazure and A. Grothendieck, *Schémas en groupes*. III (SGA3), Lecture Notes in Math., vol. 153, Springer-Verlag, Berlin, 1970.

4. P. Gérardin (a) *Représentations de groupes de Chevalley* $\mathfrak{p}$*-adiques*, C. R. Acad. Sci. Paris **275** (1972), 1159–1162.

(b) *Groupes d'Heisenberg et groupes diamants sur les corps finis*, Sém. Delange-Pisot-Poitou, groupes d'études de théorie des nombres, Paris 1972/73, no. 9.

(c) *Sur les représentations du groupe linéaire général sur un corps* $\mathfrak{p}$*-adique*, Sém. Delange-Pisot-Poitou, théorie des nombres, Paris, 1972/73, no. 12.

5. Harish-Chandra, *Harmonic analysis on reductive p-adic groups*, Proc. Sympos. Pure Math., vol. 26., Amer. Math. Soc., Providence, R.I., 1974, pp. 167–192.

6. R. Howe, *Kirillov theory on p-adic compact groups* (preprint).

7. J. A. Shalika, *Representations of the* 2×2 *unimodular group over local fields*, Seminar on Representations of Lie Groups, Institute for Advanced Study, Princeton, N. J., 1966, Exp. 2.

8. T. Shintani, *On certain square integrable irreducible unitary representations of some p-adic linear groups*, J. Math. Soc. Japan **20** (1968), 522–565. MR **38** #2252.

9. A. J. Silberger, PGL_2 *over the p-adics: its representations, spherical functions and Fourier analysis*, Lecture Notes in Math., vol. 166, Springer-Verlag, Berlin and New York, 1970. MR **44** #2891.

10. R. Steinberg, *Lectures on Chevalley groups*, Notes prepared by J. Faulkner and R. Wilson, Yale University, New Haven, Conn., 1967.

11. S. Tanaka, *On irreducible unitary representations of some special linear groups of the second order.* I, II, Osaka J. Math. **3** (1966), 217–242. MR **36** #6541.

12. J. Tits, *Sur les constantes de structure et le théorème d'existence des algèbres de Lie semi-simples*, Inst. Hautes Études Sci. Publ. Math. No. 31 (1966), 21–58. MR **35** #5487.

13. A. Weil, *Sur certains groupes d'opérateurs unitaires*, Acta Math. **111** (1964), 143–211. MR **29** #2324.

UNIVERSITÉ PARIS VII
75005 PARIS, FRANCE

THE STEINBERG CHARACTER AS A TRUE CHARACTER

W. CASSELMAN

Let k be a $\mathfrak{p}$-adic field, G the k-rational points of a connected reductive group defined over k, A_ϕ the k-rational points of a maximal split subtorus, P_ϕ the k-rational points of a minimal parabolic subgroup containing A_ϕ. For each subset Θ of the associated set Δ of positive simple roots, let $P_\Theta = M_\Theta N_\Theta$ be the corresponding standard parabolic subgroup. Let A_Θ be the center of M_Θ. Let δ_Θ be the modulus function of P_Θ.

I shall use a normalized notation for induced representations: If H_1 is any $\mathfrak{p}$-adic group, H_2 a closed subgroup, π a smooth representation of H_2, let $\operatorname{Ind}(\pi \mid H_2, H_1)$ be the right-regular representation of H_1 on the space of all locally constant functions f on H_1 with values in the space of π and of compact support modulo H_2 such that

$$f(h_2 h_1) = \delta_1^{-1/2}(h_2)\, \delta_2^{1/2}(h_2)\, \pi(h_2)\, f(h_1)$$

for all $h_1 \in H_1$, $h_2 \in H_2$ (where δ_1, δ_2 are the modulus functions of H_1, H_2).

Let π_Θ be the representation $\operatorname{Ind}(\delta_\Theta^{-1/2} \mid P_\Theta, G)$. In [**2**], Harish-Chandra has defined the Steinberg character of G to be the class function

$$\boldsymbol{St} = \sum_\Theta (-1)^{\operatorname{card}\Theta}\, \boldsymbol{ch}(\pi_\Theta).$$

(This is my own notation. For any representation π, $\boldsymbol{ch}(\pi)$ is its character.) Borel and Serre have recently given a proof (not yet published) that, in analogy with a well-known result for algebraic groups over finite fields (see Solomon [**6**], for example), this is the character of an irreducible square-integrable representation

AMS (MOS) subject classifications (1970). Primary 22D10, 22D12, 22E35, 22E50; Secondary 20C15.

of G. Their proof depends on their investigation of the cohomology of the Bruhat-Tits building and its compactification. I shall sketch here another proof, more direct even if not as elegant. I am indebted to Harish-Chandra for several remarks which improved my version of Theorem 3 considerably.

The result may be formulated more suggestively. For any subsets $\Omega \subseteq \Theta$ of Δ there is a canonical embedding of π_Θ into π_Ω, hence one of each π_Θ into π_ϕ. If Ω and Θ are any two subsets of Δ, then the intersection of the images in π_ϕ of π_Θ and π_Ω is $\pi_{\Theta \cup \Omega}$. Consider the representation π on the sum (not the direct sum) of all the images of the π_Θ, $\Theta \neq \phi$. By induction and the fact that $\boldsymbol{ch}(\pi_1+\pi_2)=\boldsymbol{ch}(\pi_1)+\boldsymbol{ch}(\pi_2)-\boldsymbol{ch}(\pi_1 \cap \pi_2)$, one has that

$$\boldsymbol{ch}(\pi) = -\sum_{\Theta \neq \phi} (-1)^{\operatorname{card} \Theta}\, \boldsymbol{ch}(\pi_\Theta)$$

and hence that the character of the quotient $\pi_\phi/\pi=\sigma$ is precisely $\boldsymbol{St}$. One-half the result mentioned above becomes

THEOREM 1. *The representation σ is irreducible.*

The cornerstone of the proof is one of the basic results in the analysis of the principal series representations of G, which will not be proven here.

THEOREM 2. *If χ is any character of M_ϕ (not necessarily unitary) then*

(a) $\operatorname{Ind}(\chi \mid P_\phi, G)$ *has a finite Jordan decomposition;*

(b) *if ϱ is any irreducible subquotient, then for some element w of the Weyl group $W=N(A_\phi)/M_\phi$, the representation ϱ has an embedding into* $\operatorname{Ind}(\chi^w \mid P_\phi, G)$.

There are stronger versions of this, not necessary here. Part (a) is a corollary of Roger Howe's general result on finitely generated admissible representations, but is also proven in the course of the proof of the theorem. Part (b) is due independently to Harish-Chandra and myself. Proofs will presumably appear soon.

The precise form of this result that we need is (superficially) weaker. Recall that if ϱ is any smooth representation of a $\mathfrak{p}$-adic unipotent group U, then $\varrho(U)$ is the subspace of all vectors v such that for some open compact subgroup U_0 of U (and there are arbitrarily large ones) we have

$$\int_{U_0} \varrho(u)\, v\, du = 0.$$

If ϱ is any G-stable subrepresentation of $\operatorname{Ind}(\chi \mid P_\phi, G)$ for some character χ, it is elementary that $\varrho \neq \varrho(N_\phi)$. Theorem 2 therefore implies (and is essentially implied by, in fact)

COROLLARY. *If ϱ is as in Theorem 2, then $\varrho \neq \varrho(N_\phi)$.*

Some more lemmata:

PROPOSITION 1. *If U is a $\mathfrak{p}$-adic unipotent group and*

$$0 \to W_1 \to W \to W_2 \to 0$$

an exact sequence of smooth U-spaces, then the sequence

$$0 \to W_1/W_1(U) \to W/W(U) \to W_2/W_2(U) \to 0$$

is exact as well.

PROOF. The definition implies immediately that $W_1(U) = W_1 \cap W(U)$.

PROPOSITION 2. *Let χ be the character $\delta_\Theta^{-1/2}$ of M_ϕ (defined since $M_\phi \subseteq M_\Theta$). Then $\pi_\Theta/\pi_\Theta(N_\phi)$ as an M_ϕ-space is isomorphic to the direct sum $\oplus_w \chi^w$ where the sum is over all $w \in W$ such that $w^{-1}(\alpha) > 0$ for all $\alpha \in \Theta$.*

PROOF. The decomposition of G into double cosets $P_\Theta w P_\phi$ gives a normal series for π_Θ as an N_ϕ-space, to which one may apply Proposition 1 and induction. (Details will appear with the proof of Theorem 2.)

PROOF OF THEOREM 1. By Proposition 2, the M_ϕ-representation $\pi_\phi/\pi_\phi(N_\phi)$ is the direct sum over all $w \in W$, $\oplus_w \chi^w$ (where $\chi = \delta_\phi^{-1/2}$). By Propositions 1 and 2 and the same reasoning as that which obtained the character of π, one can see that $\pi/\pi(N_\phi)$ is the direct sum $\oplus_w \chi^w$ over all w such that $w(\alpha) > 0$ for some simple positive root α. But there is only one w such that $w(\alpha) > 0$ for no such α, the largest element of W. We deduce that $\sigma/\sigma(N_\phi)$ has dimension one. If σ were reducible, we would have a nontrivial exact sequence

$$0 \to \sigma_1 \to \sigma \to \sigma_2 \to 0$$

and hence by Proposition 1 an exact sequence

$$0 \to \sigma_1/\sigma_1(N_\phi) \to \sigma/\sigma(N_\phi) \to \sigma_2/\sigma_2(N_\phi) \to 0.$$

But by the Corollary to Theorem 2, each of the outer terms is nontrivial, which implies that $\sigma/\sigma(N_\phi)$ has dimension $\geqq 2$, a contradiction. The proof of Theorem 1 is finished.

Now I proceed to prove

THEOREM 3. *The representation σ is $L^{1+\varepsilon}$ for each $\varepsilon>0$.*

PROOF. Let $G=K\cdot A_\phi^-\cdot\Omega\cdot K$ be the Cartan decomposition of G asserted by Harish-Chandra [**1**, p. 16], where $A_\phi^-=\{a\in A \mid \alpha(a)\leqq 1 \text{ for all } \alpha\in\Delta\}$. First, one establishes

PROPOSITION 3. *If v, $\tilde{v}$ are elements in the spaces of σ, $\tilde{\sigma}$, then there exists $C>0$ such that*

$$\langle\sigma(k_1 a\omega k_2)\, v, \tilde{v}\rangle \leqq C\cdot\delta_\phi(a)$$

(*factorization according to the Cartan decomposition*).

PROOF OF THE PROPOSITION. (1) For any subset Θ of Δ, $\sigma/\sigma(N_\Theta)$ is an irreducible M_Θ-space: Since σ is a quotient of a principal series representation π of G, $\sigma/\sigma(N_\Theta)$ is a quotient of $\pi/\pi(N_\Theta)$. But in the proof of Theorem 2 (not given here), the Jordan components of $\pi/\pi(N_\Theta)$ are all shown to be principal series representations of M_Θ. Hence, by Theorem 2, any irreducible subquotient ϱ of $\sigma/\sigma(N_\Theta)$ satisfies $\varrho\neq\varrho(N_\phi)$, and since for $\varrho=\sigma/\sigma(N_\Theta)$ itself we have $\varrho/\varrho(N_\phi)\cong\sigma/\sigma(N_\phi)$ of dimension one, it must be irreducible.

(2) This implies that A_Θ acts as scalar multiplication on $\sigma/\sigma(N_\Theta)$, and it is not hard to see that $a\in A_\Theta$ acts as $\delta_\phi(a)$. Hence, for all $a\in A$, all v in the space of σ, we have $(\sigma(a)\, v-\delta_\phi(a)\, v)\in\sigma(N_\Theta)$.

(3) Apply this to the particular case $\Theta=\Delta-\{\alpha\}$. Recall the fact (see, for example, [**3**, p. 147]) that for $v\in\sigma(N_\Theta)$, $\tilde{v}$ in the space of $\tilde{\sigma}$, if we choose a compact subgroup U_0 of N_Θ such that

$$\int_{U_0} \sigma(u)\, v\, du=0,$$

a compact subgroup U_1 of N_Θ leaving v fixed, and an element $a\in A$ such that $aU_0a^{-1}\subseteq U_1$, then $\langle\sigma(a)\, v, \tilde{v}\rangle=0$. Let w and $\tilde{w}$ be nontrivial K-stable subspaces of σ and $\tilde{\sigma}$. Define Φ to be the function on G with values in the dual of $W\otimes\tilde{W}$, taking g to be the function which takes $w\otimes\tilde{w}$ to $\langle\sigma(g)\, w, \tilde{w}\rangle$. The function Φ behaves covariantly with respect to right and left multiplication by elements of K, hence (Cartan decomposition) is determined by its values on $A_\phi^-\cdot\Omega$. Let a_0 be a generator for A_Θ modulo $Z\cdot(K\cap A_\Theta)$. Choose $U_0\subseteq N_\Theta$ such that $mU_0m^{-1}\subseteq U_0$, for all $m\in M_\phi$, such that $a_0U_0a_0^{-1}\subseteq U_0$, and such that for each $w\in W$ we have

$$\int_{U_0} \sigma(u)\,(\sigma(a_0)-\delta_\phi(a_0))\, w\, du=0.$$

Choose $U_1 \subseteq N_\Theta$ fixing each $\tilde{w} \in \tilde{W}$. Then the remark above implies that if $a \in A^-$ is such that $aU_0a^{-1} \subseteq U_1$, and $\omega \in \Omega$, then $\Phi(a\omega a_0) = \delta_\phi(a_0)\, \Phi(a\omega)$. Further, since if $aU_0a^{-1} \subseteq U_1$ then $(aa_0)\, U_0(aa_0)^{-1} \subseteq U_1$ as well, this implies that for each $a \in A_\Phi^-$ there exists an integer M such that for $m \geqq M$ we have $\Phi(a\omega a_0^m) = \delta_\phi(a_0^{m-M}) \cdot \Phi(a\omega a_0^M)$.

This is the desired asymptotic estimate for matrix coefficients in the Θ-direction. If we now apply this for all $\alpha \in \Delta$, we obtain Proposition 3 without too much trouble.

Theorem 3 follows from Proposition 3, together with Theorems 25 and 26 of [2]. (Harish-Chandra pointed this out to me, thus improving as well as correcting my original argument, which led only to a weaker result.)

If G is simply connected, B an Iwahori subgroup, and $K \supseteq B$ a good maximal compact, one can show by considering the restriction of σ to K that the subspace of vectors in σ fixed by B has dimension one (one needs for this the result in Solomon [6] on finite BN groups). One can combine this with the asymptotic estimate above to conclude

THEOREM 4. *If G is simply connected, then σ coincides with the Steinberg representation of Matsumoto* [4] *and Shalika* [5].

REFERENCES

1. Harish-Chandra, *Harmonic analysis on reductive p-adic groups*, Lecture Notes in Math., vol. 162, Springer-Verlag, Berlin, 1970.

2. ———, *Harmonic analysis on reductive p-adic groups*, Proc. Sympos. Pure Math., vol. 26, Amer. Math. Soc., Providence, R.I., 1974, pp. 167–192.

3. H. Jacquet, *Représentations des groupes linéares p-adiques*, Theory of Group Representations and Fourier Analysis (C.I.M.E., II Ciclo, Montecatini Terme, 1970), Edizioni Cremonese, Rome, 1971, pp. 119–220. MR **45** #453.

4. H. Matsumoto, *Fonctions sphériques sur un groupe semi-simple p-adique*, C. R. Acad. Sci. Paris Sér. A-B **269** (1969), A829-A832. MR **41** #8576.

5. Joseph Shalika, *On the space of cusp forms of a p-adic Chevalley group*, Ann. of Math. (2) **92** (1970), 262–278. MR **42** #423.

6. Louis Solomon, *The Steinberg character of a finite group with BN-pair*, The Theory of Finite Groups, (Sympos., Harvard Univ., Cambridge, Mass., 1968), Benjamin, New York, 1969, pp. 213–221. MR **40** #220.

UNIVERSITY OF BRITISH COLUMBIA
VANCOUVER 8, BRITISH COLUMBIA, CANADA

HARMONIC ANALYSIS ON TREES

P. CARTIER

1. Introduction. Following the method of Bruhat and Tits in [1] one associates to a reductive p-adic group G of rank one a simplicial complex of dimension 1 which appears to be a tree. The set of vertices of this tree is the discrete space $G_K/G_{\mathfrak{O}}$ (K a local field, $\mathfrak{O}$ its ring of integers). Serre used this tree in [10] and [11] to study the arithmetical properties of the subgroup of integral points $G_{\mathfrak{O}}$, namely congruence subgroups and amalgamation. We sketch here a method to deal with the representation theory of G_K using the combinatorics of the tree $G_K/G_{\mathfrak{O}}$. A detailed exposition will be submitted to the *Journal of Combinatorial Theory*.

2. Geometry of a tree ([3], [4], [10]). We consider a tree X as defined by its *vertex* set S and a set A of two-element subsets of S, the *edges*. By definition of a tree, for given vertices s and s' there exists a unique chain joining s to s', namely a sequence $[s_0, s_1, \ldots, s_m]$ of distinct vertices such that $s_0 = s$, $s_m = s'$ and $\{s_{j-1}, s_j\} \in A$ for $1 \leqq j \leqq m$. The integer m is the distance from s to s', written as $d(s, s')$. We assume moreover that X is *homogeneous of degree* $q+1$; that is, each vertex is adjacent to exactly $q+1$ edges ($q \geqq 2$).

By adjunction of a set S_∞ of *boundary points* to S, one gets a compact space $\hat{S}$ containing S as a discrete open dense subspace. A (one-sided) *geodesic* issued from the vertex s is an infinite sequence $[s_0, s_1, \ldots]$ of distinct vertices such that $s_0 = s$ and $\{s_{j-1}, s_j\} \in A$ for any $j \geqq 1$. Given s there is a bijective correspondence between the geodesics $[s_0, s_1, \ldots]$ issued from s and the boundary points b, expressed by the relation $b = \lim_{m \to \infty} s_m$ in $\hat{S}$.

Let b be a boundary point. Let s and s' be two vertices, $[s_0, s_1, \ldots]$ the geodesic from s to b, and $[s'_0, s'_1, \ldots]$ the geodesic from s' to b. There exist integers $r \geqq 0$

AMS (MOS) subject classifications (1970). Primary 22E35; Secondary 05C05, 60C05.

and $r' \geqq 0$ such that $[s_0, s_1, \ldots, s_r = s'_{r'}, s'_{r'-1}, \ldots, s'_1, s'_0]$ is the chain from s to s' and $s_{r+n} = s'_{r'+n}$ for any $n \geqq 0$. One has, therefore, $d(s, s') = r + r'$ and one puts $\delta_b(s, s') = r - r'$. From the identity $\delta_b(s, s'') = \delta_b(s, s') + \delta_b(s', s'')$ one deduces the existence of a partition $(H_n)_{n \in \boldsymbol{Z}}$ of S such that $\delta_b(s, s') = m - m'$ for s in H_m and s' in $H_{m'}$. This partition is unique up to a shift in the index. The sets H_n are the *horocycles* of center b. The shift V in the set of horocycles is defined by $V(H_n) = H_{n+1}$.

The space of all horocycles provided with a natural topology appears as a principal bundle P of basis S_∞ and structure group $\boldsymbol{Z}$ (acting via the powers of V). For each vertex s the set of horocycles going through s is a continuous cross section of P over S_∞.

3. The tree associated to $PGL_2(K)$ **[10].** Let K be a field complete under a discrete valuation, $\mathfrak{O}$ its ring of integers and $\mathfrak{p}$ the nonzero prime ideal in $\mathfrak{O}$. One assumes that the residue field $k = \mathfrak{O}/\mathfrak{p}$ is finite with q elements. One lets $\mathfrak{O}^\times = \mathfrak{O} - \mathfrak{p}$. The image of a matrix $\left(\begin{smallmatrix} a & b \\ c & d \end{smallmatrix}\right)$ in $PGL_2(K)$ shall be denoted by $\left[\begin{smallmatrix} a & b \\ c & d \end{smallmatrix}\right]$.

Let V be the vector space of column vectors $\left(\begin{smallmatrix} x \\ y \end{smallmatrix}\right)$ with coordinates taken from K. By a lattice in V we mean any $\mathfrak{O}$-submodule of V generated by two linearly independent vectors. Two lattices Λ and Λ' are called equivalent if there exists a nonzero element x in K such that $\Lambda' = x \cdot \Lambda$. The set S consists of the equivalence classes of lattices. An edge is a pair $\{s, s'\}$ of vertices with representative lattices Λ and Λ' such that $\Lambda \supset \Lambda'$ and $(\Lambda : \Lambda') = q$. One gets a tree X, homogeneous of degree $q+1$, whose boundary S_∞ can be canonically identified with the projective line $\boldsymbol{P}^1(K)$.

We can now interpret some of the standard homogeneous spaces of $G_K = PGL_2(K)$. As usual the *Borel subgroup* B_K has elements $\left[\begin{smallmatrix} a & b \\ 0 & 1 \end{smallmatrix}\right]$, the *split torus* H_K has elements $\left[\begin{smallmatrix} a & 0 \\ 0 & 1 \end{smallmatrix}\right]$, the subgroup $G_\mathfrak{O}$ consists of the integral elements $\left[\begin{smallmatrix} a & b \\ c & d \end{smallmatrix}\right]$ with a, b, c, d in $\mathfrak{O}$ and $ad - bc$ in $\mathfrak{O}^\times$, while the *Hecke subgroup* $\Gamma_0(\mathfrak{p}^m)$ is the subgroup of $G_\mathfrak{O}$ defined by the congruence $c \equiv 0 \bmod \mathfrak{p}^m$. Then S is isomorphic to $G_K/G_\mathfrak{O}$ and S_∞ to G_K/B_K while $G_K/\Gamma_0(\mathfrak{p}^m)$ is the space of pairs of vertices at the distance m and G_K/H_K is the set of pairs of distinct boundary points, in bijective correspondence with the two-sided geodesics in the tree X. Similarly, the space P of horocycles is isomorphic to G_K/M where M consists of the elements $\left[\begin{smallmatrix} a & b \\ 0 & 1 \end{smallmatrix}\right]$ with a in $\mathfrak{O}^\times$ and b in K.

4. Spherical functions [7]. We denote by $X = (S, A)$ any homogeneous tree of degree $q+1$ and Γ its full automorphism group. The vector space $C_c^\infty(S)$ consists of the complex-valued functions on S with finite support. The *Hecke algebra* $\mathscr{H}$ is the commuting algebra of the natural representation of Γ in $C_c^\infty(S)$. One describes as follows a basis $(\Theta_0, \Theta_1, \ldots)$ of the complex vector space $\mathscr{H}$: The operator Θ_m transforms a function f into the function whose value at s is the sum of the values of f at the vertices at distance m from s. The coefficient of Θ_p in the product $\Theta_m \cdot \Theta_n$ is the number of "triangles" *stu* in S with given basis *su* of length p and two other sides of lengths m and n. In particular, one gets the relations

$$\Theta_1^2 = \Theta_2 + (q+1)\,\Theta_0, \tag{1}$$

$$\Theta_1 \cdot \Theta_m = \Theta_{m+1} + q\Theta_{m-1} \qquad (\text{for } m \geqq 2), \tag{2}$$

conveniently collected into the generating series

$$\sum_{m=0}^{\infty} \Theta_m u^m = \frac{1-u^2}{1-u\Theta_1 + qu^2}. \tag{3}$$

In particular, $\mathscr{H}$ is the polynomial algebra $\boldsymbol{C}[T]$ with $T = \Theta_1$ the Hecke operator.

Fix a vertex s and let Γ_s be the stabilizer of s in Γ. We denote by $\mathscr{K}$ the Hilbert space consisting of the functions on S invariant under Γ_s, with the norm $[\sum_{t \in S} |f(t)|^2]^{1/2}$. One describes as follows an orthonormal basis $e_0, e_1, \ldots$ for $\mathscr{K}$: The function e_0 is equal to 1 at s and vanishes elsewhere, and

$$e_m = [q^{m-1}(q+1)]^{-1/2}\,\Theta_m e_0 \quad \text{for } m \geqq 1.$$

The Hecke operator T acts on $\mathscr{K}$ with a matrix given by

$$\begin{pmatrix} 0 & (q+1)^{1/2} & 0 & 0 & 0 & \cdots \\ (q+1)^{1/2} & 0 & q^{1/2} & 0 & 0 & \cdots \\ 0 & q^{1/2} & 0 & q^{1/2} & 0 & \cdots \\ 0 & 0 & q^{1/2} & 0 & q^{1/2} & \cdots \\ 0 & 0 & 0 & q^{1/2} & 0 & \cdots \\ \cdot & \cdot & \cdot & \cdot & \cdot & \cdots \end{pmatrix}.$$

Our next concern will be the spectral decomposition of T.

Let t be any complex number, which we write in the form $t = q^{1/2}(\lambda + \lambda^{-1})$. There exists on S a unique eigenfunction of T with the eigenvalue t, invariant under the group Γ_s and normalized by taking the value 1 at s. For s' at the distance $m \geqq 1$ from s, its value $F_t(s, s')$ at s' is equal to $\chi_t(\Theta_m)/q^{m-1}(q+1)$ where χ_t is the algebra homomorphism from $\mathscr{H}$ into $\boldsymbol{C}$ taking T to t. From (3) one deduces the explicit formula

$$F_t(s, s') = \frac{q(\lambda^{m+1} - \lambda^{-m-1}) - (\lambda^{m-1} - \lambda^{-m+1})}{q^{m/2}(q+1)(\lambda - \lambda^{-1})} \tag{4}$$

where $m = d(s, s')$.

The *Plancherel measure* is the unique probability measure μ on the real line $\boldsymbol{R}$ such that $\int \chi_t(\Theta_m)\, d\mu(t) = 0$ for any $m \geqq 1$. This moment problem is solved by the classical method; namely μ is the weak limit for $\varepsilon \to 0$ of

$$\pi^{-1}\,\mathrm{Im}\,\langle e_0|(T-t-i\varepsilon)^{-1}|e_0\rangle.$$

The resolvent $(T-z)^{-1}$ is easily computed from (3) and one gets

$$d\mu(t)=\frac{((4q-t^2)_+)^{1/2}}{(q+1)^2-t^2}\,dt. \tag{5}$$

The *Plancherel formula for spherical functions* reads as follows:

$$\int_{-2q^{1/2}}^{2q^{1/2}}\frac{(4q-t^2)^{1/2}}{(q+1)^2-t^2}F_t(s,s')\,dt=1 \quad \text{for} \quad s=s', \tag{6}$$

$$=0 \quad \text{otherwise.}$$

The following table collects information about the spherical functions.

Spectrum of T	$\lvert\lambda\rvert=1$	$-2q^{1/2}\leqq t\leqq 2q^{1/2}$
F_t bounded	$q^{-1/2}\leqq\lvert\lambda\rvert\leqq q^{1/2}$	$\left(\frac{\mathrm{Re}\,t}{q+1}\right)^2+\left(\frac{\mathrm{Im}\,t}{q-1}\right)^2\leqq 1$
F_t positive-definite	$\lvert\lambda\rvert=1$ *or* λ real and $q^{-1/2}\leqq\lvert\lambda\rvert\leqq q^{1/2}$	$-q-1\leqq t\leqq q+1$

Whenever X is the tree associated to $PGL_2(K)$, the previous results agree with the known results due mainly to Mautner [7].

5. Principal and supplementary spherical series. Let t be a real number in the interval $[-2q^{1/2}, 2q^{1/2}]$. We let $\mathscr{H}_t$ be the Hilbert space of functions on S with reproducing kernel $F_t(s,s')$. Any function f in $\mathscr{H}_t$ satisfies $Tf=tf$. The automorphism group Γ acts on $\mathscr{H}_t$ and one gets in this way an irreducible unitary continuous[1] representation of Γ. This provides the construction of the *principal spherical series* of representations of Γ. The *supplementary spherical series* obtains for t real in either of the intervals $]2q^{1/2}, q+1[$ and $]-q-1, -2q^{1/2}[$. Finally, for $t=q+1$ or $t=-q-1$, one gets one-dimensional representations of Γ. Whenever the tree X is associated to $PGL_2(K)$ one obtains by restriction from Γ to $PGL_2(K)$ the familiar series of representations. Notice that our proof of the fact that F_t is positive-definite for t real in $[-q-1, q+1]$ is the same in all cases and provides therefore a unified construction of the principal and supplementary series parametrized by characters of $K^\times$ trivial on the unit group $\mathfrak{O}^\times$.

Let us explain briefly how the *method of horocycles* works in our case (see [6] for an exposition of the classical case). On the space P of horocycles there

[1] The group Γ is locally compact and separable in the natural topology of pointwise convergence.

exists a unique measure ν giving the mass 1 to the set of horocycles going through a fixed vertex. This measure is invariant under the automorphism group Γ and the shift V multiplies ν by q^{-1}. One gets, therefore, a continuous representation of Γ in the Hilbert space $L^2(P, \nu)$ and one defines by $\Lambda f(\gamma)=q^{1/2}f(V\gamma)$ a unitary operator Λ in $L^2(P, \nu)$ commuting with the operators from Γ.

On the other hand, one has the natural representation of Γ in the Hilbert space $L^2(S)$ (each point of S being given the mass 1) and the Hecke operator T in $L^2(S)$. The *Radon transform* Jf of a function f in $C_c^\infty(S)$ is defined by $Jf(\gamma) =\sum_{s\in\gamma} f(s)$ for γ in P. One establishes the following facts:

(a) *The previous map J extends to an isometry J from $L^2(S)$ into $L^2(P, \nu)$ commuting with the action of Γ.*

(b) *One has $JT=q^{1/2}(\Lambda+\Lambda^{-1})\,J$.*

(c) *There exists a unitary operator W in $L^2(P, \nu)$ such that $W^2=1$, $W\Lambda=\Lambda^{-1}W$ and the image of J consists of the functions f with $Wf=f$.*

It is clear how to make explicit the spectral decomposition of Λ in $L^2(P, \nu)$: For any complex number λ of modulus 1, the suitably normalized functions f on P such that $q^{1/2}f(V\gamma)=f(\gamma)$ make a Hilbert space $\mathscr{H}'_\lambda$ which is the carrier of an irreducible unitary continuous representation of Γ. For λ and λ^{-1} one gets equivalent representations. Using J backwards, one recovers the previous results about the spectral decomposition of T.

As one expects, the supplementary series of representations can be realized in spaces of functions on P, but the description of the invariant scalar product is a little more delicate. It is also easy to extend to our case the results of Mautner [**8**] about the integral representation for the eigenfunctions of the Hecke operator.

6. Other series of representations. So far we have been able to extend the construction of the principal and supplementary series of representations to the automorphism group of a tree with an arbitrary degree $q+1\geqq 3$. We shall be very brief about the construction of the other series.

The solutions of the equation $Tf=(q+1)f$ are the harmonic functions on S, namely the functions whose value at any vertex s is the arithmetic mean of the values at the immediate neighbours of s. One can specialize to a homogeneous tree our results in [**3**]. The harmonic functions are the carrier of the *special representation*, in accordance with the general results of Borel and Serre in [**2**]. The equation $Tf=-(q+1)f$ is treated similarly.

According to a result of Silberger [**9**] made more precise by Casselman [**5**], there exists in any irreducible (admissible) representation of $PGL_2(K)$ a vector invariant under $\Gamma_0(\mathfrak{p}^m)$ for m suitably large. This suggests studying the natural representation of Γ in the space of functions $f(s, s')$ where the vertices s and s' are at a fixed distance m. One can make the decomposition of this representation into irreducible components by combinatorial methods. In this way, one can construct, for instance, irreducible representations for Γ which upon restriction to $PGL_2(K)$ break into finitely many irreducible representations of the discrete series. This provides a *combinatorial construction of the supercuspidal representations of $PGL_2(K)$*.

References

1. F. Bruhat and J. Tits, *Groupes réductifs sur un corps local*. I. *Données radicielles valuées*, Inst. Hautes Études Sci. Publ. Math., No. 41 (1972).

2. A. Borel and J.-P. Serre, *Special representations and harmonic functions on a building* (in preparation).

3. P. Cartier, *Fonctions harmoniques sur un arbre*, Symposia Mathematica **9** (1972), 203–270.

4. ———, *Géométrie et analyse sur les arbres*, Séminaire Bourbaki, 24e année, 1971/1972, Exposé 407, 18 pp.

5. W. Casselman, *On some results of Atkin and Lehner*, Math. Ann. **201** (1973), 301–314.

6. S. Helgason, *Functions on symmetric spaces*, Proc. Sympos. Pure Math., vol. 26, Amer. Math. Soc., Providence, R.I., 1974, pp. 101–146.

7. F. I. Mautner, *Spherical functions over p-adic fields*. I, II, Amer. J. Math. **80** (1958), 441–457; **86** (1964), 171–200. MR **20** #82; **29** #3582.

8. ———, *Fonctions propres des opérateurs de Hecke*, C. R. Acad. Sci. Paris Sér. A-B **269** (1969), A940-A943; ibid. **270** (1970), A89-A92. MR **41** #7031a, b.

9. A. J. Silberger, *PGL_2 over the p-adics; its representations, spherical functions and Fourier analysis*, Lecture Notes in Math., vol. 166, Springer-Verlag, Berlin and New York, 1970. MR **44** #2891.

10. J.-P. Serre, *Arbres, amalgames et SL_2*, Collège de France, 1968/69, Mimeographed set of notes written in collaboration with H. Bass.

11. ———, *Le problème des groupes de congruence pour SL_2*, Ann. of Math. (2) **92** (1970), 489–527. MR **42** #7671.

Institut des Hautes Etudes Scientifiques
F-91440 Bures sur Yvette, France

WHITTAKER MODELS FOR ADMISSIBLE REPRESENTATIONS OF REDUCTIVE $\mathfrak{p}$-ADIC SPLIT GROUPS

FRANÇOIS RODIER

1. Admissible representations. Let G be a locally compact, totally disconnected (l.c.t.d.) group. A *representation* π of G is a homomorphism of G into the group of linear automorphisms of a complex vector space E. A representation π of G is said to be *smooth* if, for any $x \in E$, the stabilizer of x in G is an open subgroup of G. A representation π of G is said to be *admissible* if π is smooth and, for any open subgroup G_1 of G, the set of all elements x in E stabilized by G_1 is finite dimensional.

Let π be an admissible representation of G in the space E. Define the representation π^* of G in the dual E^* of E by

$$\pi^*(g) \cdot f = f \circ \pi(g^{-1}) \quad \text{if } g \in G \text{ and } f \in E^*.$$

The subspace $\check{E}$ of all elements in E^* invariant under the action of an open subgroup of G is invariant under π^* and the representation $\check{\pi}$ of G in $\check{E}$ is admissible. $\check{\pi}$ is called the *contragredient* of π.

If λ and μ are two representations of G in the spaces E and F, we denote by $\mathrm{Hom}_G(\lambda, \mu)$ the space of all intertwining operators from λ to μ. An *intertwining form* of λ and μ is a bilinear form on $E \times F$ invariant by G.

Let H be a closed subgroup of G and π be a smooth representation of H in the space E. Let δ be the module of H. We denote by $\mathscr{S}(G; E; \pi)$ the space of functions $f: G \to E$ such that

$$f(hg) = \delta(h)^{-1/2}\, \pi(h)\,(f(g)) \quad \text{if } h \in H \quad \text{and} \quad g \in G.$$

f is locally constant on G, with compact support modulo H.

AMS (MOS) subject classifications (1970). Primary 22E50.

G operates by right translations on this vector space. This defines a smooth representation of G, called the *induced representation* of π, denoted by $\mathrm{Ind}_H^G \pi$. Let $\mathscr{S}(G; E)$ be the space of locally constant functions $f: G \to E$ with compact support. Let p be the map $\mathscr{S}(G; E) \to \mathscr{S}(G; E; \pi)$ defined by

$$p(f)(g) = \int_H \delta(h)^{1/2} \pi(h)^{-1} f(hg)\, dh \tag{1.1}$$

where $f \in \mathscr{S}(G; E)$ and dh is a Haar measure of H. Then p is surjective and is compatible with the right action of G on $\mathscr{S}(G; E)$ and $\mathscr{S}(G; E; \pi)$.

2. Whittaker models. Let K be a locally compact, totally disconnected, nondiscrete field. Let $\boldsymbol{G}$ be a reductive split connected group over K and G be the set of all K-rational points of $\boldsymbol{G}$. G is then a l.c.t.d. group. We define the following subgroups of $\boldsymbol{G}$: let $\boldsymbol{H}$ be a maximal K-split torus of $\boldsymbol{G}$, $\boldsymbol{B}$ a Borel subgroup containing $\boldsymbol{H}$, and $\boldsymbol{U}$ be the unipotent radical of $\boldsymbol{B}$. We denote by H, B, U the intersections of G with $\boldsymbol{H}$, $\boldsymbol{B}$, $\boldsymbol{U}$. A *character* of U is a continuous homomorphism from U into the group of complex numbers of absolute value 1. The group H operates on the set of all characters of U. If h is in H and θ is a character of U, we define the character θ^h of U by

$$\theta^h(u) = \theta(huh^{-1}) \quad \text{if } u \in U.$$

The character θ of U is said to be *principal* if the set of all h in H such that $\theta^h = \theta$ is the center Z of G.

Let θ be a principal character of U. We denote by $\mathscr{W}_\theta$ the set of all functions f on G with complex values such that

$$f(ug) = \theta(u) f(g) \quad \text{if } u \in U \text{ and } g \in G.$$

Then G operates on $\mathscr{W}_\theta$ by right translations, which defines a representation r_θ of G in $\mathscr{W}_\theta$.

Let π be a representation of G. A subspace E of $\mathscr{W}_\theta$ is called a *Whittaker model* of π with respect to θ if E is invariant under r_θ and the restriction of r_θ to E is equivalent to π. It is clear that a Whittaker model can be defined by an injective intertwining operator from π to r_θ.

Then two problems arise:

(1) Given π an admissible representation of G and θ a principal character of U, does there exist a Whittaker model of π with respect to θ?

(2) Suppose π admits a Whittaker model with respect to θ, how many models are there?

I. M. Gel'fand and D. A. Kajdan gave a partial answer to the first problem when G is the group $GL(n, K)$ [**2**].

THEOREM 1. *Let θ be a principal character of U, π an irreducible admissible supercuspidal representation of $GL(n, K)$. Then π admits a Whittaker model with respect to θ.*

Furthermore one can reduce the problem to the supercuspidal case. By a theorem of H. Jacquet and Harish-Chandra [**3**], any admissible irreducible representation of G is a subrepresentation of the induced representation $\mathrm{Ind}_P^G \tau$ where P is a parabolic subgroup of G containing B and τ a supercuspidal representation of P/V (V is the unipotent radical of P). The following theorem gives a "heredity" property of Whittaker models.

THEOREM 2. *Let P be a parabolic subgroup of G containing B and M a Levy subgroup of P containing H. Let τ be an admissible irreducible representation of M, extended to $P=MV$ by assuming that it is trivial on V. Let π' be the induced representation* $\mathrm{Ind}_P^G \tau$.

If w_0 is an element in the normalizer of H such that $B\cap w_0 B w_0^{-1}=H$, then $M\cap w_0 U w_0^{-1}$ is a maximal unipotent subgroup of M. Let θ_0 be the principal character of $M\cap w_0 U w_0^{-1}$ defined by $\theta_0(u)=\theta(w_0 u w_0^{-1})$. Let $\mathscr{W}'_{\theta_0}$ be the Whittaker space of M with respect to θ_0 and r'_{θ_0} the representation of M in $\mathscr{W}'_{\theta_0}$.

Then $\mathrm{Hom}_G(\pi', r_\theta)$ *is isomorphic to* $\mathrm{Hom}_M(\tau, r'_{\theta_0})$.

The answer to the second problem is known.

THEOREM 3. *Let θ be a principal character of U. Let π be an admissible irreducible representation of G. Then π admits at most one Whittaker model with respect to θ.*

I. M. Gel'fand and D. A. Kajdan gave a proof of this theorem in some special cases [**2**].

3. Distributions. We first need some facts about distributions on a l.c.t.d. space.

Let X be a l.c.t.d. topological space, and let E be a complex vector space. We call an *E-distribution* on X a linear form on the space $\mathscr{S}(X; E)$ of functions $f: X\to E$ which are locally constant with compact support. If T is an E-distribution on X we denote $\int_X f(x)\, dT(x)$ its value on the element f of $\mathscr{S}(X; E)$. A *distribution* on X is a $\boldsymbol{C}$-distribution.

If A is an endomorphism of the vector space E, we denote by $T\circ[A]$ the E-distribution on X given by

$$\int_X f(x)\, dT\circ[A](x) = \int_X A(f(x))\, dT(x).$$

If X is a group, the convolution product of an E-distribution on X with a distribution on X with compact support is an E-distribution on X given by

$$\int_X f(x)\, dT * S(x) = \int_X \left(\int_X f(xy)\, dS(y) \right) dT(x).$$

If $x \in X$ we denote by $\varepsilon(x)$ the Dirac distribution in x.

The intertwining forms of two induced representations are related to the distributions by the following theorem due to F. Bruhat [**1**].

THEOREM 4. *Let G be a l.c.t.d. unimodular group. Let H_1 and H_2 be two closed subgroups of G, and δ_i the module of H_i. Let τ_i be a smooth representation of H_i in the vector space E_i, π_i be the induced representation $\mathrm{Ind}_{H_i}^G \tau_i$ in the space $V_i = \mathcal{S}(G; E_i; \tau_i)$ and p_i be the canonical map $\mathcal{S}(G; E_i) \to V_i$ defined in* (1.1).

Then the space of all intertwining forms I of π_1 and π_2 is isomorphic to the space of $E_1 \otimes E_2$-distributions T on G such that

$$\varepsilon(h_1) * T * \varepsilon(h_2^{-1}) = (\delta_1(h_1)\, \delta_2(h_2))^{1/2}\, T\circ[\tau_1(h_1) \otimes \tau_2(h_2)]$$

where $h_i \in H_i$. The correspondence between I and T is given by

$$I(p_1(f_1), p_2(f_2)) = \int_G dg_2 \int_G f_1(g_1 g_2) \otimes f_2(g_2)\, dT(g_1)$$

where $f_i \in \mathcal{S}(G; E_i)$ and dg is a Haar measure on G.

4. Proof of Theorem 2. Let λ_θ be the induced representation $\mathrm{Ind}_U^G \theta$ in the space L_θ.

One needs first the following proposition:

PROPOSITION. *Let π be a smooth representation of G in E.*

(i) *Let $W \in \mathrm{Hom}_G(\pi, r_\theta)$ and A_W be the mapping from L_θ to $\check{E}$ given by*

$$\langle x, A_W(f)\rangle = \int_{U\backslash G} f(g)\, W(x)(g)\, dg$$

where $f \in L_{\bar\theta}$, and dg is an invariant measure on $U\backslash G$. Then A_W is an intertwining operator from $\lambda_{\bar\theta}$ to $\check\pi$.

(ii) *The mapping $W \to A_W$ from* $\mathrm{Hom}_G(\pi, r_\theta)$ *to* $\mathrm{Hom}_G(\lambda_{\bar\theta}, \check\pi)$ *is an isomorphism.*

Let λ'_{θ_0} be the induced representation $\mathrm{Ind}^M_{M\cap w_0Uw_0^{-1}}\theta_0$ in the space L'_{θ_0}. Then to prove Theorem 2 it is enough to prove that $\mathrm{Hom}_G(\lambda_{\bar\theta}, \check\pi)$ is isomorphic to $\mathrm{Hom}_M(\lambda'_{\bar\theta_0}, \check\tau)$. Clearly $\mathrm{Hom}_G(\lambda_{\bar\theta}, \check\pi')$ is isomorphic to the space of all intertwining forms of $\lambda_{\bar\theta}$ and π', and by Theorem 4 to the space of F-distributions T on G such that

$$\varepsilon(u) * T * \varepsilon(p^{-1}) = \delta_P(p)^{1/2}\, \bar\theta(u)\, T \circ [\tau(p)] \tag{4.1}$$

where $p \in P$ and $u \in U$. F is the space of the representation τ and δ_P is the module of P. If S is a distribution on the double coset UgP which verifies (4.1), then $S=0$ unless $UgP = Uw_0P$. As Uw_0P is open in G, this proves that the distribution T is determined by its restriction to Uw_0P. Then consider Uw_0P as a homogeneous space under the group $U \times P$. Let T' be the distribution on $U \times P$ given by

$$\int_{U\times P} \varphi(u, p)\, dT'(u, p) = \int_{Uw_0P} \left(\int_{U\cap w_0Pw_0^{-1}} \varphi(uu_1, w_0^{-1}u_1^{-1}w_0p)\, d\dot u_1 \right) dT(uw_0p)$$

where $\varphi\ \mathscr{S}(U \times P; F)$ and $d\dot u$ is a Haar measure on $U \cap w_0Pw_0^{-1}$. Then T' decomposes as a tensor product

$$dT'(u, mv) = \bar\theta(u)\, du \otimes \delta_P(m)^{-1/2}\, dQ(m)\, dv$$

where du, dv are Haar measures on U, V, and Q verifies

$$\varepsilon(u) * Q * \varepsilon(m^{-1}) = \bar\theta_{w_0}(u)\, Q \circ [\tau(m)], \qquad u \in w_0Uw_0^{-1} \cap M,\ m \in M.$$

By Theorem 4, Q corresponds to an element in $\mathrm{Hom}_M(\lambda'_{\bar\theta_0}, \check\tau)$. The correspondence we have defined between $\mathrm{Hom}_G(\lambda_{\bar\theta}, \check\pi')$ and $\mathrm{Hom}_M(\lambda'_{\bar\theta_0}, \check\tau)$ is actually a bijection.

5. Proof of Theorem 3. Let E be the space of the representation π.

Step 1. With the same hypothesis as in Theorem 3, one has the following relation:

$$\dim \mathrm{Hom}_G(\lambda_\theta, \pi) \cdot \dim \mathrm{Hom}_G(\lambda_{\bar\theta}, \check\pi) \leqq 1.$$

Suppose the product on the left-hand side is not zero and let a (resp., a') be a nonzero element in $\mathrm{Hom}_G(\lambda_\theta, \pi)$ (resp., $\mathrm{Hom}_G(\lambda_{\bar\theta}, \check\pi)$). Then the bilinear form on $L_\theta \times L_{\bar\theta}$ given by

$$I(v, v')=\langle a(v), a'(v')\rangle, \qquad v\in L_\theta,\ v'\in L_{\bar\theta},$$

is an intertwining form of λ_θ and $\lambda_{\bar\theta}$. By Theorem 4 it corresponds to a distribution T on G such that

$$\varepsilon(u_1)*T*\varepsilon(u_2^{-1})=\theta(u_1u_2)\,T \quad \text{where } u_i\in U. \tag{5.1}$$

There exists one and only one automorphism σ of G such that $\sigma(U)=U$, $\theta\circ\sigma=\theta$, and $\sigma(h)=w_0hw_0^{-1}$ if $h\in H$. By studying distributions on the double cosets BgB, which verify (5.1), one can prove that T is invariant under σ. If $v\in L_\theta$ let $\hat v$ be the element of $L_{\bar\theta}$ such that $\hat v(g)=v(\sigma(g^{-1}))$. The invariance of T under σ implies that $I(v_1, \hat v_2)=I(v_2, \hat v_1)$. From this relation one gets that $a(x)=0$ if and only if $a'(\hat v)=0$. Thus the kernel of a is actually independent of the choice of a in $\mathrm{Hom}_G(\lambda_\theta, \pi)$. As π is irreducible, one concludes by Schur's lemma that the dimension of $\mathrm{Hom}_G(\lambda_\theta, \pi)$ is one. In the same way one gets $\dim \mathrm{Hom}_G(\lambda_{\bar\theta}, \check\pi)=1$.

Step 2. (Proof of Theorem 3 when π is supercuspidal). Assume in this step that π is supercuspidal, i.e., the matrix coefficients of π are functions on G with compact support modulo Z. Let ω be the central exponent of $\pi: \omega(z)\cdot x=\pi(z)\cdot x$ if $z\in Z$ and $x\in E$. ω can be extended in a unique way into a homomorphism χ of G into the multiplicative group of the complex numbers. Then $\check\pi$ is equivalent to the representation $g\mapsto\bar\pi(g)\,|\chi(g)|^{-2}$ where $\bar\pi$ is the complex conjugate of π. From this relation it is easy to see that $\dim \mathrm{Hom}_G(\lambda_\theta, \pi)=\dim \mathrm{Hom}_G(\lambda_{\bar\theta}, \check\pi)$. Then the theorem follows in this case by Step 1 and the Proposition.

Step 3. (Proof of Theorem 3). In the general case π is a subrepresentation of $\mathrm{Ind}_P^G\tau$ where P is a parabolic subgroup of G containing H, with unipotent radical V, and τ is a supercuspical representation of P/V. Then the theorem follows by Step 2 applied to τ and Theorem 4.

References

1. F. Bruhat, *Sur les représentations induites des groupes de Lie* (Thèse), Bull. Soc. Math. France **84** (1956), 97–205. MR **18**, 907.

2. I. M. Gel'fand and D. A. Kajdan, *Representations of the group GL(n, K) where K is a local field*, Moscow, 1971; Funkcional. Anal. i Prilozen. **6** (1972), 13–44.

3. Harish-Chandra, *Harmonic analysis on reductive p-adic groups*, Proc. Sympos. Pure Math., vol. 26, Providence, R.I., 1974, pp. 167–192.

Ecole Normale Supérieure
Paris 5ᵉ, France

SPECTRA OF DISCRETE SUBGROUPS

RAMESH GANGOLLI*

1. Problems concerning the spectrum of Γ. Let G be a connected semisimple Lie group with finite center, Γ a discrete subgroup with G/Γ compact, U the (unitary) representation of G on $L_2(G/\Gamma)$ and $[U]$ the equivalence class of U. For each class $\omega \in \mathscr{E}(G)$, the unitary dual of G, let $n_\Gamma(\omega)$ be the multiplicity of ω in $[U]$. It is well known [**6**] that $[U] = \sum_{\omega \in \mathscr{E}(G)} n_\Gamma(\omega)\,\omega$, a discrete direct sum, where $n_\Gamma(\omega) < \infty$ for all ω, and is >0 only for a countable number of ω. The function n_Γ may be called the *spectral* (multiplicity) *function* of Γ. Knowledge of n_Γ is equivalent to knowledge of $[U]$. The following problems arise naturally.

PROBLEM 1. *For a given Γ determine those $\omega \in \mathscr{E}(G)$ for which $n_\Gamma(\omega) > 0$.*

If, for example, ω is described by parameters involving the structure of G, one seeks conditions on these parameters that ensure $n_\Gamma(\omega) > 0$.

PROBLEM 2. *Compute $n_\Gamma(\omega)$ in terms of the parameters that describe ω.*

Obviously, Problem 2 is a quantitative version of Problem 1.

PROBLEM 3. *To what extent does n_Γ determine Γ?*

For example, if Γ, Γ' are two subgroups of G such as above, then does $n_\Gamma = n_{\Gamma'}$ imply that Γ is isomorphic to Γ'? More generally, what group theoretic invariants

AMS (MOS) subject classifications (1970). Primary 22E45, 22E55.

Key words and phrases. Representations of semisimple Lie groups, harmonic analysis, discrete subgroups, spectra, uniform subgroups, multiplicity formulas, trace formula, spherical function, automorphic form.

* This work was supported by the National Science Foundation.

of Γ are determined by n_Γ? This problem is analogous to the duality theorems of Pontryagin and Tanaka.

Our purpose in this article is to report on the current status of these problems.

2. The Selberg trace formula. The trace formula of Selberg ([**18**], [**19**]) has proved to be the major tool for the study of these problems. Let $f \in L_1(G)$. We say f is *admissible* if (i) the series $\sum_{\gamma \in \Gamma} f(x\gamma y^{-1})$ converges everywhere on $G \times G$ to a continuous function $K_f(x, y)$ of the pair (x, y), and (ii) the operator $U(f) = \int_G f(x)\, U(x)\, dx$, which is an integral operator on $L_2(G/\Gamma)$ with kernel K_f, is of trace class on $L_2(G/\Gamma)$. When f is such a function, the trace of $U(f)$ can be computed in two different ways. The equality of the two expressions is the trace formula of Selberg. On the one hand,

$$\operatorname{Trace} U(f) = \sum_{\omega \in \mathscr{E}(G)} n_\Gamma(\omega) \operatorname{Trace} U_\omega(f),$$

where U_ω is a representation of class ω. On the other hand,

$$\operatorname{Trace} U(f) = \int_D K_f(x, x)\, dx$$

where D is a fundamental domain for Γ in G. As in [**18**], [**19**], one rewrites this last integral to get the trace formula

$$\sum_{\omega} n_\Gamma(\omega)\, \Theta_\omega(f) = \sum_{\gamma \in C_\Gamma} \operatorname{vol}(D_\gamma)\, J_\gamma(f). \tag{1}$$

Here, we have put $\Theta_\omega(f) = \operatorname{Trace} U_\omega(f)$, C_Γ is a complete set of representatives in Γ for the G-conjugacy classes of elements of Γ, $J_\gamma(f) = \int_{G/G_\gamma} f(x\gamma x^{-1})\, d\dot{x}$, G_γ being the centralizer of γ in G, $d\dot{x}$ the invariant measure on G/G_γ. Finally, D_γ is a fundamental domain for the centralizer Γ_γ of γ in G_γ, $\operatorname{vol}(D_\gamma)$ its volume. The measures dx_γ on G_γ and $d\dot{x}$ on G/G_γ are normalized so $dx = d\dot{x}\, dx_\gamma$.

The use of (1) depends on one's ability to compute $\Theta_\omega(f)$ and $J_\gamma(f)$ for admissible f. Naturally, the larger the class of admissible functions f, the more effectively one can use the formula.

3. Multiplicities of the discrete series. We use the notation of [7]. Let $\mathscr{E}_2(G)$ be the discrete series for G, and assume $\mathscr{E}_2(G) \neq \emptyset$. Let $\mathscr{E}_1(G) \subset \mathscr{E}_2(G)$ be the set of integrable classes. Let T be a compact Cartan subgroup of G, K a maximal compact subgroup containing T. By the work of Harish-Chandra [**8**], we know that $\mathscr{E}_2(G)$ is parametrized by the orbits of the Weyl group W_G of (G, T) acting on the lattice of regular integral linear forms on the Lie algebra of T. Given $\omega \in \mathscr{E}_2(G)$, one wishes to compute $n_\Gamma(\omega)$ in terms of the linear form Λ attached to ω via this correspondence. Letting f be a K-finite matrix-coefficient of ω, so that $f \in \mathscr{C}(G)$, the idea is to apply (1) to such an f.

For such f, it follows from the Schur orthogonality relations and the Harish-Chandra-Plancherel theorem that $\Theta_{\omega'}(f)=0$ for any tempered class $\omega' \neq \omega$. Thus on the left side of (1), $\Theta_{\omega'}(f)=0$ for every discrete class ω', as well as for any ω' that occurs in the Plancherel formula for G. But $\Theta_{\omega'}(f)$ may not vanish for an exceptional (i.e., nontempered) class ω. On the right side of (1), we see as a consequence of the Selberg principle [**8**] that $J_\gamma(f)=0$ unless γ is elliptic. Thus the sum on the right is finite. If one can show that f is admissible, and compute $J_\gamma(f)$ for elliptic γ, and $\Theta_{\omega'}(f)$ for exceptional ω', one has a handle on $n_\Gamma(\omega)$.

Langlands has shown ([**13**], [**14**]) that if $\omega \in \mathscr{E}_1(G)$ (and not merely $\omega \in \mathscr{E}_2(G)$), then $\Theta_{\omega'}(f)=0$ for every class $\omega' \neq \omega$. Also, $J_\gamma(f)$ can be computed for such f in terms of the parameter Λ attached to ω. Finally, $\Theta_\omega(f)$ is computable in terms of the formal degree d_ω of ω. Thus $n_\Gamma(\omega)$ is computable in these terms provided $\omega \in \mathscr{E}_1(G)$. It follows from this formula that *if Γ has no elliptic elements then* $n_\Gamma(\omega) = \operatorname{vol}(G/\Gamma)\, d_\omega$ for each $\omega \in \mathscr{E}_1(G)$. In particular, then every such class occurs in U. Thus both Problems 1 and 2 are answered for such ω. Note, however, that Problem 1 is here answered by first answering Problem 2. A direct qualitative criterion on ω which implies $n_\Gamma(\omega) > 0$ seems to be unknown.

If $\omega \in \mathscr{E}_2(G) - \mathscr{E}_1(G)$, one must, in general, expect that some terms $\Theta_{\omega'}(f)$ arising from exceptional classes will be nonzero on the left side of (1). For general G, precise knowledge about exceptional classes is not available at present to deal with these terms. However, if G is $SL(2, \boldsymbol{R})$, one knows enough about the exceptional classes to push this method through. This has been done by Langlands (unpublished) to evaluate $n_\Gamma(\omega)$ for each $\omega \in \mathscr{E}_2(G)$.

By using a different method, viz. the Atiyah-Bott-Lefschetz fixed point formula applied to an elliptic complex arising from a suitable vector bundle based on the space $T \backslash G/\Gamma$, Schmid [**17**] has shown that the multiplicity formula of Langlands is valid under the hypothesis that Λ is "sufficiently far away" from the walls of the Weyl chamber. Schmid's method does not tell us how far one must go (see, however, [**16**]). Results of Trombi and Varadarajan [**22**] show that "Λ sufficiently far away" $\Rightarrow \omega \in \mathscr{E}_1(G)$. Thus it is not clear a priori that Schmid's results extend the domain of validity of the Langlands formula. The sharpest results using this approach are those of Hotta and Parthasarathy [**11**]. They give a simple condition of Λ ensuring the validity of the Langlands formula for ω. It follows from their work that for certain G there can exist classes $\omega \in \mathscr{E}_2(G) - \mathscr{E}_1(G)$ for which the *same* formula holds.

If ω is not in $\mathscr{E}_2(G)$, answers to Problems 1 and 2 are substantially unknown. If $G = SL(2, \boldsymbol{R})$, one can show, using the Paley-Wiener theorem, *that only a finite number of exceptional classes can occur in U, and infinitely many principal series classes do occur.* Neither of these facts can be proved in general at present, though one expects them to be true. As to Problem 2, nothing is known for $\omega \notin \mathscr{E}_2(G)$ even in the case $G=SL(2, \mathbf{R})$. (See, however, the paper of Gelbart [**4**], and also these PROCEEDINGS.)

4. The type 1 spectrum. Instead of studying the entire spectrum, one may study certain pieces of it. For any $\delta \in \mathscr{E}(K)$, let $\mathscr{E}(G; \delta)$ be the subset of $\mathscr{E}(G)$ consisting of classes of type δ, and let $n_{\Gamma,\delta}$ be the restriction of the function n_Γ to $\mathscr{E}(G; \delta)$. The data $\{\omega, n_\Gamma(\omega)\}$ with ω running over $\mathscr{E}(G; \delta)$ such that $n_{\Gamma,\delta}(\omega)>0$ may be called the type δ spectrum of Γ. An important special case is when $\delta=1$, the trivial class in $\mathscr{E}(K)$. We then speak of the type 1 spectrum. Because of [**8**], no class $\omega \in \mathscr{E}_2(G)$ can occur in the type 1 spectrum.

For $\delta \in \mathscr{E}(K)$ let ξ_δ be the corresponding idempotent in the group algebra of K. Let f be admissible, and assume that $\bar{\xi}_\delta * f = f * \bar{\xi}_\delta = f$. Then $\Theta_\omega(f)=0$ unless ω is of type δ. Thus the left side of (1) collapses to $\sum_{\omega \in \mathscr{E}(G;\delta)} n_\Gamma(\omega)\, \Theta_\omega(f)$.

The Harish-Chandra-Fourier transform $\Theta_\omega(f)$ may sometimes be computed when f is restricted as above. This makes it feasible to apply (1) to such f. For example, such is patently the case when f is spherical. If ω is of type 1, let $\phi_{\lambda_\omega}(x) = \int_G \exp(\lambda_\omega - \varrho)(H(xk))\, dk$ be the elementary positive definite spherical function attached to the class ω. (This parametrization is that of [**21**].) For admissible spherical f, $\Theta_\omega(f)$ can be shown to equal $\hat{f}(\lambda_\omega) = \int_G f(x)\, \phi_{-\lambda_\omega}(x)\, dx$, so (1) reduces to

$$\sum_{\omega \in \mathscr{E}(G;1)} n_\Gamma(\omega) \hat{f}(\lambda_\omega) = \sum_{\gamma \in C} \operatorname{vol}(D_\gamma)\, J_\gamma(f). \tag{2}$$

The advantage of (2) is that the spherical transform $f \to \hat{f}$ is well understood for spherical f, due to the work of Harish-Chandra [**9**] and Trombi-Varadarajan [**21**].

Let Ω be the Casimir operator of G and let Ω_ω be the scalar by which it acts in any representation of the class ω. One can show that if ω is type 1, then Ω_ω is real and $\leqq 0$. By using the results of [**21**], one can show [**3**] that *there exists an integer* $d>0$ *such that* $\sum_{\omega \in \mathscr{E}(G;1)} n_\Gamma(\omega)\,(1-\Omega_\omega)^{-d} < \infty$. In particular, the real numbers Ω_ω have no finite point of accumulation as ω ranges over the type 1 classes occurring in U.[1]

This result has the following consequence. *If the rank of G/K is* 1, *then only finitely many exceptional classes ω can occur in the type* 1 *spectrum, i.e.*, $\{\omega;\ \omega$ *exceptional*, $n_{\Gamma,1}(\omega)>0\}$ *is finite*. More generally, (if rank $G/K>1$) one can only con-

[1] At the conference we have learned from Harish-Chandra a different, simpler proof of these results. Harish-Chandra's proof applies to the type δ situation as well.

clude that the type 1 spectrum of Γ contains only finitely many ω such that λ_ω is real valued. The point is that if rank $G/K=1$ then every exceptional class ω of type 1 has this property, due to Kostant [**12**].

Let $\mathscr{I}^1(G)$ be the space of spherical functions in $\mathscr{C}(G)\cap L_1(G)$. It is shown in [**21**] that $\mathscr{I}^1(G)$ is a Fréchet space under the seminorms $\nu_{D,r}: f\to \nu_{D,r}(f)$, where

$$\nu_{D,r}(f)=\sup_{x\in G}|Df(x)|\ \Xi(x)^{-2}(1+\sigma(x))^r.$$

Here D is any invariant differential operator, r any integer, and Ξ, σ are as in [**7**]. One can show [**3**] that *every* $f\in\mathscr{I}^1(G)$ *is admissible, and the map* $f\to \operatorname{Trace} U(f)$ *is continuous in this topology*. In particular, each $f\in C_c^\infty(G)\cap\mathscr{I}^1(G)$ is admissible. The functional $f\to\operatorname{Trace} U(f)$ is not continuous on $\mathscr{C}(G)$, and its continuity properties are tied up with the presence of elliptic elements in Γ.

The above results give partial information regarding Problem 1. Our understanding of Problem 2 is very meager. We may note in closing the following result which may be deduced from McKean [**15**]. *If* $G=SL(2,\boldsymbol{R})$ *and if* Γ *has no elliptic elements, then* $n_\Gamma(\omega)=0$ *for every exceptional class* ω *except the trivial class in* $\mathscr{E}(G)$. This is due to the fact that in this case, every exceptional class is spherical.

5. The spectrum and the structure of Γ. Let $r\geqq 0$, and define the Weyl function N_Γ by $N_\Gamma(r)=\sum n_\Gamma(\omega)$, the sum running over $\{\omega\in\mathscr{E}(G;1), |\Omega_\omega|\leqq r\}$. We have seen in §4 that $N_\Gamma(r)$ is finite for each r. In [**5**], Gel'fand conjectured that as $r\to\infty$, $N_\Gamma(r)\sim\operatorname{vol}(G/\Gamma)\, C_G r^{n/2}$ where $n=\dim(G/K)$, and C_G is a constant depending only on G.

When $G=SL(2,\boldsymbol{R})$, this was verified by Tanaka [**20**] who deduced from it: *If* Γ, Γ' *are discrete subgroups of* $SL(2,\boldsymbol{R})$ *with compact quotients, then* $n_{\Gamma,1}=n_{\Gamma',1}$ *implies that* Γ *and* Γ' *are isomorphic as abstract groups*. Indeed, $n_{\Gamma,1}=n_{\Gamma',1}$ implies that $\operatorname{vol}(G/\Gamma)=\operatorname{vol}(G/\Gamma')$, and $\operatorname{vol}(G/\Gamma)$ determines, via Gauss-Bonnet applied to $K\backslash G/\Gamma$, the genus of this space and the orders of the elements of Γ of finite order.

The asymptotic formula $N_\Gamma(r)\sim\operatorname{vol}(G/\Gamma)\, C_G r^{n/2}$ is known to be correct if G is complex [**2**], and if G is classical with rank $G/K=1$ (cf. Eaton [**1**]). If Γ has no elliptic elements, the author and G. Warner have shown (unpublished) that the asymptotic formula is correct for any G.

The validity of the asymptotic formula clearly implies that if $n_{\Gamma,1}=n_{\Gamma',1}$ then $\operatorname{vol}(G/\Gamma)=\operatorname{vol}(G/\Gamma')$. The equality of the volumes has some interesting consequences. For example, if $K\backslash G$ is Hermitian, and Γ acts freely on it, then because of the Hirzebruch proportionality principle [**10**, p. 162], the manifolds $K\backslash G/\Gamma$ and $K\backslash G/\Gamma'$ have the same χ_y-characteristic, and, for each r, the dimension of the space of Γ-automorphic forms of weight r is equal to the corresponding dimension of Γ'-automorphic forms. Thus in this case we can conclude that a knowledge of $n_{\Gamma,1}$ determines $n_\Gamma(\omega)$ for $\omega\in\mathscr{E}_1(G)$!

REFERENCES

1. T. Eaton, Thesis, University of Washington, Seattle, Wash., 1973.

2. R. Gangolli, *Asymptotic behaviour of spectra of compact quotients of certain symmetric spaces*, Acta Math. **121** (1968), 151–192. MR **39** #360.

3. R. Gangolli and G. Warner, *On Selberg's trace formula*, Japan. J. Math. (to appear).

4. S. Gelbart, *Notes on Jacquet-Langlands' theory*, Cornell University, Ithaca, N.Y., 1972.

5. I. M. Gel'fand, *Automorphic functions and the theory of representations*, Proc. Internat. Congr. Mathematicians (Stockholm, 1962), Inst. Mittag-Leffler, Djursholm, 1963, pp. 74–85. MR **31** #273.

6. I. M. Gel'fand, et al., *Generalized functions*, Vol. 6. *Theory of representations and automorphic functions*, "Nauka", Moscow, 1966; English transl., Saunders, Philadelphia, Pa., 1969. MR **36** #3725; **38** #2093.

7. Harish-Chandra, *Harmonic analysis on semisimple Lie groups*, Bull. Amer. Math. Soc. **76** (1970), 529–551. MR **41** #1933.

8. ———, *Discrete series for semisimple Lie groups*. II. *Explicit determination of the characters*, Acta Math. **116** (1966), 1–111. MR **36** #2745.

9. ———, *Spherical functions on a semisimple Lie group*. I, II, Amer. J. Math. **80** (1958), 241–310, 553–613. MR **20** #925; **21** #92.

10. F. Hirzebruch, *Neue topologische Methoden in der algebraischen Geometrie*, Ergebnisse der Mathematik und ihrer Grenzgebiete, Heft 9, Springer-Verlag, Berlin, 1956; English transl., Die Grundlehren der math. Wissenschaften, Band 131, Springer-Verlag, New York, 1966. MR **18**, 509; **34** #2573.

11. R. Hotta and R. Parthasarathy, *A geometric meaning of the multiplicity of integrable discrete classes in* $L_2(G/\Gamma)$ (to appear).

12. B. Kostant, *On the existence and irreducibility of certain series of representations*, Bull. Amer. Math. Soc. **75** (1969), 627–642. MR **39** #7031.

13. R. P. Langlands, *The dimension of spaces of automorphic forms*, Amer. J. Math. **85** (1963), 99–125. MR **27** #6286.

14. a) ———, *The volume of the fundamental domain for some arithmetical subgroups of Chevalley groups*, Algebraic Groups and Discontinuous Subgroups, Proc. Sympos. Pure Math., vol. 9, Amer. Math. Soc., Providence, R.I., 1966, pp. 143–148. MR **35** #4226.

b) ———, *Dimension of spaces of automorphic forms*, Algebraic Groups and Discontinuous Subgroups, Proc. Sympos. Pure Math., vol. 9, Amer. Math. Soc., Providence, R.I., 1966, pp. 253–257. MR **35** #3010.

15. H. P. McKean, Comm. Pure Appl. Math. **25** (1972), 225–246.

16. M. S. Narasimhan and K. Okamoto, *An analogue of the Borel-Weil-Bott theorem for hermitian symmetric pairs of non-compact type*, Ann. of Math. (2) **91** (1970), 486–511. MR **43** #419.

17. W. Schmid, *On a conjecture of Langlands*, Ann. of Math. (2) **93** (1971), 1–42. MR **44** #4149.

18. A. Selberg, *Harmonic analysis and discontinuous groups in weakly symmetric Riemannian spaces with applications to Dirichlet series*, J. Indian Math. Soc. **20** (1956), 47–87. MR **19**, 531.

19. T. Tamagawa, *On Selberg's trace formula*, J. Fac. Sci. Univ. Tokyo Sect. I **8** (1960), 363–386. MR **23** #A958.

20. S. Tanaka, *Selberg's trace formula and spectrum*, Osaka J. Math. **3** (1966), 205–216. MR **36** #312.

21. P. Trombi and V. S. Varadarajan, *Spherical transforms on semisimple Lie groups*, Ann. of Math. (2) **94** (1971), 246–303. MR **44** #6913.

22. ———, *Asymptotic behaviour of eigenfunctions on a semisimple Lie group; the discrete spectrum*, Acta Math. **129** (1972), 237–280.

SUNY AT ALBANY

UNIVERSITY OF WASHINGTON, SEATTLE

AN EXAMPLE IN THE THEORY OF AUTOMORPHIC FORMS

STEPHEN GELBART*

Let K denote a separable quadratic extension of the global field k. Using the Weil representation one can construct a class of automorphic forms for $GL(2, k)$ indexed by appropriate grossencharacters of K (see Jacquet-Langlands [**2**] and Shalika-Tanaka [**4**]). This construction generalizes some classical results of Hecke and Maass (cf. [**3**]), and it is roughly the purpose of this note to explain how. More precisely, we shall prove that the construction for *real* quadratic fields yields explicit examples of principal series representations of $SL(2, \boldsymbol{R})$ which occur discretely in $L^2(\Gamma\backslash SL(2, \boldsymbol{R}))$ for Γ some congruence subgroup of $SL(2, \boldsymbol{Z})$.

Fix k equal to $\boldsymbol{Q}$, and let $\boldsymbol{A}$ denote the adeles of $\boldsymbol{Q}$. If K is any real quadratic field $\boldsymbol{Q}(d^{1/2})$, we let v denote an arbitrary prime of K. To each grossencharacter $\omega=\prod_v \omega_v$ of K we attach a representation $\pi(\omega)=\bigotimes_p \pi_p$ of $GL(2, \boldsymbol{A})$ as follows. Suppose p is a prime of $\boldsymbol{Q}$ which splits in K with divisors v and v'. Then we set π_p equal to the representation of $GL(2, \boldsymbol{Q}_p)$ induced from the corresponding characters ω_v and $\omega_{v'}$ of $\boldsymbol{Q}_p^\times$. On the other hand, if p lies under a single prime v of K, then K_v is a quadratic extension of $\boldsymbol{Q}_p$, and ω_v determines an irreducible component $\pi(\omega_v)$ of the Weil representation of $GL(2, \boldsymbol{Q}_p)$ defined by K_v. In this case we set π_p equal to $\pi(\omega_v)$. In either case π_p is almost always a class 1 representation, since ω_v and K_v are almost always unramified. Therefore the tensor product $\bigotimes_p \pi_p$ indeed determines an irreducible admissible representation of $GL(2, \boldsymbol{A})$.

Suppose now η is the grossencharacter of $\boldsymbol{Q}$ determined by the restriction of $\pi(\omega)$ to the center of $GL(2, \boldsymbol{A})$. Then let $L_0^2(\eta)$ denote the Hilbert space of measurable functions on $GL(2, \boldsymbol{Q})\backslash GL(2, \boldsymbol{A})$ such that

(i) $\varphi(zg)=\eta(a)\,\varphi(g)$ for $z\in Z_{\boldsymbol{A}}=\{\left[\begin{smallmatrix} a & 0 \\ 0 & a\end{smallmatrix}\right]\}$;

AMS (MOS) subject classifications (1970). Primary 10D05; Secondary 22E55.

* Supported by NSF grant GP-28251.

(ii) $\int_F |\varphi(g)|^2 \, dg < \infty$, where $F = Z_A \cdot GL(2, Q)\backslash GL(2, A)$; and

(iii) $\int_{Q\backslash A} \varphi(\left[\begin{smallmatrix} 1 & x \\ 0 & 1 \end{smallmatrix}\right] g)\, dx = 0, g \in GL(2, A)$.

It is well known [2] that the natural representation of G_A on $L_0^2(\eta)$ given by right translation is the direct sum of irreducible admissible representations of $GL(2, A)$ each occurring with multiplicity one. What is remarkable is that $\pi(\omega)$ is one of these distinguished representations *provided* there is no grossencharacter μ of Q such that $\omega = \mu \circ N_{K/Q}$. (Cf. [2, Proposition 12.2]; the crucial point is that the L-function attached to $\pi(\omega)$ coincides with the Hecke L-function belonging to K and ω; $N_{K/Q}$ denotes the idele norm map from K to Q.)

To make our explicit computation as simple as possible we fix $K = Q(2^{1/2})$. With this choice of discriminant only the prime $p = 2$ ramifies in K. Moreover, since K is real, the infinite component of $\pi(\omega)$ will be a principal series representation for ω *any* grossencharacter of K. However, to insure that $\pi(\omega)$ occurs in $L_0^2(\eta)$ we must construct ω so that it does not arise from a grossencharacter of Q. Thus we set

(1) $$\omega(a) = \omega((a_v)) = \prod_{j=1}^{2} |a_{\infty_j}|^{it_j} \prod_{v \text{ finite}} |a_v|_v^{is_v}$$

with $t_1 \neq t_2$, and ∞_j the place at infinity corresponding to the imbedding $a + b(2^{1/2}) \mapsto a + (-1)^{j+1} b(2^{1/2})$. A simple computation then shows that this character is trivial on K^x only if, for some $k \in Z$,

(2) $$s = t_2 - t_1 = 2k\pi/\log(2^{1/2} - 1).$$

On the other hand, having chosen t_1, t_2 to satisfy (2), there does exist a grossencharacter ω of K of the form (1).

Henceforth, we fix $k = 1$ in (2), and assume $t_1 + t_2 = 0$. The conclusion is that $\pi(\omega) = \pi_{it_1, it_2} \otimes (\bigotimes_p \pi_p)$ occurs in $L_0^2(\eta)$. Moreover, since $t_1 + t_2 = 0$, π_{it_1, it_2} is trivial on the center of $GL(2, R)$.

We now wish to conclude that the representation π_{is} of $SL(2, R)$ occurs discretely in $L^2(\Gamma\backslash SL(2, R))$ for Γ some Fuchsian group. We observe first that by our choice of ω and K the representation π_p is class 1 for all $p \neq 2$. On the other hand, π_2 is admissible. Therefore it follows (by Theorem 1 of [1] or by direct verification) that the restriction of π_2 to $K_2^N = \{\left[\begin{smallmatrix} a & b \\ c & d \end{smallmatrix}\right] \in K_p : c = 0 \pmod N\}$ contains the identity representation for N sufficiently large. Consequently the subspace of the representation space of $\pi(\omega)$ consisting of functions invariant on the right by $K_0 = K_2^N \times \prod_{p \neq 2} K_p$ is nonempty and equivalent (as a $GL(2, R)$-module) to a *finite* number of copies of π_{it_1, it_2} (the finiteness follows from admissibility). Pick one such copy and call its space $H(t_1, t_2)$. The restriction of π_{it_1, it_2} to $SL(2, R)$ is irreducible and equivalent to the induced representation π_{is}, where

$$s = 2\pi/\log(2^{1/2} - 1).$$

Then since $G_A = GL(2, \boldsymbol{Q})\, GL^+(2, \boldsymbol{R})\, K_0$, $H(t_1, t_2)$ is naturally isomorphic to an irreducible subspace of $L^2(\Gamma\backslash SL(2, \boldsymbol{R}))$ with

$$\Gamma = GL(2, \boldsymbol{Q}) \cap GL^+(2, \boldsymbol{R})\, K_0 = \Gamma_0(N).$$

Thus we have

THEOREM. *Suppose* $s = 2\pi/\log(2^{1/2} - 1)$ *and* π_{is} *is the representation of* $SL(2, \boldsymbol{R})$ *induced from the character* $\left[\begin{smallmatrix} a & b \\ 0 & a^{-1} \end{smallmatrix}\right] \to |a|^{is}$. *Then* π_{is} *occurs discretely in* $L^2(\Gamma_0(N)\backslash SL(2, \boldsymbol{R}))$ *with* N *as above.*

Concluding remarks. (1) From the computation just sketched it seems likely that π_{is} can occur *more* than once in $L^2(\Gamma\backslash SL(2, \boldsymbol{R}))$. For $\Gamma = SL(2, \boldsymbol{Z})$ this is conjectured *not* to be the case.

(2) The occurrence of π_{is} in $L^2(\Gamma\backslash SL(2, \boldsymbol{R}))$ is equivalent to the existence of functions $g(z) = g(x, y)$ defined in the upper half-plane and such that

(i) $g((az+b)/(cz+d)) = g(z)$, for $\left[\begin{smallmatrix} a & b \\ c & d \end{smallmatrix}\right] \in \Gamma$;

(ii) $-y(\partial^2/\partial x^2 + \partial^2/\partial y^2)\, g(x, y) = (s^2 + \frac{1}{4})\, g(x, y)$; and

(iii) $g(x+iy) = g(x, y) \to 0$ uniformly in x as $y \to +\infty$.

Such functions were first studied systematically by Maass in [3]. In fact, without using the machinery of adeles and representation theory, Maass seems to have constructed examples of such functions for special choices of s and Γ.

(3) It goes without saying that it would be of interest to construct similar functions corresponding to *complementary* series representations of $SL(2, \boldsymbol{R})$.

REFERENCES

1. W. Casselman, *On some results of Atken and Lehner*, Math. Ann. **201** (1973), 301–314.

2. H. Jacquet and R. P. Langlands, *Automorphic forms on GL(2)*, Lecture Notes in Math., vol. 114, Springer-Verlag, Berlin and New York, 1971.

3. H. Maass, *Über eine neue Art von nicht analytischen automorphen Funktionen und die Bestimmung Dirichletscher Reihen durch Funktionalgleichungen*, Math. Ann. **121** (1949), 141–183. MR **11**, 163.

4. J. Shalika and S. Tanaka, *On the construction of a certain class of automorphic forms*, Amer. J. Math. **91** (1969), 1049–1076.

INSTITUTE FOR ADVANCED STUDY
CORNELL UNIVERSITY

SPECTRAL ANALYSIS OF AUTOMORPHIC FORMS ON RANK ONE GROUPS BY PERTURBATION METHODS

GILLES LACHAUD

Introduction. Let G be a connected simple linear Lie group of rank one over the field of real numbers, K a maximal compact subgroup of G, and Γ a discrete subgroup of G such that the invariant volume of G/Γ is finite. Let π be the regular representation of G into the Hilbert space $L^2(G/\Gamma)$; we extend it in the usual manner to a representation of $L^1(G)$. If we restrict ourselves to the subspace of $L^2(G/\Gamma)$ where the restriction of π to K is trivial (that is to say, to the closed subspace $L^2(K\backslash G/\Gamma)$), it is immediate that the restriction of π to the subalgebra $L^1(G, K)$ of functions on G bi-invariant under K, which is a commutative and involutive Banach convolution algebra, acts on $L^2(K\backslash G/\Gamma)$. It is thus known, by a well-known general theorem of functional analysis, that there exists a spectral measure m on the spectrum of $L^1(G, K)$, which is a closed subset of the complex plane $\mathbf{C}$, such that we have

$$(\pi(\tilde{F})\, f, g) = \int \tilde{F}(\lambda)\ dm_{f,g}(\lambda)$$

where F denotes the Gel'fand transform of $F \in L^1(G, K)$, and $dm_{f,g}$ is the measure associated with the spectral measure m and the elements f, g of $L^2(K\backslash G/\Gamma)$.

Selberg, in [**6**], and after him Langlands [**5**], have proved that the Eisenstein series, which are, in their convergence domain, eigenfunctions of the $\pi(F)$, admit a meromorphic continuation to the whole complex plane, and that their values on the spectrum of $L^1(G, K)$ enable computation of the continuous part (in the sense of Lebesgue decomposition) of the measures $dm_{f,g}$. The purpose of this article is to sketch a proof of this result based on perturbation theory of continuous

AMS (MOS) subject classifications (1970). Primary 32N15, 43A90, 47A40; Secondary 20H10, 43A85.

spectra of selfadjoint operators. This idea, known by Selberg and Gel'fand, has been worked by Faddeev [**1**] in the case of the group $PSL(2, \boldsymbol{R})$.

1. Notations. Let $G=K\cdot A\cdot N$ be an Iwasawa decomposition of G, and let $M=Z(A)\cap K$, where $Z(A)$ is the centralizer of A. We have $Z(A)=M\cdot A$, and we denote by $P=M\cdot A\cdot N$ the canonical minimal and maximal parabolic subgroup. Let α be one of the two indivisible roots [1]; we put, as usual, $2\varrho=[m(\alpha)+m(2\alpha)]\ \alpha$ where $m(\alpha)$ is the multiplicity of the root α, and the same for $m(2\alpha)$. We write

$$\varphi_\lambda(x)=\int_K a(xk)^{i\lambda-\varrho}\, dk$$

for the spherical function of index $\lambda\in\mathfrak{a}_c^*$. (If $x\in G$ we write $x=ka(x)\, n$ with trivial notations, and we agree that $\mathfrak{a}_c^*$ acts exponentially on A.) It is easy to see that $\varphi_\lambda=\varphi_\mu$ if and only if $\lambda=\pm\mu$; on the other hand Helgason and Johnson [**4**] have proved that φ_λ is bounded if and only if $|\operatorname{Im}\lambda|<|\varrho|$. Whenever this integral makes sense, we put

$$\tilde{F}(\lambda)=\int_G F(x)\,\varphi_{-\lambda}(x)\, dx$$

for $F\in L^1(G, K)$. It is well known that every character of $L^1(G, K)$ is of the form $F\to\tilde{F}(\lambda)$, with $|\operatorname{Im}\lambda|<|\varrho|$; and such a character is Hermitian if $\lambda\in\boldsymbol{R}$ or if $i\lambda\in\boldsymbol{R}$; from this we conclude that the spectrum of $L^1(G, K)$ may be identified with the space C/W, where $C=\mathfrak{a}^*\cup i\mathfrak{a}_\varrho^*$, with $\mathfrak{a}_\varrho^*=\{\lambda\in\mathfrak{a}^* \mid |\lambda|<|\varrho|\}$ and $W=\{\pm 1\}$.

Let p denote the canonical map: $G\to K\backslash G/\Gamma$, and if $t\in A$ we note

$$A_t=\{a\in A \mid \alpha(a)<\alpha(t)\}.$$

Garland and Ragunathan have given in [**2**] the precise form of a fundamental domain for G/Γ, under the conditions quoted in the Introduction. They have proved that it was, as in the classical case, the union of a compact set and of a finite number of cusps. If there is only one cusp (a situation to which we confine ourselves here, only in order to simplify the exposition, because the same method holds with minor modification when there are several cusps) we have the following:

THEOREM. *With the preceding notations:* (1) *The subgroup $\Gamma\cap P$ is included in $M\cdot N$.*

(2) *If we define $F=N/\Gamma\cap N$ and $L=\Gamma\cap P/\Gamma\cap N$, then L is finite, F is compact and L acts freely and properly on F by inner automorphisms.*

---

[1] We denote by $\mathfrak{a}$ the Lie algebra of A, by $\mathfrak{a}^*$ its dual, and by $\mathfrak{a}_+^*$ the Weyl chamber $\boldsymbol{R}^+\alpha$.

(3) *Let $E=L\backslash F$; then, if t is sufficiently large, there exists a map v such that the following diagram is commutative:*

$$\begin{array}{ccc} & K\cdot A_t\cdot N & \\ \swarrow & & \searrow^{p} \\ A_t\times E & \xrightarrow{\ v\ } & K\backslash G/\Gamma \end{array}$$

and such that v is a diffeomorphism onto its image S. Furthermore the complementary set of S is compact.

Another notation: If $x\in S$, we denote by $z(x)$ the unique element of A_t such that $v(z(x)\,n)=x$ for some $n\in N$.

2. The representation π'. The method of perturbation theory consists of approaching some operators whose spectral analysis is unknown by another whose spectral analysis is explicitly known. We are going to apply this principle to the operators $\pi(F)$; the first thing we have to do is to find some simple operators "near" to it.

When $F\in C_c(G, K)$, the operator $\pi(F)$ is given by a kernel on G/Γ; in fact we have

$$\pi(F)\,f(x)=\int_G F(xy^{-1})\,f(y)\,dy=\int_{G/\Gamma} F^{\Gamma}(x, y)\,f(y)\,dy$$

for every $x\in G$ and $f\in L^2(K\backslash G/\Gamma)$, where

$$F^{\Gamma}(x, y)=\sum_{\Gamma} F(x\gamma y^{-1}).$$

Now we have the following two lemmas, which can be found in the paper [**3**] of Godement. We recall the Bruhat decomposition: G is the disjoint union of P and PwP.

LEMMA 1. *Let $F\in C_c(G)$. Then the function*

$$(x, y)\;\mapsto\;\sum_{\Gamma\cap PwP} F(x\gamma y^{-1}),$$

viewed as a function on $(A_t\times E)\times(A_t\times E)$, has compact support.

LEMMA 2. *Let $F\in C_c^{\infty}(G)$. Then for every $p\in N$ we have*

$$\left|\sum_{\Gamma\cap P} F(x\gamma y^{-1})-F^N(x, y)\right|\ll\alpha(a_x)^p\,\chi(a_x a_y^{-1})$$

for x and y in $A_t\times E$, where

$$F^N(x, y) = \int_N F(xny^{-1})\, dn$$

and χ is the characteristic function of a compact subset of A_t, if we choose the invariant measure of N such that the volume of E is one.

This last lemma is proved with the help of the Poisson summation formula.

Thus we see that in the cusp our kernel F^Γ behaves approximately like the kernel F^N, which is much simpler. But unfortunately the map

$$F \mapsto F^N \big|_{(A_t \times E) \times (A_t \times E)}$$

is not a representation of the algebra $C_c(G, K)$ (it is something like the Wiener-Hopf operator on the half-line), and in order to use the perturbation machinery, we need an algebra of operators. So we put, for $F \in C_c(G, K)$ and $f \in L^2(K\backslash G/\Gamma)$,

$$\pi'(F)\, f(x) = \int_S [F^N(z(x), z(y)) + (tz(x)^{-1})^{2\varrho}\, F^N(t^2 z(x)^{-1}, z(y)^{-1})]\, f(y)\, dy \quad \text{if } x \in S,$$

$$= 0 \quad \text{elsewhere.}$$

So $\pi'(F)\, f$ does not depend on the variable in E. We want to prove that this is a representation. To this end we define $f_t^0 \in L^2(A)$ by putting

$$f_t^0(a) = a^\varrho \int_E f(an)\, dn \quad \text{if } a \in A_t,$$

$$= 0 \qquad \text{elsewhere.}$$

If we introduce the Harish-Chandra transform

$$H_F(a) = a^\varrho \int_N F(an)\, dn \qquad (a \in A)$$

we obtain, for $a \in A_t$,

$$\pi'(F)\, f(a) = a^{-\varrho} \int_A [H_F(ab^{-1}) + H_F(t^2 a^{-1} b^{-1})]\, f_t^0(b)\, db.$$

Since

$$F^N(a, b) = (ab)^{-\varrho}\, H_F(ab^{-1}) \quad \text{and} \quad \tilde{F}(\lambda) = \int_A H_F(a)\, a^{-i\lambda}\, da,$$

if we put

$$\hat{f}_t^0(\lambda) = \int_A f_t^0(a)\, a^{-i\lambda}\, da,$$

we obtain, because Mellin transform changes convolution into a product,

$$(*) \qquad \pi'(F)\, f(x) = \int_{\mathfrak{a}^*} \tilde{F}(\lambda)\, \hat{f}_t^0(\lambda)\, E'(\lambda, x)\, d\lambda/2\pi,$$

where $a \in A_t$, and where we have put

$$\begin{aligned} E'(\lambda, x) &= z(x)^{i\lambda-\varrho} + t^{2i\lambda} z(x)^{-i\lambda-\varrho} && \text{if } x \in S, \\ &= 0 && \text{elsewhere.} \end{aligned}$$

In the formula $(*)$, we suppose that the Haar measure of A is normalized in such a way that

$$\int_{a^\alpha < 1} a^{s\alpha}\, da = 1/s \qquad (s > 0).$$

Furthermore one sees, taking the average of the integrand over W in the formula $(*)$, that we have also

$$\pi'(F)\, f(x) = \int_{\mathfrak{a}^*/W} \tilde{F}(\lambda)\, E'(\lambda, x)\, E'(\lambda, f)\, d\lambda/2\pi$$

where

$$E'(\lambda, f) = \hat{f}_t^0(\lambda) + t^{-2i\lambda} \hat{f}_t^0(-\lambda) = \int_{G/\Gamma} E'(-\lambda, x)\, f(x)\, dx.$$

Now, in order to prove that π' is really a representation, it is sufficient to prove that the map $U' : f \mapsto E'(\cdot, f)$ from $L^2(K\backslash G/\Gamma)$ into $L^2(\mathfrak{a}^*/W)$ satisfies $U'U'^* = \mathrm{id}$, denoting by U'^* the adjoint of U'; and this is easily reduced to the Fourier-Mellin inversion formula.

3. Compactness of $\pi(F) - \pi'(F)$. Let X be the Banach space of the functions f defined on $K\backslash G/\Gamma$, with values in $\mathbf{C}$, and such that: (1) f is continuous in S and in its complementary C; (2) we have

$$f_0 \nu(an) \ll a^{2\varepsilon} \quad \text{on } A_t \times E$$

with some ε fixed once for ever. Let Y be the Hilbert space of functions defined on $K\backslash G/\Gamma$, and such that

$$\int_C |f(x)|^2 \, dx + \int_{A_t \times E} |f(an)|^2 \, a^{2\varrho+3\varepsilon} \, da \, dn < +\infty .$$

We denote now $L^2(K\backslash G/\Gamma)$ by H. We have $X \subset H \subset Y$ with dense injections.

PROPOSITION. *Let $F \in C_c(G, K)$ and $V = \pi(F) - \pi'(F) \in L(H)$. The endomorphism V applies H into X and admits a (unique) continuation as a compact application $\hat{V}$ from Y to X.*

The proof of this proposition uses the two preceding lemmas of §2, and the Ascoli theorem.

4. Continuation of the resolvent of π'. Let B be the strip $|\mathrm{Im}\,\lambda| < |\varrho + \varepsilon|$ of $\mathfrak{a}_c^*$; we call B_n the rectangle $\lambda \in B$, $|\mathrm{Re}\,\lambda| < n$, and B_n^+ is the set of $\lambda \in B_n$ with positive imaginary part.

Thanks to the Paley-Wiener theorem for spherical functions, one sees that there exists, for each natural number n, a function $F_n \in C_c^\infty(G, K)$ such that (1) $\tilde{F}_n$ is real valued on $\mathfrak{a}^*$; (2) $\tilde{F}_n$ is injective on $\mathfrak{a}_+^*$; (3) $\tilde{F}_n$ induces a bijection of B_n/W onto its range. We define, if $\mu \in B_n^+$,

$$R_n'(\mu) = [\pi'(F_n) - \tilde{F}_n(\mu)]^{-1},$$

and we consider the restriction of $R_n'(\mu)$ to X, with values in Y, denoted by the same letters.

PROPOSITION. *The map $\mu \mapsto R_n'(\mu)$ from B_n^+ to $L(X, Y)$ admits an analytic continuation $R_n'(\mu)_{X,Y}$ to the whole of B_n.*

One proves this using the properties of F_n, the spectral representation $(*)$ of $R_n'(\mu)$, and the residue theorem.

5. Continuation of the resolvent of π. We put, if $\mu \in B_n^+$,

$$R_n(\mu) = [\pi(F_n) - \tilde{F}_n(\mu)]^{-1} \in L(H),$$

$$G_n(\mu) = [\pi(F_n) - \tilde{F}_n(\mu)]\,[\pi'(F_n) - \tilde{F}_n(\mu)]^{-1} = 1 + [\pi(F_n) - \pi'(F_n)]\, R_n'(\mu).$$

It is also useful to define

$$L_n(\mu) = G_n(\mu)^{-1} = [\pi'(F_n) - \tilde{F}_n(\mu)]\, R_n(\mu),$$

so that

$$R_n(\mu)=R'_n(\mu)\,L_n(\mu).$$

In order to obtain the analytic continuation of R_n, we need the following well-known result:

LEMMA. *Let U be a connected open subset of $\mathbf{C}$, and T a holomorphic application of U into the space of compact operators of a Banach space; if $1-T(z_0)$ is invertible for some $z_0 \in U$, then the map $[1-T(z)]^{-1}$ is a meromorphic function on U.*

Now we put, for every $\mu \in B_n$,

$$G_n(\mu)_X = 1+\hat{V}R'_n(\mu)_{X,Y} \in L(X).$$

It is clear that $\hat{V}R'_n(\mu)$ is compact on X, and that $G_n(\mu)_X$ is invertible if $\mu \in B_n^+$. So, applying the preceding lemma, we see that there exists a meromorphic function $L_n(\cdot)_X$, defined on B_n, such that

$$L_n(\mu)_X = G_n(\mu)_X^{-1} \in L(X).$$

We denote by D_n the set of poles of $L_{n,X}$, and we put

$$R_n(\mu)_{X,Y} = R'_n(\mu)_{X,Y}\,L_n(\mu)_X \in L(X, Y)$$

if $\mu \in B_n - D_n$. Using the formulas at the beginning of this section, one then proves

PROPOSITION. *The map $R_n(\cdot)_{X,Y}$ is a meromorphic continuation to B_n of the map $R_n(\cdot)$ from B_n^+ to $L(X, Y)$.*

6. The spectral density. The form $(h, g) \mapsto \int h(x)\,(g(x))^- \,dx = \langle h, g\rangle$ is defined for $h \in Y$ and $g \in X$, and so places X and Y in duality. We first define, if f and g are in X,

$$e_\lambda^n(f, g) = \frac{1}{iF'_n(\lambda)} \langle [R'_n(\lambda)_{X,Y} - R_n(-\lambda)_{X,Y}]\,f, g\rangle .$$

This function of λ is meromorphic in B_n.

PROPOSITION. (1) *If $n<n'$ one has $e_\lambda^n = e_\lambda^{n'}$ if $\lambda \in B_n$; so there exists a map $\lambda \to e_\lambda$ defined on B which coincides with e_λ^n on B_n; one has $e_\lambda = e_{-\lambda}$.*

(2) *Let D be the union of the D_n. This is a locally finite subset of B, and if $\psi \in C_c(C-D/W)$, one has, for f and g in X,*

$$\int_{C/W} \psi(\lambda)\,dm_{f,g}(\lambda) = \int_{\mathfrak{a}^*/W} \psi(\lambda)\,e_\lambda(f, g)\,d\lambda/2\pi .$$

We call the map $\lambda \mapsto e_\lambda$ the *spectral density* of the spectral measure associated to π. The proof of this proposition rests on a principle of spectral analysis, and which is the following:

FORMULA OF TITCHMARSH-KODAIRA. *Let H be a Hilbert space, and T a self-adjoint operator in H; if $R(z)=[T-z]^{-1}$ is the resolvent of T, and if $dm_{f,g}$ is the spectral measure associated to T and the elements f, g of H, one has*

$$\int_{\boldsymbol{R}} \psi(\lambda)\, dm_{f,g}(\lambda) = \lim_{\varepsilon\to 0;\, \varepsilon>0} \frac{1}{2i\pi} \int_{\boldsymbol{R}} [R(\lambda+i\varepsilon)\, f - R(\lambda - i\varepsilon)\, f, g]\, \psi(\lambda)\, d\lambda$$

for every $\psi \in C_c(\boldsymbol{R})$.

The proof of this is well known.

There are two consequences of the last proposition: First, the map $\lambda \mapsto e_\lambda$ has no poles on $\mathfrak{a}^*$, and only a finite number on the part of the imaginary axis which is in the strip.

7. The theorem. We put, for $\lambda \in B_n - D$ and $f \in X$,

$$E_n(\lambda, f) = E'(\lambda, L_n(\lambda)_X f)$$

and one sees that

$$e_\lambda(f, f) = |E_n(\lambda, f)|^2 = |E_n(-\lambda, f)|^2.$$

This implies, if $F \in C_c(G, K)$,

$$E_n(\lambda, \pi(F) f) = \tilde{F}(\lambda)\, E_n(\lambda, f),$$

and we get also, using general von Neumann theory, a function c_n defined on $\mathfrak{a}^*_+$, measurable and of absolute value one, such that

$$E_n(\lambda, f) = c_n(\lambda)\, E_n(-\lambda, f).$$

The $E_n(\lambda, \cdot)$ are distributions, but we have the following:

LEMMA. *Let* $\operatorname{Im}\lambda < 0$. *There exists at most one distribution T_λ on $K\backslash G/\Gamma$ satisfying the following conditions:*

(1) *The distribution $T_\lambda - E'_\lambda$ is in $L^2(K\backslash G/\Gamma)$.*

(2) *One has $\omega T_\lambda = \tilde{\omega}(\lambda)\, T_\lambda$, where $\tilde{\omega}(\lambda) = -(|\lambda|^2 + |\varrho|^2)$; and then it is an analytic*

function on G such that

$$T^0_\lambda(x)=a(x)^{i\lambda-\varrho}+ca(x)^{-i\lambda-\varrho},$$

with $c\in \mathbf{C}$.

This lemma is proved using Bernstein's theorem and the principle of quick decrease of K-finite automorphic eigenfunctions of the Casimir operator in a Siegel set, when constant term is removed (see [**3**]).

We then obtain some functions $E(\lambda, x)$ such that

$$E_n(\lambda, f)=\int (E(\lambda, x))^- f(x)\,dx \qquad (\lambda\in B_n)$$

and also an analytic continuation of the c_n functions, using part (2) of the lemma.

Now the Eisenstein series

$$(*) \qquad \sum_{\Gamma/\Gamma\cap P} a(x\gamma)^{i\lambda-\varrho}$$

converges if $\operatorname{Im}\lambda>\varrho^*$ and satisfies the conditions (1) and (2) of the preceding lemma. It is then equal to $E(\lambda, x)$ if $\varrho<\operatorname{Im}\lambda<\varrho+\varepsilon$, and we obtain in this manner a function which is meromorphic in the complex plane. Watching the constant term, one obtains also the continuation of the function c.

We then obtain the following:

THEOREM. *The Eisenstein series* (∗) *admits a meromorphic continuation in the complex plane, except perhaps on zero, with respect to* λ, *and poles independent of* x, *analytic on* G *and such that:*

(1) *If* $f\in X$, *the integral*

$$E(\lambda, f)=\int_{K\backslash G/\Gamma} f(x)\,(E(\lambda, x))^-\,dx$$

converges for every f *in the half-space* $\operatorname{Im}\lambda>-\varrho$ *and satisfies*

$$E(\lambda, \pi(F)\, f)=\tilde{F}(\lambda)\, E(\lambda, f)$$

for every $F\in C_c(G, K)$.

(2) *If we denote by* $L^2_1(K\backslash G/\Gamma)$ *the orthogonal supplement of the closed linear span of the eigenfunctions of the* $\pi(F)$ *in* $L^2(K\backslash G/\Gamma)$, *we have the Plancherel formula*

$$\int_{K\backslash G/\Gamma} |f(x)|^2\,dx=\int_{\mathfrak{a}^*/W} |E(\lambda, f)|^2\,d\lambda/2\pi$$

for every $f\in L^2_1(K\backslash G/\Gamma)$.

(3) *In the strip* $|\mathrm{Im}\,\lambda| < |\varrho|$, *the poles of E are located either on the vertical segment (and only a finite number of them lie there), or on the strip* $-\varrho < \mathrm{Im}\,\lambda < 0$.

(4) *There exists in* $C - \{0\}$ *a meromorphic function c, with the same poles as the Eisenstein series, such that the following formulas hold:*

$$E(\lambda, x) = c(\lambda)\, E(-\lambda, x);$$
$$c(\bar{\lambda})\,(c(\lambda))^{-} = c(\lambda)\, c(-\lambda) = 1;$$
$$E^0(\lambda, x) = a(x)^{i\lambda - \varrho} + c(\lambda)\, a(x)^{-i\lambda - \varrho}.$$

Using simple arguments from the theory of elliptic differential operators, it is possible to see that if f and the distributions $\omega^n f$ belong to $L_1^2(K\backslash G/\Gamma)$, one has the inversion formula corresponding to the Plancherel formula of the part (2), namely

$$f(x) = \int_{\mathfrak{a}_+^*} E(\lambda, f)\, E(\lambda, x)\, d\lambda/2\pi.$$

Before finishing, let me say that the same method seems to be yet fruitful in the case where G is an algebraic group defined over $\boldsymbol{Q}$, and of rank one over this field, if we take for Γ an arithmetic subgroup of G; note only that some preliminary decomposition of the space $L^2(K\backslash G/\Gamma)$ is necessary, due to the fact that if we take a maximal parabolic subgroup P of G, defined over $\boldsymbol{Q}$, the reductive part of a Levi subgroup of G is no longer compact, but only anisotropic.

REFERENCES

1. L. D. Faddeev, *Expansion in eigenfunctions of the Laplace operator on the fundamental domain of a discrete group on the Lobačevskiĭ plane*, Trudy Moskov. Mat. Obšč. **17** (1967), 323–350 = Trans. Moscow Math. Soc. **1967**, 357–386. MR **38** #5062.

2. H. Garland and M. S. Ragunathan, *Fundamental domains for lattices in* (**R**-) *rank one semi-simple Lie groups*, Ann. of Math. (2) **92** (1970), 279–326. MR **42** #1943.

3. R. Godement, *The spectral decomposition of cusp-forms*, Proc. Sympos. Pure Math., vol. 9, Amer. Math. Soc., Providence, R.I., 1966, pp. 225–234. MR **35** #1713.

4. S. Helgason and K. Johnson, *The bounded spherical functions on symmetric spaces*, Advances in Math. **3** (1969), 586–593. MR **40** #2787.

5. R. P. Langlands, *Eisenstein series*, Proc. Sympos. Pure Math., vol. 9, Amer. Math. Soc., Providence, R.I., 1966, pp. 235–252. MR **40** #2784.

6. A. Selberg, *Discontinuous groups and harmonic analysis*, Proc. Internat. Congr. Math. (Stockholm, 1962), Inst. Mittag-Leffler, Djursholm, 1963, pp. 177–189. MR **31** #372.

UNIVERSITÉ PARIS VII
PARIS, FRANCE

INDEXES

AUTHORS

Richard Askey, Department of Mathematics, University of Wisconsin, Madison, Wisconsin 53706

Robert J. Blattner, Department of Mathematics and Statistics, University of Massachusetts, Amherst, Massachusetts 01002

J. Brezin, Department of Mathematics, University of North Carolina, Chapel Hill, North Carolina 27514

P. Cartier, 24 rue Ronsard, F-91470-Limours en Hurepoix, France

W. Casselman, Department of Mathematics, University of British Columbia, Vancouver 8, British Columbia, Canada

Ronald R. Coifman, Department of Mathematics, Washington University, St. Louis, Missouri 63130

Leon Ehrenpreis, Department of Mathematics, Yeshiva University, New York, New York 10033

Pierre Eymard, Departement de Mathématique, Université de Nancy 1, Case officielle no. 140, 54000 Nancy, France

Charles Fefferman, Department of Mathematics, University of Chicago, Chicago, Illinois 60637

M. Flensted-Jensen, Matematisk Institut, Universitetsparken 5, DK-2100 København, Denmark

G. B. Folland, Department of Mathematics, University of Washington, Seattle, Washington 98195

Harry Furstenberg, Institute of Mathematics, Hebrew University, Jerusalem, Israel

Ramesh Gangolli, Department of Mathematics, University of Washington, Seattle, Washington 98195

Stephen Gelbart, Department of Mathematics, Cornell University, Ithaca, New York 14850

Paul Gerardin, U.E.R. de Mathématiques, Université Paris VII, 75005 Paris, France

Kenneth I. Gross, Department of Mathematics, University of North Carolina, Chapel Hill, North Carolina 27514

Harish-Chandra, School of Mathematics, Institute for Advanced Study, Princeton, New Jersey 08540

Sigurdur Helgason, Department of Mathematics, Massachusetts Institute of Technology, Cambridge, Massachusetts 02139

Takeshi Hirai, Department of Mathematics, Faculty of Science, Kyoto University, Kyoto, Japan

Roger Howe, Department of Mathematics, State University of New York, Stony Brook, New York 11790

Hervé Jacquet, Department of Mathematics, Graduate School, City University of New York, 33 West 42nd Street, New York, New York 10036

Kenneth Johnson, Department of Mathematics, Rutgers University, New Brunswick, New Jersey 08903

Adam Kleppner, Department of Mathematics, University of Maryland, College Park, Maryland 20742

A. W. Knapp, Department of Mathematics, Cornell University, Ithaca, New York 14850

Tom Koornwinder, Mathematisch Centrum, 2e Boerhaavestraat 49, Amsterdam, The Netherlands

Ray A. Kunze, Department of Mathematics, University of California, Irvine, California 92664

Gilles Lachaud, 33, Avenue du Bac, 94210-La Varenne Saint-Hilaire, France

J. Lepowsky, Department of Mathematics, Yale University, New Haven, Connecticut 06520

Lars-Åke Lindahl, Matematiska Institutionen, Sysslomansgatan 8, S-752 23 Uppsala, Sweden

Ronald Lipsman, Department of Mathematics, University of Maryland, College Park, Maryland 20742

H. Lee Michelson, Department of Mathematics, Bar-Ilan University, Ramat-Gan, Israel

Willard Miller, Jr., Department of Mathematics, University of Minnesota, Minneapolis, Minnesota 55455

Calvin C. Moore, Department of Mathematics, University of California, Berkeley, California 94720

L. Pukanszky, Department of Mathematics, University of Pennsylvania, Philadelphia, Pennsylvania 19104

François Rodier, U.E.R. de Mathématiques, Université Paris VII, 2 Place Jessieu, 75005 Paris, France

H. Rossi, Department of Mathematics, Brandeis University, Waltham, Massachusetts 02154

P. J. Sally, Jr., Department of Mathematics, University of Chicago, Chicago, Illinois 60637

Allan J. Silberger, Mathematisches Institut der Universität, 53 Bonn, Federal Republic of Germany

T. A. Springer, Mathematisch Instituut der Rijksuniversiteit, Universiteitscentrum "de Uithof", Budapestlaan 6, Utrecht, The Netherlands

E. M. Stein, Department of Mathematics, Princeton University, Princeton, New Jersey 08540

R. Strichartz, Department of Mathematics, Cornell University, Ithaca, New York 14850

P. C. Trombi, Department of Mathematics, University of California, Los Angeles, California 90024

V. S. Varadarajan, Department of Mathematics, University of California, Los Angeles, California 90024

M. Vergne, 11 Rue de Quatrefages, Paris 5[e], France

Nolan R. Wallach, Department of Mathematics, Rutgers University, New Brunswick, New Jersey 08903

Guido Weiss, Department of Mathematics, Washington University, St. Louis, Missouri 63130

Joseph A. Wolf, Department of Mathematics, University of California, Berkeley, California 94720

REFERENCED AUTHORS

Roman numbers refer to pages on which a reference is made to an author or work of an author.

Italic numbers refer to pages on which a complete reference to a work by the author is given.

Boldface numbers indicate the first page of the articles in the book.

SUBJECTS